P9-EEG-496

All IP in 3G CDMA Networks

All IP in 3G CDMA Networks

The UMTS Infrastructure and Service Platforms for Future Mobile Systems

Jonathan P. Castro
Orange Communications SA/AG, Switzerland

Telephone (+44) 1243 779777

Email (for orders and customer service enquiries): cs-books@wiley.co.uk
Visit our Home Page on www.wileyeurope.com or www.wiley.com

Other Wiley Editorial Offices

John Wiley & Sons Inc., 111 River Street, Hoboken, NJ 07030, USA

Jossey-Bass, 989 Market Street, San Francisco, CA 94103-1741, USA

Wiley-VCH Verlag GmbH, Boschstr. 12, D-69469 Weinheim, Germany

John Wiley & Sons Australia Ltd, 33 Park Road, Milton, Queensland 4064, Australia

John Wiley & Sons (Asia) Pte Ltd, 2 Clementi Loop #02-01, Jin Xing Distripark, Singapore 129809

John Wiley & Sons Canada Ltd, 22 Worcester Road, Etobicoke, Ontario, Canada M9W 1L1

Wiley also publishes its books in a variety of electronic formats. Some content that appears in print may not be available in electronic books.

British Library Cataloguing in Publication Data

A catalogue record for this book is available from the British Library

ISBN 0-470-85322-0

Typeset in 10/12pt Times by Thomson Press (India) Limited, New Delhi
Printed and bound in Great Britain by Antony Rowe Ltd, Chippenham, Wiltshire
This book is printed on acid-free paper responsibly manufactured from sustainable forestry
in which at least two trees are planted for each one used for paper production.

To:

My family for their endurance, and to my dear friends and colleagues for their understanding, because while putting together this book I stole too many precious moments from them.

And

Once again to all the 3GPP contributors for their dedication to make the UMTS specifications a reality and for continuing to work on its evolution, without whom some of the contents of this book would never have been possible.

Contents

Preface

At the time of this writing many mobile network operators in Europe have already started, while others brace to launch, wideband mobile services enabled by Universal Mobile Telecommunication System (UMTS).

Therefore, aims to ensure a rapid growth in traffic volume and a rise in new services will indeed change the configuration and structure of wireless network. As a result, future mobile communications systems with highly integrated services, flexibility and higher throughput will not only be a common trend.

Yet, to support such trends and features while sustaining the evolution of UMTS building blocks coupled with the efficient use of spectrum and optimum management of radio resources remains a challenge.

To meet these challenges the 3GPP carries on with the consolidation and enhancements of Wide-band Code Division Multiple Access (WCDMA) and the hybrid Time Division-CDMA as the radio techniques for the Universal Mobile Telecommunication Systems (UMTS). Hence, UMTS conceived at the eve of this new millennium will progress with its large impact on the landscape of future wideband mobile networks and serve as the leading platform for wireless multimedia communications.

The new specification extracts in this edition intend to provide selected updates on the new developments of the UMTS specification. At the same time, they also provide examples of prospective implementations. Thus, it is hoped that the synthesis presented in this new edition will continue to serve as a practical reference.

In this context, to stay as a valuable source on UMTS air-interface and network issues, application platform examples, complementary technologies etc. this book keeps the concise and integrated description of principles, methods and technology used in the standard specification. As such, again the 3GPP recommendations structure and contents are followed closely to comply entirely with the concept, terminology, approach and style, and not only the technical essence.

In drafting the standards or UMTS technical specifications, the experts have tried to reduce the risk of ambivalent interpretations and, not necessarily, ease the understanding of a common reader. The logic, constructive discussions and consensus behind the choices or equivalent solutions have not always been retained in the final specifications. Therefore, in some ways this book attempts to present an objective unified view of the key aspects of UMTS, its evolution and complementary technologies. On the other hand, the areas of UMTS are vast in content and details, and not all are within the scope of these writings; the same applies to its evolution and complementary enablers. Nevertheless, this edition will still enable anyone who desires to understand what UMTS has in the specifications and get a good grasp of its design and selected new features, but the 3GPP technical documents remain the official specifications with all their appropriate ownership and origin of contribution.

Since this volume has been introduced in good faith as a useful reference for the UMTS technology and its evolution along with complementary technologies, and with the impressions of the writers alone, the author does not take the responsibility for any misuse or error while dealing with the information provided. Then again, if for some reasons some representations could have been omitted due to time constraints or the dynamic changing process of the specifications, they will be revised and corrected for later reprints. Therefore, the author welcomes any comments and suggestions for improvements or changes that could enhance further this contribution to UMTS and its changing pace now under full deployment phase.

The chapters in this book cover specific design details of the building blocks in the UMTS air interface, in particular the physical layer. It addresses the technical part of the specification for both the FDD and TDD modes. On the other hand, it also introduces the key criteria for network dimensioning and deployment of 3G systems assumption and evolution from 2G mobile networks from the provider's point of view. To illustrate the progressive steps of UMTS standards such as the evolution towards predominant packet switching oriented communications, this edition focusses more on the 'All IP' enabling concept for efficient advanced mobile multimedia service offering.

A brief summary of the chapters can be outlined as follows:

Chapter 1 deals with concrete requirements for 3G mobile systems after summarising the rapid growth of wireless communications and the Internet, as well as the evolution of its building blocks and the overall convergence trends based on IP technology. It also outlines briefly enhancing technologies such as capacity increasing antennas, Multi-User Detection Techniques and Software Radio applications.

Chapter 2 presents the fundamentals of system analysis, e.g. multiple access options, which considers narrow-band and wide-band digital channels, as well as the background for the UTRA FDD and TDD modes. It covers signal processing aspects describing principles of spread-spectrum, modulation and spreading, the CDMA performance, PN sequences, power control and handovers. It presents the communication environments envisaged for UMTS operation and deployment. It also describes the channel models used to verify and justify the performance for the selected operating scenarios. It provides a summary of the mathematical formulation for the performance analysis results seen in the forthcoming chapters.

Chapter 3 describes the UMTS service components. It covers the UMTS bearer architecture, concepts in Quality of Service (QoS) for 3G systems (including e2e QoS for the IP Multimedia Subsystem), multimedia transmission and traffic classes in UMTS. The classes include conversational, streaming, interactive and background types. Sensitivity to IP transmission impairments are also covered here. To provide an overview of potential applications in UMTS this chapter also summarises service offerings and selected area of service technology.

Chapter 4 describes the UTRA physical layer design and configuration, where we introduce all the building blocks in detail with their respective technical description and requirements. It covers dedicated common transport channels, configuration of FDD and TDD physical channels in the uplink and downlink with their spreading and coding characteristics. Spreading and modulation, including scrambling, multiplexing and channel coding are also the contents of this chapter. The chapter presents the aforementioned characteristics for the FDD and TDD separately for each mode or unified when the case applies to both.

Chapter 5 describes the UMTS development platform. It introduces its architecture top-down identifying the core and access network domains. It defines the UTRA identifiers and functions, e.g. system access control, radio channel ciphering and deciphering, mobility functions and radio resource management and control functions. It presents also mobility management with its signalling connections and impacts of mobility handling. Chapter 4 presents also the UTRAN sychronisation and UTRAN interfaces besides pointing out 2G/3G network co-existing issues. It introduces the radio interface protocol architecture with its structure in terms of services and function layers. This chapter thus outlines the most relevant elements, which require technical description for design and implementation.

Chapter 6 introduces the future, predominantly PS, domain networks based on the IP-Multimedia Subsystem (IMS). It covers the conceptual architecture of UMTS Release 4 and 5. It starts with the evolution of R99 and discusses briefly the long-term view and vision of the UMTS architecture. Then it describes the components of IMS with their corresponding interfaces or reference points. This chapter also presents the introduction and considerations of mobility management, registration aspects, multimedia signaling, service platforms, QoS aspects and transport issues like the basic differences of Ipv4 and Ipv6. As a practical approach to the co-existing IMS and current networks, it illustrates the integration of IMS to current UMTS networks now under full deployment stage.

Chapter 7 introduces the factors that influence 3G-network dimensioning. It discusses coverage and capacity trade-off in the FDD mode pointing out impacts from soft handover, power control and orthogonality deviations. It covers the analysis of parameters for multi-service traffic in PS and CS. It establishes service models starting from capacity projections and service strategy. Cellular coverage planning issues, i.e. the coverage concept, radio network parameter assumptions for CS and PS and characteristics of CDMA cells (with its theoretical capacity and cell loading effects) constitute the essential parts of this chapter. The chapter covers link budget principles for the forward and reverse links and their respective formulation. In the latter part, these principles are applied for a field study. For completeness the chapter also describes briefly the dimensioning of the RNC in the UTRAN side. Chapter 7 also presents the dimensioning of the Core Network and Transmission systems. In the last part, results of the field study are provided using hypothetical parameters to illustrate the concepts end-to-end. The illustrations correspond to dimensioning exercises carried out while optimising 3G networks. However, the input and output values in this chapter do not necessarily reflect actual values that may be used directly while dimensioning a future UMTS network. Finally, to complete the assessment of UMTS network deployment within 2G networks like GSM, this chapter discusses briefly co-location and site sharing, as well as co-location of antenna systems.

Chapter 8 introduces the High-Speed Downlink Packet Access (HSDPA) as the enhancement of the UMTS transmission rates in the evolution of the radio access. HSDPA increases the network capacity without major new infrastructure change but simply SW enabling in the Node B in most cases. The chapter provides the technical background of the transmission rate enhancing features and illustrates the performance of some of the key building blocks of HSDPA. It also describes the initial deployment phase of the prospective operational solutions.

Chapter 9 introduces the UTRA transmission system starting from the spectrum allocation, i.e. the UTRA frequency bands. It presents the radio transmission and reception aspects, describing transmitter and receiver characteristics for the User Equipment (UE) and the Base Station (BS). It describes the maximum output power and output power dynamics, out of synchronisation output power handling and transmit On/Off power. Details on the output RF spectrum emissions, such as occupied bandwidth and out band emission, spectrum emissions, adjacent channel leakage power ratio, spurious emissions and transmit modulation and inter-modulation are discussed in this chapter. The summary of examples includes a review of simulation scenarios for the co-existence of FDD/FDD when analysing ACIR with macro-to-macro and macro-to-micro cases. Before presenting results the chapter also reviews propagation models.

Chapter 10 summarises the main application platforms as the services enabler entities delivering content and managing distribution according to user profiles. It describes the key characteristics of the streaming, downloading and location positioning solutions. These platforms are elements, which can greatly exploit the wideband capabilities of the UMTS radio-access bearers.

Chapter 11 presents issues on resource and network management. It covers radio resource management and signalling, i.e. managing power (fast and low). The conceptual aspects of network management are covered from the Network Management System point of view. It also covers initial considerations for network optimisation.

Finally, Chapter 12 covers the complementary technologies, which do/will co-exist with UMTS. Here we describe primarily EDGE as the intermediate evolution of the conventional TDMA-based radio access from GSM, WLAN as an extension of capacity in hot spots and WIMAX as the forthcoming wireless broadband backbone for UMTS, including the fixed segment.

The last chapter also describes additional features of the 'ALL IP' and more intelligent Node B options introduced in Chapter 1. Thus identifying the benefits of packet optimised solutions in the overall scheme of network evolution and convergence with the mobile/fixed networks. Hence, this new print complements the earlier edition with the key building blocks of the UMTS evolution towards an end-to-end packet network evolving within an all-purpose telecommunications environment and co-existing with the classical mobile networks.

Jonathan P. Castro

Abbreviations

2G 3G	Second Generation Third Generation
3GPP	Third Generation Partnership Project
4G	Fourth Generation
8-PSK	8-Phase Shift Keying
AAL2	ATM Adaptation Layer Type 2
AC	Access Controller
ACI	Adjacent Channel Interference
ACIR	Adjacent Channel Interference Power Ratio
ACK	Acknowledgement
ACLR	Adjacent Channel Leakage Power Ratio
ACP	Adjacent Channel Protection
ACS	Adjacent Channel Selectivity
AI	Acquisition Indicator
AICH	Acquisition Indicator Channel
AMR	Adaptive Multi-Rate
AP	Access Preambles, Access Points
API	Access Preamble Acquisition Indicators
ARQ	Automatic Repeat Request
AS	Application Servers, Access Slots
ASC	Access Service Classes
ATM	Asynchronous Transfer Mode
AVI	Actual Value Interface
AWGN	Additive White Gaussian Noise
BTS	Base Transceiver Station
BB	Base Band
BCCH	Broadcast Control Channel
BCH	Broadcast Channel
BER	Bit Error Rate
BLER	Block Error Rate
BMC	Broadcast and Multicast Control
BPSK	Binary Phase Shift Keying
BSC	Base Station Controller
BSIC	Base Station Identity Code
BSS	Base Station Subsystem
BYE	Session Termination
CC	Call Control, Convolutional Coding
CCCH	Common Control Channel

CCH	Control Channels
CCPCH	Common Control Physical Channel
CCTrCH	Coded Composite Transport Channel
CD	Collision Detection
CD/CA-ICH	Collision Detection/Channel Assignment-Indicator Channel
CD-DSMA	Collision Detection-Digital Sense Multiple Access
CDF	Cumulative Density Function, Cumulative Distribution Function
CDMA	Code Division Multiple Access
CI	Commercial/Industrial
C/I	Carrier-to-Interference Ratio
CN	Core Network
COST	European Cooperation in the Field of Scientific and Technical Research
CPCH	Common Packet Channel
CPICH	Common Pilot Channel
CRC	Cyclic Redundancy Check
CS	Coding Scheme, Circuit Switched
CSICH	CPCH Status Indicator Channel
CTCH	Common Traffic Channel
DCA	Dynamic Channel Allocation
DCCH	Dedicated Control Channel
DCH	Dedicated Channel
DHCP	Dynamic Host Client Protocol
DiffServ	Differentiated Services
DL	DownLink
DNS	Domain Name Server
DPCCH	Dedicated Physical Control Channel
DPCH	Dedicated Physical Channel
DPDCH	Dedicated Physical Data Channel
DRNC	Drifting RNC
DS	DiffServ, Dual Stack
DSCH	Downlink Shared Channel
DSMA	Digital Sense Multiple Access
DTCH	Dedicated Traffic Channel
DTX	Discontinuous Transmission
El	Standard 2 Mbps Transmission Line
EDGE	Enhanced Data Rate for GSM Evolution
EFR	Enhanced Full Rate
EGPRS	Enhanced GPRS
EIRP	Equivalent Isotropic Radiated Power
ETSI	European Telecommunications Standards Institute
FACH	Forward Access Channel
FAUSCH	Fast Uplink Signalling Channel
FBI	Feedback Information
FCS	Fast Cell Selection, Fast Cell Switching

FDD	Frequency Division Duplex
FDMA	Frequency Division Multiple Access
FEC	Forward Error Correction
FER	Frame Erasure Rate, Frame Error Rate
FN	Frame Number
FP	Frame Protocol
FTP	File Transfer Protocol
GGSN	Gateway GPRS Supporting Node
GMSK	Gaussian Minimum Shift Keying
GoS	Grade-of-Service
GP	Guard Period
GPRS	General Packet Radio Service
GPS	Global Positioning System
GSM	Global System for Mobile Communications
GW	Gateway
HCS	Hierarchical Cell Structure
HO	Handover
HSCSD	High-Speed Circuit-Switched Data
HTML	Hypertext Markup Language
HTTP	Hypertext Transfer Protocol
HW	Hardware
ID	Identifier
IE	Information Element
IEE	Institution of Electrical Engineers
IEEE	Institute of Electrical and Electronics Engineers
IETF	Internet Engineering Task Force
IF-HO	Intermediate-frequency-HO
IMD	Inter-Modulation
IMEI	International Mobile Equipment Identities
IMSI	International Mobile Subscriber Identity
IntServ	Integrated Services
IP	Internet Protocol
IPv4	IP version 4
IPv6	IP version 6
IR	Incremental Redundancy
IS	Interim Standard (US)
IS-136	North American TDMA
IS-54	North American TDMA Digital Cellular
IS-95	North American Version of the CDMA Standard
ISCP	Interference Signal Code Power
ISDN	Integrated Services Digital Network
ISP	Internet Service Provider
ITU	International Telecommunication Union

Iu	Interconnection Point Between an RNC and a Core Network
Iub	Interface Between an RNC and a Node B
Iur	Logical Interface Between Two RNCs
JTACS	Japan TACKS
kbps	kilo bits per second
ksps	kilo symbols per second
L1	OSI Layer 1: physical layer
L2	OSI Layer 2: radio data link layer
L3	OSI Layer 3: radio network layer
LA	Link Adaptation
LAC	Location Area Code
LAN	Local Area Network
LCS	Location Communication Services
LDAP	Lightweight Directory Access Protocol
MAC	Medium Access Control, Message Authentication Code
MCC	Mobile Country Code
MCL	Minimum Coupling Loss
Mcps	Mega chips per second
MCS	Modulation and Coding Scheme
MHA	Mast Head Amplifier
MM	Mobility Management
MMUSIC	Multiparty Multimedia Session Control
MNC	Mobile Network Code
MPLS	Multi-protocol Label Switching
MRC	Maximal Ratio Combining
MS	Mobile Station
MSC	Mobile-services Switching Centre
MT	Mobile Terminal, Mobile Terminated, Mobile Termination
MTU	Maximum Transfer Unit
NAS	Non-Access Stratum
NB	Narrow Band
NBAP	Node B Application Part
NCx	Network Control
NE	Network Elements
NF	Noise Fig.
NMS	Network Management System
NRT	Non-real Time
NSS	Networking Subsystem
NW	Network
OFDM	Orthogonal Frequency Division Multiplexing
OSA	Open Service Architecture

OSI	Open Systems Interconnection
OTDOA	Observed Time Difference of Arrival
OVSF	Orthogonal Variable Spreading Factor
PC	Power Control
PCCH	Paging Control Channel
P-CCPCH	Primary-CCPCH
PCH	Paging Channel
PCPCH	Physical CPCH
P-CPICH	Primary-CPICH
PCM	Pulse Code Modulation
PCS	Personal Communications Systems
PCU	Packet Control Unit
PDC	Pacific Digital Cellular
PDH	Plesiochroneouw Digital Hierarchy
PDCP	Packet Data Convergence Protocol
PDP	Packet Data Protocol, Policy Decision Point
PDSCH	Physical DSCH
PDU	Packet Data Unit
PI	Paging Indicator
PICH	Paging/Page Indicator Channel
PLMN	Public Land Mobile Network
PN	Pseudo-random Noise
PPP	Point-to-Point Protocol
PRACH	Physical RACH
PS	Packet Switched
PSC	Primary Synchronisation Code
PSK	Phase Shift Keying
PUSCH	Physical Uplink Shared Channel
QoS	Quality of Service
QPSK	Quadrature Phase Shift Keying
R99	Release 1999
R00	Release 2000
RA	Routing Area, Radio Access
RAB	Radio-Access Bearer
RAC	Routing Area Code
RACH	Random-Access Channel
RAKE	Special Receiver Type Used in CDMA
RAM	Radio Access Mode
RAN	Radio-Access Network
RANAP	Radio-Access Network Application Part
RAT	Radio-Access Technique
RAU	Rate Adaptation Unit
RB	Radio Bearer
RF	Radio Frequency

RL	Radio Link
RLB	Radio Link Budget
RLC	Radio Link Control
RM	Resource Manager
RNC	Radio Network Controller
RNS	Radio Network Subsystem
RNSAP	Radio-Network Subsystem Application Part
RNTI	Radio Network Temporary Identity
RR	Round Robin
RRC	Radio Resource Control, Root-Raised Cosine
RRM	Radio Resource Management
RRU	Radio Resource Utilisation
RSCP	Received Signal Code Power
RSSI	Received Signal Strength Indicator
RSVP	Resource Reservation Protocol
RT	Real-Time
RTCP	RTP Control Protocol
RTP	Real-Time Transport Protocol
RTT	Radio Transmission Technology, Round-Trip Time
RU	Resource Unit
RX	Receive
RXD	Receive Diversity
SAI	Service Area Identifier
SAP	Service Access Point
S-CCPCH	Secondary-CCPCH
SCH	Synchronisation Channel
S-CPICH	Secondary-CPICH
SCTP	Stream Control Transmission Protocol
SDH	Synchronous Digital Hierarchy
SDP	Session Description Protocol
SDU	Service Data Unit
SF	Spreading Factor
SFN	System Frame Number
SGSN	Serving GPRS Support Node
SHO	Soft HO
SIGTRAN	Signalling Transport
SIM	Subscriber Identity Module
SIP	Session Initiation Protocol
SIR	Signal-to-Interference Ratio
SLA	Service Level Agreement
SM	Session Management, Service Management
SMG	Special Mobile Group
SMS	Short Message Services
SNR	Signal-to-Noise Ratio
SR	Software Radio
SRNC	Serving RNC

SRTT	Smoothed Round-Trip Time
SS	Spruad Spectrum
SSC	Secondary Synchronisation Code
S-SCH	Secondary-SCH
SSDT	Site Selection Diversity Transmission
STM	Synchronous Transport Module
STTD	Space Time Transmit Diversity
T1	1.544 Mbps Transmission Link
TCH	Traffic Channel
TCP	Transmission Control Protocol
TDD	Time Division Duplex
TDM	Time Division Multiplex
TDMA	Time Division Multiple Access
TE	Terminal Equipment
TFC	Transport Format Combination
TFCI	Transport Format Combination Indicator
TFCS	Transport Format Combination Set
TFI	Transport Format Indicator
TFS	Transport Format Set
TMSI	Temporary Mobile Station Identity
TN	Termination Node, Transit Network
TOM	Telecom Operations Map
TPC	Transmit Power Control
TR	Technical Recommendation
TrCH	Transport Channel
TRAU	Transcoding and Rate Adaptation Unit
TS	Training Sequences
TSG	Technical Specification Group
TSTD	Time Switched Transmit Diversity
TTI	Transmission Time Interval
TID	Transmission Intermodulation Distortion
TTP	Traffic Termination Point
UBR	Unspecified Bit Rate
UDP	User Datagram Protocol
UE	User Equipment
UEP	Unequal Error Protection
UL	UpLink
Um	Radio Interface for GSM BSS
UM	Unacknowledged Mode
UMTS	Universal Mobile Telecommunication System/Services
URA	UTRAN Registration Area
U-RNTI	UTRAN Radio Network Temporary Identity
USIM	UMTS Subscriber Identity Module, User Services Identity Module
VAS	Value Added Service

VBR	Variable Bit Rate
VCI	Virtual Circuit Identifier
VHE	Virtual Home Environment
VLR	Visitor Location Register
VMS	Voice Mail System
VoIP	Voice over IP
VPI	Virtual Path Identifier
WAP	Wireless Application Protocol
WCDMA	Wide-band Code Division Multiple Access
WDP	Wireless Data Protocol
WI	Walfisch-Ikegami
WML	Wireless Markup Language
WiMAX	Worldwide Inter-operability Micro-wave Access
WLAN	Wireless Local Area Network

1

Evolving Mobile Networks

While the history of mobile communications is long [1–3], and the background of mobile networks thereby is also long, in this chapter we focus on the historic evolution in terms of network architecture and services starting with Second Generation (2G) mobile systems. In particular, we consider the development of the architecture of Global System for Mobile Communications (GSM), since it is by far the most widespread mobile system in the world today. This will provide the basis to cover the introduction of Universal Mobile Telecommunication Services (UMTS) in relation to its Core Network (CN) and radio architectures. The latter will in turn serve as the platform to present UMTS radio-access technology, which is one of the aims of this book.

1.1 THE GROWTH OF MOBILE COMMUNICATIONS

Today wireless voice service is one of the most convenient and flexible means of modern communication. GSM technology has been at the leading edge of this wireless revolution. It is the technology of choice in over 120 countries and for more than 200 operators worldwide. Figure 1.1 illustrates how current estimates indicate that by the year 2005 there will be around 1.4 billion wireless subscribers (i.e. wireless access network access – mobile telephone users), out of which more than 50% will depend on GSM technology and its evolution.

As the wireless revolution has been unfolding, the Internet has also shown a phenomenal growth simultaneously. The advent of the World Wide Web and web browsers has propelled TCP/IP protocols into the main stream, and the Internet is widespread not only in the corporate environment but also in households. Large number of consumers have embraced the Internet and use it today to access information online, for interactive business transactions and e-commerce as well as electronic mail.

The success of mobile communications, i.e. the ubiquitous presence it has established, and the emergence of the Internet point towards a tremendous opportunity to offer integrated services through a wireless network.

One of the main market segments for wireless services besides corporate intranet/Internet access is the consumer sector. The availability of intelligent terminals[1] or multi-purpose wireless telephones is already ushering a new era of the information age, where subscribers

[1]For example, WAP terminals.

All IP in 3G CDMA Networks J. Castro
 ISBN: 0-470-85322-0

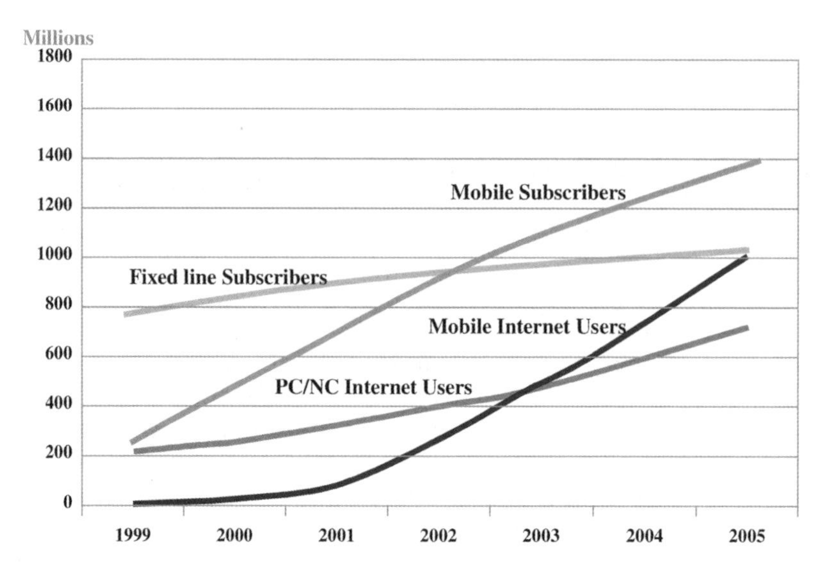

Figure 1.1 The growth of mobile and Internet services.

can receive the following directly through GSM/GPRS: news, sport updates, stock quotes, etc. However, the progress of audio-visual techniques and the support for a web-like interface in a new generation of terminals will push consumers to a new era of multimedia communications with a focus on services rather than technology. Figure 1.2 illustrates the 3G subscriber growth, which will enhance and accelerate multimedia communications in the mobile arena.

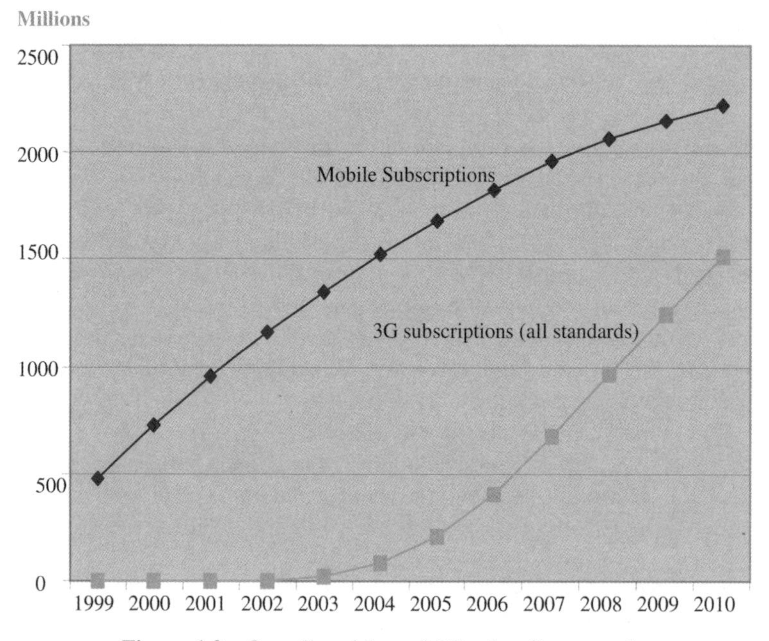

Figure 1.2 Overall mobile and 3G subscriber growth.

1.2 ROADMAP TO BROADBAND WIRELESS MULTIMEDIA

Broadband Wireless Multimedia (BWM) has started long ago, however, only recently it has been in the main stream of the development within mobile networks. The ubiquitous presence of Radio Frequency (RF), network, terminal and software technology has motivated the expansion of wide-band wireless applications in the cellular network environment. As these technologies blend, the number of applicational possibilities grow higher and higher, and so does the system intricacy.

Thus, to a great extent the evolution and success of BWM depend on the efficient blending of technologies and making it available to users outside its design sophistication.

Today UMTS stands as the platform to make BWM a practical reality, it offers a framework with open standards, which harmonises the building blocks and assures consistency in its evolution. Contrary to what it may appear, it does not only offer an advanced radio interface, UMTS also incorporates all the building blocks to make BWM easy towards the user by masking the unavoidable complexity through a well-defined architectural structure. It comprises the contribution of all network domains and terminal technologies. It sets not only specifications for infrastructure elements but also the technical recommendations to implement services and applications, as well as network management.

This book starting with this first chapter aims to expand the key building blocks of UMTS and thereby offers a concise perception of its capabilities and characteristics to enable BWM through the next decade.

To support the growth of Internet-type services[2] and future demands for wireless services, ETSI SMG, 3GPP and other standards bodies[3] have completed or are now completing specifications to provide a transition platform or evolution path for wireless networks like GSM. Figure 1.3 illustrates the wireless data technology options, which could be summarised as follows:

- 14.4 kilo bits per second (kbps) allows GSM data calls with a rate of 14.4 kbps per time slot, resulting in a 50% higher data throughput compared to the current maximum speed of 9.6 kbps.
- High-Speed Circuit-Switched Data (HSCSD) aggregates symmetrically or asymmetrically several circuit channels, e.g. 28.8 kbps for two time slots (2 + 2) or 43.2 kbps for three time slots (3 + 1).
- General Packet Radio Service (GPRS) enables GSM with Internet access at high spectrum efficiency by sharing time slots between different users. It affords data rates of over 100 kbps to a single user while offering direct IP connectivity.
- Enhanced Data Rate for GSM Evolution (EDGE) modifies the radio link modulation scheme from GMSK to 8QPSK, thereby increasing by three times the GSM throughput using the same bandwidth. EDGE in combination with GPRS (E-GPRS) will deliver single-user data rates of over 300 kbps.
- UMTS as Third Generation T G (3G) wireless technology utilises a wide-band CDMA or TD/CDMA transceiver. Starting with channel bandwidths of 5 MHz it will offer data rates

[2]Including voice or IP as a new trend.
[3]In the USA – T1P1, in Japan – ARIB, in Korea – TTA and in China – CWTS.

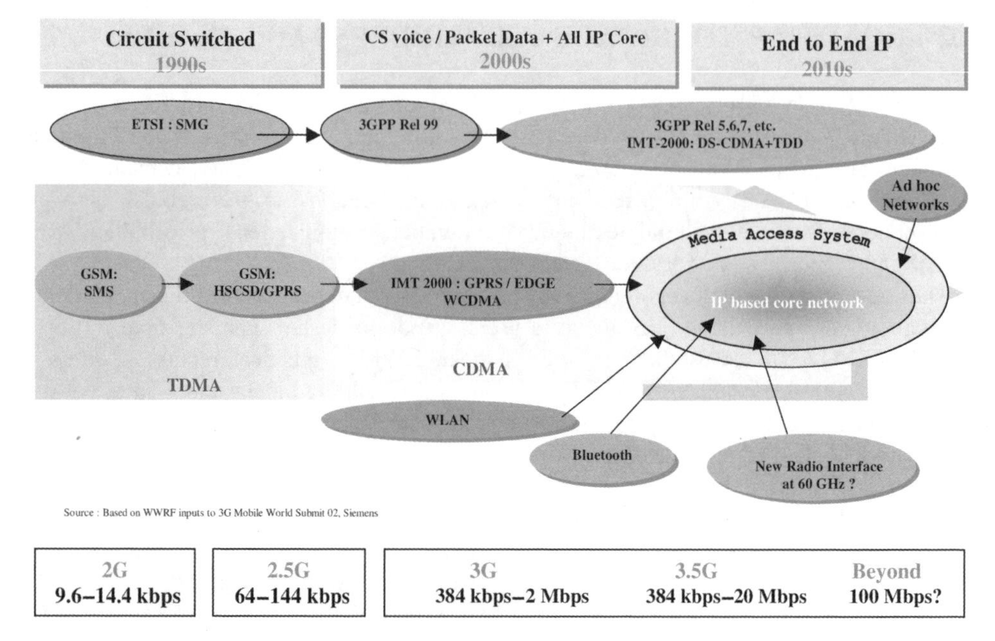

Figure 1.3 Evolution for wireless network,[4] e.g. GSM.

up to 2 Mbps. The introduction of UMTS with Release 1999 (R99) will use new spectrum and new radio network configurations while using the GSM core infrastructure.

Although the circuit-switched enhancements such as HSCSD has increased transmission rates, it is packet-switched enhancements which will meet the challenges or demands posed on current wireless networks. Thus, GPRS and UMTS with EDGE as an intermediate solution will provide the platform to support integrated services of voice and data including multimedia.

As seen in Figure 1.3, after the introduction of R99, where the main innovation took place in the radio side with the incorporation of WCDMA, subsequent releases of UMTS (e.g. R5, R6, etc.) will bring evolution in the core by adding 'All IP' features. Thus, by 2005 UMTS (CNs) will support real-time packet switching to enhance VoIP and streaming, for example. By this time, continuing innovation in the radio will also start expanding transmission rates up to 20 Mbps with High-Speed Packet Downlink Access (HSPDA)[5] and intelligent or Adaptive Antennas (AA).

Figure 1.3 illustrates also the 2010s period, where UMTS will aim to offer end-to-end IP transport (i.e. RAN and core) plus new radio technologies and push transmission rates up by 100 Mbps. Therefore, it seems reasonable to think that UMTS and its seamless complementary access technologies (e.g. WLAN and Bluetooth) will serve as the BWM platform for future mobile network evolution.

While GPRS and UMTS meet the demands for Internet (IP) features and higher bandwidths in mobile networks, another evolution step is taking place in the network

[4] IS-136 has adopted EDGE as its air–interface expansion.
[5] Dedicated chapters will cover HSPDA and AA.

infrastructure. This is the convergence of single networks into a multi-purpose backbone network. The next section covers this step, which will also have an impact on the implementation of UMTS radio-access technology.

1.2.1 Convergence of Fixed and Mobile Networks

Convergence, i.e. the closer inter-working between fixed and mobile telecommunications, although has long been a buzzword in the telecom market, is now coming into reality. As Ericsson puts it [4], fixed and mobile convergence includes everything from new services to the integration of nodes, networks and operating systems. The user may have, e.g. the same voice mailbox for fixed and mobile telephony, while the operator can also use the large sections of the network in a co-ordinated manner for different types of access. Thus, convergence is now a new frontier in communications, where UMTS will evolve [5–14].

Figure 1.4 illustrates how single service networks will evolve into multi-purpose networks with multi-level access points. With IP becoming more pervasive in the backbone, the challenge of integrating voice and data services in the fixed and mobile environments become more formidable.

It boils down to the transformation of the telecom, computer and media industry, resulting into the converged industry as illustrated in Table 1.1.

Clearly then, UMTS will be part of the convergent industry with a trend towards multi-services within integrated infrastructures.

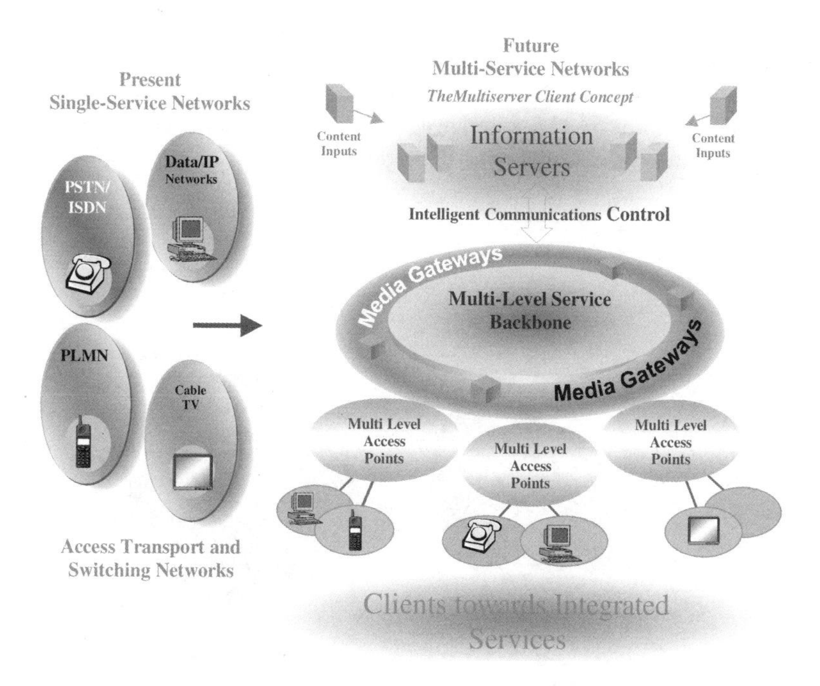

Figure 1.4 Multi-service network.

Table 1.1 The converging industry in telecommunications, computers and media

Telecom industry (New telecome)	Computer industry (Converged industry)	Media industry (New battle grounds)
PSTN	Main frames	–
PTN	Desktop computing	
ISDN	PL-LAN PC-Servers	–
GSM	PC-WAN	–
GSM+GPRS	Internet/intranet WWW	Electronic publication
UMTS mobile	Electronic commerce	Interactive entertainment
UMTS WLANs	–	–

1.2.2 The Next Decade of UMTS

Predominant standards will drive the evolution of mobile networks in the context of convergence and growth of the wireless Internet. For example, Figure 1.5 shows technology trends, in which GSM and WCDMA will set the path for expansion into fully BWM systems. The global evolution will result into one seamless network, where GSM radio operating at 800, 900, 1800 or 1900 MHz spectrum with nationwide coverage will share CN, transport, sites, terminals and network management with WCDMA radio operating at 2 GHz frequency with urban and sub-urban coverage at the beginning.

Thus, an extension of coverage and service beyond 2005 will to a great degree follow the UMTS path, using WCDMA and its enhancements as the radio-access technology. On the other hand, EDGE or other technologies will still serve as an alternative for expansion, as it is happening in the US with some operators.

However, in terms of throughput and broadband capabilities, WCDMA offers more. Figures 1.6 and 1.7 illustrate a reference comparison of packet performance, where WCDMA supports more users and higher throughput than concurrent access techniques, e.g.

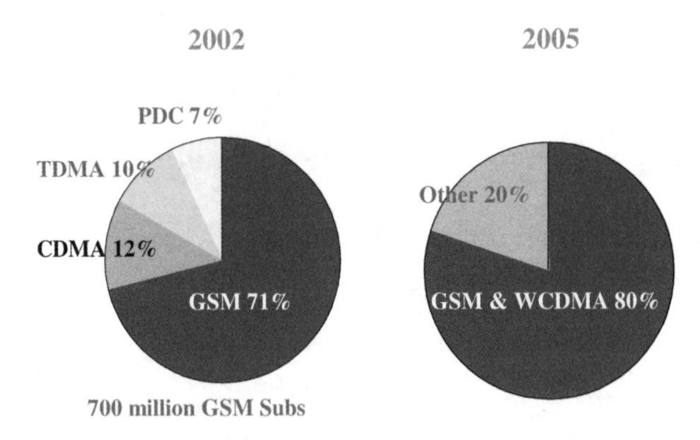

Figure 1.5 Technology trends of mobile digital subscribers.

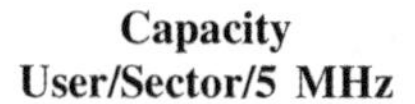

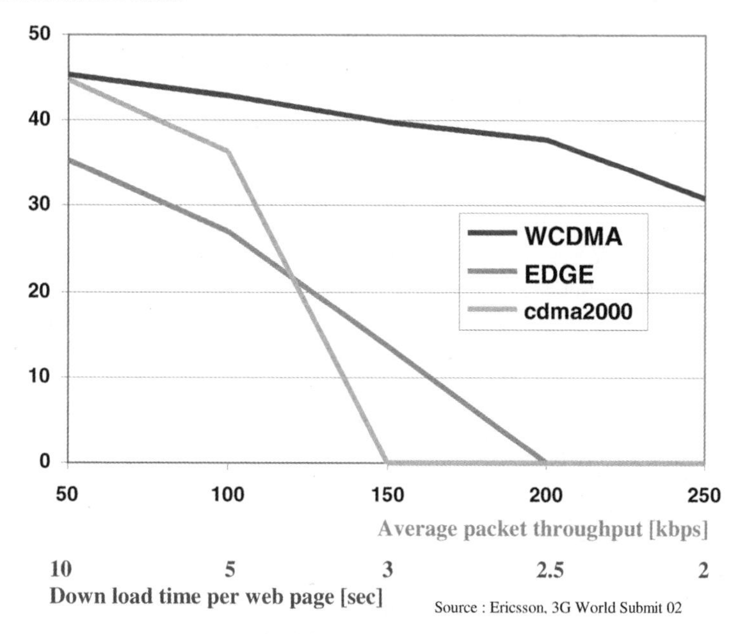

Figure 1.6 Packet data performance in WCDMA.

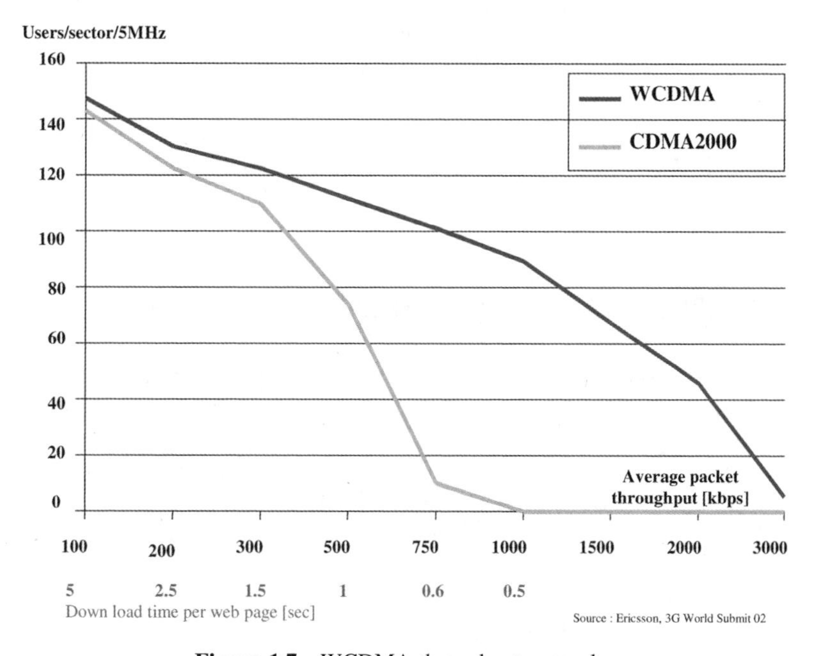

Figure 1.7 WCDMA throughput example.

CDMA2000 and EDGE. The latter utilises existing 200 kHz GSM spectrum carrier as a direct evolution in the air-interface side, while CDMA2000 uses 1.25 MHz carrier also as direct radio evolution of IS-95.

UMTS works with new and existing spectrum using a 5 MHz carrier built on GSM core network. Therefore, there will be coverage with seamless handover between WCDMA and GSM/GPRS networks even if UMTS starts covering primarily urban areas at its introduction.

UMTS, taking advantage in most countries from GSM covering nationwide with either 800, 900, 1800 or 1900 MHz, one core GSM/UMTS (utilising one transport, sharing as many sites as possible, exploiting GSM/UMTS multi-mode terminals) and one network management system, stands as the seamless global network of the future.

Figure 1.7 illustrates how as a whole, WCDMA radio in UMTS offers the capabilities to expand in terms of throughput and co-exists with evolving network technologies. WCDMA offers more traffic per radio carrier, thereby higher capacity and lower cost per user. For example, within 200 kHz GSM carrier, we can support 8 voice users, while with 5 MHz WCDMA carrier we go up to 76 voice users.

Furthermore, as seen in Figure 1.7 projections indicate that later releases of UMTS will support more users than CDMA2000 and EDGE with higher QoS and throughput. Although such comparisons are relative and depend on assumptions selected as well as the measurement criteria, the potential of WCDMA is there due to its larger bandwidth.

Finally, the optimisation potential of UMTS through capacity enhancing techniques are yet to appear in the field. Hence, it remains to exploit improvements in:

- basic system algorithms, e.g. reduce operating E_o/N_o in the baseband,
- radio network, e.g. optimise admission and congestion control,
- power management, e.g. self-optimised power control for more efficient SHO,
- transmission diversity improvements in the downlink,
- hierarchical cell structure to maximise service options,
- adaptive antennas, beam-forming methods, optimised modulation, etc.

1.3 UMTS PERFORMANCE ENHANCING TECHNOLOGIES

Information and material science technology as well as design and manufacturing techniques continue advancing with positive impacts on telecommunications, and thereby also in UMTS. Hence, in the following we will aim to gather key applications, which will increase the overall performance of UMTS, e.g. as noted above. Since the scope for progress can be large, this section will only summarise main solutions and outputs but not necessarily cover in depth all aspects studied already or under study in related technical literature. The main areas of interest include means to increase capacity, efficient transport and spectrum efficiency. While the options to achieve these objectives may vary from contribution to contribution, here we will attempt to pick up the main stream, which is somehow also supported by the standard bodies.

1.3.1 Drivers To Rise Output

The first set of key drivers to maximise output and performance from current and forthcoming mobile infrastructure solutions include the following:

- Spectrum availability and cost-limited frequency ranges for UMTS coupled with speculation for high profits has made ownership extremely expensive and now it must be exploited by all means.
- Mixed-media wireless data applications will require essentially more bandwidth to enrich and motivate mobile users. After all, UMTS is meant to enhance consumer experience at all levels and make wireless multimedia ubiquitous.
- Packing more bits within an allocated frequency implies improving spectral efficiency, which in turn allows more users to be served at larger rates. This provides a vital means for cost reduction.
- Maximum coverage besides broadband transmission, i.e. high rate services, has also become an important concern to augment and speed up 'Return of Investments' (ROI).

Before we list the different technologies maturing to provide a great opportunity to achieve substantial increases in spectral efficiency, coverage and overall system cost reduction, we shall bring the criteria to apply these into commercial solutions.

1.3.2 Applying New Technologies in Evolving UMTS Networks

Above all, technologies are to facilitate and make easier the provision of new services and not make it more complex, needless to say more costly. Thus, we have to see how services will flow in the infrastructure:

- Broadband data services will be asymmetric. Thus, the downlink will represent greater challenge assuming that traffic volumes will be higher downwards at first.
- Packet data, as the new main stream traffic, will require adaptive modulation and coding to maximise throughput in order to maintain flexibility and high performance. These will result in edge intelligence to enable faster MAC functions.
- Improvements shall therefore better fit if applied to both BTS and UE, e.g. multiple antennas.
- If the predominant data comes in the packet mode; logically, the access network infrastructure shall evolve into routed IP architecture, with seamless integration of any type of radio technologies.

The second set of key drivers, which will be outlined in the next chapter, concerns solutions and applications to value added services themselves.

In the next sections we will condense the main aspects of selected enhancing technologies, capacity increasing techniques and system solutions, e.g. intelligent edge nodes.

1.3.3 Capacity Increasing Antennas

By increasing the number of BS antennas, we can resolve the uplink limitation of WCDMA. However, this approach does not allow a single-step solution because many factors intervene before completing the process. These factors include propagation environment, BS configuration and environmental issues originating, e.g. from different power levels and network integration procedures with the RNC. However, here we consider first the BS configuration by looking at the antenna design. We need low correlation between the antennas achievable by adequate separation between the antennas. The beam-forming technique may exploit a uniform linear array, where the inter-antenna spacing falls near 1/2 of a carrier wavelength. Then sectors using narrow beams will have an increased antenna gain when compared to typical sector antenna.

While the pico- and micro-environments have a higher angular diversity, the macro-environment has a lower angular diversity, but higher multi-path diversity. Thus, the macro-environment can benefit from beam-forming techniques, because the latter applies more to lower angular diversity conditions. The optimum number of branches will depend on the accuracy of the channel estimation; Godara [15,16] presents more beam-forming options related to mobile applications. The result implies that there exists considerable range of options to increase capacity through adaptive antenna techniques. As Javed [17] puts it (Table 1.2), it's a complex trade-off dictated by:

- air-interface constraints,
- benefits versus complexity/cost metrics,
- propagation channel considerations (e.g. angle/temporal spread).

Table 1.2 Downlink antenna techniques

Space time techniques for downlink antenna solutions		
Beamforming 'Narrow' aperture Signal 'spatially coloured'	Diversity 'Wide' aperture Signal 'spatially white'	*Multiple Input Multiple Output (MIMO)* Exploits parallelism and MIMO channel
Planewave beamforming Fixed (switched) beam 'Bearing estimation'	*Feedback/TDD diversity* Selective transmit diversity TxAA and pre-RAKE	*Feedforward MIMO* Space time trellis Space time block codes Layered space time coding Space time turbo trellis
Non-plane beamforming Max SINR Min BER	*Feedforward diversity* Delay Diversity (DD) Code Word Diversity (CWD) Orthogonal Transmit... (OTD) Phase Sweeping Transmit... (PSTD) Amplitude Sweeping Transmit... (ASTD) Time Switched Transmit... (TSTD)	*Feedback MIMO* Eigenmode STC Water pouring STC

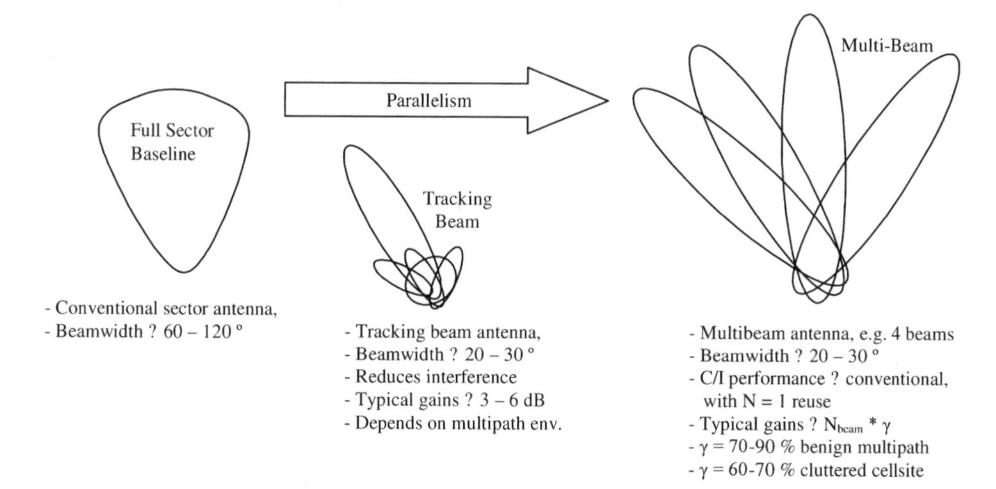

Figure 1.8 Basic beam-forming options.

If we go along the considerations of the technical specifications, e.g. [18], and exploit the power of parallelism as illustrated in Figure 1.8 (after [17]) we shall see that a multi-beam approach represents the most effective means to enhance aggregate capacity per cell site. Although antenna and feeder complexity increases, this approach takes advantage of compact antenna facet technology with integral beamforming. Beam gain relative to the full-sector system can permit increased coverage or may allow relaxation of PA requirements, and thereby reduce cell-site costs [17].

On the other hand, non-uniform densities may limit capacity benefits obtained from spatial processing; e.g. in the case of WCDMA, the terminal will not accurately assess the spectrum angle seen by the BTS, which means DL beam solutions will depend on UL measurements. Thus, very selective beamforming, which depends from the spread angle in the RF propagation channel, shall be deployed.

At the end, what an UMTS operator or service provider will expect is that capacity improvements do not mean radical replacement in the BTS or terminal side, but a modular and seamless process as the technology matures.

Other technical details on related applications shall be covered in Chapter 8, while introducing the UTRA High Speed Downlink Packet Access (HSDPA).

1.3.4 Multi-User Detection Techniques

Multi-User Detection (MUD) techniques may apply to both the UL and DL. However, initially due to processing power constraints in the MS, MUD may be exploited first in the BS. Thus, here we look at performance enhancement primarily in the UL while implementing MUD in the BS. The two UTRA modes, i.e. FDD and TDD, can benefit from MUD techniques. In fact, the joint detection algorithm is already an inherent part of the TDD mode.

Capacity within interference-limited WCDMA can improve through the use of efficient receivers. This implies that the structured multiple-access interference can be dealt with at the receiver through multi-user detectors [19]. MUD techniques have been covered at length in [20,21]. Here we aim to point out some of the promising techniques, which can apply to future releases of the WCDMA mode.

Studies on MUD techniques for WCDMA BS receivers [22–24] indicate that a multi-stage Parallel Interference Cancellation (PIC) may suite well WCDMA systems with a single Spreading Factor (SF). The parallel interference cancellation implies that interference gets cancelled from all users concurrently. MUD techniques for multi-service WCDMA with a variable spreading factor has been studied in [25], where a Group-wise Serial Interference Cancellation (GSIC) receiver [26–28] appears to be the most promising of the present receiver designs. In this technique, users with a given SF are also detected concurrently, after which the MAI[6] originated by them gets suppressed by the users having different SF.

1.3.5 Software Radio Applications

Although 3G wireless communication concepts, e.g. IMT-2000 family of networks, aim towards global standardisation to break away with multiple standards deployed in particular geographical areas, there is a need for multi-frequency transceivers operating in common hardware platforms for practical solutions in the medium and long terms.

This solution appears more realistic today through Software Radio (SR), and through the application of flexible and programmable transceivers. Thus, SR sets itself as a key technology to drive the realisation of global standards in 3G systems. The evolution of GSM to UMTS alone will benefit multi-band, multi-mode [GSM 900, 1800, 1900, GPRS, UMTS (FDD and TDD)] terminals. On the other hand, SR not only applies to terminals or Mobile Stations (MSs) but also the to the Base Stations (BS). In the sequel we cover SR as part of the enabling techniques in the MS and BS.

The main limitation of the feasibility of MUD in real commercial systems has been the disproportionate processing speeds afforded by current DSP[7] technology and the requirement of the detection and estimation algorithms. Although overall performance of DSPs has increased and keeps increasing, 3G systems also are pushing the signal processing capabilities higher and higher. Tasks such as high-data-rate signal acquisition, more accurate channel estimation for highly selective fading environments, fast signal quality estimation algorithms involved in power control and optimum combining of signals for diversity gains in space and time, demand all the power a processor can produce. These demands can be realised more rapidly through Software Defined Radio (SDR).

Thus, while compatibility between standards remains attractive, SDRs will shape into software and hardware reconfigurable radios in the RF, Intermediate Frequency (IF), as well as baseband processing stages [29–32].

1.3.6 Packet Oriented Architecture

The introduction of UMTS (Release 1999) benefited from the established GSM building blocks and architecture. Its immediate evolution through Releases 4 and 5 has even added further IP subsystems, e.g. IP Radio-Access Network (RAN) transport and IP-Multimedia

[6]Multiple access interference.
[7]Digital signal processor.

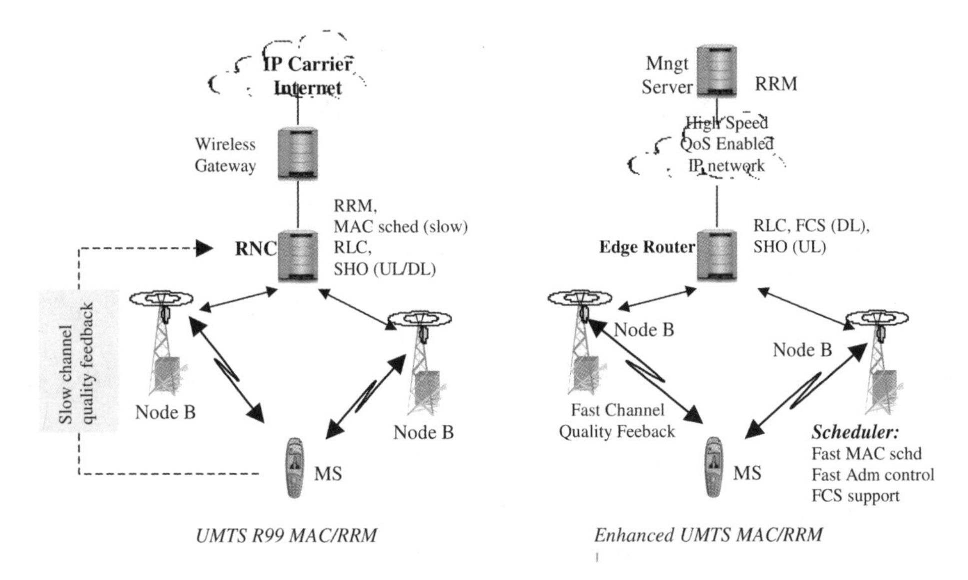

Figure 1.9 MAC/RRM in R99 and enhanced UMTS.

Subsystem (IMS). This implies to a great degree that enhancements on the UMTS architecture will come through a more efficient use of the packet mode features.

Then if we look at the Network Elements (NE) of the established mobile infrastructure illustrated in Figure 1.9 (left), we shall notice that Node B's have limited functions, which make the channel quality feedback slow. As outlined in [17] this architecture has:

- circuit-oriented design inappropriate and inefficient for busty traffic,
- scheduler at RNC–slow resource allocation,
- slow updates for SHO during cell selection.

A packet oriented MAC model illustrated also in Figure 1.9 (right after [17]) would allow a more dynamic interaction of core and Mobile Station (MS), resulting in

- a spectrum efficient packet-by-packet fast scheduling,
- Fast Resource Allocation (i.e. code, modulation and bandwidth) with the scheduler at the BTS,
- Fast Cell Switching (FCS) due to the optimisation for packet data delivery, capacity, coverage and high data rate at the cell edge:

The design of the enhanced UMTS MAC/RRM will permit to go from a hierarchical to a routed (edge-intelligent) network design. In the first case

- each Node B gets virtually connected point-to-point to one and only one RNC, where each RNC performs control/bearer plane functions, as well as terrestrial switching interactivity functions,
- traffic flows are strictly constrained up/down the hierarchy,

- HO between RNCs implies full transfer control, bearer and/or switching functions.

Within routed network architectures, following the approach and terminology in [17], we can split or separate the RNC functions as follows:

- Control plane functions would go to control or computing servers, e.g. RNC-C.
- Bearer plane functions would go to radio link processor pool, e.g. RNC-B.
- Traffic aggregation and switching functions would get delegated to routed network infrastructure.
- In this design, any Node B can communicate with any RNC-B, i.e. strict hierarchy gets eliminated.
- Likewise, any Node B can communicate directly (through a routed network) with any concerned RNC-C.

Figure 1.10 illustrates the evolution of routed network architectures. Some as in [17] consider the basis for fourth generation networks. However, if technologies mature early enough, the UMTS performance enhancement can be deployed gradually without having to wait for a big wave to improve capacity, for example. In fact, the last thing the cellular telecom industry needs today is an ambiguous perception of technology capabilities.

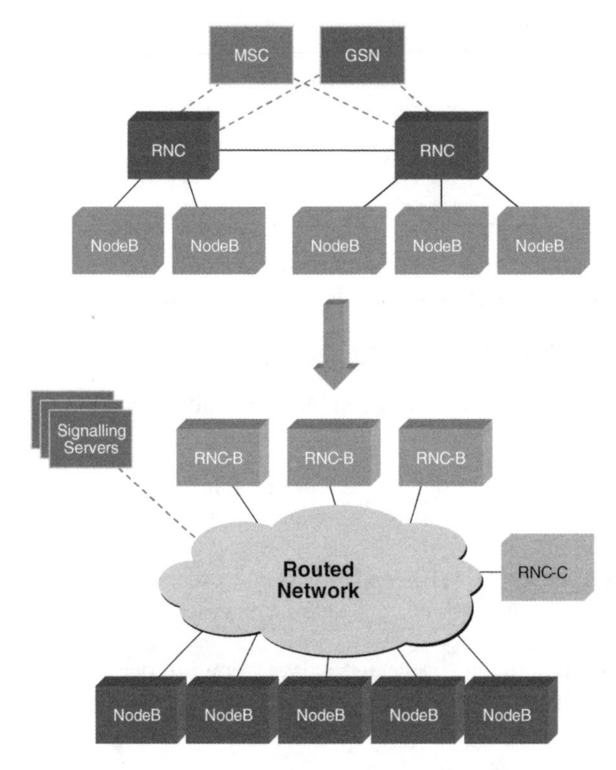

Figure 1.10 Towards routed networks.

1.3.7 Implementation and Integration Aspects

Research studies aiming to improve the overall performance of multiple-access techniques such as WCDMA or TDCDMA have provided interesting and applicable methods. However, these results may not necessarily be part of the first UTRA commercial systems in the next 2 years. Thus, it will be some time before techniques such as software radio, adaptive antennas and multi-user detection enhance capacity, coverage and increase system stability.

Implementation and integration appear as key limitations to bring these advanced techniques into operating systems or near future[8] exploitable networks. Processing power demands, for example, do not allow rapid implementation of the above methods. Furthermore, integrating such techniques into smaller components is a great challenge. This means that while less optimum supporting techniques like system on a chip, maximising power consumption or operating at very low power come into place, the aforementioned improvements will remain academic.

At present, while UMTS frequency licensing is becoming big business for governments, operators seem to have fallen into the spin of supremacy and consolidation for market share and have somehow forgotten the timeliness of technology. Manufacturers are finding themselves in a race to supply plain vanilla solutions and are incapable of implementing true breakthroughs in multiple-access or radio-access techniques.

Thus, it seems reasonable to think that it may be for the benefit of the industry as a whole and governments themselves to concentrate on putting more resources into the realisation of new communication technologies than just coping with spectrum allocation and acquisition to offer services with higher transmission rates. Such an approach will make UMTS a clear platform for advanced technology from the start and not just one more alternative to provide new mobile applications.

1.4 UMTS NETWORK REQUIREMENTS

The first release of UMTS (R99) introduced major evolution in the air interface, while subsequent releases, e.g. R4 and R5, brought newer functions and new building blocks to the core side. In addition, it also has an impact on the service configuration options. Hence, the 3G, or more specifically UMTS requirements in this section cover selected portions of four main areas, i.e. services, air interface, core-network access and the enhancements for long-term evolution.

1.4.1 UMTS Services Aspects

The scope of services can be largely focused on different issues like service management, charging and billing, terminals, network management, quality of service and security. Here, however, we will be looking at services from the principle side in order to establish a framework to present the UMTS air interface. An extract of the service principles outlined in

[8]Recent evaluation on end-to-end industrial solutions do not yet show these techniques as a part of the product.

the ETSI specifications UMTS services aspects – service principles and UMTS services [33] and service capabilities [16] – can be summarised as follows:

> UMTS is the realisation of a new generation of mobile communications technology for a world in which personal communication services should allow person-to-person calling, independent of location, the terminal used, the means of transmission (wired or wireless) and the choice of technology.

UMTS shall therefore be in compliance with the following objectives:

a. To provide a single integrated system in which the user can access services in an easy-to-use and uniform way in all environments.

b. To allow differentiation between service offerings of various serving networks and home environments.

c. To provide a wide range of telecommunication services including those provided by fixed networks and requiring user bit rates of up to 2 Mbps/s as well as services specific to mobile communications. These services should be supported in residential, public and office environments, and in areas of diverse population densities. These services are provided with a quality comparable with that provided by fixed networks such as ISDN.

d. To provide services via hand held, portable, vehicular mounted, movable and fixed terminals (including those which normally operate connected to fixed networks), in all environments (in different service environments – residential, private domestic and different radio environments) provided that the terminal has the necessary capabilities.

e. To provide support to roaming users by enabling users to access services provided by their home environment in the same way even when roaming.

f. To provide audio, data, video and particularly multimedia services.

g. To provide for the flexible introduction of telecommunication services.

h. To provide the capability to support Universal Personal Telecommunications (UPT).

i. To provide within the residential environment the capability to enable a pedestrian user to access all services normally provided by fixed networks.

j. To provide within the office environment the capability to enable a pedestrian user to access all services normally provided by PBXs and LANs.

k. To provide a substitute for fixed networks in areas of diverse population densities, under conditions approved by the appropriate national or regional regulatory authority.

l. To provide support for interfaces which allow the use of terminals normally connected to fixed networks.

In addition, UMTS aims:

- to enable users to access a wide range of telecommunication services, including many that are today undefined as well as multimedia and high data rates,

- to facilitate the provision of small, easy to use, low cost terminals with long talk time and long standby operation,
- to provide an efficient means of using network resources (particularly radio spectrum).

Based on the above objectives, specific requirements related to services are outlined in the ETSI specifications [4,34,35]. These requirements are primarily concerned with items such as quality of service, security and charging, service management, etc.

1.4.2 UMTS Terrestrial Radio-Access Aspects

The UMTS Terrestrial Radio-Access (UTRA) system requirements are based on the service requirements. The latter sets the demands which UTRA specification aims to meet. Table 1.3 summarises key (selected) requirements identified for the UTRA concept from [36].

By looking at the bearer capabilities in Table 1.3 we can see that evolution towards higher rates will initially apply mainly to indoor rates. In this environment convergence will also have higher impact. In addition, UTRA will not only prevent seamless HO between different operators or access networks, but also support HO between 2G and 3G systems, e.g. GSM and UMTS.

UTRA will support key technologies, like ATM, IP, B-ISDN, as well as GSM, when it comes down to CN transports. This will consolidate the trend of 2G CN towards integrated circuit-switched and packet-switched services.

1.4.3 IP-Multimedia CN Subsystem (IMS) Requirements

The requirements in the preceding section cover primarily R99. Here we update requirements of key components in R5, in particular the ones for IMS based on the inputs from [37].

IMS comprises all CN elements for the provision of IP-multimedia applications over IP-multimedia sessions. An application handles one or more media simultaneously, such as speech, audio, video and data (e.g. chat text and shared whiteboard), in a synchronised mode in the user side. A multimedia application may involve multiple parties and multiple connections, while adding or deleting resources within a single IP-multimedia session. A user may invoke concurrent IP-multimedia applications in an IP-multimedia session.

An IMS session implies a set of multimedia senders and receivers and the data streams flowing from senders to receivers. These sessions are supported by the IP-multimedia CN subsystem and are enabled by IP connectivity bearers (e.g. GPRS as a bearer). A user may invoke concurrent IP-multimedia sessions.

Some of the high level requirements for IP-multimedia applications can be summarised as follows:

- Negotiable QoS in IMS sessions both before establishment and during the session.
- Negotiable QoS for individual media components in IMS sessions both before establishment and during activity.

Table 1.3 UTRA high level requirements

Key requirements	Description
	Bearer capabilities
Maximum user bit rates	*Rural outdoor*:[a] at least 144 kbps (goal to achieve 384 kbps), maximum speed: 500 km/h *Sub-urban outdoor*:[b] at least 384 kbps (goal to achieve 512 kbps), maximum speed: 120 km/h *Indoor/low range outdoor*:[c] at least 2 Mbps, maximum speed: 10 km/h The UTRA definition should allow evolution towards higher bit rates
Flexibility	Negotiation of bearer service attributes (bearer type, bit rate, delay, BER, etc.) Parallel bearer services (service mix), real-time/non-real-time communication modes, etc. Circuit and packet oriented bearers Support scheduling (and pre-emption) of bearers (including control bearers) within priority Adaptability of link to quality, traffic and network load, as well as radio conditions Wide range of bit rates should be supported with sufficient granularity Variable bit rate real-time capabilities should be provided Bearer services appropriate for speech shall be provided
Handover	Provide seamless (to user) handover between cells of one operator The UTRA should not prevent seamless HO between different operators or access networks Efficient handover between UMTS and second generation systems, e.g. GSM, should be possible
	Operational requirements
Compatibility with services provided by present core transport networks	ATM bearer services GSM services Internet Protocol (IP) based services B/N-ISDN services
Radio-access network planning	If radio resource planning is required, automatic planning shall be supported
Public network operators	It shall be possible to guarantee pre-determined levels of QoS and quality to public UMTS ops
Private and residential operators	The radio-access scheme should be suitable for low cost applications where range, mobility and user speed may be limited Multiple unsynchronised systems should be able to successfully co-exist in the same environment It should be possible to install base stations without co-ordination Frequency planning should not be needed
	Efficient spectrum usage
Spectrum efficiency	High spectrum efficiency for typical mixtures of different bearer services Spectrum efficiency at least as good as GSM for low bit rate speech
Variable asymmetry of total band usage	Variable division of radio resource between uplink and downlink resources from a common pool (NB: this division could be in either frequency, time or code domains)

Table 1.3 (*Continued*)

Key requirements	Description
Spectrum utilisation	Allow multiple operators to use the band allocated to UMTS without co-ordination[d] It should be possible to operate the UTRA in any suitable frequency band that becomes available such as first and second generation system's bands
Coverage/capacity	The system should be flexible to support a variety of initial coverage/capacity configurations and facilitate coverage/capacity evolution Flexible use of various cell types and relations between cells (e.g. indoor cells and hierarchical cells) within a geographical area without undue waste of radio resources Ability to support cost effective coverage in rural areas
Mobile terminal viability	Hand-portable and PCM-CIA card sized UMTS terminals should be viable in terms of size, weight, operating time, range, effective radiated power and cost
Network complexity and cost	The development and equipment cost should be kept at a reasonable level, taking into account cell-site cost, cross-connect, signalling load and traffic overhead (e.g. due to handovers)
Mobile-station types	It should be possible to provide a variety of mobile-station types of varying complexity, cost and capabilities in order to satisfy the needs of different types of users
	Requirements from bodies outside SMG
Alignment with IMT-2000	UTRA shall meet at least the technical requirements for submission as a candidate technology for IMT-2000 (FPLMTS)
Minimum bandwidth allocation	It should be possible to deploy and operate a network in a limited bandwidth (e.g. 5 MHz)
Electro-magnetic Compatibility (EMC)	The peak and average power and envelope variations have to be such that the degree of interference caused to other equipment is not higher than in today's systems
RF radiation effects	UMTS shall be operative at RF emission power levels, which are in line with the recommendations related to electro-magnetic radiation
Security	The UMTS radio interface should be able to accommodate at least the same level of protection as the GSM radio interface does
Co-existence with other systems	The UMTS terrestrial radio access should be capable of co-existing with other systems within the same or neighbouring band depending on systems and regulations Multi-mode implementation capabilities It should be possible to implement dual mode UMTS/GSM terminals cost effectively

[a]The specified bit rate will be available throughout the operator's service area, with the possibility of large cells.
[b]The specified bit rate will be available with complete coverage of a suburban or urban area, using micro-cells or smaller macro-cells.
[c]The specified bit rate will be available indoors and localised coverage outdoors.
[d]NB: the feasibility of spectrum sharing requires further study.

- End-to-end QoS for voice comparable to the circuit-switched (e.g. AMR codec based) one.
- Automatically negotiable QoS for current/new service capabilities during roaming.
- Operator IP policy control implementing capabilities for IMS applications.
- IMS sessions shall support different varieties of identifiable and inter-operable media types (e.g. default codec selection and header compression).
- Each IMS session shall support one or more IP-multimedia applications.
- IMS applications shall not risk reduction in privacy, security or authentication compared to corresponding GPRS and circuit-switched services.
- *IMS shall support inter-working between the packet- and circuit-switched services, and with PSTN and ISDN networks, as well as inter-work with Internet.*
- IMS shall support basic voice calls between IMS and CS domain/PSTN-style networks, and be able to convey the following service associated information: CLIP/CLIR; call forwarding, call barring, call waiting/hold, MPTY; when roaming: home environment and serving network services.
- IMS shall support access independence and serve subscribers regardless[9] of how they obtain an IP connection (e.g. GPRS, fixed lines, LAN, etc.). It shall also enable operators to limit network topology visibility to authorised entities.
- IMS shall support Internet-application sessions developed outside the 3GPP community, and enable operators to limit network topology visibility to authorised entities.
- CAMEL [38], MExE [39], SAT [40] and OSA [41], which are the identified service capabilities of VHE in 22.121 [42], should evolve to support IMS, e.g. additions to APIs, service capability features, service capability servers, user profile etc.
- *Both telecom and Internet numbering and addressing schemes shall be supported. IMS communication establishment (both mobile originating and terminating) depending on originator shall be able to be based on E.164 (e.g. +1 23 456 789) or SIP URL (sip: my.name@company.org)* [43].
- It shall be possible for the network operator or service provider to use the following.

  The same E.164 number for IP-multimedia sessions and CS speech telephony (TS11) [44].

  A different E.164 number if desired for IP-multimedia sessions.

"This allows customers who originally had only an E164 MSISDN to retain the same number for receiving communications in the IMS domain and also in the CS domain when outside IMS coverage" [37].

[9]Assuming that all authentication and security measures have passed.

1.5 CONCLUSIONS

Chapter 1 has presented a window to perceive the environment into which UMTS[10] will develop. It has set the background to introduce *UMTS radio-access technology*, the aim of this book. From the impressive growth of GSM and the Internet, as well as the UMTS air-interface specification requirements, UMTS Terrestrial Radio Access (UTRA) is well positioned to play the key role in the convergence of telecommunications towards integrated services. Therefore, the contents of future chapters describe in more detail some of the key elements shown generically in this chapter.

REFERENCES

1. Mehrotra, A., *Cellular Radio–Analog and Digital Systems*, Chapter 1. Norwood, MA: Artech House, 1994.
2. Young, W.R., Advanced mobile phone service: introduction, background, and objectives, *Bell System Tech. J.*, **58**(1), 1–14, 1979.
3. Macdonald, V.H., Advanced mobile phone service: the cellular concept, *Bell System Tech. J.*, **58**(1), 15–41, 1979.
4. Ericsson, Connection No. 2 June 1999.
5. UMTS 22.25, Quality of Service and Network Performance.
6. UMTS 22.05, Service Capabilities.
7. UMTS 33.20, Security Principles for UMTS.
8. UMTS 22.15, Security and Charging.
9. UMTS 22.24, New charging and Accounting Mechanisms.
10. UMTS 22.70, Virtual Home Environment.
11. UMTS 22.71, Automatic Establishment of Roaming Agreements.
12. UMTS 23.05, Network Principles.
13. TG24 Requirements for Charging, Billing, Accounting, Tariffing.
14. UMTS 22.20, Service Management.
15. Godara, L.C., Application of antenna arrays to mobile communications, Part I: Performance improvement, feasibility, and system considerations, *Proc. IEEE*, **85**(7), 1031–1060, 1997.
16. Godara, L.C., Application of antenna arrays to mobile communications, Part II: Beam-forming and direction-of-arrival considerations, *Proc. IEEE*, **85**(8), 1195–1245, 1997.
17. Javel, A., Nortel Networks "Evolving Enabling Technologies for 4G", Third Generation (3G) Mobile World Summit, Tokyo, Japan, 15–18th Jan 2002.
18. 3GPP TR 25.848 V4.0.0 (2001-03), Physical layer aspects of UTRA High Speed Downlink Packet Access, Release 4.
19. Verdú, S., Minimum probability of error for asynchronous Gaussian multiple-access channels, *IEEE Trans. Inform. Theory*, **32**(1), 85–96, 1986.
20. Verdú, S., *Multiuser Detection*, Cambridge, UK: Cambridge University Press, 1998.
21. Juntti, M. and Glisic, S., Advanced CDMA For Wireless Communications, in *Wireless Communications: TDMA Versus CDMA* (Eds. S. G. Glisic and P. A. Leppänen), Dordrecht: Kluwer, 1997, Chapter 4, pp. 447–490.
22. Ojanperä, T., Prasad, R. and Harada, H., Qualitative comparison of some multiuser detector algorithms for wideband CDMA, *Proc. IEEE Vehic. Tech. Conf.*, **1**, 46–50, 1998.

[10] In particular, UTRA.

23. Correal, N.S., Swanchara, S.F. and Woerner, B.D., Implementation issues for multiuser DS-CDMA receivers, *Int. J. Wireless Inform. Networks*, **5**(3), 257–279, 1998.
24. Juntti, M. and Latva-aho, M., Multiuser receivers for CDMA systems in Rayleigh fading channels, *IEEE Trans. Vehic. Tech.*, in press.
25. Wijting, C.S., Ojanperä, T., Juntti, M.J., Kansanen, K. and Prasad, R., Groupwise Serial Multiuser Detectors for Multirate DS-CDMA, *Proc. IEEE Vehic. Tech. Conf.*, in press.
26. Juntti, M., Performance of multiuser detection in multirate CDMA systems, *Wireless Pers. Commun.*, **11**(3), 293–311, 1999.
27. Juntti, M., Performance of multiuser detection in multirate CDMA systems, *Wireless Pers. Commun.*, **11**(3), 293–311, 1999.
28. Juntti, M., Multiuser detector performance comparisons in multirate CDMA systems, *Proc. VTC'98*, Ottawa, Canada, 1998, 36–40.
29. Seskar, I. and Mandayam, N., Software-defined radio architectures for interference cancellation in DS-CDMA systems, *IEEE Pers. Commun.*, **6**(4), 26–34, 1999.
30. Tsurumi, H. and Suzuki, Y., Broadband RF stage architecture for software defined radio in handheld terminal applications, *IEEE Commun. Mag.*, **37**(2), 90–95, 1999.
31. Walden, R.H., Performance trends of analog-to-digital converters, *IEEE Commun. Mag.*, **37**(2), 96–101, 1999.
32. Chester, D.B., Digital IF filter technology for 3G systems: An introduction, *IEEE Commun. Mag.*, **37**(2), 102–107, 1999.
33. UMTS 22.01, Service Aspects – Service Principles.
34. UMTS 22.25, Quality of Service and Network Performance.
35. TG32 UMTS – Radio Requirements.
36. *High Level Requirements Relevant for the Definition of the UTRA Concept*, V3.0.1, 1998–2010.
37. 3GPP TS 22.228 Service requirements for the IP-Multimedia, Core Network Subsystem (Stage 1), (Release 5), V5.4.0 (2001–2012).
38. 3GPP TS 22.078: Customised Applications for Mobile network Enhanced Logic (CAMEL); Service definition – Stage 1.
39. 3GPP TS 22.057: Mobile Execution Environment (MExE); Service description, Stage 1.
40. 3GPP TS 22.038: Technical Specification Group Services and System Aspects; USIM/SIM Application Toolkit (USAT/SAT); Service description; Stage 1.
41. 3GPP TS 22.127: Technical Specification Group Services and System Aspects; Stage 1 Service Requirement for the Open Service Access (OSA).
42. 3GPP TS 22.121: Technical Specification Group Services and System Aspects; The Virtual Home Environment.
43. RFC2543: SIP: Session Initiation Protocol.
44. 3GPP TS 22.003: CS Teleservices supported by a PLMN.

2

System Analysis Fundamentals

2.1 FUNDAMENTALS OF SYSTEM ANALYSIS

Third generation systems focus on providing a universal platform to afford multifarious communication options at all levels, i.e. the radio as well as the core network* sides. This implies the application of optimum techniques in multiple access and interworking protocols for the physical and upper layers, respectively. This chapter discusses the background of the multiple access or radio part of the UMTS specification. Several sources [1–5] have already covered all types of fundamentals related to the air interface. Thus, we focus only on the communication environment to access the radio link performance for coverage analysis and network dimensioning in forthcoming chapters.

2.1.1 Multiple-Access Options

The access technologies utilised in UTRA are unique because of the type of implementation and not because they are new. The combination of CDMA and TDMA techniques in one fully compatible platform make UTRA special. The WCDMA and hybrid TDMA/CDMA allow the FDD and TDD modes to co-exist seamlessly to meet the UMTS services and performance requirements. In the sequel we cover the fundamental characteristics for each access technique which serves as a building block for the UTRA modes.

2.1.1.1 Narrow-Band Digital Channel Systems

The two basic narrow-band techniques include FDMA (using frequencies) and TDMA (using time slots). In the first case, frequencies are assigned to users while guard bands maintain between adjacent signal spectra to minimise interference between channels. In the second case, data from each user takes place in time intervals called slots. The advantages of FDMA lie on the efficient use of codes and simple technology requirements. But the

*Includes Value Added Services – VAS solutions.

All IP in 3G CDMA Networks J. Castro

ISBN: 0-470-85322-0

drawbacks of operating at a reduced signal-to-interference ratio and the inhibiting flexibility[1] of bit-rate capabilities outweigh the benefits. TDMA allows flexible rates in multiples of basic single channels and sub-multiples for low bit rate broadcast transmission. It offers frame-by-frame signal management with efficient guard band arrangements to control signal events. However, it requires substantial amounts of signal processing resources to cope with matched filtering and synchronisation needs.

2.1.1.2 Wide-Band Digital Channel Systems

Some of the drawbacks and limitations in the narrow-band channel systems made room for wide-band channel system designs. In wide-band systems the entire bandwidth remains available to each user, even if it is many times larger than the bandwidth required to convey the information. These systems include primarily Spread Spectrum (SS) systems, e.g. Direct-Sequence Spread Spectrum (DSSS) and Frequency Hopping Spread Spectrum (FHSS). In DSSS, emphasised in this book, the transmission bandwidth exceeds the coherent bandwidth, i.e. the received signal after de-spreading resolves into multiple time-varying delay signals that a RAKE receiver can exploit to provide an inherent time diversity receiver in a fading environment. In addition, DSSS has greater resistance to interference effects when compared to FDMA and TDMA. The latter greatly simplifies frequency band assignment and adjacent cell interference. In addition, capacity improvements with DSSS (or more commonly referred to as DS-CDMA,[2] resulting from the voice activity factor) cannot apply effectively to FDMA or TDMA. With DS-CDMA, e.g. adjacent micro-cells share the same frequencies, whereas interference in FDMA and TDMA does not allow this. Other benefits and features can be found in [6–8]. Here we focus on the WCDMA or FDD mode and TDMA/CDMA or TDD mode of the UTRA solution.

2.1.1.3 The UTRA FDD Mode: WCDMA

Figure 2.1 illustrates some of the UTRA Frequency Division Duplex (FDD) characteristics. This mode uses wide-band Direct-Sequence Code Division Multiple Access (DS-CDMA), denoted as WCDMA. To support bit rates up to 2 Mbps, it utilises a variable spreading factor and multi-code links. It supports highly variable user data rates through the allocation of 10 ms frames, during which the user data rate remains constant, although the latter may change from frame to frame depending on the network control. It realises a chip rate of 3.84 Mcps within 5 MHz carrier bandwidth, although the actual carrier spacing can be selected on a 200 kHz grid between approximately 4.4 and 5 MHz, depending on the interference situation between the carriers.

The FDD has a self-timing point of reference through the operation of asynchronous BSs, and it uses coherent detection in the uplink and downlink based on the use of pilot reference symbols. Its architecture allows the introduction of advanced capacity and coverage enhancing CDMA receiver techniques, e.g. multi-user detection and smart adaptive

[1]The maximum bit per channel remains fixed and low.
[2]Direct-sequence code division multiple access.

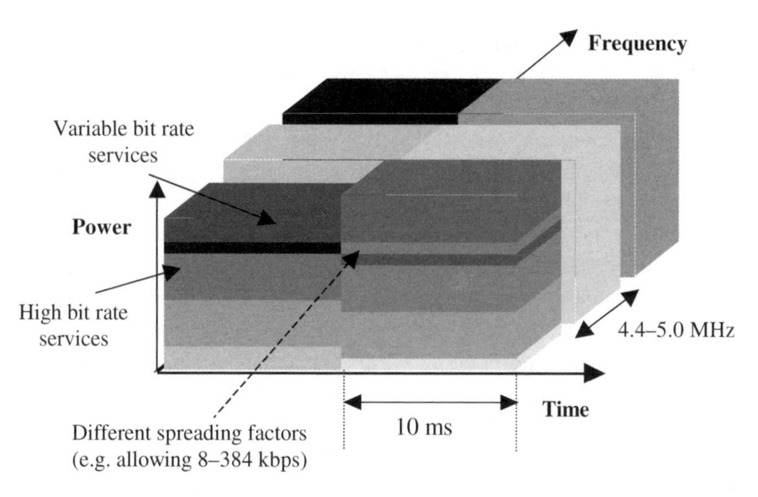

Figure 2.1 The UTRA WCDMA or FDD mode characteristics.

antennas. In addition, it will seamlessly co-exist with GSM networks through its inter-system handover functions of WCDMA.

2.1.1.4 The UTRA TDD Mode: TD/CDMA

The second UTRA mode results from the combination of TDMA and FDMA and exploits spreading as part of its CDMA component. It operates in time division duplexing using the same frequency channel.

In this mode, the MSs can only access a Frequency Division Multiplexing (FDM) channel at specific times and only for a specific period of time. Thus, if a mobile gets one or more

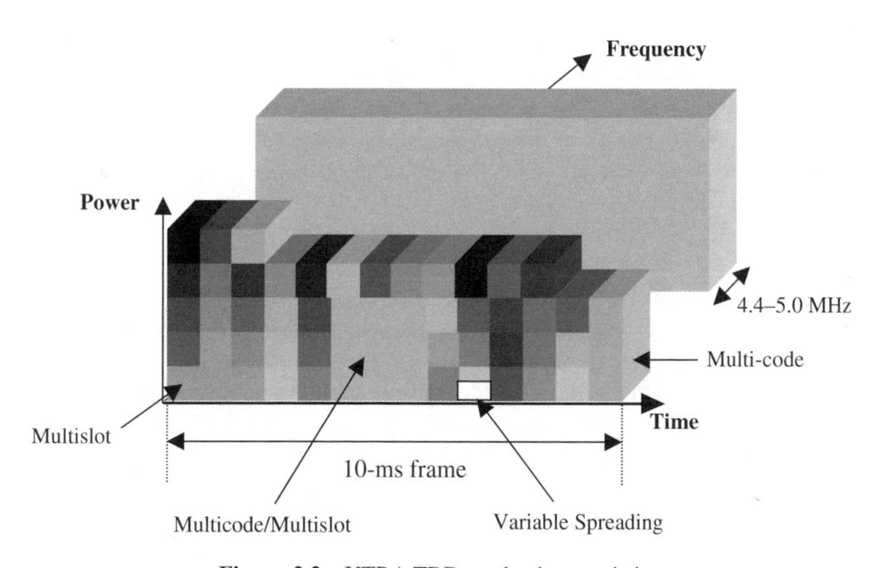

Figure 2.2 UTRA TDD mode characteristics.

Time Slots (TSs) allocated, it can periodically access this set of TSs throughout the duration of the frame. Spreading codes described in Chapter 4 separates user signals within one or more slots. Hence, in the TDD mode we define a physical channel by a code, a TS and a frequency, where each TS can be assigned to either the uplink or the downlink depending on the demand. Users may obtain flexible transmission rates by occupying several TSs of a frame as illustrated in Figure 2.2, without additional processing resources from the transceiver hardware. On the other hand, when more than one frequency channel gets occupied, utilisation of transceiver resources will increase if the wide-band transmission cannot prevent it. We achieve variable data rates through either multi-code transmission with fixed spreading or through single code with variable spreading. In the first case, a single user or users may get multiple spreading codes within the same TS; while in the second case, the physical channel spreading factor may vary according to the data rate.

2.1.2 Signal Processing Aspects

In the following, we review signal processing characteristics for the WCDMA as well as TD/CDMA as a base to describe key functions of the UTRA FDD and TDD modes. These include spreading aspects, modulation and coding.

2.1.2.1 The Spread Spectrum Concept

Digital designs of communication systems aim to maximise capacity utilisation. We can, for example, increase channel capacity by increasing channel bandwidth and/or transmitted power. In this context, CDMA operates at much lower S/N ratios as a result of the extra channel bandwidth used to achieve good performance at low signal-to-noise ratio. From Shannon's channel capacity principle [9] expressed as

$$C = B \log_2\left[1 + \frac{S}{N}\right] \tag{2.1}$$

where B is the bandwidth (Hz), C is the channel capacity (bps), S is the signal power and N is the noise power, we can find a simple definition of the bandwidth as

$$B = \frac{C}{1.44} \times \frac{N}{S} \tag{2.2}$$

Thus, for a particular S/N ratio, we can achieve a low information error rate by increasing the bandwidth used to transfer information. To expand the bandwidth here, we add the information to the spreading spectrum code before modulation. This approach applies, for example, to the FDD mode, which uses a code sequence to determine RF bandwidth. The FDD mode has robustness to interference due to higher system processing gain[3] (G_p). The latter quantifies the degree of interference rejection and can be defined as the ratio of RF

[3]Reference processing gains for spread spectrum systems have been established between 20 and 50 dB.

bandwidth to the information rate

$$G_p = \frac{B}{R} \tag{2.3}$$

From [10] in a spread-spectrum system, thermal noise and interference determine the noise level. Hence, for a given user, the interference is processed as noise. Then, the input and output S/N ratios can relate as

$$\left(\frac{S}{N}\right)_o = G_p\left(\frac{S}{N}\right)_i \tag{2.4}$$

Relating the S/N ratio to the E_b/N_o ratio,[4] where E_b is the energy per bit and N_o is the noise power spectral density, we get

$$\left(\frac{S}{N}\right)_i = \frac{E_b \times R}{N_o \times B} = \frac{E_b}{N_o} \times \frac{1}{G_p} \tag{2.5}$$

From the preceding equations we can express E_b/N_o in terms of the S/N input and output ratios as follows

$$\frac{E_b}{N_o} = G_p \times \left(\frac{S}{N}\right)_i = \left(\frac{S}{N}\right)_o \tag{2.6}$$

2.1.2.2 Modulation and Spreading Principles

In wide-band spread-spectrum systems like the FDD mode, the entire bandwidth of the system remains available to each user. To such systems, the following principles apply: first, the spreading signal has a bandwidth much larger than the minimum bandwidth required to transfer desired information or baseband data. Second, data spreading occurs by means of a code spreading signal, where the code signal is independent of the data and is of a much higher rate than the data signal. Lastly, at the receiver, de-spreading takes place by the cross-correlation of the received spread signal with a synchronised replica of the same signal used to spread the data [10].

2.1.2.2.1 Modulation

If we view Quadrature Phase Shift Keying (QPSK) as two independent Binary Phase Shift Keying (BPSK) modulations, then we can assume the net data rate doubles. We now provide the background for QPSK to serve as background to the applications in UTRA presented in Chapter 4.

For all practical purposes we start with M-PSK, where $M = 2^b$ and $b = 1$, 2 or 3 (i.e. 2-PSK or BPSK, 4-PSK or QPSK and 8-PSK). In the case of QPSK modulation the phase of the carrier can take on one of the four values 45°, 135°, 225° or 315° as we shall see later.

[4]Unless otherwise specified, here we assume N_o thermal and interference noise.

The QPSK power spectral density (V^2/Hz) could be then defined as

$$S(f) = A^2 T_s \left\{ \frac{\sin[\pi T_s(f - f_c)]}{[\pi T_s(f - f_c)]} \right\} \tag{2.7}$$

where f_c is the un-modulated carrier frequency, A is the carrier amplitude and T_s is the symbol interval. When T_b is the input binary bit interval, T_s may be expressed as

$$T_s = T_b \log_2 M \tag{2.8}$$

The power spectral density of an un-filtered M-PSK signal occupies a bandwidth which is a function of the symbol rate, $r_s = (1/T_s)$. Thus, for a given transmitter symbol, the power spectrum for any M-PSK signal remains the same regardless of the number M of symbol levels used. This implies that BPSK, QPSK and 8-PSK signals each have the same spectral shape if T_s remains the same in each case.

Spectral Efficiency For an M-PSK scheme each transmitted symbol represents $\log_2 M$ bits. Hence, at a fixed input bit rate, as the value of M increases, the transmitter symbol rate decreases, which means that there is an increase in spectral efficiency for larger M.

Thus, for any digital modulation the spectral efficiency (η_s), i.e. the ratio of the input data rate (r_b) to the allocated channel bandwidth (B), is given by

$$\eta_s = \frac{r_b}{B} \quad \text{(bps Hz)} \tag{2.9}$$

the 8-PSK spectral efficiency will be three times greater than that for BPSK. However, this will be achieved at the expense of the error probability.

Now allocating the RF bandwidth of an M-PSK signal we should remember that its spectrum rolls off relatively slowly. Therefore, it is necessary to filter the M-PSK signal so that its spectrum is limited to a finite bandpass channel region avoiding adjacent channel interference. Using *Nyquist filtering* or *raised cosine filtering* prevents the *adjacent channel interference*, as well as the *Intersymbol Interference* (ISI) due to filtering. The raised-cosine spectra are characterised by a factor α_B, known as the *excess bandwidth factor.* This factor lies in the range 0–1, and specifies the excess bandwidth of the spectrum compared to that of an ideal bandpass spectrum ($\alpha_B = 0$) for which the bandwidth would be $B = r_s$. Typical values of α_B used in practice are 0.3–0.5 [11].

Thus, for M-PSK transmission using the Nyquist filtering with roll-off (α_B) the required bandwidth will be given by

$$B = r_s(1 + \alpha_B) \tag{2.10}$$

Then the maximum bit rate in terms of the transmission bandwidth (B) and the roll-off factor (α_B) can be defined as

$$r_b = \frac{B \log_2 M}{1 + \alpha_B} \tag{2.11}$$

However, if we assume an M-PSK with an ideal Nyquist filtering (i.e. $\alpha_B = 0$) the signal spectrum is centred on f_c, it is constant over the bandwidth $B = 1/T_s$ and zero outside that band. Then the transmitted bandwidth for the M-PSK signal and the respective spectral efficiency are given by

$$B = \frac{1}{T_b \log_2 M} \quad \text{and} \quad \eta_s = \frac{r_b}{B} = \log_2 M \tag{2.12}$$

Bit Error Rate (BER) Performance In M-PSK modulation, the input binary information stream is first divided into b bit blocks, and then each block is transmitted as one of M possible symbols, where each symbol is a carrier frequency sinusoid having one of M possible phase values [11]. Among the M-PSK schemes, BPSK and QPSK are the most widely used. Nevertheless, here we review only the QPSK scheme. In QPSK each transmitted symbol (Figure 2.3) represents two input bits as follows:

Input bits	*Transmitted Symbols*
00	$A\ \cos(w_c t + 45°)$
01	$A\ \cos(w_c t + 135°)$
11	$A\ \cos(w_c t + 235°)$
10	$A\ \cos(w_c t + 315°)$

The conversion from binary symbol to phase angles is done using Gray coding. This coding permits only one binary number to change in the assignment of binary symbols to adjacent phase angles, thereby minimising the de-modulation errors, which in a digital receiver result from incorrectly selecting a symbol adjacent to a correct one.

Figure 2.3 illustrates a block diagram frequently used for any form of M-PSK modulation. For QPSK, the multiplexer basically converts the binary input stream into two parallel, half-rate signals $v_I(t)$ and $v_Q(t)$ (i.e. the in-phase and quadrature signals). These signals taking values $+A/\sqrt{2}$ or $-A/\sqrt{2}$ in any symbol interval are fed to two balanced modulators with

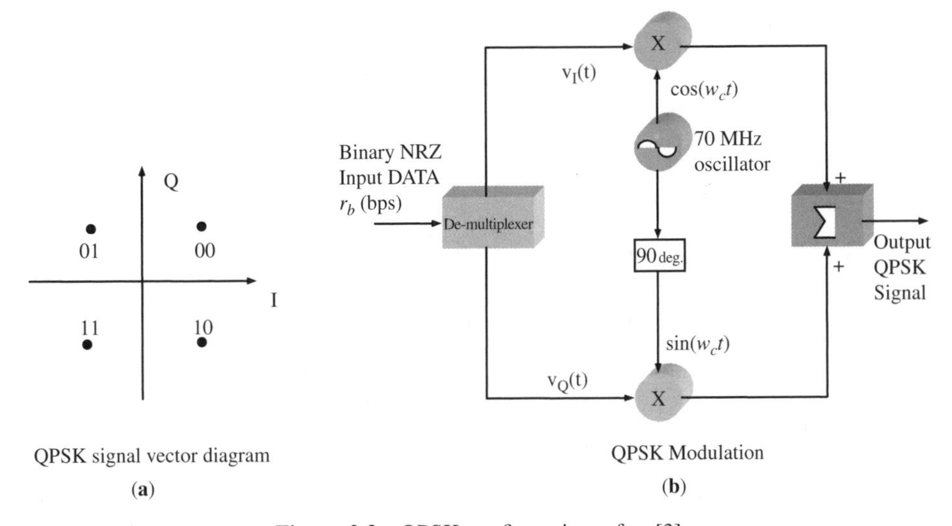

Figure 2.3 QPSK configuration, after [3].

input carriers or relative phases 0° and 90°, respectively. Then the QPSK signal could be given by

$$s(t) = v_I(t)\cos w_c t + v_Q(t)\sin w_c t \tag{2.13}$$

If we assume T_s is the time interval and $v_I = +A/\sqrt{2}$ and $v_Q = -A/\sqrt{2}$, it can be shown that the output $s(t)$ is

$$s(t) = A\cos\left(w_c - \frac{\pi}{4}\right) \tag{2.14}$$

Assuming a *coherent de-modulator*, the latter includes a *quadrature detector* consisting of two balanced multipliers with carrier inputs in phase quadrature, followed by root-Nyquist filter in the output I and Q arms. Then, the resultant I and Q signals are sampled at the centre of each symbol to produce the de-modulator output I and Q signals, which in turn are delivered to the decoder [11].

Generally, an M-PSK modulator produces symbols with one of M phase values spaced $2\pi/M$ apart. Then each signal is de-modulated correctly at the receiver when the phase is within π/M rad of the correct phase at the de-modulator sampling instant. If noise is present, evaluation of the probability of error requires a calculation of the probability that the received phase lies outside the angular segment within π/M rad of the transmitted symbol at the sampling instant.

Therefore, the probability that a de-modulator error occurs can be referred to as the *symbol error probability* (P_s). In the context of the M-ary modulation scheme with $M = 2^b$ bits, each symbol represents b bits. The most probable symbol errors are then those that choose an incorrect symbol adjacent to the correct one. When using Gray coding, only one bit error results from a symbol error. Thereupon, the bit error probability (P_b) is related to the symbol error probability by

$$P_b = \frac{P_s}{m} \tag{2.15}$$

In the case of QPSK, symbol errors occur when the noise pushes the received phasor into the wrong quadrant as illustrated in Figure 2.4. In this figure it is assumed that the

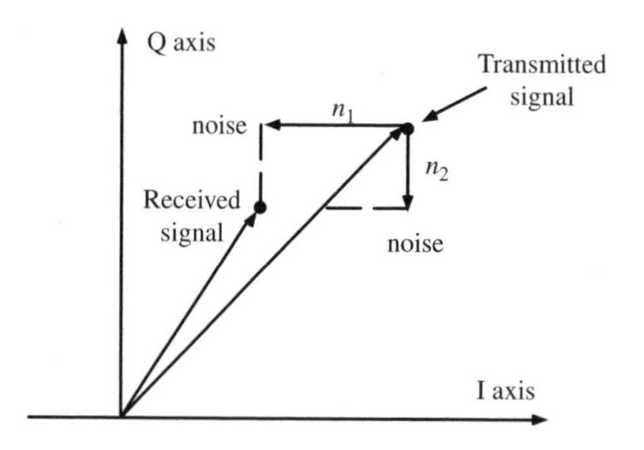

Figure 2.4 Transmitted and received signal vectors [11].

transmitted symbol has a phase of $\pi/4$ rad, corresponding to the de-modulator I and Q values of $v_I = V$ and $v_Q = V$ volts (i.e. noise-free case). Thus, if we consider that the noise phasors (n_1 and n_2) are pointing in directions that are most likely to cause errors, then a symbol error will occur if either n_1 or n_2 exceeds V.

Now, if for simplicity we also assume that a QPSK signal is transmitted without Nyquist filtering and de-modulated with hard decisions, the probability of a correctly de-modulated symbol value is equal to the product of the probabilities that each de-modulator low-pass filter output lies in the correct quadrant. Then the probability that the de-modulated symbol value is correct is given by

$$P_c = (1 - P_{e1})(1 - P_{e2}) \tag{2.16}$$

where P_{e1} and P_{e2} are the probabilities that the two filter output sample values are in the wrong quadrant. When showing that the low-pass filters are equivalent to integrators, which is the optimum choice if Nyquist filtering is not used, P_{e1} and P_{e2} can be expressed as

$$P_{e1} = P_{e2} = Q\left(\sqrt{\frac{E_s}{N_o}}\right) \tag{2.17}$$

where $E_s = A^2 T_s/2$ is the energy per symbol, $N_o/2$ is the two-sided noise power noise spectral density (in V^2/Hz) at the de-modulator input and the function $Q(x)$ is the complementary integral Gaussian function. The error function erf(x) given by

$$\mathrm{erf}(x) = \frac{2}{\sqrt{\pi}} \int_0^x \exp(-y^2)\mathrm{d}y \tag{2.18}$$

and complementary error function erfc(x) expressed as

$$\mathrm{erfc}(x) = 1 - \mathrm{erf}(x) \tag{2.19}$$

are not fully identical to the integral Gaussian function $G(x)$, and the complementary integral $G_c(x)$ or $Q(x)$ in our case. Now if we assume $G_c(x) = Q(x)$, we can use the following function to evaluate our error probabilities:

$$Q(x) = \frac{1}{2}\mathrm{erfc}\left(\frac{x}{\sqrt{2}}\right) \tag{2.20}$$

Then since $P_{e1} = P_{e2}$, the *symbol error probability* could be written as

$$P_s = 1 - P_c = 2P_{e1} - P_{e1}^2 \tag{2.21}$$

which at $P_{e1} \leq 1$ becomes

$$P_s \approx 2P_{e1} \tag{2.22}$$

Substituting P_{e1} from eqn (2.17) into eqn (2.22), the QPSK *symbol error probability* can be given by

$$P_s \approx 2Q\left(\sqrt{\frac{E_s}{N_o}}\right) \tag{2.23}$$

Now, for QPSK $E_s = 2E_b$, where E_b is the energy per bit; then making use of eqn (2.23) we get the bit error rate probability (P_{BER}) for the QPSK system as follows:

$$P_{\text{BER}} = Q\left(\sqrt{\frac{2E_b}{N_o}}\right) \tag{2.24}$$

Here we found the P_{BER} assuming that no Nyquist filtering was present. However, according to [11], this P_{BER} also holds when root-Nyquist filters are used at the transmitter and receiver under the assumption that the de-modulator input energy (E_b) and the noise power density (N_o) are the same for both cases.

2.1.2.3 CDMA System Performance

As noted earlier, CDMA systems tolerate more interference than typical TDMA or FDMA systems. This implies that each additional active radio user coming into the network increases the overall level of interference to the cell-site receivers receiving CDMA signals from mobile-station transmitters. This depends on its received power level at the cell site, its timing synchronisation relative to other signals at the cell site and its specific cross-correlation with other CDMA signals. Consequently, the number of CDMA channels in the network will depend on the level of total interference that the system can tolerate. As a result, the FDD mode behaves as an interference-limited system, where technical design will play a key role in the overall quality and capacity performance. Thus, despite advanced techniques such as multi-user detection and adaptive antennas, a robust system will still need a good bit error probability with a higher level of interference.

When we consider that at the cell site all users receive the same signal level assuming Gaussian noise as interference, the modulation method has a relationship that defines the bit error rate as a function of the E_b/N_o ratio. Therefore, if we know the performance of the signal processing methods and tolerance of the digitised information to errors, we can define the minimum E_b/N_o ratio for a balanced system operation. Then, if we maintain operation at this minimum E_b/N_o, we can obtain the optimum performance of the system. From [11–23] we can define the relationship between the number of mobile users (M), the processing gain (G_p) and the E_b/N_o ratio as follows:

$$M \approx \frac{G_p}{(E_b/N_o)} \tag{2.25}$$

On the other hand, the E_b/N_o performance can be seen better in relation with Shannon's limit in AWGN,[5] which simplified can be presented as

$$\frac{C}{B} < \frac{1}{\log_e 2}\left(\frac{S}{N}\right) \quad \text{and} \quad \frac{C}{B} < \frac{1}{\log_e 2}\left(\frac{E_b}{N_o}\right)\left(\frac{C}{B}\right) \tag{2.26}$$

then

$$\frac{E_b}{B} \geq \log_e 2 = 0.69 = -1.59\,\text{dB} \tag{2.27}$$

provides error-free communications. Then for Shannon's limit the number of users can be projected from

$$M = \frac{G_p}{0.69} = 1.45G_p \tag{2.28}$$

Shannon's theoretical limit implies that a WCDMA system can support more users per cell than classical narrow-band systems limited by the number of dimensions. On the other hand, this limit in practice has $E_b/N_o = 6$ dB as a typical value. However, due to practical limitations, accommodating as many users in a single cell as indicated by Shannon's limit is not possible in a CDMA system, and this applies also to the UTRA FDD. Thus, cell capacity depends upon many factors (e.g. receiver-modulation performance, power-control accuracy and inter-system interference), and the upper-bound theoretical capacity of an ideal noise-free CDMA channel has also limitations by the processing gain (G_p) [10].

Multiple transmissions in neighbouring CDMA cells using the same carrier frequency cause interference, is denoted by β factor. This event will cause reduction of the number of users in a cell, because the interference from users in other cells has to be added to the interference generated by the other mobiles in the user's cell. β may range from 0.4 to 0.55. In addition to the interference factor, we also introduce the imperfect power control or power control accuracy factor α, which ranges from 0.5 to 0.9. Interference can be reduced by the voice activity factor ν ranging from 0.45 to 1.0. If we use directional antennas at the base station, the sectorised cell will have a sectors, each of the antennas used at the cell will radiate into a sector of $(360/a)°$, resulting in an interference improvement factor λ. Average values for β, α, ν and λ (three-sector cell) are 0.5, 0.85, 0.6 and 2.55, respectively [10]. Then incorporating all the preceding factors the user capacity equation becomes

$$M \approx \frac{G_p}{E_b/N_o} \times \frac{1}{1+\beta} \times \alpha \times \frac{1}{\nu} \times \lambda \tag{2.29}$$

In the next section we also review pseudo-random sequences as part of the signal processing aspects relevant for the operation of the UTRA modes.

[5]Additive Gaussian white noise.

2.1.2.4 Pseudo-Random Sequences

Pseudo-random Noise (PN), i.e. deterministic periodic sequences in WCDMA, performs the following tasks: bandwidth spreading of the modulated signal to wider transfer bandwidths and signal discrimination among users transmitting in the same bandwidth of multiple-access methods.

The characteristics of these sequences are: 1/2 relative frequencies of zero and one; for zeroes or ones, half of all run lengths are of length 1; one-quarter are of length 2, one-eighth are of length 3, etc. When a PN sequence shifts by any non-zero number of elements, the resulting sequence will have an equal number of agreements and disagreements with respect to the original sequence.

We generate PN sequences by combining feedback shift register outputs. This register consists of consecutive two-state memory or storage stages and feedback logic. Binary sequences shift through the shift register in response to clock pulses. We logically combine the contents of the stages to produce the input to the first stage. The initial contents of the stages and feedback logic determine the successive contents of the stages. We call a feedback shift register and its output linear when the feedback logic consists of entirely mod-2 adders.

The output sequences get classified as either maximal length or non-maximal length. The first ones are the longest sequences that can be generated by a given shift register of a given length, while all other sequences besides maximal length sequences are non-maximal length sequences. In the binary shift register sequence generators, the maximal length sequence has $2^n - 1$ chips, where n is the number of stages in the shift registers. A property of the maximal length sequences implies that for an n-stage linear feedback shift register, the sequence repetition period in clock pulses is $T_o = 2^n - 1$. When a linear feedback shift register generates a maximal sequence, then all its non-zero output sequences result in maximal sequences, regardless of the initial stage. A maximal sequence contains $2^{n-1} - 1$ zeroes and 2^{n-1} ones per period.

Other characteristics of PN sequences (e.g. properties of maximal length PN sequences, auto-correlation, cross-correlation and orthogonal functions) are described in [10]. In the following we review additional WCDMA characteristics, such as power control and soft handover.

2.1.2.5 Power-Control Characteristics

Accurate and *fast power* control becomes imperative in WCDMA. It increases network stability and prevents near-far effect (UL) or cell blocking by overpowered MSs. Open-loop or slow power control would not cope with the highly non-correlated fast fading between UL and DL as a consequence of the large frequency separation. Chapter 4 describes the technical details of fast power control. The latter applies to both the UL and DL. In the first case, the BS balances the MS's power after comparing the received Signal-to-Interference Ratio (SIR) to a SIR_{target}. In the second case, we aim to provide sufficient additional power to MSs at the cell edges in order to minimise other-cell interference.

The *outer-loop* or slow power control adjusts the BS's reference SIR_{target} based on the needs of a single or independent radio link. It aims to maintain constant quality established by the network through a target BER or FER, for example. The RNC handles the command steps to lower or increase the reference SIR_{target}.

2.1.2.6 *Soft Handover Characteristics*

While there is hard handover for carrier change or hierarchical cell transition, and inter-system handover to pass from FDD to TDD or GSM, in WCDMA two types of soft handovers characterise the cell transition process. These include *softer* and *soft* handovers. In the first case, an MS finds itself in the overlapped cell coverage area of two adjacent sectors of a BS. The MS communicates simultaneously with BS through two channels (2 DL codes) corresponding one to each sector. The MS's RAKE receives and processes the two signals, where its fingers generate the necessary de-spreading codes for each sector. The UL process occurs in the BS, where the BS receives the MS's channel in each sector and routes them to the same RAKE receiver for the typical maximal ratio combining process under one active power-control loop per connection.

In the second case, i.e. soft handover, an MS finds itself in the overlapping cell coverage area of two sectors corresponding to different BSs. Communications between MS and BS occur simultaneously through two channels, one from each BS. In the DL, the MS receives both signals for maximal ratio combining. In the UL, the MS code channel arrives from both BS, and is routed to the RNC for combining, in order to allow the same frame reliability indicator provided for outer-loop power control when selecting the best frame. Two active power-control loops participate in soft handover, i.e. one for each BS.

While softer handover may occur only in about 10% of links, soft handover may occur in about 30% of the links. Thus, for the latter provision in terms of extra power, RAKE processing, RNC transmission lines will be essential.

2.2 THE 3G COMMUNICATIONS ENVIRONMENT

This section provides dedicated reference models for the test environments cited in the forthcoming chapters, in particular the deployment contents presented in Chapter 7. These test environments aim to cover the range of UMTS operating environments. Thus, the necessary parameters to identify the reference models include the test propagation environments, traffic conditions and user information rate for reference voice and data services. It also presents some performance objectives and criteria for each operating environment.

The test operating environments are direct extracts from the recommendations considered for the evaluation process of the Radio Transmission Technologies (RTTs) submitted to ETSI and ITU as UTRA candidate solutions. Thus, the contents bring together or are based entirely on the specifications outlined in [11–14].

2.2.1 Mapping High Level Requirements onto Test Environments

This chapter maps high level service requirements summarised in Chapter 1 onto test environments described in the next sections. The mapping identifies the maximum user bit rate in each test environment, together with the maximum speed, expected range and associated wide-band channel model. Table 2.1 illustrates the suggested reference values.

Table 2.1 Radio transmission test environments [13]

High level description	Maximal bit rate (kbps)	Maximal speed (km/h)	Test environment channel models	Cell coverage
Rural outdoor	144	500	Vehicular channels A and B	Macro-cell
Sub-urban outdoor	384	120	Outdoor-to-indoor and pedestrian channels A and B	Micro-cell
			Vehicular channel A	Macro-cell
Indoor/low range outdoor	2048	10	Indoor channels A and B	Pico-cell
			Outdoor-to-indoor and pedestrian channel A	Micro-cell

2.2.1.1 *Reference Services*

The UMTS minimum set of services to appropriately characterise bearers include ranges of supported data rates, BER requirements, one-way delay requirements, activity factor and traffic models. The forthcoming section covers traffic models and Table 2.2 provides example values for access reference services, such as speech, Low Delay Data (LDD), Long Delay Circuit-Switched Data (LCD) and Un-restricted Delay Data (UDD). The latter corresponds to connectionless data for packet services; 12.2 kbps corresponds to AMR rates not necessarily part of the early test recommendations [13].

Table 2.2 Reference data rates

Test environments	Indoor office	Outdoor to indoor and pedestrian	Vehicular 120 km/h	Vehicular 500 km/h
Speech (kbps)	8, 12.2	8, 12.2	8, 12.2	8, 12.2
BER	$\leq 10^{-3}$	$\leq 10^{-3}$	$\leq 10^{-3}$	$\leq 10^{-3}$
Delay (ms)	20	20	20	20
Activity (%)	50	50	50	50
LDD data (kbps)	144, 384, 2048	64, 144, 384	32, 144, 384	32, 144
BER	$\leq 10^{-6}$	$\leq 10^{-6}$	$\leq 10^{-6}$	$\leq 10^{-6}$
Delay (ms)	50	50	50	50
Activity (%)	100	100	100	100
UDD data (packet) Connection-less information types	See Section 2.2.8.1 and Table 2.7	See Section 2.2.8.1 and Table 2.7	See Section 2.2.8.1 and Table 2.7	See Section 2.2.8.1 and Table 2.7
LCD data (kbps)	144, 384, 2048	64, 144, 384	32, 144, 384	32, 144
BER	$\leq 10^{-6}$	$\leq 10^{-6}$	$\leq 10^{-6}$	$\leq 10^{-6}$
Delay (ms)	300	300	300	300
Activity (%)	100	100	100	100

2.2.2 Channel Types

As a global standard, UMTS aims for a broad range of environment characteristics, e.g. large and small cities, tropical, rural and desert areas. Reference parameters describing the propagation models for these areas include

1. time delay spread with its structure and its statistical variability (e.g. probability distribution of time delay spread),
2. geometrical path loss rule (e.g. R^{-4}) and excess path loss,
3. shadow fading and multi-path fading characteristics (e.g. Doppler spectrum, Rician vs. Rayleigh) for the envelope of channels,
4. operating radio frequency.

Characterisation of rapid fading variation occurs by the channel impulse response, where response modelling takes place using a tapped-delay-line implementation. The Doppler spectrum characterises the tap variability. These environments are represented in terms of propagation from [13] by indoor office, outdoor-to-indoor and pedestrian, vehicular and mixed.

2.2.3 Indoor Office

This environment has small cells and low transmit powers, where both BSs and pedestrian users remain indoors, with path loss rule varying due to scatter and attenuation by walls, floors and metallic structures, e.g. partitions and filing cabinets, all producing some type of shadowing effects. These effects include log-normal shadow fading with standard deviation of 12 dB, and fading ranges from Rician to Rayleigh, with Doppler frequency offsets set by walking speeds.

The indoor office path loss is based on the COST[6] 231 model; this low increase of path loss versus distance is a worst case from the interference point of view and is defined as follows:

$$L = L_{\mathrm{FS}} + L_c + \sum k_{\mathrm{wi}} L_{\mathrm{wi}} + n^{((n+2)/(n+1)-b)} * L_f \tag{2.30}$$

where L_{FS} is the free space between transmitter and receiver, L_c is the constant loss, k_{wi} is the number of penetrated walls of type i, n is the number of penetrated floors, L_{wi} is the loss of wall type i, L_f is the loss between adjacent floors and b is the empirical parameter. L_c normally is set to 37 dB. $n = 4$ is the average for an indoor office environment. For capacity calculations in moderately pessimistic environments, the model can be modified to $n = 3$ (Table 2.3).

Under the simplifying assumptions of the office environment the indoor path loss model has the following form:

$$L = 37 + 30\ \log_{10}(R) + 18.3n^{((n+2)/(n+1)-0.46)} \tag{2.31}$$

[6]COST 231 Final Report (e.g. propagation environments), Commission of the European Communities.

Table 2.3 Weighed average for loss categories indoor environment [13]

Loss category	Description	Factor (dB)
L_f	Typical floor structures (i.e. offices) Hollow pot tiles Reinforced concrete Thickness type < 30 cm	18.3
L_{w1}	Light internal walls Plasterboard Walls with large numbers of holes (e.g. windows)	3.4
L_{w2}	Internal walls Concrete, brick Minimum number of holes	6.9

where r is the transmitter–receiver separation given in m and n is the number of floors in the path. L shall in no circumstances be less than free space loss. A log-normal shadow fading standard deviation of 12 dB can be expected.

2.2.3.1 *Physical Deployment*

The specific assumptions about the indoor physical deployment environment can be summarised from [13] as area per floor $= 5000\ \text{m}^2$, number of floors $= 3$, room $= 20 \times 10 \times 3\ \text{m}^3$, corridor $= 100 \times 5 \times 3\ \text{m}^3$, log-normal standard deviation $= 12$ dB and MS velocity $= 3$ km/h. Figure 2.5 illustrates a default deployment scheme, where base stations use omni-directional antennas. For spectrum efficiency evaluation, quality statistics should only be collected in the middle floor. See the mobility model in [13].

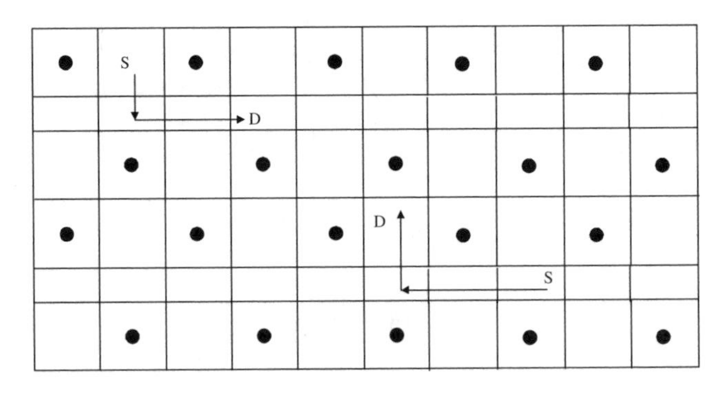

Figure 2.5 Indoor office model and deployment scheme.

2.2.4 Outdoor-to-Indoor and Pedestrian

We also characterise this environment by small cells and low transmit power, where BSs with low antenna heights stay outdoors while pedestrian users may remain in the streets and/or inside buildings and residences. A geometrical path loss rule of R^{-4} may satisfy; however, a wider range would serve better. When the path has a line of sight on a canyon-like street, the path loss follows a R^{-2} rule with Fresnel zone clearance. In the absence of Fresnel zone clearance, a path loss rule of R^{-4} will apply, but a range up to R^{-6} may occur due to trees and other obstructions along the path. Log-normal shadow fading with a standard deviation of 10 dB applies to outdoors and 12 dB to indoors. Building penetration loss averages 12 dB with a standard deviation of 8 dB. Walking speeds set Rayleigh and/or Rician fading rates, but not faster fading due to reflections from moving vehicles.

Generally, the total transmission loss L (dB) between isotropic antennas for outdoor transmission loss equals the sum of:

- free space loss, L_{fs},
- the diffraction loss from rooftop to the street, L_{rts},
- the reduction due to multiple screen diffraction past rows of buildings, L_{msd},

where L_{fs} and L_{rts} are independent of the BS antenna height, while L_{msd} depends on whether the base-station antenna is at, below or above building heights. Then L is given as

$$L(d) = L_{\text{fs}} + L_{\text{rts}} + L_{\text{msd}} \tag{2.32}$$

Given an MS-to-BS separation R, the free space loss (L_{fs}) between them is given by

$$L_{\text{fs}} = -10\log_{10}\left(\frac{\lambda}{4\pi d}\right)^2 \tag{2.33}$$

The diffraction from the rooftop down to the street level gives the excess loss to the mobile station

$$L_{\text{rts}} = -10\log_{10}\left[\frac{\lambda}{2\pi^2 r}\left(\frac{1}{\theta} - \frac{1}{2\pi + \theta}\right)^2\right] \tag{2.34}$$

where

$$\theta = \tan^{-1}\left(\frac{|\Delta h_m|}{x}\right) \quad \text{and} \quad r = \sqrt{(\Delta h_m)^2 + x^2} \tag{2.35}$$

Δh_m is the difference between the mean building height and the mobile antenna height; x is the horizontal distance between the mobile and the diffracting edges [13].

For the general model, the multiple screen diffraction loss from the base antennas due to propagation past rows of buildings is

$$L_{\text{msd}} = -10\log_{10}(Q_M^2) \tag{2.36}$$

where Q_M is a factor dependent on the relative height of the base-station antenna as being either at, below or above the mean building heights [15,16].

In this case the base-station antenna height is near mean rooftop level, then

$$Q_M = \frac{d}{R} \tag{2.37}$$

The total transmission loss for the near rooftop case then becomes

$$L = -10\log_{10}\left(\frac{\lambda}{2\sqrt{2}\pi R}\right)^2 - 10\log_{10}\left[\frac{\lambda}{2\pi^2 r}\left(\frac{1}{\theta} - \frac{1}{2\pi + \theta}\right)^2\right] - 10\log_{10}\left(\frac{d}{R}\right)^2 \tag{2.38}$$

When $\Delta h_b = -5$ m, $\Delta h_m = 10.5$ m, $x = 15$ m and $b = 80$ m, as typical in an urban and sub-urban environment, the above path loss expression reduces to a simple function of the transmitter-to-receiver distance R (km) and frequency f (MHz),

$$L = 40\log_{10}(R) + 30\log_{10}(f) + 49 \tag{2.39}$$

where R is the base-station–mobile-station separation in kilometres and f is the carrier frequency of 2000 MHz for UMTS band application. L shall in no circumstances be less than free space loss. This model applies to the Non-Line-of-Sight (NLOS) case only and describes worse case propagation assuming log-normal shadow fading with a standard deviation of 10 dB for outdoor users and 12 dB for indoor users. The average building penetration loss is 12 dB with a standard deviation of 8 dB [13].

A more detailed model uses a recursive approach [17] that calculates the path loss as a sum of LOS and NLOS segments. The shortest path along the streets between the BS and the MS has to be found within the Manhattan environment and the path loss (dB) is given by

$$L = 20\log_{10}\frac{4\pi d_n}{\lambda} \tag{2.40}$$

where d_n is the illusory distance, λ is the wavelength and n is the number of straight street segments between BS and MS (along the shortest path).

The illusory distance is the sum of the street segments obtained recursively using the expressions $k_n = k_{n-1} + d_{n-1}c$ and $d_n = k_n s_{n-1} + d_{n-1}$ where c is a function of the street crossing, e.g. for a 90° street crossing $c = 0.5$. Furthermore, s_{n-1} is the length in metres of the last (straight path) segment. We set the initial values as $k_0 = 1$ and $d_0 = 0$, and we get the illusory distance as the final d_n when the last segment has been added. When extending the model to cover the micro-cell dual slope behaviour, we express L as

$$L = 20\log_{10}\left(\frac{4\pi d_n}{\lambda} D\left(\sum_{j=1}^{n} s_{j-1}\right)\right), \quad \text{where} \quad D(x) = \begin{cases} x/x_{\text{br}}, x > x_{\text{br}} \\ 1, \leq x_{\text{br}} \end{cases} \tag{2.41}$$

Before the break point x_{br} the slope is 2, after the break point it increases to 4. The break point x_{br} is set to 300 m. x is the distance from the transmitter to the receiver.

When taking into account propagation effects going above rooftops, path loss calculation occurs according to the shortest geographical distance using the COST Walfish–Ikegami model and with antennas below rooftops, i.e.

$$L = 24 + 45\log(d + 20) \tag{2.42}$$

where d is the shortest physical geographical distance from the transmitter to the receiver in metres. The final path loss results from the minimum between the street path loss value and the path loss based on the shortest geographical distance, i.e.

$$\text{Path loss} = \min(\text{Manhattan path loss},\ \text{macro-path loss})$$

This path loss model applies only to micro-cell coverage with antenna located below the rooftop. When the urban structure has macro-cell coverage, the first path loss case applies [13].

2.2.4.1 *Physical Deployment*

The same test service requirements of the preceding section (indoor) apply for the outdoor-to-indoor pedestrian environment. The specific assumptions about the physical deployment environment include the following:

- Indoor: building penetration loss of 12 dB, standard deviation of 8, log-normal standard deviation of 10 and MS velocity of 3 km.
- Outdoor: building penetration loss is NA, log-normal standard deviation is of 10 and MS velocity of 3 km.

A Manhattan-like structure defined for the outdoor-to-indoor and pedestrian environment can be applied with following assumptions: 6.5 km^2, 200×200 m^2 block, 30 m street width and 10 m BS height. Figure 2.6 illustrates the default deployment scheme with BS using omni-directional antennas, where expected quality statistics would only arise from among cells marked with a T. See the mobility model in [13].

2.2.5 Vehicular

We characterise the vehicular environment by larger cells and higher transmit power. The recommendations imply a geometrical path loss rule of R^{-4} and log-normal shadow fading with 10 dB standard deviation in urban and sub-urban areas. Rural areas with flat terrain will have lower path loss than that of urban and sub-urban areas. In mountainous areas we can apply a path loss rule closer to R^{-2} assuming that BS locations do not suffer from blocking. Vehicle speeds set Rayleigh fading rates.

The vehicular environment applies to scenarios in urban and sub-urban areas outside the high rise core where the buildings have nearly uniform height. In this model, the BS has

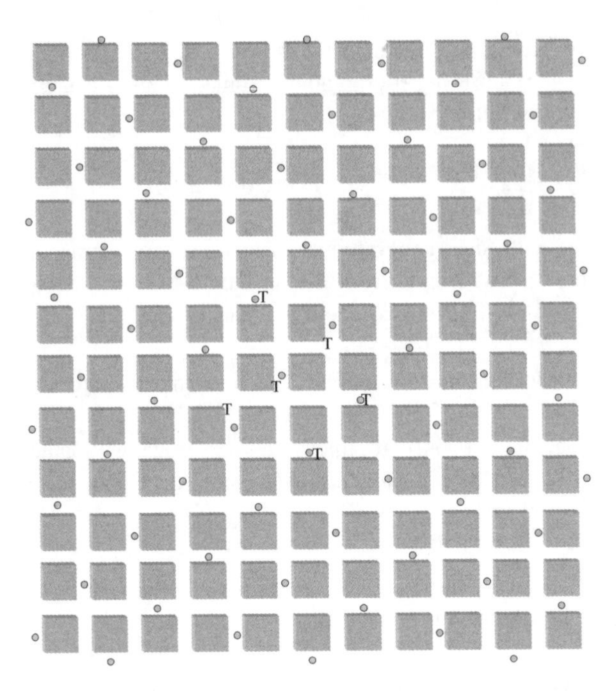

Figure 2.6 Manhattan-like urban model example.

antenna height above rooftop level with Q_M as the factor depending on the relative height of the BS antenna,

$$Q_M = 2.35\left(\frac{\Delta h_b}{d}\sqrt{\frac{b}{\lambda}}\right)^{0.9} \tag{2.43}$$

where Δh_b is the height difference between BS antenna and the mean building rooftop height and b is the average separation between rows of buildings. Then the total transmission loss for the above rooftop case becomes

$$\begin{aligned} L = &-10\log_{10}\left[\left(\frac{\lambda}{4\pi R}\right)^2\right] - 10\log_{10}\left[\frac{\lambda}{2\pi^2 r}\left(\frac{1}{\theta} - \frac{1}{2\pi+\theta}\right)^2\right] \\ &- 10\log_{10}\left[(2.35)^2\left(\frac{\Delta h_b}{R}\sqrt{\frac{d}{\lambda}}\right)^{1.8}\right] \end{aligned} \tag{2.44}$$

In the building environment, measurements [18] showed that the path loss slope behaves as a linear function of the base-station antenna height relative to the average rooftop (Δh_b).

Then, the above path loss equation can be defined as

$$L = -10\log_{10}\left[\left(\frac{\lambda}{4\pi R}\right)^2\right] - 10\log_{10}\left[\frac{\lambda}{2\pi^2 r}\left(\frac{1}{\theta} - \frac{1}{2\pi+\theta}\right)^2\right] - 10\log_{10}\left[(2.35)^2\frac{\left(\Delta h_b\sqrt{d/\lambda}\right)^{1.8}}{R^{2((1-4)\times 10^{-3}\Delta h_b)}}\right] \tag{2.45}$$

where

$$\theta = \tan^{-1}\left(\frac{|\Delta h_m|}{x}\right), \quad r = \sqrt{(\Delta h_m)^2 + x^2} \tag{2.46}$$

Δh_m is the difference between the mean building height and the mobile-antenna height and x is the horizontal distance between the mobile and the diffracting edges.

If $\Delta h_m = 10.5$ m, $x = 15$ m and $b = 80$ m, typical urban and sub-urban environment values with average four-storey building heights, then the above path loss expression L reduces to a simple function of the transmitter-to-receiver distance R (km). We measure the BS antenna height from the average rooftop Δh_b in metres, and frequency (f) in MHz

$$L = [40(1-4)\times 10^{-3}\Delta h_b]\log_{10}(R) - 18\log_{10}(\Delta h_b) + 21\log_{10}(f) + 80 \quad \text{(dB)} \tag{2.47}$$

When we assume a fixed BS antenna height of 15 m above the average rooftop, i.e. $\Delta h_b = 15$ m, and a carrier frequency of 2000 MHz, the vehicular path loss L becomes

$$L = 128.1 + 37.6\log_{10}(R) \tag{2.48}$$

L shall in no circumstances be less than the free space loss. This model applies to the NLOS case only and describes the worse case propagation. Log-normal shadow fading with 10 dB standard deviation are assumed in both urban and sub-urban areas [13]. The path loss model is valid for a range of Δh_b from 0 to 50 m.

2.2.5.1 Physical Deployment

When assuming a cell radius of 2000 m, services up to 144 kbps apply; and with a cell radius of 500, services above 144 kbps (e.g. 384 kbps) apply. If the BS antenna height remains above the average rooftop height of 15 m, a hexagonal cell layout with distances between base stations equal to 6 km can serve as reference. Figure 2.7 illustrates this type of tri-sectored cells using the GSM based antenna pattern shown in Figure 2.8.

2.2.5.2 Mobility Model

The vehicular reference mobility model uses a pseudo-random mobility model with semi-directed trajectories, the mobile's position gets updated according to the de-correlation

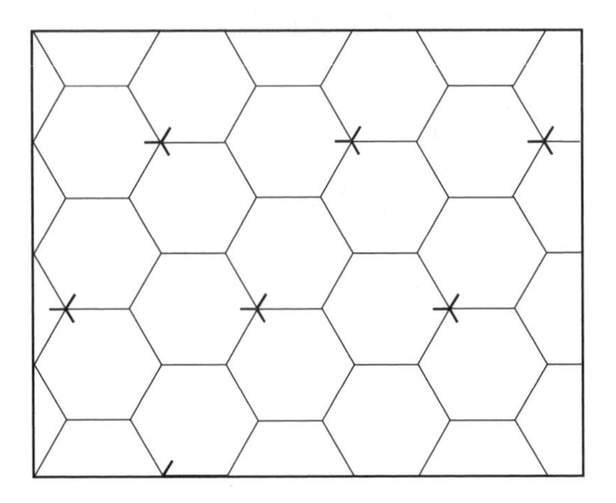

Figure 2.7 Vehicular deployment model.

length and direction can change at each position update following a given probability within a sector. For a reference example, we can assume constant mobile speeds of 120 km/h, with a direction probability change at position update of 0.2, and a maximal angle for direction update of 45°. Mobiles get uniformly distributed on the map and their direction randomly chosen at initialisation [13].

2.2.6 Mixed

Here we illustrate a mix environment by a vehicular (macro-cells) and an outdoor-to-indoor (micro-cells) environment taking place in the same geographical area. In this area, fast

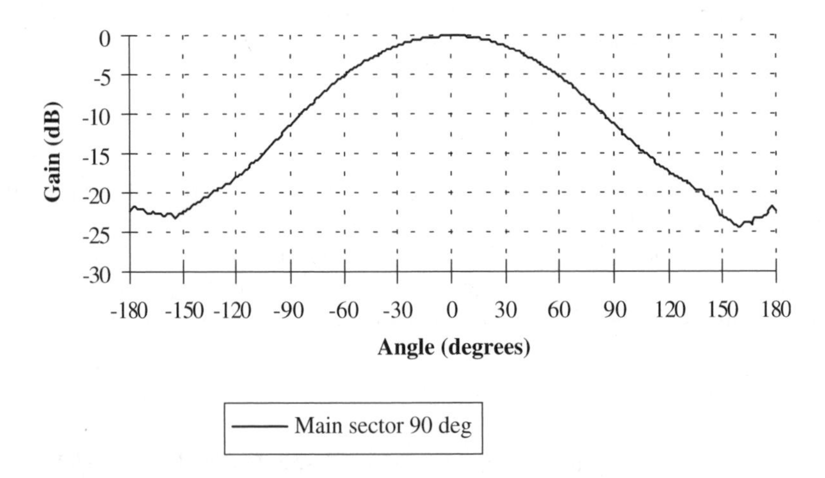

Figure 2.8 Horizontal antenna pattern example based on GSM three-sectored antenna [13].

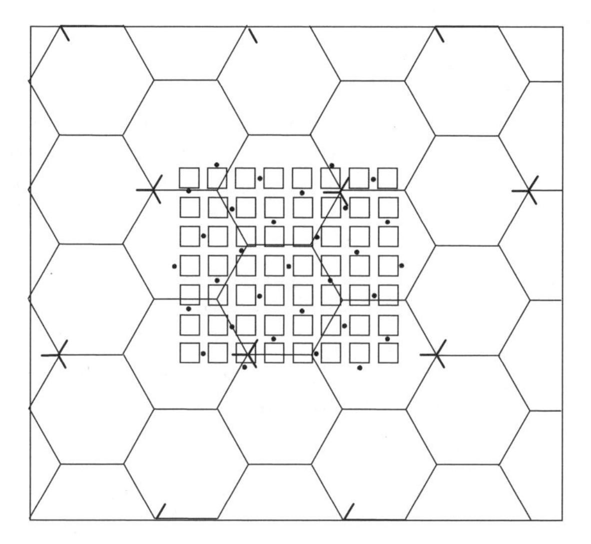

Figure 2.9 Mixed physical environment and proposed deployment model.

moving terminals (e.g. vehicles and trains) will most likely connect to the macro-cells to reduce the hand-off rate (number of hand-offs per minute) and slow moving terminals (pedestrians and boats on a shore) will probably connect to the micro-cells to achieve high capacity. The reference assumptions [13] about combined outdoor and vehicular physical deployment environments can be as follows: the log-normal standard deviations are 10 dB for both outdoor and vehicular environments, mobile speeds are 3 and 80–120 km/h for outdoor and vehicular environment, respectively. The proportions of users are 60% and 40% for outdoor and vehicular environment, respectively.

Mobility Model The mobility model will follow outdoor and vehicular patterns allowing appropriate handover between macro-cells and micro-cells for all users (Figure 2.9).

2.2.6.1 Long-Term Fading De-correlation Length

We characterise the log-normal fading in the logarithmic scale around the mean path loss, L (dB), by a Gaussian distribution with zero mean and standard deviation. In this context, due to the slow fading process versus distance Δx, adjacent fading values correlate. Then, the normalised auto-correlation function $R(\Delta x)$ can be described with sufficient accuracy by an exponential function [19]

$$R(\Delta x) = \exp\left(-\frac{|\Delta x|}{d_{\text{cor}}}\ln 2\right) \tag{2.49}$$

where the de-correlation length d_{cor} depends on the environment. From this principle, we may assume 20 m for the de-correlation in the vehicular, and 5 m for the outdoor-to-indoor pedestrian environment. For the latter the evaluation of de-correlation length may not be fully valid.

2.2.7 Channel Impulse Response

The environments described in the preceding sections have a channel impulse response based on a tapped-delay-line model. The number of taps, the time delay relative to the first tap, the average power relative to the strongest tap and the Doppler spectrum of each tap characterise the model. Most of the time delay spreads are relatively small, but occasionally, there are worst case multi-path characteristics that lead to larger delay spreads. Here we consider the worst case. Two multi-path channels capture this delay spread better than a single tapped delay line. The reference simulation channel model can use a discrete Wide Sense Stationary Un-correlated Scattering (WSSUS) channel model, where the sum of delay replicas represents the received signal of the input signal weighed by independent zero-mean complex Gaussian time-variant processes. Hence, if $z(t)$ and $w(t)$ denote the complex low pass representations of the channel input and output, respectively, then [13]

$$w(t) = \sum_{n=1}^{N} \sqrt{p_n} g_n(t) z(t - \tau_n) \tag{2.50}$$

where p_n is the strength of the nth weight and $g_n(t)$ is the complex Gaussian process weighing the nth replica.

The Doppler spectrum of the nth path or the power spectrum of $g_n(t)$ controls the fading rate due to the nth path. Therefore, to define this channel model we can specify only the Doppler spectra of the tap weights $\{P_n(\upsilon); n = 1, \ldots, N\}$, the tap delays $\{\tau_n; n = 1, \ldots, N\}$ and the tap weight strengths $\{p_n(\upsilon); n = 1, \ldots, N\}$. Interpreting the process $g_n(t)$ as the superposition of un-resolved multi-path components arriving from different angles and in the vicinity of the delay interval, we have

$$\left(\tau_n - \frac{1}{2W} < \tau < \tau_n + \frac{1}{2W}\right) \tag{2.51}$$

where W is the bandwidth of the transmitted signal.

Generally, each ray has a different Doppler shift corresponding to a different value of the cosine of the angle between the ray direction and the velocity vector, which means that we can assume the following: First, very large number of receive rays arrive uniformly distributed in azimuth at the MS and at zero elevation for each delay interval for outdoor channels.[7] At the BS in general the received rays arrive in a limited range in azimuth. Second, for indoor channels[8] a very large number of receive rays arrive uniformly distributed in elevation and azimuth for each delay interval at the BS.

The first assumption matches the ones made in by Clarke [20] and Jakes [21] in narrow-band channel modelling. Thus, the same Doppler spectrum will result, i.e.

$$P_n(\upsilon) = P(\upsilon) = \frac{1}{\pi} \frac{1}{\sqrt{(V_{\mathrm{mob}}/\lambda)^2 - \upsilon^2}}, \quad |\upsilon| < \frac{V}{\lambda} \tag{2.52}$$

[7]Also, the antenna pattern is assumed to be uniform in the azimuthal direction.
[8]Assuming that the antenna is either a short or half-wave vertical dipole.

Table 2.4 Indoor office reference environment tapped-delay-line parameters [13]

Tap	Channel A		Channel B		Doppler spectrum
	Relative delay (ns)	Average power (dB)	Relative delay (ns)	Average power (dB)	
1	0	0	0	0	Flat
2	50	−3.0	100	−3.6	Flat
3	110	−10.0	200	−7.2	Flat
4	170	−18.0	300	−10.8	Flat
5	290	−26.0	500	−18.0	Flat
6	310	−32.0	700	−25.2	Flat

where V_{mob} is the velocity of the mobile and λ is the wavelength at the carrier frequency. The term 'classic' is used to identify this Doppler spectrum [13].

The second assumption results in a Doppler spectrum that is nearly flat, and the choice of a flat spectrum has been made, i.e.

$$P_n(v) = P(v) = \frac{\lambda}{2V_{\text{mob}}}, \quad |v| < \frac{V_{\text{mob}}}{\lambda} \tag{2.53}$$

Hence, this Doppler spectrum is referred to as 'flat' [13].

Tables 2.4–2.6 describe the tapped-delay-line parameters for each of the environments introduced in the preceding sections. Three parameters characterise each tap of the channels, i.e. the time delay relative to the first tap, the average power relative to the strongest tap and the Doppler spectrum of each tap. A ±3% variation in the relative time delay allows channel sampling rate matching some in simulations [13].

Table 2.5 Outdoor-to-indoor and pedestrian reference environment tapped-delay-line parameters [13]

Tap	Channel A		Channel B		Doppler spectrum
	Relative delay (ns)	Average power (dB)	Relative delay (ns)	Average power (dB)	
1	0	0	0	0	Classic
2	110	−9.7	200	−0.9	Classic
3	190	−19.2	800	−4.9	Classic
4	410	−22.8	1200	−8.0	Classic
5	NA	NA	2300	−7.8	Classic
6	NA	NA	3700	−23.9	Classic

Table 2.6 Vehicular reference environment, high antenna, tapped-delay-line parameters [13]

	Channel A		Channel B		
Tap	Relative delay (ns)	Average power (dB)	Relative delay (ns)	Average power (dB)	Doppler spectrum
1	0	0.0	0	−2.5	Classic
2	310	−1.0	300	0	Classic
3	710	−9.0	8900	−12.8	Classic
4	1090	−10.0	12 900	−10.0	Classic
5	1730	−15.0	17 100	−25.2	Classic
6	2510	−20.0	20 000	−16.0	Classic

2.2.8 Traffic Types and Propagation Models

We can represent real time services (e.g. speech and CS data services) by generating calls according to a Poisson process assuming a mean call duration of 120 s. The speech would be an on-off model, with activity and silent periods being generated by an exponential distribution. The mean value for active and silence periods is 3 s and independent of the uplink and downlink, and both are exponentially distributed. For circuit-switched data services, we can assume a traffic model with constant bit rate model and 100% activity.

2.2.8.1 Packet or Non-real Time Services

We can represent non-real time services by a WWW browsing session consisting of a sequence of packet calls. We only consider the packets from a source, which may be at either end of the link but not simultaneously. A subscriber may initiate a packet call when requesting an information entity. During this call several packets may be generated, which means that the packet call constitutes a sequence of packet bursts [22,23]. Figure 2.10 illustrates the bursts during the packet call typically seen in fixed network packet transmissions.

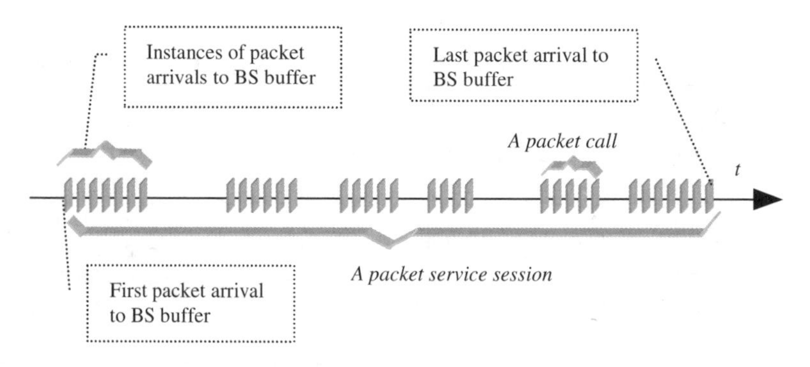

Figure 2.10 Packet service session characteristics.

The behaviour of Figure 2.10 can be modelled through the following parameters:

- Session arrival process – Poisson process.
- Number of packet calls per session – $N_{pc} \in \mathrm{Geom}(\mu_{Npc})$.
- Reading time between packet calls – $D_{pc} \in \mathrm{Geom}(\mu_{Dpc})$.
- Number of datagrams within a packet call – $N_d \in \mathrm{Geom}(\mu_{Nd})$.
- Inter-arrival time between datagrams (within a packet call) – $D_d \in \mathrm{Geom}(\mu_{Dd})$.
- Size of a datagram, S_d – Pareto distribution with cut-off:

$$\begin{aligned} f_x(x) &= \frac{\alpha k^\alpha}{x^{\alpha+1}}, \quad x \geq k \\ F_x(x) &= 1 - \left(\frac{k}{x}\right)^\alpha, \quad x \geq k \\ \mu &= \frac{k\alpha}{\alpha - 1}, \quad \alpha > 1 \\ \sigma^2 &= \frac{k^2\alpha}{(\alpha-2)(\alpha-1)^2}, \quad \alpha > 2 \end{aligned} \tag{2.54}$$

Packet size is defined by the following formula:

$$\text{Packet size} = \min(P, m) \tag{2.55}$$

where P is the normal Pareto distributed random variable ($\alpha = 1.1$, $k = 81.5$ bytes) and m is the maximum allowed packet size, $m = 66\,666$ bytes. The PDF of the packet size becomes

$$f_n(x) = \begin{cases} \frac{\alpha k^\alpha}{x^{\alpha+1}}, & k \leq x < m \\ \beta, & x = m \end{cases} \tag{2.56}$$

where β is the probability that $x > m$. It can easily be calculated as

$$\beta = \int_m^\infty f_x(x)\mathrm{d}x = \left(\frac{k}{m}\right)^\alpha, \quad \alpha > 1 \tag{2.57}$$

Then it can be calculated as

$$\mu_n = \int_{-\infty}^{\infty} x f_n(x)\mathrm{d}x = \int_k^{m-} x\frac{\alpha k^\alpha}{x^{\alpha+1}}\mathrm{d}x + m\left(\frac{k}{m}\right)^\alpha = \cdots = \frac{\alpha k - m(k/m)^\alpha}{\alpha - 1} \tag{2.58}$$

with the parameters above the average size: $\mu_n = 480$ bytes [13,22,23] indicates that according to the values for α and k in the Pareto distribution, the average packet size μ is 480 bytes. The average requested file size is $\mu_{Nd} \times \mu = 25 \times 480$ bytes $\approx$ 12 kbytes. The inter-arrival time is adjusted in order to get different average bit rates at the source level. Table 2.7 illustrates characteristics of connectionless information rates for WWW from [22].

Table 2.7 Characteristics of connection-less information types [22]

Packet based information rates, e.g. Internet services	Average number of packet calls in a session	Average reading time between packet calls (s)	Average amount of packets within a packet call	Average inter-arrival time between packets (s)[a]	Parameters for packet size distribution
WWW surfing UDD 8 kbps	5	412	25	0.5	$k = 81.5$ $\alpha = 1.1$
WWW surfing UDD 32 kbps	5	412	25	0.125	$k = 81.5$ $\alpha = 1.1$
WWW surfing UDD 64 kbps	5	412	25	0.0625	$k = 81.5$ $\alpha = 1.1$
WWW surfing UDD 144 kbps	5	412	25	0.0277	$k = 81.5$ $\alpha = 1.1$
WWW surfing UDD 384 kbps	5	412	25	0.0104	$k = 81.5$ $\alpha = 1.1$
WWW surfing UDD 2048 kbps	5	412	25	0.00195	$k = 81.5$ $\alpha = 1.1$

[a]The different inter-arrival times correspond to average bit rates of 8, 32, 64, 144, 384 and 2048 kbps.

2.3 CONCLUDING REMARKS

This chapter summarises the essential background to investigate the UTRA physical layer and its impact on its architecture. It has provided the reference models to represent the communication environments and the signal processing issues. It was not the aim of the author to cover the different topics in depth but to set them as review points for further study when required.

REFERENCES

1. Haykin, S., *Communications Systems*. New York: Wiley, 1983.
2. Shanmugam, K.S., *Digital and Analog Communication Systems*. New York: Wiley, 1979.
3. Schwartz, M., *Information Transmission, Modulation, and Noise*. New York: McGraw-Hill, 1970.
4. Steele, R., *Mobile Radio Communications*. Piscataway, NJ: IEEE Press, 1994.
5. Yang, S.C., *CDMA RF System Engineering*. Norwood, MA: Artech House, 1998.
6. Viterbi, A., *Principles of Spread Spectrum Communication*. Reading, MA: Addison-Wesley, 1997.
7. Cooper, G. and McGillem, C., *Modern Communications and Spread Spectrum*. New York: McGraw-Hill, 1998.
8. Dixon, R., *Spread Spectrum Systems with Commercial Applications*. New York: Wiley, 1994.
9. Shannon, C.E., Communications in the presence of noise, *Proc. IRE*, **37**, 10–21, 1949.
10. Garg, V.K., Smolik, K. and Wilkes, J.E., *Applications of CDMA in Wireless/Personal Communications*. New Jersey: Prentice-Hall, 1997.
11. TS 101 111, UMTS 21.01, Universal Mobile Telecommunication System (UMTS); Overall Requirements on the Radio Interface(s) of the UMTS.
12. TG32 UMTS–Radio Requirements.
13. High Level Requirements Relevant for the Definition of the UTRA Concept, V3.0.1, 1998–2010.

14. Draft New Recommendation ITU-R M, FPLMTS.REVAL, Guidelines for Evaluation of Radio Transmission Technologies for IMT-200/FPLMTS.
15. Xia, H.H. and Bertoni, H.L., Diffraction of cylindrical and plane waves by an array of absorbing half screens, *IEEE Trans. Antennas Propagation*, **40**(2), 170–177, 1992.
16. Maciel, L.R., Bertoni, H.L. and Xia, H.H., Unified approach to prediction of propagation over buildings for all ranges of base station antenna height, *IEEE Trans. Vehicular Technol.*, **42**(1), 41–45, 1993.
17. Berg, J.E., A Recursive Method For Street Microcell Path Loss Calculations, PIMRC '95, Vol. 1, 1995, 140–143.
18. Xia, H.H. *et al.*, Microcellular propagation characteristics for personal communications in urban and sub-urban environments, *IEEE Trans. Vehicular Technol.*, **43**(3), 743–752, 1994.
19. Gudmundson, M., Correlation model for shadow fading in mobile radio systems, *Electron. Lett.*, **27**(23), 2145–2146, 1991.
20. Clark, R.H., A Statistical Theory of Mobile Reception, *Bell System Tech. J.*, **47**, 957–1000, 1968.
21. Jakes, W.C. (ed.), *Microwave Mobile Communications*. New York: Wiley, 1974.
22. Anderlind, E. and Zander, J., A traffic model for non-real-time data users in a wireless radio network, *IEEE Commun. Lett.*, **1**(2), 1997.
23. Miltiades, E. *et al.*, A multiuser descriptive traffic source model, *IEEE Trans. Commun.*, **44**(10), 1996.

3
UMTS Service Components

3.1 THE SERVICE CREATION ENVIRONMENT

Above all, UMTS concerns *services* with the ultimate goal to enhance and enrich the user experience by providing faster, efficient and more flexible service access with a large variety of choices and preferences. The complexity does not concern the subscribers and it shall also appear masked to service providers.

All the hype regarding business projections with phenomenal and voluminous profits in the past have only undermined the user perception of UMTS. The speculation of the media coupled with greediness of certain governments (e.g. UK and Germany) made the spectrum cost in some regions unnecessarily costly.

But despite the unexpected events, UMTS will progress into the service platform for future information systems; the whole communication industry is behind it. Technology will no longer be perceived as an issue but as the driver for innovation and service variety and differentiation. As illustrated in Chapter 1, UMTS will stand as the natural evolution of a majority of today's mobile network infrastructures. With some patience and consistent availability of solutions, UMTS will deliver the means to implement the key elements of tomorrow's communication network and service designs. After all, Rome was not built in a day, and UMTS too will take some time as it took for GSM (e.g. 5+ yrs). With this in mind, this chapter will aim to contribute towards the main components of UMTS services before the different building blocks, and these are described in the forthcoming chapters.

3.1.1 UMTS Service Characteristics

By now in the mobile telecom community we all know that there is not one 'killer data application' for GPRS or UMTS as it has been the voice for GSM. Although the challenges appear dauntless, perseverance and conventional wisdom will lead to establish the sets of services for the main stream of subscribers, because there is not only one, but 'many killer applications' as there are many types of users.

On the other hand, it is a matter of allowing mobile usage trends to expand to all social strata and generate new lifestyles, which embrace the new forms of data communication and information sharing. This on its own will bring new waves of demand to make UMTS-type services the ideal vehicle to meet multifarious and ubiquitous needs.

All IP in 3G CDMA Networks J. Castro
 ISBN: 0-470-85322-0

Hence, giving new services other than voice their corresponding attributes, e.g. making them Relevant, Efficient, Affordable and Localised (i.e. REAL), they will appeal to all those to whom they are due. Because everyone needs to communicate and share information through the most practical way depending what is at stake, putting UMTS services to everyone's reach will simply fill all the gaps to interact between persons or even between man to machine, or trigger machine to machine communication for that.

Satisfying mobility (anywhere), convenience (anytime), immediacy and flexibility (anything to anyone) through making services personal and available in real time, UMTS will cover all spectra of lives.

3.1.1.1 Service Examples

What are then those new services that will reach every consumer? They are not necessarily all the new ones, but also current ones offered in a more dynamic and optimised environment, i.e. focusing on the user needs. Table 3.1 illustrates a *service classification example* to use it as a reference in the forthcoming discussions. Clearly, this is a non-exhaustive list used in the industry with different variations, to which one may add or integrate service types or classes.

To bring services to a new dimension through UMTS or to realise services shown in Table 3.1 under a new dynamics and optimised surrounding, next we will present a common arrangement, which will facilitate the blending of all factors leading to enrich the user experience.

3.1.2 Application Development Principles

Services and application may appear to convey the same thing; however, they do differ on their function, although sometimes they may work as one. *Applications enable services and services are the bundling of applications offered to the subscribers.* The first appeals more to operators or service providers, while the second to developers or manufacturers packaging network solutions. *One application may serve several services or one service may use several applications.* Both services and applications depend on each other and their realisation requires accurate specifications.

Thus, in the following we will attempt to summarise the guidelines to arrive at the ideal process of service creation. We shall outline the key areas, e.g. market development, development platforms and tools, solution architectures with corresponding inter-operability and security aspects and facilitating application development and testing communities, finally deploying successful, highly performing and reliable applications to enable attractive and interesting services with large variety and flexibility.

3.1.2.1 New Product and Solutions

- Getting *innovation* into the mobile market becomes critical today. To foster this development the creation of an ecosystem, i.e. an environment leading to new telecommunication services, is essential.

Table 3.1 Service classification example

Thick: can work off-line → intelligent client			Thin: needs on-line link → partially intelligent client		
Vertical applications	Information services	Enhanced communications	M-commerce Internet browsing	Leisure	Corporate and residn . . . services
Machine-to-machine - Data retrieval - Remote surveillance - Device management - Remote operation	*Location Based Services (LBS)* - Local Info - Traffic info - Positioning - Navigation	*Speech* - Voice (2G and 3G) - Video telephony - Video conferencing - Show me sessions	*E-payment financial services* - Electronic wallet - Micro-payment - C-to-C payment Stock trading	*Games* - Interactive games - Games on demand - Multi-player games - 3D games	*Remote access* - Corporate access - Private access - May require services from domains 1–5
Automotive/tracking - Transport monitoring - Vehicle tracking - Aggression protect - Vehicle alarms - Automatic assistance	- *Directory* - W/yellow pages - Info kiosks - Transport schedule - City guide - Administrations	*Downloading education material* - Instructions - Service Indications - Directions - Office info - Interest rates	*Banking* - Automatic payment - Currency rates - Personal account info - Credit checks - Video clips, etc.	*Music/video +* *- downloading* - Music streaming - Video streaming - Ringtones, radio - Imaging, E-books - Office term agendas	*Corporate VPNs* - Wireless office - Own server E-mail - Corp video conf. - Corp bulletin boards
Home automation - Device control management - Surveillance - Fire alarms, etc.	*New* - Dedicated news - General news - Special events - Sports results	*Streaming video* - Live instructions - Direct Media access - Others	*Shopping* - Electronic mall - Bookings - Auctions - Orders	*Gambling* - Lottery - Casino	*Family VPNs* - Social net bulletin - Own Server PDAs - Own E-mail server - Access family clubs

Table 3.1 *(Continued)*

Thick: can work off-line → intelligent client			Thin: needs on-line link → partially intelligent client		
Vertical applications	Information services	Enhanced communications	M-commerce Internet browsing	Leisure	Corporate and residn . . . services
Medical equipment - Emergency alarms - Inventory supply - Devise connection - Control/maintenance	*Tourism* - Vacation info - Hotels - Restaurants - Sightseeing info	*Messaging* - Voice mail, SMS - E-mail - Unified messaging - MM messaging	*Ticketing* - Electronics ticket - Ticket purchasing - Ticket transfer	*Entertainment* - Children - Teenagers - Young people - Adults - Special groups	
		Active community - Chat sessions - Instant messaging - Meeting circles - Personal portals	*Direct marketing* - Vouchers - Fidelity programs - Classifieds - Auctions	*Lifestyles* - Hobbies - Parties - Fashion - Gourmet	
		Personal Info Mngt - Synchro. . . agenda - Directory, info, etc.		*Dating Services* - Partner finder - Meeting specials	

- Building an appropriate environment implies a venture creation, which starts with the synthesis, discovery, design, launch and growth. This means that an ecosystem will include market needs, technology, human capital and financial capital. A smooth interaction of all these factors will bring in new successful services.

3.1.2.2 User Friendly Applications

1. Stay away of 'slow response times' systems[1] with unfathomable user interfaces. If your grandmother does not understand the user interface, do not deploy the service. A service should be up and running within three keystrokes at most. Do not be fooled, although it may be worth 1000 words it may take long.
2. Remember, mobile cellular Internet *is different* from World Wide Web Internet running on wired systems where bandwidth is not always critical. Thus, 3G wireless Internet will not be browsing, but reaching precise information just as a surgeon uses the scalpel to get to a specific part.
3. In the application design process, one will think as the customers and identify who they are while trying to understand their problems and communication needs; and will address all of them appropriately, i.e. 'first time users, regular users and power users'. Segment here may no longer play a major role as in voice-only services era. Personalisation has to reach practical needs and not only as superfluous user desires.
4. A service provider or service developer will break down the user scenarios, minimise the tasks to solve a problem or to get an information, will remove un-necessary tasks, prioritise remaining ones and will test applications in user-like set-ups and with the most likely devices to reach a true user experience.
5. To multiply the power of mobile Internet, e.g. an application will proactively bring intelligence to the user's hands. For example, contrary to the traditional WAP, it will *push*[2] information to increase session interactivity and speed up the process to get the information.
6. Understand that the application life cycle in the wireless Internet environment may not exceed 12 months; this may be too short for developers. Thus, they need to act fast with the appropriate platforms and the right testing environment besides a precise set of requirements.
7. This means that operators or service providers need to change their mentality of service creation. While in the traditional wired Internet application distribution is easier, in the mobile Internet without relevant applications or services there is no chance of usage expansion.

Then it boils down to giving value and choice and not just quantity of *superficial applications.* Ringtones and logos, although they increase traffic, will not justify all the infrastructure sophistication forthcoming for UMTS. Operators need to nurture

[1]Avoid the Microsoft phenomenon, e.g. as the computing power increases the SW fatness also increases.
[2]For example, available with WAP V2.0.

developers by providing the best possible environment for end-to-end testing and not just a set of requirements. Go beyond NTT DoComo team, who drove the whole convergence of application and service creation for their mobile Internet.

A portal, for example, will address the mobile user and not just subscriber in general, based on the wired Internet solutions. Wireless transmission optimisation for the handset demands real testing and mastering of constraints.

8. For the first time, operators are the integrators of a whole set of technologies in order to offer wireless data services. Hence, they will need to understand their role or suffer on their business. Classical management approaches or traditional managers will need to change or let those who have a better approach and blend closer with engineering do the task. It is as simple as that.
9. Likewise, developers will need to produce with measure and specificity and not expect that one product will meet all demands. Flexibility, modularity and inter-operability will be a must.

3.1.2.3 Application Attributes and Network Optimisation

As Feige and Cisco [1] put it, delivering WEB and/or multimedia content through the mobile Internet is not the same as doing it through the wired Internet. And rightly so, the constraints in the cellular environment are higher and very often underestimated by the Internet community. Larger bandwidth availability in the fixed Internet may mask the non-optimised overhead and make delays invisible to the user, while in the mobile Internet it translates into slow response and sluggish service highly perceived by the user.

The challenges stand in optimising, e.g. the transmission environment, network, applications platforms, etc. But one critical challenge stands on perceiving the nature of the application. Thus, in the following we will highlight the relationship of the network and delivering mobile content.

1. Data through the cellular environment today has limitations from non-optimised Internet backbones, traditional cross-Internet connections (middle mile), limited air-interface bandwidth (the last mile) and in many cases non-ideal or non-appropriate edge devices.
2. The problems and solutions for cellular data offering are then:[3]

   optimise TCP to prevent inefficient transmission link in the transmission interface → overhead due to BER,

   maximise image and text compression when transmitting fat or rich text (e.g. HTML) in a limited bandwidth,

   transcode with more than one markup language (e.g. HTML, SGML, XHTML, CHTML, WML and AnyML) for multiple devices,

   store multiple instances of same content after first request to minimise latency and increase scalability → catch to reduce CPU load in the backend and save bandwidth,

[3]For example, looking from Cisco's perception.

build multifunction dedicated mobile data portals capable to interface with all types of application platforms (e.g. personalised, special workflows, push messaging, synchronised data bases, etc.) and developer's systems.

3. In conclusion, implement mobile multi-service architecture from the beginning. Understand the constraints and do not just adapt an old style wired Internet portal.
4. Do not just follow a common trend, learn the lessons from the real world; *set a new trend*:

   Wireless content tends to be static today delivered through WTP/WSP (WAP) or HTTP/ TCP (i mode). Yet, transport protocols should not be mistaken with markup languages.

   Even though XML seems to be the winner in markup languages with the others as subset, and TCP is the major transport protocol.

   See that TCP was designed for a wired infrastructure. It assumes packet loss due to network congestion, constant bandwidth, stable Round-Trip Time (RTT) and TCP handshake overhead.

   Yet the cellular wireless environment challenges these assumptions. What then? Modify your transport layer dynamically depending on the environment and the application. Hence, use the right TCP stack.
5. Access the behaviour of streaming type applications through your cellular network environment. See, e.g.:

   Video transmission is time sensitive. Hence, through HTTP (i.e. TCP based) may not be as efficient as with RTP (UDP based) where visual impact is minimised with a network loss.
6. Finally, device awareness content will maximise the service experience and minimise transmission flows. Device sensitive servers will channel the right information at the appropriate time and apply the correct policies to adapt the content.

3.1.2.4 Testing Mobile Applications

As mentioned earlier, to successfully introduce useful and meaningful application we need complete environment. Such arrangement by all means will also include the testing capabilities. 'Short time to market' implies following common procedures end-to-end, i.e. all players shall align to contribute to satisfy all value chain requirements from development to targeted user, which means that all players have non-negligible responsibilities, e.g.:

- identify and communicate application requirements to developers early,
- address specific usability on time – identify the user experience expected,
- simplify the features to minimise complexity – leave room for evolution,
- set testing criteria and 'end-to-end' evaluation environment,
- classify testing domains, e.g. bearer services + components, management, etc.
- engineering conformance to requirements,

- marketing conformance and readiness to requirements,
- verification of user perception to service objectives,
- compliance with 'return of investments' projections.

If we notice the list above, all points are common sense, yet we often tend to underestimate the co-ordination of all concerned parties and expect that someone else will do it. Developers assume that one product or application will apply to most cases and imply that testing corresponds primarily to the service provider; they have not idea of end-to-end performance and focus only on product features ignoring optimisation. Service providers expect that all application will be plug and play; marketing neglects the time required to validate engineering aspects. The result comes down to delayed and poorly performing products with minimum impact in the public or user perception, basically not meeting their needs or arising new interest.

Who are then the main parties involved, or what are the key testing building blocks for successful applications? While the answer may vary depending on the application, a minimum set of conditions can be summarised as follows:

1. Use a common testbed, preferably the service provider's one
   Incorporate all main NEs, including billing, load and security testing.
2. Developers shall assume continuous service assurance
   during functional and operational tests,
   own application performance responsibility across boundaries,
   i.e. infrastructure providers and third party solutions,
   support content tests through authentication and expanded geography,
   benchmark solutions through meaningful comparisons,
   distinguish access and user segments, e.g. radio and subscriber types.
3. Service providers shall integrate and runs of all service building blocks
   will discriminate internal and external network testing;
   i.e. determine own network dependencies and roaming ones,
   provide an optimised bearer path for all applications and services,
   outline inter-operability constraints consistent with early requirements,
   identify load levels in real operations and apply them during testing,
   define application platform dependencies while testing new services,
   contribute with terminals to test compliance to earlier requirements.
4. Network infrastructure providers shall also be active participants during testing
   Nominal compliance with standard specifications will not be enough.
   Shall provide open description of functional performance in NEs early.
   Will support end-to-end application and services testing.
   Will provide realistic simulation toolkits.
   Will contribute to verification of simulated tests in real time.
   Will facilitate automated testing of new applications and services.

Will support multiple or diverse terminal testing, not just a selection,
provide open infrastructure to all device types in advance.

Clearly, we need timely co-ordination of three key players, i.e. developers, operators or service providers and infrastructure suppliers. We can no longer underestimate the learning curve. Together, we need to define testing profiles and make it available before applications development and verify afterwards well in advance of commercial introduction.

3.1.2.5 Implementing Converging Services

UMTS network and service enabling capabilities are evolving continuously. For example, in this edition we present the evolution of R99 to R4 and R5 as part of an integrated enhancement, i.e.

- packet capabilities for GSM came in with class B GPRS terminals,
- simultaneous voice and data get enabled with class A GPRS terminals,
- broadband bearers + QoS get enabled with WCDMA in UMTS R99,
- we obtain CN optimisation and introduce OSA with UMTS R4,
- R5 enables all IP Core features supporting SIP based applications and services.

These evolving features inter-operate or are to do so seamlessly at the network and device level. Logically, application development and service creation will be expected to do likewise.

This means that implementation of new applications will assume a converged service environment, where, e.g. Internet, multimedia, E-mail, presence, instant messaging and m-commerce co-exists flawlessly with voice services, and perform automatically as the *network* and *device* enablers allow.

Backward and forward compatibility shall be then a practical requirement and not a voluntary option. As users, we all focus on the service performance and not the infrastructure or application technology.

Without a doubt, such requirement has challenging implications, but they are not impossible to reach, because today standards or technical specifications for all domains and release phases are out or under completion. Besides, full compatibility does not need to happen all once or in one device or network. By making pragmatic roadmaps we can also build flexible architectures for transcending applications, i.e. applications that can adapt to evolve as enablers come into place.

3.1.3 UMTS Service Features

UMTS services will not only offer mobile services supported by second generation systems such as GSM, but will also expand these services to higher rates and greater flexibility. The services evolving in the GSM platform through its Circuit Switched (CS) and Packet Switched (PS) will continue in UMTS while new services get introduced in its new domains, e.g. IMS.

Table 3.2 Range of transmission rates

High level description	Maximal bit rate (kbps)	Maximal speed (km/h)	Cell coverage
Rural outdoor	144s	500	Macro-cell
Sub-urban outdoor	384	120	Micro-cell Macro-cell
Indoor/Low range outdoor	2048	10	Pico-cell Micro-cell
HSDPA	8000 (D4)	NA	NA

Thus, future UMTS services will have user transmission rates from low bit up to 8[4] Mbps. Although, high rates will occur primarily within indoor environments, these will be substantial. Table 3.2 illustrates this increase. For completeness, here we already add the progress of the downlink through HSDPA[1] in Releases 5 and 6. This table does not show yet the performance increase due to the evolution of the new UMTS components presented in Chapter 1 for the long-term evolution.

Then the question of the coverage range for UTMS is no longer just what transmission rates, but what type of services, when and where. It is no longer 'communications anywhere anytime', but 'what I want when I want wherever I want'.

Practically, the exploitation of wider transmission rates will facilitate the expansion of data traffic. As illustrated in Table 3.3 there exists a clear trend for the convergence of IP protocol to wireless, or to what we now call *wireless IP*. The latter will consolidate the *wireless Internet*, where more than 200 million Internet and 300 million mobile subscribers will merge into 1 billion wireless Internet users.

As discussed already in the preceding section, 'non-voice services' will make demands not only on manufacturers and operators but also from supporting industries, creating a need of applications for new service enablers. Such a demand will also introduce new challenges and the need for pragmatic integration of services and devices, as well as new data processing and managing techniques. These demands can be summarised as needs as illustrated in Table 3.4.

Table 3.3 Convergence of Internet Protocol (IP) to wireless

Computer mobility: high speed terminals	Telecommunication mobility: integrated wide area services	Media mobility: choice of personal services
Constant Internet access	ISDN services	Streaming audio
Easy electronic mail	Video telephony	Video on demand
Instant real time images	Wide-band data services	Interactive video services
Ubiquitous multimedia	Location services coupled with application servers	TV/radio/data contribution and distribution
Wireless broadband IP-multimedia capabilities	SIP based IP communications in cellular network environments	Multi-session mixed media transmission

[4]Referring to the advent of High Speed Downlink Packet Access (HSDPA).

Table 3.4 Needs for service providers and technology enablers

Needs for service providers	Needs for technology enablers
Strategy for innovative services	Well integrated CS, PS and IMS subsystems
Economic and spectrum efficiency data pipe	Advanced value added platforms (e.g. WAP, IS, LCS, unified messaging, OSA, (etc.)
Standard interface-to-phone display	Power efficient handsets
Dynamic management control points	Effective yet very light devices (OSs)
New and flexible billing systems	Text → speech
Perception of market needs	Speech → text
Personalisation	Intelligent voice recognition
Addressing all user segments	Multi-band terminals exploiting software radio
New data processing and management techniques	Synchronisation
Cost efficient and secure terminals and devices	Pragmatic user interfaces (e.g. efficient portals)
Clear requirements and testing environment	Secure application techniques

Clearly, the challenges cover all main areas of SW/HW and management technology. In the forthcoming sections and next chapters we will see how UMTS addresses these needs and outlines the main approaches and requirements to meet the challenges.

3.2 THE UMTS BEARER ARCHITECTURE

As illustrated in Figure 3.1 [2], UMTS proposes a layered bearer service architecture, where each bearer service on a specific layer offers its individual services based on lower layers. Thus, the UMTS bearer service architecture serves as an ideal platform to offer end-to-end service quality.

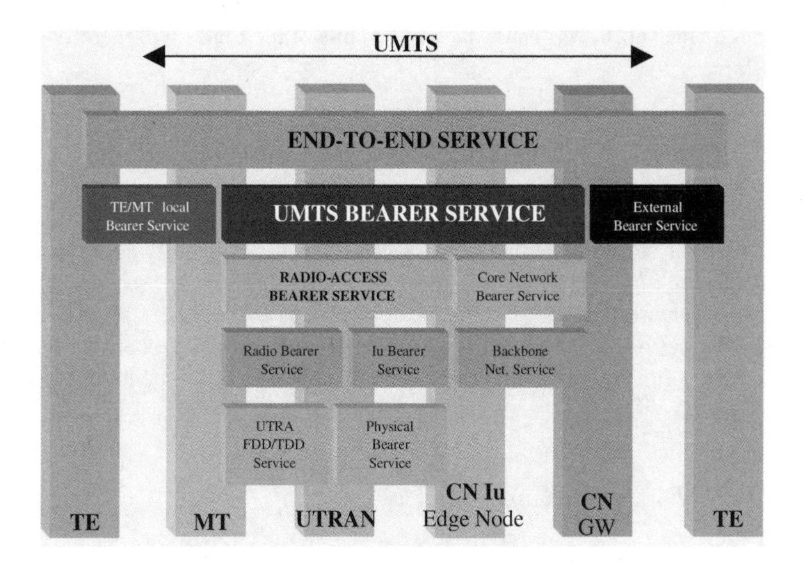

Figure 3.1 UMTS bearer service architecture after [2].

End-to-end services imply solutions from a Terminal Equipment (TE) to another TE, having a certain Quality of Service (QoS) offered to a user of a network, where the latter decides whether he is satisfied with the provided QoS or not. A QoS Bearer Service (BS) defines clearly the characteristics and functionality required from the source to the destination of a service. A BS includes all aspects to enable the provision of a contracted QoS, e.g. control signalling, user plane transport and QoS management functionality.

Because of its layered-bearer service architecture, UMTS permits users or applications to negotiate, re-negotiate or change appropriate bearer characteristics to carry their information. Negotiations take place based on application needs, network resource availability and demands of QoS.

3.2.1 Radio Access – Core – Backbone and Iu Bearer Relations

Figure 3.1 also illustrates the main UMTS bearer services, i.e. the Radio-Access Bearer (RAB) service and the Core Network Bearer (CNB) service. Both services enable the optimised realisation of UMTS bearer services in the cellular network environment reflecting mobility aspects and mobile subscriber profile requirements. The examples are given below.

3.2.1.1 The Radio-Access Bearer Service

The RAB service provides confidential transport of signalling and user data between Mobile Terminal (MT) and Core Network (CN) Iu edge node with the appropriate QoS to meet the negotiated UMTS bearer service or with the default QoS for signalling. The service maintains the characteristics of the radio interface for a moving MT, e.g. unequal error protection for the expected Service Data Unit (SDU) through lower RABs. In R99, a RAB unequal error protection applies only for services using a codec integrated in the core network, which implies that UMTS bearer service cannot use the attribute *SDU format information* to define sub-flows. Thus, the payload bits of the SDUs will therefore be equally protected [2].

3.2.1.2 The Core Network and Backbone Bearer Service

The CN bearer service of the UMTS core network connects the UMTS CN Iu edge node with the CN gateway to the external network. This service efficiently controls and utilises the backbone network in order to provide the contracted UMTS bearer service, where packet core network supports different backbone bearer services for a variety of QoS [2].

The CN bearer service uses generic backbone network services, where the latter covers layer 1 and layer 2 functionality selected according to QoS requirements in the CN bearer service.

3.2.1.3 The Radio and Iu Bearer Services

Radio Bearer Service (RBS) realises the RAB service. The role of the RBS is to cover all the aspects of the radio interface transport while using the UTRA FDD/TDD physical layer.

To support unequal error protection, UTRAN and MT shall have the ability to segment/ reassemble the user flows into the different sub-flows requested by the RAB service. The segmentation/ reassemble is given by the SDU payload format signalled at the RAB establishment. The RBS handles the part of the user flow belonging to one sub-flow, according to the reliability requirements for that sub-flow.

The Iu bearer service together with the physical bearer service provides the transport between UTRAN and CN. Iu bearer services for packet traffic shall provide different bearer services for a variety of QoS [2].

3.2.2 Management and Allocation of QoS Functions

Relations between QoS internal functions and network nodes to maintain a specified QoS are implementation specific. Allocation of these functions to the UMTS entities indicates the specific entity requirement to enforce the QoS commitments negotiated for the UMTS bearer service.

The QoS management functions of all UMTS entities together ensure the provision of the negotiated service between the UMTS bearer service access points. The translation and mapping with UMTS external services provide the end-to-end QoS.

3.2.2.1 Control Plane Management Functions

Table 3.5 illustrates the main functions of the control plane. Notice how these functions perform distinct tasks to meet and maintain QoS request of the different UMTS bearer services. The admission capability control by discriminating the different network entity adequacy to serve or meet QoS requests does also perform service retention.

Figure 3.2 shows the flow sequence of control plane management functions to regulate the UMTS bearer services. These functions support establishment and modification of UMTS bearer services

- through signalling and negotiation with the UMTS external services, and by
- establishment or modification of all UMTS internal services providing the required characteristics.

3.2.2.2 Control Plane Allocation of QoS Management Functions

The following function descriptions are direct extracts from [2] outlined here for completeness.

As seen in Figure 3.2, the translation functions (Trans.) in the *MT* and the *Gateway* convert between external service signalling and internal service primitives, include the translation of the service attributes. The packet oriented translation function in the gateway is FFS.

The *MT-UMTS BS* manager, *CN EDGE* and *Gateway* signal between each other and via the translation function with external instances, to establish or modify a UMTS bearer service. Each UMTS BS manager interrogates its associated admission/capability control

Table 3.5 Control plane management functions

Service manager	Translation function	Admission Capability Control (ACC)	Subscription control
Co-ordinates functions to establish, modify and maintain the service it is responsible for. It provides all user plane QoS management functions with the relevant attributes. It offers services to other instances, e.g. it signals with peer service managers and uses services provided by other instances. May perform an attribute translation to request lower layer services. May interrogate other control functions to receive permission for service provision.	Converts between UMTS service primitive bearer service control and the various service control protocols interfacing external networks. Translation includes converting between UMTS bearer service attributes and QoS attributes of the external network service control protocol (e.g. between IETF-TSPEC and UMTS service attributes). The service manager may include a translation function to convert between its service attributes and the attributes of a lower layer service it is using.	Maintains information about all available network entity resources and about all resources allocated to UMTS bearer services. It determines whether resources provided by an entity can meet requirements for each UMTS bearer service request or modification. Then, it reserves these resources if allocated to the UMTS bearer service. Checks also Network entity capability to provide the requested service. The resource control performed by the ACC supports also the service retention.	Checks the administrative rights of the UMTS bearer service user to use the requested service with the specified QoS attributes.

whether the network entity supports the specific requested service and whether the required resources are available. In addition, the CN EDGE UMTS BS manager verifies with the subscription control the administrative rights to use the service.

The MT-UMTS BS manager translates the UMTS bearer service attributes into attributes for the local bearer service and requests this service from the local BS manager.

The UMTS BS manager of the CN EDGE translates the UMTS bearer service attributes into RAB service attributes and Iu bearer service attributes. It also translates UMTS bearer service attributes into CN bearer service attributes and requests its Iu BS manager, its CN BS manager and the RAB manager in the UTRAN to provide the required services. Consequently, it triggers the action on all corresponding managers.

The RAB manager through its admission/capability control verifies whether UTRAN supports the specific requested service and whether the required resources are available. It translates the RAB service attributes into radio bearer service and Iu bearer service attributes

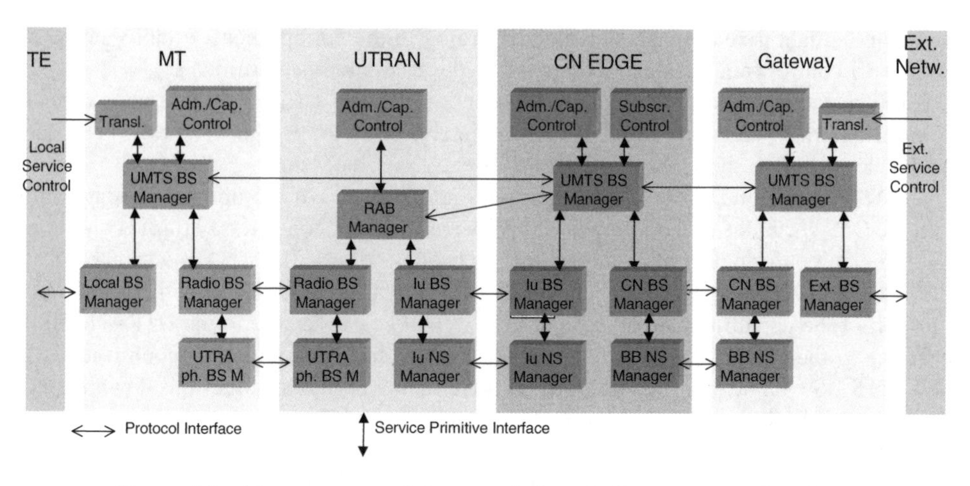

Figure 3.2 Flow sequence of the control plane QoS management functions.

and requests the radio BS manager and the Iu BS manager to provide bearer services with the required attributes.

The gateway UMTS BS manager translates the UMTS bearer service attributes into CN bearer service attributes and requests its CN BS manager to provide the service. In addition, it translates the UMTS bearer service attributes into the external bearer service attributes and requests this service from the external BS manager.

In the radio side the Iu and CN BS managers use services provided by lower layers as illustrated in Figure 3.2.

3.2.2.3 User Plane QoS Management Functions

Figure 3.3 illustrates the user plane QoS management functions of the UMTS BS. These functions, provided by the QoS management control elements with relevant characteristics,

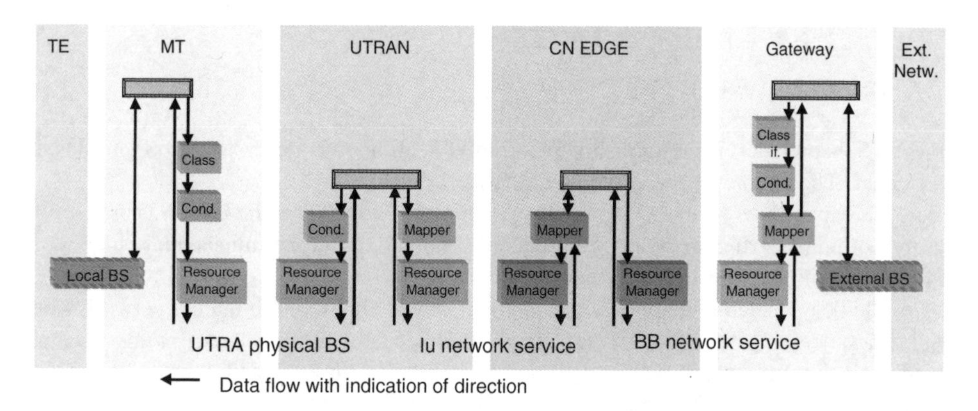

Figure 3.3 User plane QoS management functions.

maintain the data transfer characteristics according to the commitments established by the UMTS BS control functions and expressed by the bearer service attributes.

The FFS gateway and MT classification function (Class.) assign user data units received from the external bearer service or the local bearer service to the appropriate UMTS bearer service according to the QoS requirements of each user data unit.

The MT traffic conditioner (Cond.) provides conformance of the uplink user data traffic with the QoS attributes of the relevant UMTS bearer service. The gateway traffic conditioner may provide conformance of the downlink user data traffic with the QoS attributes of the relevant UMTS bearer service; i.e. on a per PDP context basis. Because the packet oriented transport of the downlink data units from the external bearer service to the UTRAN and the buffering in the UTRAN may result in bursts of downlink data units non-conformant with the UMTS BS QoS attributes, a traffic conditioner in the UTRAN frames this downlink data unit traffic according to the relevant QoS attributes.

The resource manager may act as a traffic conditioner, e.g. it can provide conformance with the relevant QoS attributes by appropriate data unit scheduling, or implicitly condition the traffic when fixed resources get dedicated to one bearer service. Thus, traffic conditioning does not imply separate functions.

Mapping function marks each data unit with the specific QoS indication related to the bearer service performing the transfer of the data unit.

Each network-entity Resource Manager (RM) is responsible for a specific resource, e.g. to distribute its resources between all bearer services requesting transfer of data units on these resources. Hence, the RM aims to provide the QoS attributes required for each individual bearer service.

3.3 QoS ATTRIBUTES IN UMTS BEARER SERVICES

To appropriately support asymmetric uplink/downlink channels, UMTS offers uni-directional and bi-directional bearer services; for the latter, attributes such as maximum bit rate, guaranteed bit rate and transfer delay are exploited. These attributes describe the service, which the UMTS network provides to the user of the UMTS bearer service. A set of QoS attributes generally stored in the HLR as QoS profile, characterise a given service.

3.3.1. Source of Attributes

At UMTS bearer service establishment or modification, the necessary QoS profiles get considered. The source of the profiles may arise from [2]:

The UE capabilities form a QoS profile, which may not fully use the UMTS bearer service features offered by the network. For example, not all initial terminals may support PS 384 kbps downlink even if the network will.

It would be expected that applications using the UE within the terminating network would request a specific QoS profile to the UE at UMTS bearer establishment or modification. However, if the application request does not do so, the UE may use a QoS profile configured within the UE (e.g. by AT commands manually), which may lead the Terminal Equipment (TE) to derive a QoS profile outside the scope of UMTS.

In principle, a QoS profile in the UMTS subscription will describe the upper limits for the provided service, if the service user requests specific values within the profile range. Otherwise, the QoS profile will serve as the default QoS service profile requested by the user.

Generally, an operator will configure and store default QoS profile(s) for the UMTS bearer services as its network supports. Thus, when unsupported QoS requests occur, negotiation and modification of QoS will take place at the UMTS bearer service establishment.

3.3.2 UMTS and Radio-Access Bearer Service Attributes

Table 3.6 summarises the main UMTS bearer service attributes, which include the traffic classes. In the radio access it applies to both CS and PS domains.

Table 3.6 UMTS and radio-access bearer service attribute characteristics

Characteristics	Description (direct extracts from [3])
Traffic classes	– Conversational, streaming, interactive and background type of application for which the UMTS or radio-access bearer is optimised. It serves the UMTS traffic source or UTRAN buffer allocation.
Maximum bit rate (kbps)	– Maximum number of bits delivered by UMTS and to UMTS or by UTRAN and to UTRAN at a service access point within a period of time, divided by the duration of the period. – The traffic is conformant with the maximum bit rate as long as it follows a token bucket algorithm, where token rate equals maximum bit rate and the bucket size equals maximum SDU size. – The maximum bit rate is the upper limit a user or application can accept or provide. All UMTS or RAB bearer attributes may be fulfiled for traffic up to the maximum bit rate depending on the network conditions.
Guaranteed bit rate (kbps)	– It is the number of bits delivered by UMTS or by UTRAN to UTRAN at a SAP within a period of time (provided that there is data to deliver), divided by the duration of the period. – Delay and reliability attributes are guaranteed, for the traffic exceeding the guaranteed bit rate the UMTS bearer service attributes are not guaranteed.
Delivery order (y/n)	– It indicates whether the UMTS bearer shall provide in-sequence SDU delivery or not. – Whether out-of-order SDU are dropped or re-ordered depends on the specified reliability.
Maximum SDU size (bytes)	– The maximum allowed SDU size – It is used for admission control and policing

Table 3.6 (*Continued*)

Characteristics	Description (direct extracts from [3])
SDU format information (bits)	– List of possible exact sizes of SDUs – If a RAB service uses unequal error protection, the SDU format information defines the exact sub-flow (+bit rates) of the SDU payload – In the RAB it also serves to operate RLC protocol mode transparently for higher spectral efficiency and lower bearer cost and better error protection.
SDU error ratio	– Indicates the fraction of SDUs lost or detected erroneous whilst only for conforming traffic. – In UMTS and RAB the SDU error ratio performance with reserved resources does not depend on the loading conditions as it does it without it. For example, in interactive and background classes, it is used as target value.
Residual bit error ratio	– For the RAB, which also applies to UMTS, *residual bit error ratio* indicates the undetected bit error ratio in the delivered SDUs if error detection has been requested for a sub-flow. Otherwise it indicates the Bit Error Ratio (BER) in the delivered SDUs. – For equal error protection in the RAB, we only need one value.
Delivery of erroneous SDUs (y/n)	– Indicates whether SDUs detected as erroneous shall be delivered or discarded (y is for error detection applied and error delivered, n for errors discarded. – For unequal error protection in the RAB the attribute is set per sub-flow, which may have different settings.
Transfer delay (ms)	– Indicates maximum delay for 95th percentile of the distribution of delay for all delivered SDUs during the lifetime of a bearer service, where delay for an SDU is defined as the time from a request to transfer an SDU at one SAP to its delivery at the other SAP. – It is used to specify the delay tolerated by the application. It allows UTRAN to set transport formats and ARQ parameters.
Traffic handling priority	– Specifies the relative importance for handling of all SDUs belonging to the UMTS or radio-access bearer compared to the SDUs of other bearers – Used within UMTS (UTRAN) interactive class to allow ideal traffic scheduling.
Allocation/retention priority	– Specifies the relative importance of given UMTS or radio-access bearers compared to other UMTS or radio-access bearer for allocation and retention. It is a subscription attribute, which is not negotiated from the mobile terminal. – It is used to differentiate between bearers while performing bearer allocation and retention in a resource scarce environment.
Source statistics descriptor ('speech'/'unknown')	– Specifies characteristics of the source of submitted SDUs – May serve to calculate statistical multiplex gain in admission control on the radio and the Iu interface.

3.4 UMTS QUALITY OF SERVICE CLASSES AND ATTRIBUTES

The main four classes of UMTS traffic differentiated by their delay sensitivity are conversational, streaming, interactive and background. Conversational classes have higher delay sensitivity than background classes. The first two classes correspond to real-time classes, while the second two to non-real time. The sensitivity to delay distinguishes real time classes, e.g. video telephony has the higher susceptibility; hence a conversational class will carry its data rather than a streaming class.

We distinguish interactive class from background class through the latency of response within application in which they are used; e.g. the first may apply to interactive services like web browsing with instant response demands, while the second may apply to background file download like FTP, which can operate without response constraints. Due to non-strict delay requirements, compared to conversational and streaming classes, interactive and background classes offer better error rate protection by means of channel coding and retransmission. Traffic in the interactive class has higher priority in scheduling than background class traffic; thus, background applications use transmission resources only when interactive applications do not need them, which is critical in cellular wireless environment where the bandwidth is lower compared to fixed networks [2].

The classification in UMTS attempts to take into account the cellular transmission environment primarily through the parameters in the radio interface, in order to offer a robust end-to-end channel. Table 3.7 illustrates the UMTS QoS classes.

3.4.1 Conversational Class

Typical speech over CS bearers (GSM), Voice over IP (VoIP), video conferencing and video telephony are some examples of the conversational class, which in turn represents real-time services. The latter corresponds to symmetric traffic with end-to-end delay thresholds below 399 ms. On the other hand, interactive class when working with latency delays below or the same as of those of conversational, they can also be considered real-time services.

Required characteristics for live (human) end users during conversation depend strictly on the sensibility of the corresponding human senses. Nevertheless, there are common thresholds, which will underline the minimum understanding levels. Thus, the maximum transfer delay is given by the human perception of video and audio conversation. As a result, the limit for acceptable transfer delay is very strict, as failure to provide low enough transfer

Table 3.7 QoS classes in UMTS, after [1]

Traffic class	Conversational	Streaming	Interactive	Background
Characteristics	Preserve time (variation) relation between stream information entities Conversational pattern–low delay	Preserves also time relation between information entities	Request response pattern Preserve data integrity	Connectionless, delay tolerating transmission Requires high data integrity
Applications	Voice, video telephony	Streaming video	Internet or web browsing	Download, ftp, E-mail, etc.

delay will result in unacceptable lack of quality. The transfer delay requirement is therefore both significantly lower and more stringent than the round-trip delay of the interactive traffic case [2].

Real-time conversation–fundamental characteristics for QoS:

- preserve time relation (variation) between information entities of the stream;
- conversational pattern (stringent and low delay).

3.4.1.1 UMTS Service Attributes in Conversational

When the UMTS bearer carries speech service, we can set *source statistics descriptor* to calculate the statistical multiplexing gain in the core network, UTRAN, as well as the UE and use it for admission control.

Despite the bit-rate variation of a conversational source codec, we consider conversational traffic relatively non-bursty, where a *maximum bit rate* specifies the upper limit bit rate with which a UMTS bearer delivers SDUs at the SAPs; and where a UMTS bearer does not need to transfer traffic exceeding the *guaranteed bit rate*.[5] Finally, because of the non-bursty traffic we can guarantee the *transfer delay* of an arbitrary SDU.

UTRAN realises *conversational* bearers without RLC re-transmissions by using the *SDU format information,* which is obtained through dividing the largest defined SDU format. Thereby, it transports efficiently and at lower costs than if we adapt the RLC PDU size to the UMTS bearer SDU size.

The *maximum SDU size*, used for admission control and policing, applies only when the *SDU format information* does not get specified. If specified, e.g. with one or several sizes, the SDU size is variable and each SDU shall exactly conform to one of the specified sizes.

We specify the application error rate requirement by using the *SDU error ratio*, *residual bit error ratio* and *delivery of erroneous SDU* attributes. They also serve to determine whether the application wants UMTS to detect and discard SDUs containing errors by applying an appropriate forward error correction.

3.4.1.2 Radio-Access Bearer (RAB) Service Attributes in Conversational

As in the UMTS bearer, when the RAB carries speech service, we can set *source statistics descriptor* to calculate the statistical multiplexing gain in the core network, UTRAN, as well as the UE and use it for admission control.

To support unequal error protection requests in conversational class for a given RAB (per sub-flow), we specify the attributes *delivery of erroneous SDUs*, *Residual bit error ratio* and *SDU error ratio.* The first attribute determines whether error detection shall be used and, if so, whether SDUs with error in a certain sub-flow shall be delivered or not. The second one specifies the bit error ratio for undetected delivered bits, while the last one specifies the fraction of SDUs with detected error in each sub-flow.

When the RAB service transports user data SDU payloads with unequal error protection, the SDU formats conform to possible exact sizes, where the payload bits are statically

[5]*Maximum and guaranteed* bit rate attributes are used for resource allocation within UMTS.

structured into sub-flows and the *SDU format information* attribute defines the exact sub-flow format of SDU payload.

UTRAN has a rate control protocol to manage requests of specified periodic rates with SDU format information sources. For example, it controls the rate between *guaranteed bit rate* and *maximum bit rate*, where each of these two rates corresponds to an SDU format specified in the *SDU format information*. When the SDU size is constant (e.g. CS data), the SDU format information may include a list of possible bit rates per sub-flow, to allow rate control of the sub-flows by change of inter-PDU transmission interval (IPTI) [3].

3.4.2 Streaming

Streaming implies transmitting information continuously in streams. This technique facilitates Internet browsing by allowing displays even before the completion of information transfer. It has higher tolerance for jitter to support the large asymmetry of Internet applications. Through buffering, the streaming technique smoothes out packet traffic and offers it as it becomes available. Thus, it can support video on demand as well as web broadcast. While both types of video applications can benefit from the same video compression technologies, they differ in the usage of coding, protocols, etc. Thus, we can offer two types of video applications and address or offer services to more than one type of user depending on the transmission rate or delay sensitivity. Its fundamental characteristic in real time implies the preservation of time relation between information entities of the stream.

3.4.2.1 UMTS Service Attributes in Streaming

As in the conversational class, if the UMTS bearer carries streaming, we can set *source statistics descriptor* to calculate the statistical multiplexing gain in the core network, UTRAN, as well as the UE and use it for admission control.

Again as in the conversational class, we consider streaming traffic relatively non-bursty, where a *maximum bit rate* specifies the upper limit bit rate with which a UMTS bearer delivers SDUs at the SAPs; and where a UMTS bearer does not need to transfer traffic exceeding the *guaranteed bit rate*.[6] However, since the traffic is non-bursty, it is meaningful to guarantee a *transfer delay* of an arbitrary SDU.

Typical streaming transfer delay requirements passed primarily through the *transfer delay* attribute are in a range where at least in a part of this range the Radio Link Controller (RLC) re-transmission may be used.

UTRAN can also realise *streaming* bearers without RLC re-transmissions by using the *SDU format information,* which is obtained through dividing the largest defined SDU format. Thereby, it transports efficiently and at lower costs than if we adapt the RLC PDU size to the UMTS bearer SDU size.

Also as in the conversational class, the *maximum SDU size*, used for admission control and policing, applies only when the *SDU format information* does not get specified. If

[6] *Maximum and guaranteed* bit-rate attributes are used for resource allocation within UMTS.

specified, e.g. with one or several sizes, the SDU size is variable and each SDU shall exactly conform to one of the specified sizes.

In streaming we also specify the application error rate requirement by using the *SDU error ratio*, *residual bit error ratio* and *delivery of erroneous SDU* attributes, which serve as well to determine whether the application wants UMTS to detect and discard SDUs containing errors by applying an appropriate forward error correction.

3.4.2.2 Radio-Access Bearer Service Attributes in Streaming

When the RAB carries streaming speech, we do also set the *source statistics descriptor* to enable UTRAN to calculate a statistical multiplexing gain on radio and Iu interfaces to use it for admission control.

When applying unequal error protection to streaming class, we specify *delivery of erroneous SDUs*, *residual bit error ratio* and *SDU error ratio* attributes per sub-flow, where, as in the conversational case, the first attribute determines whether error detection shall be used and, if so, whether SDUs with error in a certain sub-flow shall be delivered or not. The second one specifies the bit error ratio for undetected delivered bits, while the last one specifies the fraction of SDUs with detected error in each sub-flow.

For data SDU payloads aspects and UTRAN rate control protocol issues in *streaming* see the RAB service attributes in *conversational.*

3.4.3 Interactive

Logically, we denote interactive to be the dynamic exchange of information through a man-machine interface or machine-to-machine inter-connection. The tempo of the dynamics will depend on the application or the purpose of the device under interaction. In the context of Internet applications like web browsing, the response time will depend on the type of information requested and the quality of the link as well as protocols in use. Delay sensitive applications will demand faster interaction, e.g. emergency devices, system controls, etc. Other applications such as location services, games, passive information centres, etc. will operate within flexible round-trip delays. In the forthcoming section we cover other applications.

3.4.3.1 UMTS Bearer Service Attributes Within Interactive Class

We optimise this bursty bearer class to transport human or machine interactions with remote equipment, e.g. web browsing. The traffic conditioning includes the maximum bit rate to be able to limit the delivered data rate for applications and external networks.

Traffic handling priority allows differentiation of bearers within the interactive class. This simplifies implementation, which would be otherwise complex if we need to set absolute guarantees on delay, bit rate etc.

Through UMTS internal scheduling UMTS bearer SDUs with higher traffic handling priority supersede the priority of SDUs of other bearers within the interactive class [3–10].

Because it is basically impossible to combine attributes specifying delay, bit rate, packet loss, etc. interactive bearers do not guarantee quality, and the actual bearer quality will depend on the load of the system and the admission control policy of the network operator or service providers.

The SDU error ratio, residual bit error ratio and delivery of erroneous SDUs specify the bit integrity of the delivered data as the only additional attribute for the interactive class, but since there are not reserved resources for interactive class, we use the SDU error ratio as a target value. SDU error ratio cannot be guaranteed under abnormal load conditions [3].

3.4.4 Background

While the background class still grows with innovative solutions, it remains as one of the traditional data communication techniques. It serves for E-mail, SMS, database inquiry and information service platforms. Delay does not have critical consequence in this class, although delays of more than a minute will be highly noticeable.

But despite the non-demanding round-trip delays, accuracy becomes critical. Thus, the background users expect error-free communications. For example, control mechanisms measuring performance or monitoring actions will need a reliable accuracy when sending or transmitting information.

3.4.4.1 UMTS Service Attributes in Background Class

We optimise non-delayed sensitive background class for machine-to-machine communication, e.g. messaging services. Background applications tolerate a higher delay than applications using the interactive class, which is the main difference between the background and the interactive classes.

UMTS transfers background class SDUs only during spare capacity in the network. To guarantee bit integrity of delivered data in a background class we only need SDU error ratio, residual bit error ratio and delivery of erroneous SDU attributes. When applying traffic conditioning we use *maximum bit rate* to limit the delivered data rate for applications and external networks. Finally, as in the interactive class, because there are no reserved resources for background class, we use the *SDU error ratio* as a target value, and this ratio cannot be guaranteed under abnormal load conditions [3].

3.4.4.2 RAB Service Attributes in Interactive and Background Classes

The attributes outlined for the UMTS bearer services corresponding to these classes apply also to the RAB.

3.4.5 Summary of UMTS and RAB Service Attributes

Table 3.8 summarises the defined UMTS and radio-access bearer service attributes according to class. Notice that for both the UMTS and RAB the attributes appear basically the same.

Table 3.8 UMTS and radio-access bearer attributes defined for each bearer traffic class [3]

Traffic class Attributes	Conversational	Streaming	Interactive	Background
Maximum bit rate	X	X	X	X
Delivery order	X	X	X	X
Maximum SDU size	X	X	X	X
SDU format information	X	X		
SDU error ratio	X	X	X	X
Residual bit error ratio	X	X	X	X
Delivery of erroneous SDUs	X	X	X	X
Transfer delay	X	X		
Guaranteed bit rate	X	X		
Traffic handling priority			X	
Allocation/retention priority	X	X	X	X
Source statistics descriptor	X	X		

3.4.6 Iu and Core Network Bearer Service Attributes

The *Iu Bearer service* together with the *physical bearer service* provides the transport between UTRAN and CN.

Likewise, the UMTS packet CN will support different backbone bearer services for a variety of QoSs.

Iu bearer services for packet traffic provide different bearer services for a variety of QoSs.

The operator or service provider will choose the QoS capabilities it needs in the IP layer or QoS capabilities in the ATM layer.

IETF defined *differentiated services* apply for IP based Iu bearer services. Thus, inter-operation with IP based networks uses *differentiated services* when an operator or service provider chooses ATM-SVC as an internal dedicated transport bearer.

The operator or service provider controls the mapping from UMTS QoS classes to DiffServ code points. This mapping depends on bandwidth and provisioning of resources among the different DiffServ classes, which the operators control to satisfy their cost and performance requirements [3].

Inter-operability between operators will take place based on the use of Service Level Agreements (SLAs), which are an integral part of the defined DiffServ architecture.

3.4.7 UMTS Bearer and RAB Service Attribute Value Ranges

UMTS bearer service and radio-access bearer services have a list of finite attribute values or the allowed value range defined. The value range defines possible usable values for an attribute considering every potential service condition for Release 1999. When a service is defined as a combination of attributes, further limitations may apply; for example, the shortest possible delay may not be possible to use together with the lowest possible SDU error ratio. Service requirements, i.e. required QoS and performance for a given UMTS service, is defined in the service requirement specifications 3GPP TS 22.105 [4]. The aspect

Table 3.9 UMTS and radio access bearer services attribute value ranges summary, after [3]

Traffic class	Conversational class	Streaming class	Interactive class	Background class
Maximum bit rate (kbps)	$< 2048^{a,b}$	$< 2048^{a,b}$	< 2048 overhead[b,c]	< 2048 overhead[b,c]
Delivery order	Yes/No	Yes/No	Yes/No	Yes/No
Maximum SDU size (octets)	≤ 1500 or 1502^{d}	≤ 1500 or 1502^{d}	≤ 1500 or 1502^{d}	≤ 1500 or 1502^{d}
SDU format information	[e]	[e]		
Delivery of erroneous SDUs	Yes/No[f]	Yes/No[f]	Yes/No[f]	Yes/No[f]
Residual BER	5×10^{-2}, 10^{-2}, 5×10^{-3}, 10^{-3}, 10^{-4}, 10^{-5}, 10^{-6}	5×10^{-2}, 10^{-2}, 5×10^{-3}, 10^{-3}, 10^{-4}, 10^{-5}, 10^{-6}	4×10^{-3}, 10^{-5}, 6×10^{-8g}	4×10^{-3}, 10^{-5}, 6×10^{-8g}
SDU error ratio	10^{-2}, 7×10^{-3}, 10^{-3}, 10^{-4}, 10^{-5}	10^{-1}, 10^{-2}, 7×10^{-3}, 10^{-3}, 10^{-4}, 10^{-5}	10^{-3}, 10^{-4}, 10^{-6}	10^{-3}, 10^{-4}, 10^{-6}
Transfer delay (ms)	100 – Maximum value 80 for RAB	250 – Maximum value		
Guaranteed bit rate (kbps)	$< 2048^{a,b}$	$< 2048^{a,b}$		
Traffic handling priority			1,2,3	
Allocation/Retention priority	1,2,3	1,2,3	1,2,3	1,2,3
Source statistic descriptor	Speech/unknown	Speech/unknown		

[a]For bit rate of 2048 kbps the UTRAN operates in transparent RLC protocol mode with negligible overhead from layer 2 protocols.

[b]Despite the UMTS network capabilities to support a large number of different bit rate values, these values get limited to the necessary in order to minimise terminal and charging complexity, as well as reduce inter-working functions.

[c]It is necessary to estimate the impact from layer 2 protocols on maximum bit rate in the non-transparent RLC protocol mode.

[d]When PDP type is PPP, maximum SDU size is 1502 octets. Otherwise, maximum SDU size is 1500 octets.

[e]RAN (WG3) in 3GPP will recommend the possible values of exact SDU sizes for which UTRAN can support during a transparent RLC protocol mode.

[f]Only the MT/TE on the UMTS bearer side provides error indication when delivery of erroneous SDUs are set to 'Yes'. R99 does not support error indication signalling from the CN gateway side outside UMTS networks. *These items in footnote (f) do not apply to the RAB service attributes.*

[g]Values are derived from CRC lengths of 8, 16 and 24 bits on layer 1.

of future proof coding (beyond Release 1999) of attributes in protocol specifications is not considered in the defined value list or value range tables. See Table 3.9 for the value ranges of the UMTS bearer service attributes. The value ranges reflect the capability of UMTS network.

3.5 MAPPING QoS ATTRIBUTES

There are three paths for mapping QoS attributes:

1. From application attributes to UMTS bearer service attributes, which are operator and/or implementation dependent.
2. From UMTS bearer service attributes to CN bearer service attributes, which also depends on the operator's choice.
3. From UMTS bearer service attributes to radio-access bearer service attributes, which is outlined next.

3.5.1 From UMTS Bearer Services to RAB Service Attributes

As noted in [3], when establishing a UMTS bearer and the underlying Radio-Access Bearer (RAB) in support of a service request, some attributes on UMTS level do not have the corresponding attribute at the RAB level. For example, UMTS requested transfer delay bearer will often be larger than the requested transfer RAB delay, because the transport through the core network will already use a part of the acceptable delay.

For the following attributes/settings, the UMTS bearer attribute value will normally be the same as the corresponding attribute value for the RAB (see Table 3.10).

In the next attributes (Table 3.11), the UMTS bearer attribute value will normally not be the same as the corresponding attribute value for the RAB. The relation between the UMTS bearer service attribute values and RAB service has implementation implications and depends, for example, on network dimensioning [3].

Table 3.10 Corresponding UMTS and RAB bearer attributes

Maximum bit rate	Traffic handling priority
Delivery order	Allocation/retention priority
Delivery of erroneous[a] SDUs	Maximum SDU size
Guaranteed bit rate	SDU format[b] information

[a]If delivery of erroneous SDUs is set to 'Yes', the handling of error indications on UMTS bearer level and RAB level differs. Error indications can only be provided on the MT/TE side of the UMTS bearer [3].

[b]Exact sizes of SDUs is the same, exact SDU format payload does not exist at UMTS bearer level.

Table 3.11 Relation between UMTS bearer attributes and RAB service attributes

RAB residual BER	Reduced with bit errors introduced in the Core Network (CN), by the Core Network Bearer (CNB) service
RAB SDU error ratio	Reduced with errors introduced in the CN by CNB service
RAB transfer delay	Reduced with delay introduced in the CN, e.g. on transmission links or in a codec resident in the CN
RAB level only	
SDU format information	Exact SDU payload format is retrieved from the codec integrated in the CN
Source statistics descriptor	Set to speech if the RAB transports compressed speech generated by the codec integrated in the core network

3.6 END-TO-END QoS

End-to-end QoS covers requirements for IMS and its dependence on building blocks.

3.6.1 End-to-End QoS Negotiation and Policy Requirements

The main QoS negotiation and policy requirements can be summarised as follows (see other requirements in [5]):

The UMTS R4 and R5 QoS negotiation mechanisms used for providing end-to-end QoS will be backward compatible with UMTS Release 1999. It will not make any assumptions about the situation in external networks, which are not within the scope of 3GPP specifications, or about application layer signalling protocols and applications, which may be used on terminal equipment attached to mobile terminals.

The UMTS network shall be able to negotiate end-to-end QoS also for mobile terminals and applications, which are not able to use QoS negotiation mechanisms other than the ones provided by UMTS. Thus, no changes to non-UMTS specific QoS negotiation mechanisms will occur.

The UMTS policy mechanisms described in TS 23.060 will be used for control of the UMTS bearers, and the interaction between UMTS bearer services and IP bearer services will only occur at the translation function in the UE and GGSN.

3.6.2 QoS End-to-End Functional Architecture

To provide QoS end-to-end, we manage the QoS within each domain. Thus, we use an IP Bearer Service (BS) manager to control the external IP bearer service, which due to different techniques applied within the IP network, communicates to the UMTS BS manager through the translation function. When setting up the PDP context, the user has access to one of the following options [5]:

1. *Basic GPRS IP connectivity service*. The bearer gets established according to the user's subscription, local operator's IP bearer resource based policy, local operator's admission

control function and GPRS roaming agreements. *IP bearer resource based local policy* decision may be applied to the bearer.

2. *Enhanced GPRS based services.* We use the bearer to support an enhanced application-layer service, such as Instant Messaging (IM). Here, *service-based local policy* decisions (e.g. authorisation and policy-based control) may be also applied to the bearer.

The *IP policy control,* as a logical policy decision element, enables co-ordination between events in the application layer and resource management in the IP bearer layer, which may be also internal to the *IP BS manager* in the GGSN.[7]

Non-UMTS network resources needing to provide QoS, inter-work with an external resource manager that controls those resources as follows [5]:

a. Signalling along the flow path (e.g. RSVP and LDP).
b. Packet marking or labelling along the flow path (e.g. DiffServ and MPLS).
c. Interaction between policy control and/or resource management elements.
d. Border routers between networks enforcing service level agreements.

Then for policy control the following applies [5]:

1. The UMTS IP policy framework aims as much as possible to conform to IETF 'Internet standards', which may be used for policy decision, authorisation and control of the IP level functionality, at both user and network levels.
2. To facilitate separate function evolution, a separation between the scope and roles of the UMTS policy mechanisms and the IP policy framework exists.

3.6.2.1 IP BS Manager

IP BS manager[8] applies standard IP mechanisms to manage IP bearer services. These mechanisms may vary and have different service controlling parameters from the ones used within UMTS. The IP BS manager implementation may include the support of DiffServ edge function and the RSVP function. The *translation/mapping function* provides the inter-working between the mechanisms and parameters used within the UMTS bearer service, and those used within the IP bearer service while interacting with the IP BS manager. If needed, we map the IP QoS parameters into UMTS QoS parameters in the GGSN. In the UE, we map the QoS requirements determined from the application layer (e.g. SDP) to either the PDP context parameters or IP layer parameters (e.g. RSVP).

When an IP BS manager exists both in the UE and the gateway node, it is possible that these IP BS managers communicate directly with each other by using relevant signalling protocols.

[7]The IP policy architecture does not mandate the policy decision point to be external to the GGSN.
[8]End-to-end QoS management functions do not cover CS service cases, or an ATM gw IP service inter-working.

Table 3.12 IP BS manager capability in the UE and GGSN [5]

Capability	UE	GGSN
DiffServ edge function	Optional	Required
RSVP/IntServ	Optional	Optional
IP policy enforcement point	Optional	Required[a]

[a]Although GGSN IP policy enforcement capability is required, GGSN policy control is a network operator choice.

Table 3.12 illustrates the minimum equipment functionality expected in order to allow multiple network operators to provide inter-working between their networks for end-to-end QoS. It does not exclude other functions, e.g. over-provisioning or combinations of these mechanisms.

Table 3.12 illustrates also how the *IP BS managers* in the UE and GGSN provide the set of capabilities for the IP bearer level, where the provision of the IP BS manager is optional in the UE, and required in the GGSN.

3.6.2.2 IP Policy Control

The IP policy control stands as a logical policy decision element, which applies standard IP mechanisms to implement policy in the IP bearer layer. These mechanisms may conform, e.g. to the framework defined in IETF [RFC2573], *a framework for policy-based admission control*, where the IP policy control is effectively a Policy Decision Point (PDP). The IP policy control makes decisions regarding the network, based on IP policy rules, and communicates these decisions to the *IP BS manager* in the GGSN, which is the IP Policy Enforcement Point (PEP).

The Policy Decision Function (PDF) stands as the logical entity of the P-CSCF, which when implemented in a separate physical node, does not exist as a standard interface between the PDF and P-CSCF.

Now the Go [6] interface links the PDF and GGSN, and supports the transfer of information and policy decisions between the *policy decision point* and the *IP BS Manager* in the GGSN.

The PDF makes policy decisions using the information obtained from the P-CSCF, where we map the P-CSCF (PDF) application level parameters (e.g. SDP) into IP QoS parameters. The P-CSCF (PDF) recides in the same domain[9] as the GGSN [5].

3.6.2.3 Resource Manager

In the UMTS network various nodes perform resource management in the admission control decision, usually under the direct control of the UMTS network. Likewise, in IP networks, there also exists resource management to ensure that resources required for a service are available. Because resources for the IP bearer service to be managed are not necessarily

[9]Currently in IETF, inter-domain policy interactions are not defined yet.

owned by the UMTS network, the resource management of those resources would be performed through an external resource management function for the IP network.

Furthermore, where the UMTS network uses external IP network resources as part of the UMTS bearer service, e.g. for the backbone bearer service, it may also be necessary to inter-work with an external IP resource manager.

For the external interaction, the GGSN supports DiffServ edge functionality along other mechanisms. These are given below [5]:

1. *Signalling along the flow path*. Explicit requests are either granted or rejected through the exchange of signalling messages (per flow–RSVP or aggregated–resources changes) between network elements along the path of the IP packet flow.
2. *Interaction between network management entities*. Resource requirements are explicitly negotiated and provisioned through network management entities, and then enforced in border nodes separating DiffServ administrative domains.
3. *Service level agreements enforced by the border routers between networks*. Resources are allocated along the path based on agreements between the network operators, then provisioned with the characteristics of the aggregated traffic that is allowed to flow between systems.

3.6.2.4 GGSN and UE Translation/Mapping

Translation/mapping function, which interacts with the IP BS manager and with the UMTS BS manager in the GGSN and in the UE, provides inter-working between the mechanisms and parameters used within the UMTS bearer service and those used within the IP bearer service.

For service-based local policy, the GGSN *Translation/mapping function* maps IP bearer based policy information into UMTS bearer based policy information. The GGSN uses this mapping for service-based local policy over the UMTS network.

3.6.2.5 End-to-End IP QoS Management Functions

Figure 3.4 shows QoS management functions for controlling the external IP bearer services and the relation to the UMTS bearer service QoS management functions. It also illustrates the scenario for control of an IP service using IP Bearer Service (BS) managers in both possible locations, i.e. in the UE and gateway node and an external resource manager. In addition, it indicates the optional communication path between the IP BS managers in the UE and the gateway node.

3.6.3 Capabilities of Key End-to-End QoS Functional Elements

3.6.3.1 GGSN

GGSN *DiffServ Edge Function* complies with IETF specifications for differentiated services [7], which we use to provide QoS for the external bearer service. *DiffServ edge function*

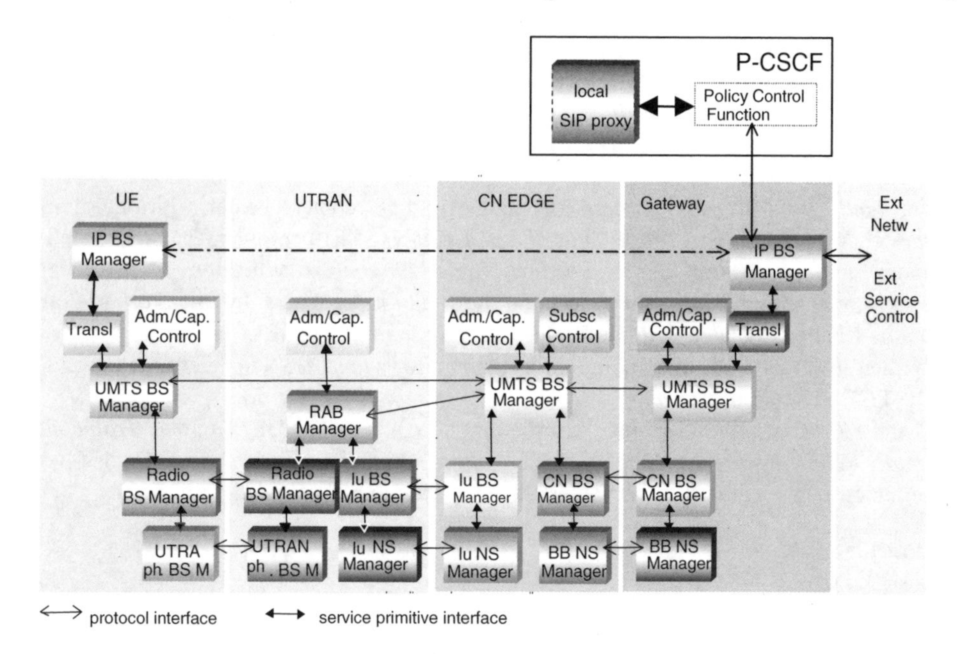

Figure 3.4 UMTS QoS management functions for control plane bearer service of an external IP server [5].

parameters (i.e. classifiers, meters and packet handling actions) may be statically configured on the GGSN, derived from PDP context parameters and/or derived from RSVP signalling. Statically configured DiffServ functions may include classifiers, meters, markers, droppers and shapers acting on uplink traffic [5].

DiffServ functions configured on the basis of PDP context parameters consist of marking user packets, where the DiffServ Code Point (DSCP) results from the PDP context parameters according to statically configured rules.

The *Service-based Local Policy* SBLP *enforcement point* controls QoS provided to a combined set of IP flows. It includes policy-based admission control applied to the bearer associated with the flows, and configuration of the policy based 'gating' functionality in the user plane. Service-based local policy decisions are either 'pushed' to or requested by the GGSN via the Go interface.

Policy-based admission control ensures that resources used by a particular set of IP flows are within the 'authorised resources' specified via the Go interface, which provide an upper bound on allocable resources for the set of IP flows, and are expressed as a maximum authorised bandwidth and QoS class.[10] The PDF generates a maximum authorised QoS class for the set of IP flows, which gets mapped by the GGSN *translation/mapping function* to give the authorised resources for UMTS bearer admission control.

A 'gate' consisting of a packet classifier and a gate status (open/closed) implemented in the GGSN defines policy enforcement in the user plane. This gate acts as policy enforcement

[10]The QoS class identifies a bearer service having a set of bearer service characteristics associated with it.

function interacting through the Go interface with PCF as the *policy decision point* for QoS resource authorisation at the IP BS level for uni-directional flow of packets (i.e. in either the upstream or downstream direction). An open gate enables packet flow acceptance and DiffServ enforcement. Otherwise, flow packets get dropped. For more details on *gate* GGSN QoS capabilities, see [5].

The *binding mechanism handling* associates the PDP context bearer with one or more *IP flows* in order to support *Service-Based Local Policy* (SBLP) enforcement, where the PDP includes binding information to associate the PDP context activation or modification messages with the SBLP policy decision information provided by the *Policy Control Function* (PCF) associated with the IP flow(s). In order to allow SBLP policy information to be 'pulled' from the PCF, the binding information allows the GGSN to determine the address of the PCF to be used [5].

If the GGSN receives binding information, it ignores any UE supplied *Traffic Flow Template* (TFT), discarding thereby all TFT filters. Otherwise, when the UE sends binding information to the network it populates the TFT filters with wildcard values.

3.6.3.2 UE

DiffServ edge function acts as a DiffServ (DS) boundary for the traffic from applications running on the UE. As specified in RFC2475, DS boundary node must be able to apply the appropriate *Per Hop Behaviour* (PHB) to packets based on the DS code point, and even sometimes perform traffic conditioning functions. When using GGSN DiffServ marking, we do not need the UE DiffServ edge function.

RSVP[11]*/IntServ*[12] *function* enables the UE to request end-to-end QoS using RSVP messages as defined in IETF standards. These messages may also serve the network to inform which DSCP will the UE use, include the authorisation token, flow identifier(s) in a policy data object when the UE carries the authorisation token and may as well serve to trigger PDP context activation/modification. The FFS acts as the inter-working between MT and TE [5].

Binding mechanism associates the PDP context bearer to the IP flow(s) to support SBLP policy enforcement in the GGSN. The binding information containing the authorisation token and flow identifier(s) provides the binding mechanism and the UE includes the PDP context activation and modification messages. We may also use the authorisation token to bind a RSVP session with a SIP session by including the authorisation token and flow identifier(s) in RSVP messages. For IMS services, the authorisation token is provided to the UE by the P-CSCF during SIP session establishment [5].

For each bi-directional media flow, the UE ensures that the 64-bit IPv6 address prefix of the source address of outgoing packets matches the prefix of the destination address supplied for incoming packets [5].

[11]Resource reservation protocol.
[12]Integrated services.

3.6.3.3 P-CSCF (PDF)

The main service-based local policy decision point functions based on [5] are as follows:

a. Session QoS resources authorisation (e.g. bandwidth in terms set of IP flows–destination address and port), using the SDP contained in the SIP signalling message to calculate the proper authorisation.

b. For bi-directional media flows, the P-CSCF (PDF), according to operator policy, assumes 64-bit IPv6 address prefix of the source address for downstream packets, and is the same as the prefix of the destination address for upstream packets of the same media flow.

The P-CSCF (PDF) enforces UE behaviour with respect to the IMS media components' assignment to the same PDP context or to separate PDP contexts.

The P-CSCF (PDF) decides whether new QoS authorisation (bandwidth, etc.) is needed due to the mid-call media or codec change.[13]

The PDF functions as a policy decision point for the service-based local policy control.

The PDF exchanges the authorisation information with the GGSN via the Go interface.

PDF provides final policy decisions controlling allocated QoS resources for the authorised media stream. The decision gets transferred from PDF to GGSN.

At IMS session release, the PDF revokes the QoS resource authorisation for the session.

3.6.3.3.1 Binding Mechanism Handling

The PDF generates an authorisation unique[14] token for each SIP session and the P-CSCF sends the authorisation token to the UE in SIP signalling.

3.6.4 Go Interface (PDF–GGSN) Functional Requirements

The Go interface enables service-based local policy and QoS inter-working information to be pushed to or requested by the GGSN from a Policy Control Function (PDF). It provides information to support the following functions in the GGSN:

- control of service-based policy gating function in GGSN,
- UMTS bearer authorisation,
- charging correlation related function.

[13] For example, a new authorisation is required when an UE resource request exceeds previous authorisation, etc.
[14] Across all PDP contexts associated with an APN.

The Common Open Policy Service (COPS), IETF compliant, protocol supports a client–server interface between the policy enforcement point in the GGSN and Policy Control Function (PDF). It allows both push and pull operations and may store policy decisions in a local policy decision point, allowing thereby the GGSN to make admission control decisions without requiring additional interaction with the PDF.

3.6.4.1 Information Elements Exchanged via Go Interface

The COPS protocol supports several messages between a client and a server. It also incorporates additional 3GPP Go-specific information elements to support the SBLP control functions as covered in the preceding sections. In the COPS framework context, 'client type' identifies the Go interface allocated for a 3GPP Go COPS client (GGSN), which we cover next for UMTS. The events specific to the UMTS or IP bearer service trigger the request messages from the GGSN PEP to the PDF.

A GGSN Request (REQ) message to the PCF allows the GGSN to request SBLP policy information for a set of IP flows identified by *binding information*. The latter information associates the PDP context to the IP flow(s) of an IMS session, and is used by the GGSN to request SBLP policy information from the PDF. The *binding information* includes:

1. an authorisation token sent by the P-CSCF to the UE during SIP signalling,
2. one or more flow identifiers used by the UE, GGSN and PDF uniquely identify the IP media flow(s).

The authorisation token is unique within the scope of the operator's domain. It conforms to relevant IETF standards on SIP extensions for media authorisation.

Flow identifiers specify an IP media flow associated with the SIP session. They are based on the *media components* ordering description structure defined by a single 'm=' line and port numbers within that media component in the SDP. A flow identifier combined with the authorisation token shall be sufficient to uniquely identify an IP media flow [5].

A *Decision* (DEC) message from the PDF to the GGSN contains decision objects, which include one of the following commands:

- Install (admit request/install configuration, commit),
- Remove (remove request/remove configuration).

These commands are used to:

- authorise QoS/revoke QoS authorisation for one or more IP flows,
- control forwarding for one or more IP flows.

The *responses* from the PEP to the PDF include an acknowledgement and/or an error response to commands received by the PEP. The following response messages shall be supported: Report state (success/failure/accounting) (RPT).

The *Delete Request State* (DRQ) message from PEP to PDF indicates that request state of a previously authorised bearer resource is no longer available/relevant at the

GGSN and removal of corresponding COPS at the PDF. It includes request deletion reason.

The Install command used to authorise QoS contains the following policy information associated with the IP flow(s):

1. Packet classifier(s). 5 Tuple: source IP address, destination IP address, source port, destination port and protocol (may be wildcarded).
2. Authorised QoS information. It provides an upper bound on reservable resources or allocated for the combined set of IP flows, includes DiffSer[15] class and data rate parameter.
3. Packet handling action. It defines the packet handling that should be accorded to packets matching the packet classifier–gate status signifies packets being passed (gate open) or silently discarded (gate closed).
4. Charging information – ICID. It enables GGSN with awareness of IMS session level charging identifier commands.

The PDF sends the ICID provided by the P-CSCF as part of the authorisation (Install) decision, and the report state contains the charging correlation information, which is used to correlate usage records (e.g. CDRs) of the GGSN with IMS session records from the P-CSCF. For this purpose, the GGSN sends the GCID of the PDP context and the GGSN address to the PDF as part of the authorisation report (RPT) [5].

Finally, the messages which revoke QoS authorisation or remove configuration information provide only information required to perform the action (e.g. the COPS handle element, which is used as a way of identifying the installed decision information).

3.6.5 Implementing End-to-End QoS

A practical solution would allow a UMTS operator or service provider to use standardised Internet QoS mechanisms across the IP core backbone network. GGSN/edge router function would connect to the IP backbone. These mechanisms would then afford the use of Service Level Agreements (SLAs) as a management tool to control resources of the IP core backbone network. The SLA would contain specifications on technical issues and administrative contractual information.

The connection from the User Equipment (UE) to the remote terminal/server can be established over different network paths. The UE accesses UTRAN over Uu radio interface. Over the Iu-PS interface UTRAN can, e.g. connect to the SGSN and multimedia gateway. The core IP backbone would transport the IP traffic between the SGSN and the GGSN (Gn interface) and between the GGSN and application servers in the operator's service network or a border router to external Internet.

For example, a subscriber flow for real time IP applications will use dedicated radio channels over Uu, and that the QoS through UMTS would be controlled per user flow in RNC, SGSN and GGSN nodes, respectively. The nodes would use the IP core backbone for

[15] Used only to identify the maximum allowed traffic class.

IP based transport over Iu and Gn. For this transport QoS would be controlled on aggregated flows using DiffServ for classification and conditioning and MPLS for the actual QoS implementation. From the GGSN to remote terminal/server the QoS would be controlled on the aggregated flows. The GGSN would support DiffServ for classification and conditioning and MPLS for the actual QoS implementation for the traffic on the Gi interface.

3.7 QoS WITHIN INTER-WORKING SCENARIOS

As part of its backward compatibility requirements, UMTS needs to inter-work existing network technologies. Thus, in the following we introduce the most common technologies that UMTS will be capable to inter-work with.

3.7.1 UMTS-GSM CS

The mapping between UMTS-GSM CS follows the GSM CS mechanisms and call control parameters with two handover functions, i.e. UMTS to GSM CS and GSM CS to UMTS. At the introduction of UMTS networks, the first will be more frequent and rarely the second one.

3.7.1.1 Handover from UMTS to GSM CS

When a UMTS call is set up in the CN, the BC IEs are mapped into QoS RAB attributes at call set-up. If the CN has to perform a handover towards GSM, the non-anchor MSC performs an assignment based on GSM specific traffic channel attributes.

Since we use the BSSMAP protocol over the E-interface and because no appropriate procedure exists to map QoS attributes into BSSMAP parameters, the anchor MSC maps the BC IEs into GSM traffic channel parameters, following existing GSM procedures for call set-up.

This requires that we code the BC IE according to GSM protocol requirements, i.e. all those parameters not applicable to UMTS should nevertheless be correctly specified by the UE in order to perform a handover to GSM according the above specified principles.

3.7.1.2 Handover from GSM CS to UMTS

When a GSM call is set up in the CN, the BC IEs are mapped into channel type parameters at call set-up.

If the GSM-CN has to perform a handover towards UMTS, the non-anchor MSC performs an assignment based on UMTS specific radio-access bearer attributes.

As in the preceding case, since we use the BSSMAP protocol over the E-interface, the non-anchor MSC uses the received channel type parameter (e.g. ‘speech or data indicator’, the type of data service (transparent/non-transparent) and user rate) to derive the QoS RAB attributes.

3.7.2 UMTS-GSM GPRS

GPRS Release 1999 (R99) QoS attributes are equivalent to the UMTS QoS attributes. Nevertheless, for inter-working purposes between different releases, we define mapping rules between GPRS Release 1997/1998 (R97/98) and GPRS Release 1999 (R99) as well as UMTS. Mapping occurs whenever the UE, the SGSN, the GGSN and the HLR nodes are of different releases, R97/98 or R99. We require mapping in PDP context activation and modification procedures and when a R99 HLR inserts subscriber data towards a R97/98 SGSN.

Here we define complete mapping rules. However, if a user requests a QoS profile which the network does not support (e.g. a low delay and a high reliability), the decision to support such attribute combination or not will depend on the admission control functionality within the PDP context activation procedure, and the QoS for such a profile may be re-negotiated by the network based on the available resources [3].

The overall principle of mapping between two profiles applied in their respective network releases implies that both require similar QoS. Notwithstanding GPRS R97/98 equipment will not support real-time services, which corresponds to R99 conversational and streaming traffic classes. Therefore, the mapping for UMTS handover to GSM/GPRS will always be to the non-real-time interactive and background traffic classes, at least the latter has been upgraded to R99 or later GPRS release.

3.7.2.1 General Guidelines

R99 air-interface session management and GTP messages contain R99 attributes as an extension of the R97/98 QoS information element; hence, we may prevent un-necessary mapping. The following cases can be illustrated from [3]:

1. When a R97/98 MS visits a GPRS R99 or UMTS SGSN and the GGSN is of R97/98 or R99, the visited SGSN does not perform any mapping of QoS attributes.
2. With a R99 GGSN, the GTP version 1 (R99) QoS profile only contains the R97/98 QoS attributes. Thus, for this PDP context we do not need a Traffic Flow Template (TFT).
3. When a R99 UE visits a GPRS R99 or UMTS SGSN (or serving PLMN) and the GGSN (or home PLMN) is of R97/98, the visited SGSN (or visited PLMN) provides bearers capable to support QoS according to R99. In this case, when a PDP context gets activated (mobile or network initiated) mapping takes place in the serving SGSN.
4. When MS initiated PDP context activations or network initiated PDP context activations occur, the home R97/98 GGSN responds to the activation request by returning the QoS negotiated profile, which contains the accepted and changed R97/98 attributes. In these cases, mapping of the changed attributes into R99 attributes takes place in serving SGSN and signalled to the UE in the activate PDP context accept message.
5. In principle,[16] returned and unchanged attributes during negotiation procedures do not get mapped a second time by serving SGSN, i.e. the unchanged R99 attributes received in the create PDP context response message gets sent to UE in QoS negotiated profile of the activate PDP context accept message.

[16]According to a general mapping rule.

6. A R99 MAP message contains R99 attributes as an extension of the R97/98 QoS information element when 'Insert Subscriber Data' message is sent to a R99 SGSN.
7. When a R99 HLR sends an 'Insert Subscriber Data' message to a R97/98 SGSN, the message contains the R97/98 QoS attributes.
8. A R99 SGSN uses the R99 attributes of subscribed QoS profile when a R99 UE requests to use subscription data in the PDP context activation.
9. The R99 SGSN uses R97/98 attributes of subscribed QoS profile when a R97/98 MS requests to use subscription data in the PDP context activation.

3.7.2.2 Determining R99 attributes from R97/98 attributes

This mapping is applicable in the following cases [3], see also Table 3.13:

Table 3.13 Rules for determining R99 attributes from R97/R98 attributes [3]

Resulting R99 attribute		Derived from R97/98 attribute	
Name	Value	Value	Name
Traffic class	Interactive	1,2,3	Delay class
	Background	4	
Traffic handling priority	1	1	Delay class
	2	2	
	3	3	
SDU error ratio	10^{-6}	1, 2	Reliability class
	10^{-4}	3	
	10^{-3}	4, 5	
Residual bit error ratio	10^{-5}	1, 2, 3, 4	Reliability class
	4×10^{-3}	5	
Delivery of erroneous SDUs	No	1, 2, 3, 4	Reliability class
	Yes	5	
Maximum bit rate (kbps)	8	1	Peak throughput class
	16	2	
	32	3	
	64	4	
	128	5	
	256	6	
	512	7	
	1024	8	
	2048	9	
Allocation/retention priority	1	1	Precedence class
	2	2	
	3	3	
Delivery order	Yes	Yes	Reordering required (Info in SGSN and GGSN PDP contexts)
	No	No	
Maximum SDU size	1500 octets	Fixed value	

- Handover of PDP context from GPRS R97/98 SGSN to GPRS R99 or UMTS SGSN.
- PDP context activation in a serving R99 SGSN with a R97/98 GGSN. When GGSN responds to the PDP context activation, mapping of the changed R97/98 QoS attributes received from the GGSN to R99 QoS attributes is performed in the serving SGSN.
- This mapping is also applicable if a R99 UE allows an application to request a PDP context activation with R97/98 QoS attributes, e.g. via AT command.

Since the reordering required attribute is not available in the MS, the MS shall set the R99 delivery order attribute to the value 'subscribed' (see 3GPP TS 24.008).

3.7.2.3 Determining R97/98 Attributes from R99 Attributes

This mapping is applicable in the following cases [3] (see also Table 3.14).

PDP context is handed over from GPRS R99 or UMTS to GPRS R97/98. Since the allocation/retention priority attribute is not available in the UE, the UE sets the R97/98 precedence class attribute to the value 'subscribed'; and in the case of asymmetric bearers, the higher value of the maximum bit rate attributes for downlink and uplink is selected and used for the maximum bit rate value.

1. When a R99 UE performs a PDP context activation in a serving R99 SGSN, while the GGSN is of R97/98. In this case the SGSN performs mapping of the R99 QoS attributes to the R97/98 QoS attributes.
2. A R99 HLR may need to map the stored subscribed QoS attributes in the HLR subscriber data to R97/98 QoS attributes that are going to be sent in the insert subscriber data message from the R99 HLR to the R97/98 and R99 SGSN. Therefore, it is recommended that R97/98 QoS attributes are stored in the HLR in addition to the R99 QoS attributes.
3. A R99 UE (except UMTS only UE) may receive a request for a PDP context activation with R99 QoS attributes, e.g. via AT command.

3.7.3 UMTS-PSTN

Because the PSTN does not have QoS mechanisms, thus QoS attribute inter-working/ mapping is not needed [3].

3.7.4 UMTS-ISDN

ISDN does not have QoS mechanisms either, hence, QoS attribute inter-working/mapping is not needed. Nevertheless, means for determining required bandwidth, delay and reliability are required, in particular for MT cases, since for MO cases it is simple [3].

Table 3.14 Rules for determining R97/98 attributes from R99 attributes [3]

Resulting R97/98 attribute		Derived from R99 attribute	
Name	Value	Value	Name
Delay class	1	Conversational	Traffic class
	1	Streaming	Traffic class
	1	Interactive	Traffic class
		1	Traffic handling priority
	2	Interactive	Traffic class
		2	Traffic handling priority
	3	Interactive	Traffic class
		3	Traffic handling priority
	4	Background	Traffic class
Reliability class	2	$\leq 10^{-5}$	SDU error ratio
	3	$10^{-5} < x \leq 5 * 10^{-4}$	SDU error ratio
	4	$> 5 \times 10^{-4}$	SDU error ratio
		$\leq 2 \times 10^{-4}$	Residual bit error ratio
	5	$> 5 \times 10^{-4}$	SDU error ratio
		$> 2 \times 10^{-4}$	Residual bit error ratio
Peak throughput class	1	< 16	Maximum bit rate (kbps)
	2	$16 \leq x < 32$	
	3	$32 \leq x < 64$	
	4	$64 \leq x < 128$	
	5	$128 \leq x < 256$	
	6	$256 \leq x < 512$	
	7	$512 \leq x < 1024$	
	8	$1024 \leq x < 2048$	
	9	≥ 2048	
Precedence class	1	1	Allocation/retention priority
	2	2	
	3	3	
Mean throughput class	Always = 31	NA	
Reordering required (Info in SGSN and GGSN PDP contexts)	yes	yes	Delivery order
	no	no	

3.7.5 UMTS-Internet

Internet applications follow Internet QoS attributes for the selection of the class and appropriate traffic attribute values. These applications do not directly use the services of UMTS but they use Internet QoS definitions and attributes, which are mapped to UMTS QoS attributes at API. Currently there are two main Internet QoS concepts, namely, integrated services and differentiated services. The mapping between Internet QoS and UMTS QoS is presented in the following clauses.

There is a support for IP based QoS models for PDP contexts, meaning both Integrated Services (IntServ) signalled by RSVP [RFC2205] and Differentiated Services (DiffServ) (6-bit QoS attribute on each IP packet). Applications residing in the TE control both mechanisms, enabling thereby different application specific QoS levels for the same PDP context. Application level IP based QoS gets mapped to UMTS packet core QoS by a network element at the border of the network, e.g. a 3G-gateway node. The RSVP support requires flow establishment, and possibly aggregation of flows, within the UMTS packet core network, while differentiated services require that there is either one QoS profile for each traffic type or alternatively the priority and traffic type information is included in the data packets [3].

3.7.6 Error in Real-Time Packet Multimedia Payloads

Here we briefly outline the error resilience of different encoded media streams when considering the support of unequal error protection for real-time packet multimedia services, and provides some indicative figures for the residual bit error rates that could be tolerated by audio-visual H.323 payloads in a 3G environment [3].

H.323 employs the H.225.0 packetisation scheme, which in turn uses UDP/IP and RTP to transport each media stream. Figure 3.5 illustrates the H.323 packet structure and its header compression.

We may compress about 40 bytes H.323 packet header into 2–4 bytes while assuming that this information will require reliable transmission, to prevent any errors in the header, which would result in the loss of the complete H.323 packet. On the other hand, for real-time multimedia streams that cannot accommodate a large delay,[17] we may use codecs that are tolerant to residual bit errors.

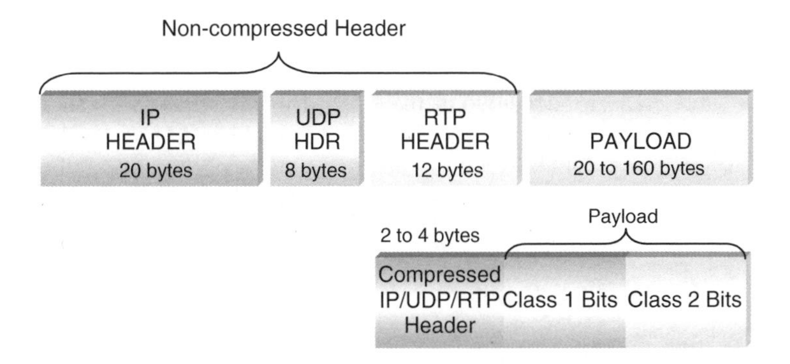

Figure 3.5 Non-compressed and compressed H.323 packet structure.

[17]And therefore packet re-transmission.

Table 3.15 AMR speech and MPEG-4 video codec attribute examples [3]

Bit rate	Delay	BER	FER
AMR speech codec payload			
4.75–12.2 kbps	End-to-end delay 100 ms[a]	10^{-4} for class 1 bits	< 0.5%[b]
		10^{-3} for class 2 bits	
		($\sim 10^{-2}$) (ok for some applications)	
MPEG-4 video payload			
24 – 28 kbps[c]	End-to-end 150 and 400 ms	10^{-6}–no visible degradation	
	Video codec delay < 200 ms	10^{-5}–little visible degradation	
		10^{-4}–some visible artefacts	
		> 10^{-3}–limited practical application	

[a]Codec frame length is 20 ms.
[b]With graceful degradation for higher erasure rates.
[c]Variable, average rate scaleable from... and higher.

3.7.6.1 Error Resilience Factors

Factors that influence error resilience include [3]:

1. the media type,
2. the Quality of Service (QoS) required,
3. the specific codec used.

Furthermore, we may also sub-divide real-time media streams into different classes on the basis of bit error sensitivity as illustrated in Table 3.15, where in some cases the most sensitive bits may be protected by in-band checksum information. We also note that in addition to the effect of residual bit errors in the media stream, the QoS may further degrade by packet loss due to errors in the H.323 header.

The values Table 3.15 are indicative of the QoS attributes required by audio and video media streams, including Bit Error Rates (BER) and Frame Erasure Rate (FER).

For practical reference, Table 3.15 illustrates the AMR speech codec and the MPEG-4 video codec.

3.7.7 Discriminating QoS Profiles

Handover executions from R99 to GPRS R97/98 networks need to determine which PDP context in a set of PDP contexts provides the highest QoS, because within a set of PDP contexts with the same APN and PDP address, all PDP contexts except the one with the highest QoS profile gets de-activated [3].

To determine which PDP context has the highest QoS we apply the ranking illustrated Table 3.16, where we maintain only the PDP context(s) with the highest QoS ranking and

Table 3.16 QoS profile ranking

QoS ranking	Traffic class	Traffic handling priority
1	Interactive	1
2	Conversational	Not applicable
3	Streaming	Not applicable
4	Interactive	2
5	Interactive	3
6	Background	Not applicable

de-activate the rest. In a second pass, if more than one PDP context remains, the PDP context with the highest value for the maximum bit rate attributes for downlink or uplink gets selected, and we de-activate all PDP contexts except the one(s) with the highest maximum bit rate selected.

If more than one PDP context remains after the second pass, all PDP contexts except the one with the lowest NSAPI gets de-activated.

3.7.7.1 Determination of Traffic Class Weights in HLR QoS Profiles

In the subscription record of a HLR, the QoS profile represents the maximum QoS per PDP context to an associated APN. Afterwards, it is possible to negotiate all QoS parameters, including an appropriate traffic class for each QoS flow. This applies to the first PDP context established, as well as the next one, i.e. this includes primary and secondary PDP context activations. The traffic classes have increasing weight according to the order background, interactive, streaming and conversational rankings also illustrated in Table 3.16.

3.8 APPLICATIONS AND SERVICE OFFERINGS

The questions arising from the exploitation of wireless networks, more in particular IP based network or non-voice services, can be summarized as follows:

- What are these services?
- Who are they targeted at?
- How much do we offer them for?
- How do we apply technology? Or what technology do we require?

3.8.1 Sensitivity to IP Transmission Impairments

To conclude the UMTS traffic classes, in the following we briefly outline some criteria for different applications in the context of the aforementioned classes.

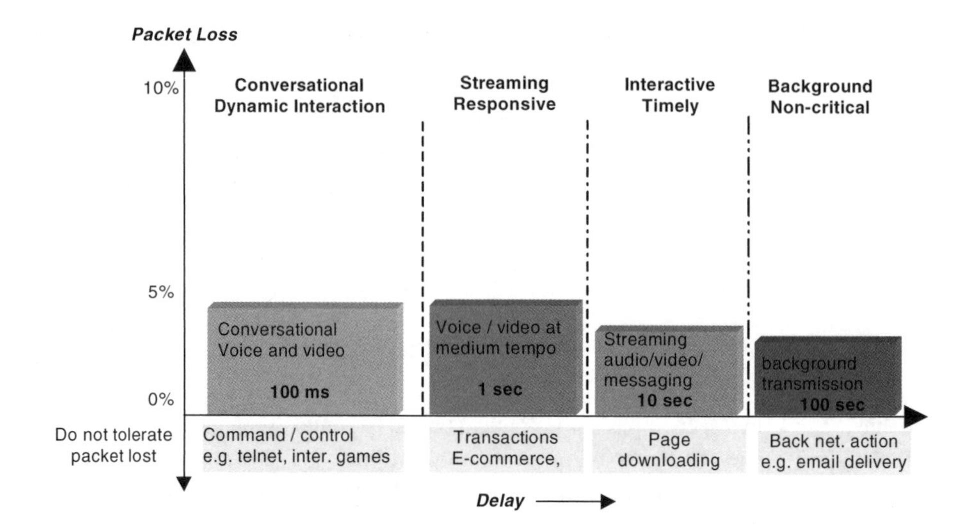

Figure 3.6 Sensitivity of applications to delay in IP environments.

From the algorithmic representation of delays (x axis) and linear scale of packet loss estimation (y axis) in Figure 3.6, we can see the sensitivity of applications to IP impairments. Clearly, entries below the vertical axis do not tolerate any type of packet loss; e.g. command/control actions in Telnet or interactive games, on-line banking, e-commerce, etc. This means that reliable service transmission will imperatively include both delay control and packet transfer integrity.

Controlling delay implies keeping end-to-end one-way delay below 250 ms, otherwise this impairment will annoy users and service quality perception will diminish. When packets get lost due to late arrival or discarded as result of congestion, the missing information degrades multimedia transmission, demanding Packet Loss Concealment (PLC) techniques in voice type transmission and error correction or re-sending on data transmission.

Although PS or IP networks have flexibility when using codecs, we still need to add encoding time to the end-to-end delay. While the delay for different types of codecs

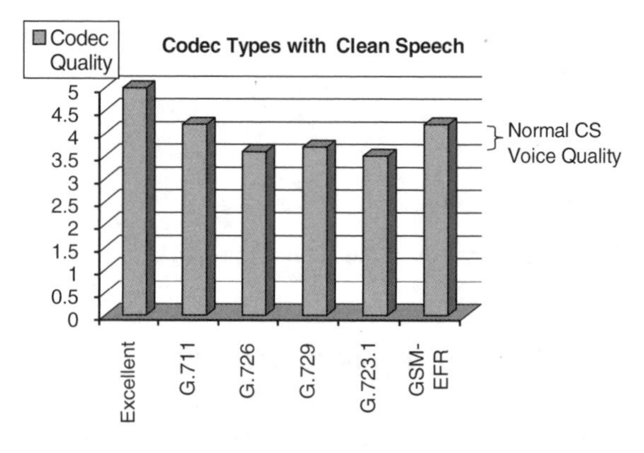

Figure 3.7 Codec quality perception.

Table 3.17 Selected codec parameters

Codec	Type	Bit rate (kbps)	Frame size (ms)	Total delay (ms)[a]
G.711	PCM	64	Based on packet size	
G.726	ADPCM	32	Based on packet size	
G.729/A	CS-ACELP	8	10	25
G.732.1	MP-MLQ	6.3	30	67.7
GSM-EFR	ACELP	12.2	20	40

[a]Total delay assumes one frame per packet.

illustrated in Figure 3.7 vary depending on their physical parameters, e.g. type, bit rate and frame size as noted in Table 3.17, all must allow normal CS voice quality.

Voice quality obtained through test methods, e.g. Mean Opinion Score (MSO) described in ITU recommendations P.800 rates the GSM-EFR codec as quite acceptable (Figure 3.7). This codec corresponds to the AMR family selected for UMTS. Hence, when it comes to delay limits for future VoIP services, e.g. 3G networks will not add un-necessary delays.

Finally for completeness, we list the ITU G.114 recommendations on delay limits. Depending on the applications delay ranges can be noted as follows:

- 50 ms: limit for processing delay will vary based on processing power,
- 0–150 ms: one way delay is acceptable,
- 150–400 ms: one way delay is acceptable depending on the applications,
- $>$ 400 ms: one way is unacceptable.

In summary, sources of delay in PS network include:

- propagation (while the signal moves through the channel),
- processing (encoding/transcoding is of 50–140 ms, packetisation of 0–60 ms and DSP functions, e.g. filtering is of 0–25 ms),
- packet loss mitigation (queuing and jitter buffers are of 20–50 ms and inter-leaving is of 5–90 ms).

3.8.2 UMTS Generic Services

The strength of UMTS services will not reside in one or two applications, but in the conjunction and complementation of a series of application and technologies, which will generate different sets of services. Figure 3.8 illustrates a generic set of non-application targets primarily for PS networks including multimedia features. In this illustration, we can see the characteristics of connectionless and connection oriented services, i.e. variable and constant bit rate.

We not only need to know to what groups we can address these services (e.g. enterprises, communication firms, telematic centres, content and location based providers, commerce

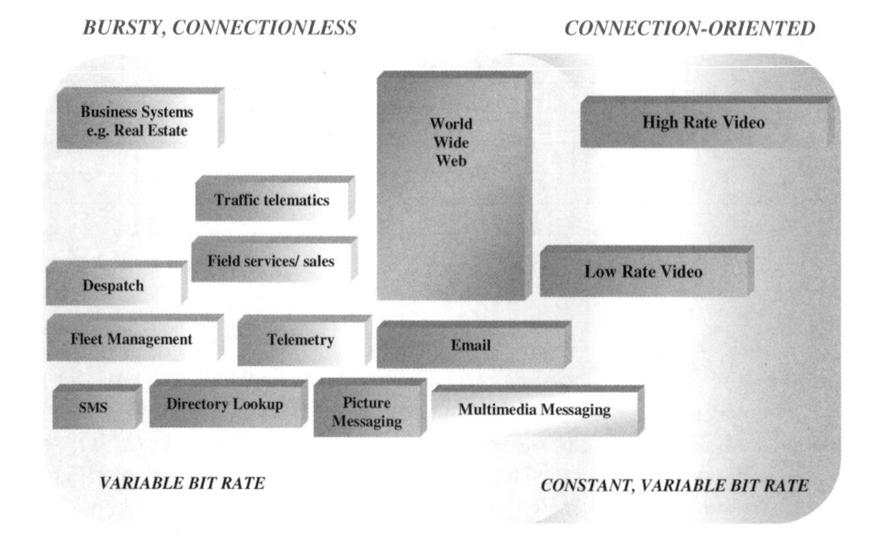

Figure 3.8 Generic range non-voice applications.

organisations and typical wireless operators aiming to minimise operational costs and churn). We also need to know where is the end user and how does he/she apply technology.

3.8.2.1 Enabling Speech

The Adaptive Multi-Rate (AMR) techniques will enable the UMTS speech codec. This codec consists of single integrated speech codec with eight source rates controlled by the RAN, i.e 12.2 (GSM-EFR), 10.2, 7.95, 7.40 (IS-641), 6.70 (PDC-EFR), 5.90, 5.15 and 4.75 kbps. The use of the average required bit rate has impacts on interference levels, thereby on capacity and battery life. Logically, lower rates will favour capacity and battery life duration, but not necessarily quality.

The AMR coder [8] works with speech frames of 20 ms, i.e. 160 samples at a sampling rate of 8000 samples/s. It may switch its bit rate at every frame through in-band signalling or through a dedicated channel. It uses Multi-rate Algebraic Code Excited Linear Prediction Coder (MR-ACELP) as a coding scheme. We extract CELP parameters at each 160 speech samples for error sensitive tests. The latter consist of three error classes (A–C), where class A has the highest sensitivity and requires strong channel coding. The AMR speech codec can tolerate about 1% Frame Error Rate (FER) of class A bits without any deterioration of the speech quality. For class B and C bits a higher FER can be allowed. The corresponding Bit Error Rate (BER) of class A bits will be about 10^{-4}.

AMR allows an activity factor of 50% (while parties have a telephone conversation) through a set of basic functions:

- Background acoustic noise evaluation on the Tx to transmit key parameters to the Rx.
- Voice Activity Detector (VAD) on the Tx.

- A Silence Descriptor (SID) frame that passes transmission comfort noise information to the Rx at regular intervals. This noise gets generated on the Rx in the absence of normal speech frames.

3.8.2.2 Enabling Circuit-Switched Video Telephony

Video telephony has higher BER requirements than speech due to its video compression features; however, it has the same delay sensitivity of speech. Technical specifications [9] in UMTS recommend ITU-T Rec. H.324M for video telephony in CS links, while at present there exists two video telephony options for PS links, i.e. ITU-T Rec. H.323 [4] and IETF SIP [10]. The H.323 has characteristics similar to H.324M.

The adapted[18] H.324 includes essential elements such as H.223 for multiplexing and H.245 for control. It also includes H.263 video codec, G.723.1 speech codec and V.8bis. I may have MPEG-4 video and AMR to better suit UMTS services as illustrated in Figure 3.9.

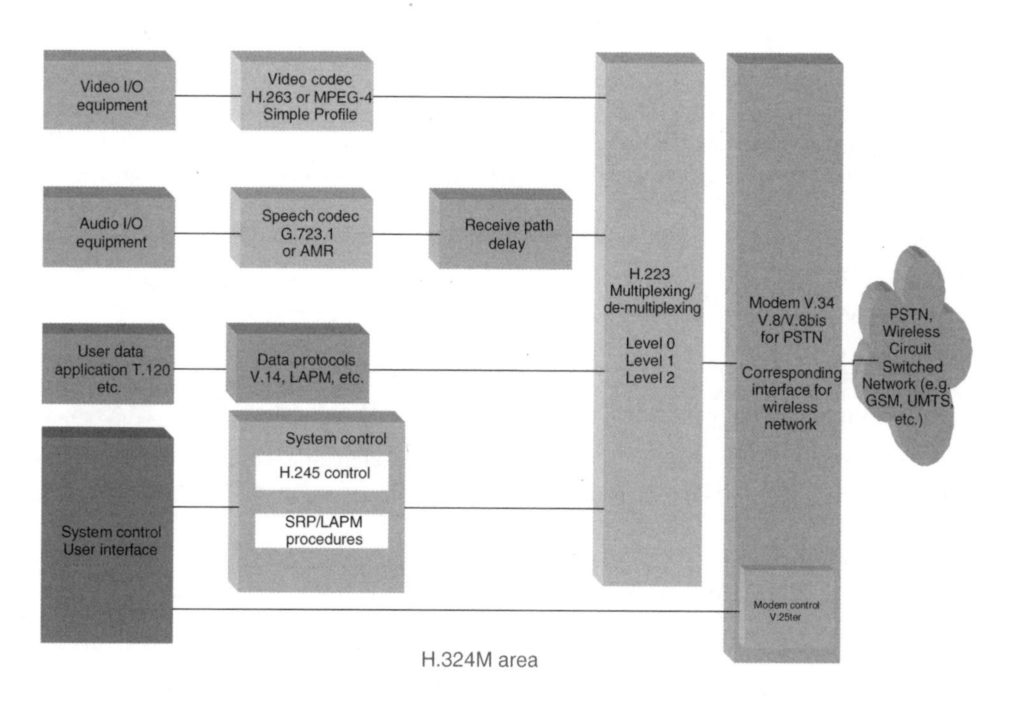

Figure 3.9 The ITU Rec. H.324 model.

Technical specifications include seven phases for a call, i.e. set-up, speech only, modem learning, initialisation, message, end and clearing. Backward compatibility occurs through level 0 of the H.223 multiplexing, which is the same as H.324 [11].

The H.324 terminal has an operation mode for use over ISDN links. Annex D in the H.324 recommendations defines this mode of operation as H.324/I [12]. H.324/I offers direct

[18]Adapted to wireless from what was originally meant for fixed networks.

inter-operability with the H.320 terminals, H.324 terminals on the GSTN, H.324 terminals operating on ISDN and voice telephones.

For seamless data communications between UMTS and PSTNs, the UMTS call control mechanism takes into account V.8bis messages. These messages get interpreted and converted into UMTS messages and V.8bis, respectively. The latter contains identification procedures and selection of common modes of operation between Data Circuit-terminating Equipment (DCE) and between Data Terminal Equipment (DTE). Essential V.8bis features include:

- flexible communication mode selection by either the calling or answering party,
- enabling automatic identification of common operating modes,
- enabling automatic selection between multiple terminals sharing common telephone channels,
- friendly user interface to switch from voice telephony to a modem based communications.

3.8.2.3 *Enabling Packet-Switched Video Telephony*

The H.323 ITU-T protocol standard for multimedia (and IP telephony) call control enables PS multimedia communications in UMTS. The standard:

- employs a peer-to-peer model in which the source terminal and/or GW is the peer of the destination terminal and/or GW,
- treats Gateways (GW) and terminals alike,
- requires GWs and terminals to provide their own call control/processing functions,
- provides multiple options for voice, data and video communications,
- it may employ a gatekeeper function to provide telephone number to IP address translation, zone admission control and other resource management functions.

Figure 3.10 illustrates the H.323 architecture, which incorporates a family of standards including H225, H245 and H450. As an international standard for conferencing over packet networks, H.323:

- acts as a single standard to permit Internet telephony products to inter-operate,
- also serves as base for standard inter-operability between ISDN- and telephony-based conferencing systems,
- has the flexibility to support different HW/SW and network capabilities.

The logical channels in H.323 get multiplexed at the destination port transport address level. The transport address results from the combination of a network address and a port identifying a transport level endpoint, e.g. an IP address and a UDP port. Packets having different payload types go to different transport address, thereby eliminating usage of separate multiplexing/de-multiplexing layer in H.225.0. The H.225 standard

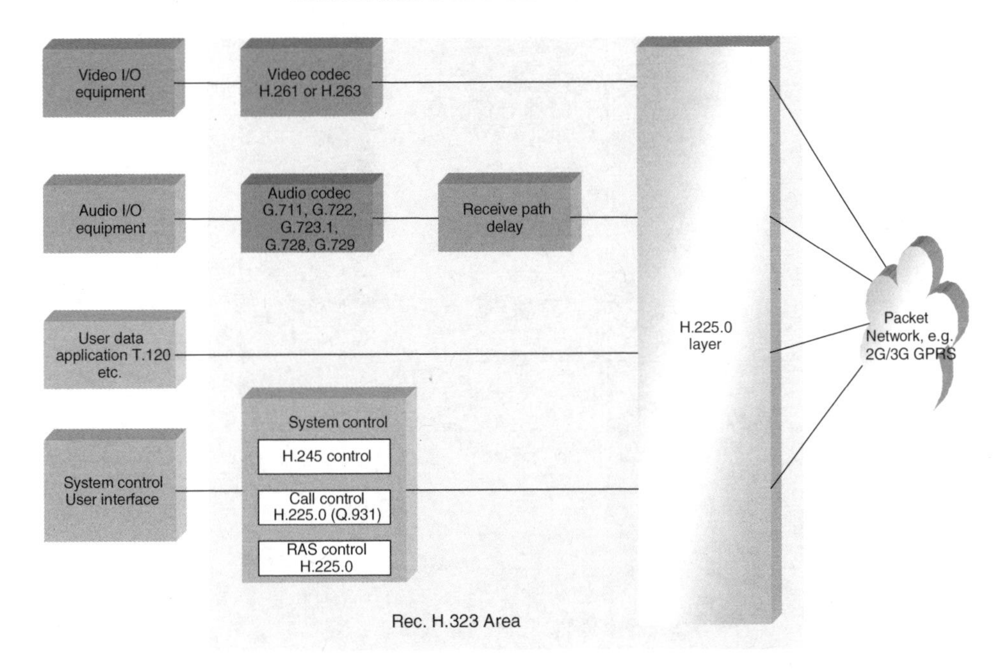

Figure 3.10 The ITU Rec. H.323 model.

uses RTP/RTCP[19] for media stream packetisation and synchronisation for supporting LANs. This usage depends on the usage of UDP/TCP/IP. BER control takes place at lower layers; thus, incorrect packets do not reach the H.225 level.

When both audio and video media act in a conference, they transmit using separate RTP sessions, and RTCP packets get transmitted for each medium using two different UDP port pairs and/or multi-cast addresses. Thus, direct coupling does not exist at the RTP level between audio and video sessions, and synchronised playback of a source's audio and video takes place using timing information carried in the RTCP packets for both sessions.

Point-to-point H.323 conference occurs with two TCP connections between the two terminals, i.e. one for call set-up connection and one for conference control and feature exchange. The first connection carries the call set-up messages defined in H.225.0, i.e. the Q.931 channel. After a first TCP connection on a dynamic port, the calling parties establish the second TCP connection to the given port, where the second connection carries the conference control messages defined in H.245. Thus, the H.245 serves to exchange audio and video features in the master/slave context.

3.8.2.4 Session Initiation Protocol (SIP)

The Session Initiation Protocol (SIP) is another alternative to enable PS video telephony. Developed in IETF by the Multiparty Multimedia Session Control (MMUSIC) group, SIP is an application layer control signalling protocol for creating/modifying and terminating sessions with one or more participants, e.g. Internet multimedia conferences, Internet telephone calls

[19]Real-time transport protocol/real-time transport control protocol.

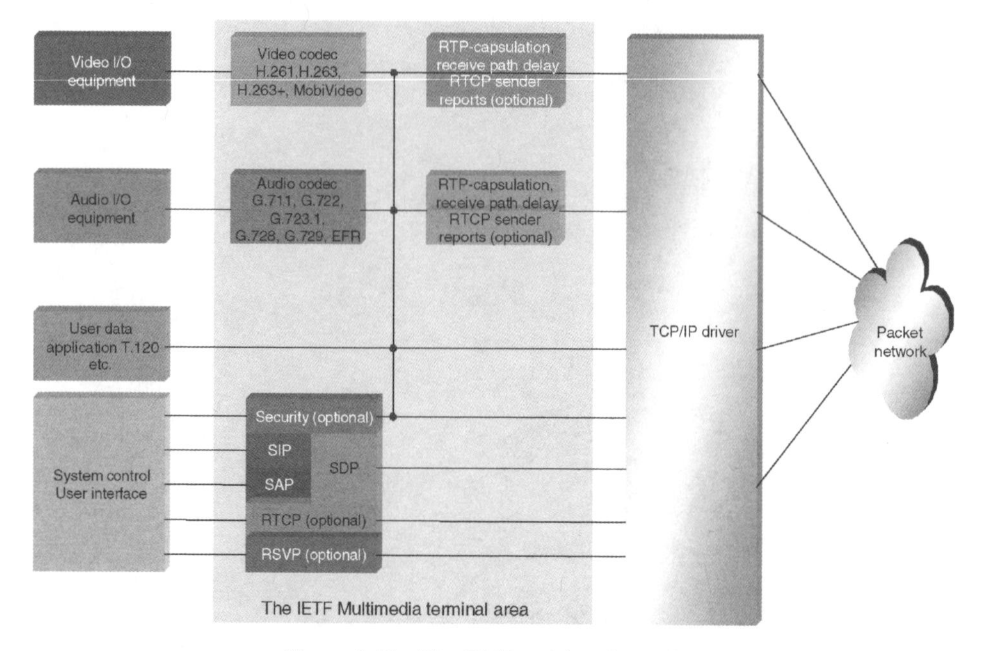

Figure 3.11 The IETF multimedia model.

and multimedia distribution. Participants in a session can communicate via multi-cast or via a mesh of uni-cast relations, or a combination of these. See Figure 3.11. SIP corresponds to:

- the overall IETF multimedia data and control architecture currently incorporating protocols such as Resource Reservation Protocol – RFC 2205 (RSVP) – for reserving network resources,
- the Real-Time Transport Protocol (RTP–RFC 1889) for transporting real-time data and providing QoS feedback,
- the real-advertising multimedia sessions via multicast and the Session Description Protocol (SDP – RFC 2327) for describing multimedia sessions.

Nevertheless, it does not depend on any of the above for its functionality and operation. SIP transparently supports name mapping and re-direction services, thereby allowing the implementation of ISDN and IN telephony subscriber services and enabling personal mobility. Technically, SIP has the following characteristics:

- called and calling peers can specify their preference of where they would like calls to be connected,
- use of user@domain as call addresses and http look-alike messages,
- only deals with tracking down users and delivering a call to an endpoint, i.e. it is orthogonal to other signalling protocols,
- uses servers for re-direction (re-direct server), user location tracking (registrar) and fork request (proxy server),

- it does not have address initiation and termination like H.323 (widely accepted),
- simple and easy to implement by IP developers.

SIP supports five phases of establishing and terminating multimedia calls:

- user location–determination of the end system for connection,
- user capabilities–determination of the media and media parameters for usage,
- user availability–determination of the willingness of the called party to engage in communications,
- call set-up–ringing establishment of call parameters at both called and calling party,
- call handling–including transfer and termination of calls.

SIP can also initiate multi-party calls using a Multi-point Control Unit (MCU) or fully meshed inter-connection instead of multi-cast (Table 3.18).

Undoubtedly, SIP is the most appropriate protocol to enable PS video telephony in UTMS. At this writing, technical bodies are debating the final outcome. From the author's point of view, it seems evident that SIP would lead to better results and widespread usage of video telephony.

3.8.2.5 *Layer Structure Enabling for Multimedia – MEGACO/H.248*

Media Gateway Control (MEGACO) or H.248 is part of the protocols that will facilitate the control of video telephony on the PS side. Megaco/248 jointly developed by ITU TG-16 and

Table 3.18 SIP vs. H.323

	H.323	SIP
Standards body	ITU TSG-16	IETF Music
Properties	Based on H.320 conferencing and ISDN Q.931 legacy Difficult to extend and update No potential beyond telephony Complex, mono-lithic design	Based on web principles (Internet friendly) Easy to extend and update Readily extensible beyond telephony Modular simplistic design
Standards status	H.450.x series provides minimal feature set (pure per approach) Adding mixed peer/stimulus approach (inefficient architecture) Slow moving	No real end-device feature standard yet Many options for advanced telephony features Good velocity
Industry acceptance	Established now, primarily system level Few, if any, H:323 base telephones End user primarily driven by Microsoft (NetMeeting), Intel, etc.	Rapidly growing industry momentum (system level) Growing interest in SIP phones and soft clients

IETF covers all gateway applications moving information streams from IP networks to PSTN, ATM and others. These include PSTN trunking, gateways, ATM interfaces, analog line and telephone interfaces, announcement servers, IP phones and many others.

The Megaco IP phone master/slave approach is entirely compatible with peer-level call control approaches such as SIP and H.323. It acts orthogonal to the last two protocols. Megaco/H.248 allows:

- profiles to be defined, i.e. allows application level agreements on GW organisation and behaviour to be made for specific application types to reducing complexity,
- allows support of multiple underlying transport types (e.g. ALF reliability layer over UDP and TCP), and both text and binary encoding; the latter enables more appropriate support for a broader range of application scales (e.g. big vs. small gateways) and more direct support for existing systems.

3.8.2.6 IETF Signalling Transport (SIGTRAN)

SIGTRAN develops an essential Stream Control Transmission Protocol (SCTP), which we view as a layer between the SCTP user application and an un-reliable end-to-end datagram service such as UDP. Thus, the main function of SCTP amounts to reliable transfer of user datagrams between peer SCTP users. It performs this service within the context of an association between SCTP nodes, where APIs exist at the boundaries.

SCTP has connection-oriented characteristics but with broad concept. It provides means for each SCTP endpoint to provide the other during association start-up with a list of transport addresses (e.g. address/UDP port combinations) by which that endpoint can be reached and from which it will originate messages. The association carries transfers over all possible source/destination combinations, which may be generated from two end lists. As a result, SCTP offers the following services:

- application-level segmentation,
- acknowledged error-free non-duplicated transfer of user data,
- sequenced delivery of user datagrams within multiple streams,
- enhanced reliability through support of multi-homing at either or both ends of the association,
- optional multiplexing of user datagram into SCTP datagrams.

3.8.3 Family of UMTS Users

In the process of identifying the potential 3G users we can segment the subscriber body based on the population distribution, e.g. business, residential and mass market. We can further break these groups down into heavy and light users. However, our interest lies in finding who does actually correspond to each group and how much traffic they generate.

3.8.3.1 Business Subscribers

Business users will follow their enterprises and set the pace according to the wealth of resources and activity intensity. While a simple distinction would fall into large, medium and small corporations, it will not identify the true nature of business users. Thus, for all practical purposes we will group (non-exhaustively) into the following:

- *Information technologist* – involved in generating or transferring all types of modern information in communications and computers, software, etc.
- *Designers and producers* – working in manufacturing, heavy industry, product lines, etc.
- *Distributors and retailers* – active in marketing, sales and product distribution.
- *Financial and legal people* – banking and financing work, legal world activities, etc.

The classification above aims to group activities while identifying the type of business the subscribers will foster. Then, based on the profile we can see the volumes of traffic and demands they will generate.

3.8.3.2 Residential Subscribers

We can characterise these subscribers by their lifestyle. The latter in turn will provide a window to the amount of traffic they will generate. To make it simple and logical we can classify them into the following:

- *Communicators* – those continuously involved in social activities, communicating at all times.
- *Always prepared* – keeping up with the trends and having all means of modern communications.
- *World travellers* – relocating often, an international citizen.
- *Well to do* – the wealthy and established pillars of the community owning the national capital.

3.8.3.3 Mass Market Subscribers

All the remaining population groups not listed in the preceding segments correspond to this category, e.g. children and young people of 5–22 years, the labour force, educational groups (i.e. university), institutions, government bodies, etc. All of us, although not classified in the above categories, may also correspond to this segment.

3.8.4 Cost and Services

Regardless of who the subscriber is or to what subscriber segment he/she belongs, a user will always be looking to cost-value investments. Costly or too sophisticated communication services will not appeal to any of the aforementioned segments. Expensive services like the ones proposed by the Iridium[20] group will not gain sufficient penetration to justify investments. Thus, 3G services will not only need to be affordable, but also efficient to generate interest in all segments. No doubt, acceptance level will vary from group to group and region to region, but affordability and utility will go before wide acceptance. Hence, the key issue is to meet the needs of whatever segment of the population group.

3.8.5 UMTS Services Technology

To meet the needs implies making available the correct tools and environment. Now, if we assume that the infrastructure arrangements will take care of the environment, it remains a big task to find a tool or user equipment device to satisfy users.

A terminal not only needs to be a smart device capable of accessing a PS network, support bandwidth on demand, audio streaming and multimedia, it will also need versatility and have multiple capabilities.

A multi-functional device will make the difference in future usage and acceptance of higher transmission rates offered through UMTS. Market penetration and widespread usage of these of multimedia services will depend on the available and affordable terminals, as well as the pragmatic applications.

Wireless device inter-connections, intelligent voice recognition, wireless E-mail, simultaneous voice and data, user defined closed user group, location services [13,14], personal profile portal, location based delivery and marketing will only occur with efficient integration and inter-working of multiple technologies.

During 2002–2003, more than 50% of terminals will be replaced ranging from low end to high end, with about 80% penetration of mobile users in some regions; today's smart-phones will be tomorrow's low end terminals.

Thus, the minimum features for a UMTS handset at the start of 3G services will consist of:

- dual mode UMTS/GSM of 900, 1800 and 1900 MHz, including GPRS and HSCSD for seamless compatibility and roaming with 2G networks;
- integrated, WAP, Bluetooth;
- voice control and intelligent voice recognition (e.g. VoxML);
- large colour display and limited multimedia features;
- simultaneous UMTS sessions from 64 kbps up to 384 kbps;
- approximately $100 \times 50 \times 18$ mm and < 100 g;
- accessories – headset, camera, GPS and all existing accessories.

Furthermore, information centric devices like PDAs will have additional options, e.g.

- advanced multimedia capabilities;

[20]Satellite mobile services offering mainly voice and low data with world coverage.

- video clip and play support with easy man–machine interface;
- video and music stream support;
- standard and open OS;
- WAP and Java application capabilities;
- HTML and XML browser, E-mail client and personal portal configuration capabilities;
- intelligent phone management features, e.g. cmd completion;
- advanced colour touch-screens;
- Bluetooth and all necessary features integrated, e.g. pull-able headset, camera, etc.

3.8.5.1 Applications

Applications may not necessarily come from the technology design. However, the final blend will depend on the available and accessible technology. Therefore, the creation and implementation of applications will require large complicity between those providing technology solutions, those generating application platforms (including SW) and those planning to offer services. For example, the quality and utilisation feature of location services will depend not only on the information services server, but also on the capabilities of the terminal to display the information.

3.9 CONCLUSIONS

The set of service components for UMTS will continue to evolve, e.g. Chapter 10 outlines further UMTS application and services enablers. Thus, this chapter points mainly the key quality control elements that we cannot neglect as we follow the trends for non-voice services over integrated 3G CS and PS networks.

On the service provider side, again, it does not matter within what segment subscribers are, at the end, with the penetration of mobile services, free-Internet and the choice[21] of service provider, users will only care about quality, price and value. Ideal platforms for service differentiation are just now coming out. The implementation process for new services exploiting full 3G capabilities is thus still evolving.

REFERENCES

1. G. Feige, Cisco, 3GSM World Congress, Cannes, France, 2002.
2. 3GPP, Technical Specification Group, QoS Concept (3GPP TS 23.107 V5.5.0, 2002).

[21]The increase in operators during the 3G licensing process in many countries will create higher competition yet.

3. 3GPP, Technical Specification Group (TSG) SA, Transparent end-to-end Packet Switching Streaming Service (PSS) General Description, Release 5, V5.0.0, (2002–2003).
4. ITU-T H.323, Packet Based Multimedia Communications Systems, 1998.
5. 3GPPTS 23.207, End-to-End QoS Concept and Architecture, V5.8.0, 2003–2006).
6. 3GPP TS 23.002: 'Network Architecture.'
7. RFC 2475: 'An Architecture for Differentiated Services (DiffServ)'
8. 3GPP, Mandatory Speech Codec Speech Processing Functions, AMR Speech Codec; General Description (3G TS 26.071, 1999).
9. Technical Specification Group, Codec for Circuit Switched Multimedia Telephony Service, General Description, 3GPP, TS 26.110, 1999.
10. Handley, M., *et al.*, SIP: Session Initiation Protocol, RFC2543, IETF, 1999.
11. 3GPP, Architecture Principles for Release 2000 (3G TR 23.821), V1.0.1, 2000–2007).
12. ITU-T H.324, Terminal For Low Bit-rate Multimedia Communication, 1998.
13. 3GPP, Technical Specification Group Services and System Aspects, Services and System Aspects, Location Services (LCS), Service description, Stage 1, 3G TS 22.071, 1999.
14. 3GPP, Technical Specification Group (TSG) RAN, Working Group 2 (WG2), Stage 2 Functional Specification of Location Services in URAN, 3G TR 25.923, 1999.

4

The UTRA Physical Layer Design

The UTRA design is composed basically of three parts, i.e. radio aspects corresponding primarily to the physical layer, radio interface aspects incorporating layers 2 and 3 and network aspects inter-working directly with the core network. This chapter describes the UTRA physical layer including both FDD and TDD modes, as well as spreading and modulation, multiplexing and channel coding, and physical layer procedures.

4.1 SUMMARY OF FEATURES

Figure 4.1 illustrates the relationship of the physical layer (L1) and the upper layers (L2 and L3). L1 interfaces the Medium Access Control (MAC) sub-layer of L2 and the Radio Resource Control (RRC) portion of L3. L1 offers different transport channels to the MAC and the MAC offers different logical channels to the Radio Link Control (RLC) sub-layer of L2. Thus, there are Service Access Points (SAPs) between the different layers/sub-layers. A transport channel is characterised by the way information is transferred over the radio interface. The type of information transferred characterises a logical channel.

Two types of physical channels are defined in L1, i.e. Frequency Division Duplex (FDD) and Time Division Duplex (TDD). The first (FDD) mode is characterised by code, frequency and in the uplink by the relative phase (I/Q); the second (TDD) mode has in addition a time slot characterisation. The Radio Resource Control (RRC) manages L1.

The data transport services offered to higher layers by L1 occur through the use of transport channels via the MAC sub-layer. Table 4.1 illustrates some of the L1 or physical layer services. Through inter-working (e.g. a UE) provision of compatible bearers is assured.

Based on the types of physical channels L1 has two multiple access techniques:

- A Direct-Sequence Code Division Multiple Access (DS-CDMA) with the information spread within 5 MHz bandwidth, also referred to as Wide-band CDMA (WCDMA).
- A Time Division Multiple Access (TDMA) + CDMA often denoted as TDMA/CDMA or TD/CDMA resulting from the extra slotted feature.

The two access schemes afford UTRA two transmission modes, i.e. Frequency Division Duplex (FDD) corresponding to WCDMA operating with pair bands, and Time Division Duplex (TDD) corresponding to TD/CDMA operating with unpaired bands. The flexibility

All IP in 3G CDMA Networks J. Castro
© 2004 John Wiley & Sons, Ltd ISBN: 0-470-85322-0

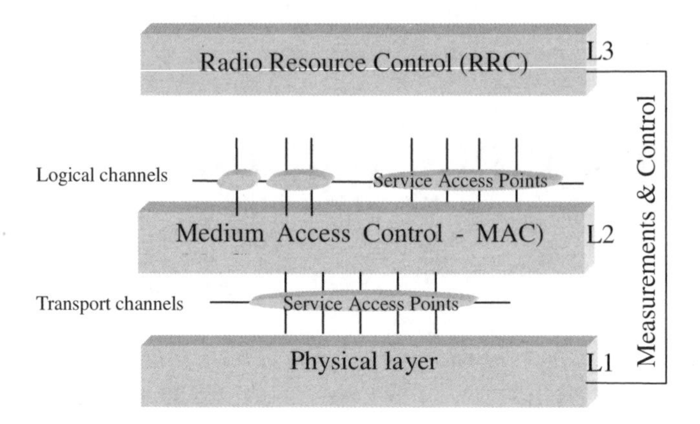

Figure 4.1 A radio interface protocol architecture around L1.

to operate in either FDD or TDD mode allows efficient spectrum utilisation within the frequency allocation in different regions, e.g. Europe, Asia, etc.

The FDD mode or WCDMA is thus a duplex method where uplink and downlink transmissions use two different radio frequencies separated, e.g. by 190 MHz. The TDD mode is a duplex method where uplink and downlink transmissions occur over the same radio frequency by using synchronised time intervals. In the TDD, time slots in a physical channel are divided into transmission and reception parts. Information on uplink and downlink are transmitted reciprocally. The UTRA has QPSK as modulation scheme. In the WCDMA or FDD mode the spreading (and scrambling) process is closely associated with modulation. The different UTRA families of codes are channelisation codes derived with a code-tree structure to separate channels from the same source, and codes to separate different cells.

Table 4.2 illustrates the harmonised parameters of the two UTRA modes.

A 10 ms radio frame divided into 15 slots (2560 chips/slot at the chip rate 3.84 Mcps) applies to two modes. A physical channel is therefore defined as a code (or number of codes) and additionally in TDD mode the sequence of time slots completes the definition of a

Table 4.1 Main functions of the UTRA physical layer

Macro-diversity distribution/combining and soft handover execution	Power weighing and combining of physical channels
Error detection on transport channels and indication to higher layers	Modulation and spreading/ de-modulation and de-spreading of Physical channels
FEC encoding/decoding of transport channels	Frequency and time (chip, bit, slot and frame) synchronisation
Multiplexing of transport channels and de-multiplexing of coded composite transport channels	Radio characteristics measurements including FER, SIR, interference power, etc., and indication to higher layers
Rate matching (data multiplexed on DCH)	Inner-loop power control
Mapping of coded composite transport channels on physical channels	RF processing

Table 4.2 UTRA FDD and TDD harmonised parameters

Parameters	UTRA TDD	UTRA FDD
Multiple access	TDMA, CDMA (inherent FDMA)	CDMA (inherent FDMA)
Duplex method	TDD	FDD
Channel spacing and carrier chip rate	5 MHz (nominal) and 3.84 Mcps	
Time slot and frame length	15 slots/frame and 10 ms	
Spreading factor	1,2,4,8,16	4, . . . , 512
Channel allocation	Slow and fast DCA supported	No DCA required
Types of burst	Traffic bursts, random access and synchronisation burst	DTX time mask defined, burst not applicable
Multi-rate concept	Multi-code, multi-slot and orthogonal variable spreading	Multi-code and orthogonal variable spreading
Forward Error Correction (FEC) codes	Convolutional coding $R = 1/2$ or 1/3 constraint length $K = 9$, turbo coding (eight-state PCCC $R = 1/3$) or service specific coding	
Inter-leaving	Inter-frame inter-leaving (10, 20, 40 and 80 ms)	
Modulation	QPSK	
Detection	Coherent, based on midamble	Coherent, based on pilot symbols
Dedicated channel power control	UL: open loop; 100 or 200 Hz DL: closed loop; rate $\leq$ 800 Hz	Fast closed loop; rate = 1500 Hz
Intra-frequency handover	Hard handover	Soft and softer handovers
Inter-frequency handover	Hard handover	
Intra-cell interference cancellation	Support for joint detection	Support for advanced receivers at base station

physical channel. The information rate of the channel varies with the symbol rate being derived from the 3.84 Mcps chip rate and the spreading factor.

We derive the symbol rate from the 3.84 Mcps chip rate and the spreading factor to obtain a variable rate in the channel. The information rate of the channel, e.g. varies with spreading factors from 256 to 4 for FDD uplink, from 512 to 4 for FDD downlink and from 16 to 1 for TDD uplink and downlink. Consequently, modulation symbol rates vary from 960 to 15 ksps (7.5 ksps) for FDD uplink (downlink), respectively, and for TDD the momentary modulation symbol rates vary from 3.84 Msps to 240 ksps, where sps stands for symbols/s.

The UTRA has QPSK as modulation scheme. In the WCDMA or FDD mode the spreading (and scrambling) process is closely associated with modulation. The different UTRA families of codes are:

- channelisation codes derived with a code-tree structure to separate channels from the same source, and codes to separate different cells;
- gold codes with 10 ms period (38 400 chips at 3.84 Mcps) used in the FDD mode, with the actual code itself of length $2^{18} - 1$ chips, and scrambling codes of length 16 used in the TDD mode;

- User Equipment (UE) separating codes: gold codes with 10 ms period, or alternatively S(2) codes 256 chip period for FDD mode, and codes with period of 16 chips and midamble sequences of different length depending on the environment for the TDD mode.

The key physical layer procedures involved with UTRA operation are:

- power control, with both inner loop and slow quality loop for FDD mode, and for TDD mode open loop in uplink and inner loop in downlink;
- cell search operation.

Measurements reported to higher layers and network containing radio characteristics like FER, SIR, interference power, etc. are:

- handover measurements within UTRA, e.g. determination of relative strength of a cell. In the FDD mode, identification of timing relation between cells to support asynchronous soft handover;
- other measurement procedures are preparation for HO to GSM 900/1800/1900; UE procedures before random access-process; and procedures for Dynamic Channel Allocation (DCA) in the TDD mode.

4.2 DEDICATED AND COMMON TRANSPORT CHANNELS

Transport channels are defined by how and with what features data are transferred over the air interface. The generic classification of transport channels includes two groups, i.e. *dedicated* and *common* channels. The first group uses inherent UE addressing, while the second uses explicit UE addressing when addressing is required.

4.2.1 Dedicated Transport Channels

There is primarily one transport Dedicated Channel (DCH) for uplink or downlink in the FDD and TDD modes, which is used to carry user or control information between the UTRAN and an UE. The DCH is transmitted over the entire cell or over only a part of the cell using, e.g. beam-forming antennas.

4.2.2 Common Transport Channels

While the intrinsic function of each common transport channel may not necessarily be identical in the FDD and TDD modes, both sets have basically the same function and acronym. Table 4.3 summarises the essential definitions for the two modes.

Both FDD and TDD have a similar number of transport channels; however, the FDD mode does not have an Uplink Shared Channel (USCH) and the TDD mode does not have a Common Packet Channel (CPCH).

Table 4.3 Summary of common transport channels

FDD mode	TDD mode
BCH – Broadcast Channel	*BCH – Broadcast Channel*
Downlink transport channel that is used to broadcast system- and cell-specific information. The BCH is always transmitted over the entire cell and has a single transport format.	
FACH – Forward Access Channel	*FACH – Forward Access Channel(s)*
Downlink transport channel used to carry control information to a mobile station when the system knows the cell location of the mobile station. In the FDD, it can be transmitted over the entire cell or over only a part of the cell using, e.g. beam-forming antennas, and it can also be transmitted using slow power control. In the TDD it may carry short user packets.	
PCH – Paging Channel	*PCH – Paging Channel*
Downlink transport channel is transmitted always over the entire cell, and used to carry control information to a mobile station when the system does not know the location cell of the mobile station. In the FDD mode transmission of the PCH is associated with the transmission of physical-layer generated paging indicators, to support efficient sleep-mode procedures.	
RACH – Random-Access Channel	*RACH – Random-Access Channel*
Uplink transport channel, always received from the entire cell, used to carry control information from the mobile station. In FDD, the RACH is characterised by a collision risk and by using open-loop power control for transmission. In TDD it may also carry short user packets.	
CPCH – Common Packet Channel	*USCH – Uplink Shared Channel*
Uplink transport channel associated with a dedicated channel on the downlink, which provides power control and CPCH control commands (e.g. emergency stop). It is characterised by initial collision risk and by using inner loop power control for transmission.	Uplink transport channel shared by several UEs carrying dedicated control or traffic data.
DSCH – Downlink Shared Channel	*DSCH – Downlink Shared Channel*
Downlink transport channel shared by several UEs carrying dedicated control or traffic data. In FDD it is associated with one or several downlink DCH(s). It may be transmitted over the entire cell or over only a part of the cell using, e.g. beam-forming antennas.	

The CPCH transport channel in FDD performs essential power control commands, which may not be required in TDD. Likewise, the USCH transport channel performs essential commands in TDD, which may not be required in FDD.

4.3 CONFIGURATION OF FDD PHYSICAL CHANNELS

Physical channels in FDD inherit primarily a layered structure of radio frames and time slots. A radio frame is a processing unit consisting of 15 slots with a length of 38 400 chips,

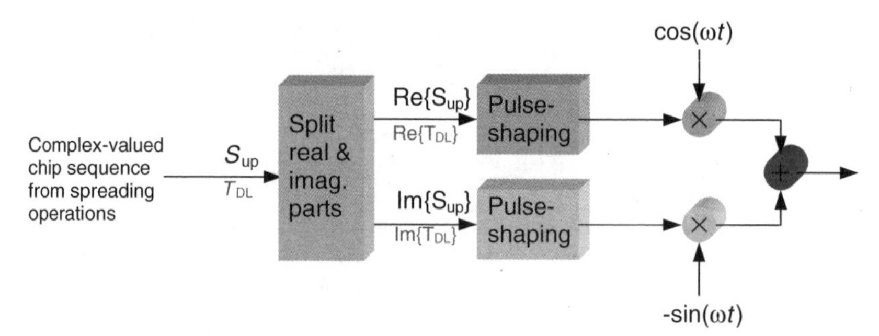

Figure 4.2 Uplink/downlink modulation process.

and slot is a unit consisting of fields containing bits with a length of 2560 chips. The slot configuration varies depending on the channel bit rate of the physical channel; thus, the number of bits per slot may be different for different physical channels and may, in some cases, vary with time. The basic physical resource is the code/frequency plane, and on the uplink, different information streams may be transmitted on the I and Q branches. Thus, a physical channel corresponds to a specific carrier frequency and code, and on the uplink there is in addition a relative phase (0 or $\pi/2$) element.

4.3.1 Uplink and Downlink Modulation

The uplink modulation uses a chip rate of 3.84 Mcps, where the complex-valued chip sequence generated by the spreading process has QPSK modulation as seen in Figure 4.2. The pulse-shaping characteristics are described in [1].

The downlink modulation also has a chip rate of 3.84 Mcps, with a QPSK modulated complex-valued chip sequence generated by the spreading process. Figure 4.2 does also represent the downlink modulation process. However, the DL pulse-shaping characteristics are described in [2].

4.3.2 Dedicated Uplink Physical Channels

The two types of uplink dedicated physical channels, i.e. Dedicated Physical Data Channel (DPDCH) and Dedicated Physical Control Channel (DPCCH) are *I*/*Q* code multiplexed within each radio frame. The uplink DPDCH carries the DCH transport channel, while the uplink DPCCH carries L1 control information such as known pilot bits to support channel estimation for coherent detection, Transmit Power Control (TPC) commands, Feedback Information (FBI) and an optional Transport Format Combination Indicator (TFCI).

The TFCI informs the receiver about the instantaneous transport format combination of the transport channels mapped to the uplink DPDCH transmitted simultaneously. There is one and only one uplink DPCCH on each radio link; however, there may be zero, one or several uplink DPDCHs on each radio link. Figure 4.3 illustrates the frame structure of the uplink dedicated physical channels, where each frame has 10 ms length split into 15 slots (T_{slot}) of 2560 chips length, corresponding to one power control period.

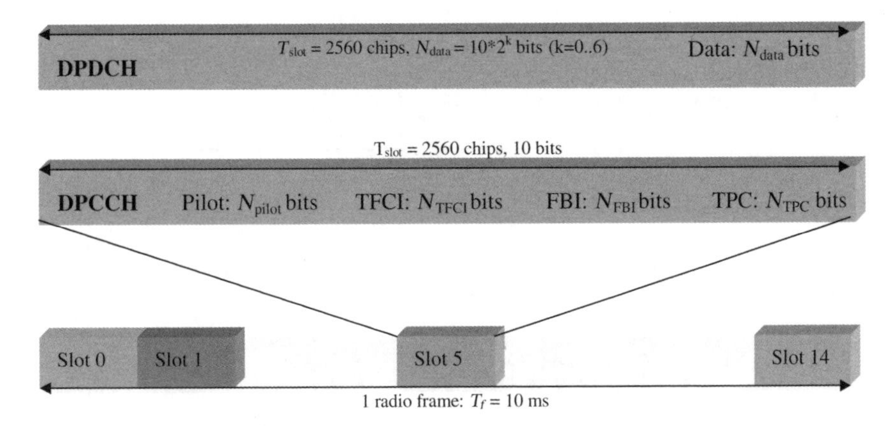

Figure 4.3 Uplink frame structure DPDCH/DPCCH.

Parameter k in Figure 4.3 determines the number of bits per uplink DPDCH slot. It is related to the spreading factor defined as $\mathrm{SF} = 256/2^k$, which may range from 256 down to 4. The SF in the uplink DPCCH is always equal to 256 corresponding to 10 bits per uplink DPCCH slot. Table 4.4 illustrates the exact number of bits in the uplink DPDCH, while Table 4.5 shows the different uplink DPCCH fields (i.e. N_{pilot}, N_{TFCI}, N_{FBI} and N_{TPC}). The pilot patterns are given in Table 4.6 and the TPC bit pattern is given in Table 4.8. Upper layers configure the slot format. The channel symbol rate and SF for all cases in Table 4.5 are 15 and 256, respectively. Channel bit and symbol rates illustrated in Tables 4.4 and 4.5 reflect rates before spreading.

The FBI bits (S field and D field) support the techniques requiring feedback from the UE to the UTRAN access point, including closed loop mode transmit diversity and Site Selection Diversity Transmission (SSDT). The open SSDT signalling uses the S field and the closed loop mode transmit diversity signalling uses the D field. The S field consists of 0, 1 or 2 bits while the D field consists of 0 or 1 bit. Table 4.5 shows the total FBI field size, i.e. the N_{FBI}. Simultaneous use of SSDT power control and closed loop mode transmit diversity requires that the S field consists of 1 bit. The use of the FBI fields is described in detail in [3].

Table 4.4 DPDCH fields

Slot form at (i)	Channel bit rate (kbps)	Channel symbol rate (ksps)	SF	Bits/frame	Bits/slot	N_{data}
0	15	15	256	150	10	10
1	30	30	128	300	20	20
2	60	60	64	600	40	40
3	120	120	32	1200	80	80
4	240	240	16	2400	160	160
5	480	480	8	4800	320	320
6	960	960	4	9600	640	640

Table 4.5 DPCCH fields

Slot format (i)	Channel bit rate (kbps)	Bits/frame	Bits/slot	N_{pilot}	N_{TPC}	N_{TFCI}	N_{FBI}	Slots/frame
0	15	150	10	6	2	2	0	15
0A	15	150	10	5	2	3	0	10–14
0B	15	150	10	4	2	4	0	8–9
1	15	150	10	8	2	0	0	8–15
2	15	150	10	5	2	2	1	15
2A	15	150	10	4	2	3	1	10–14
2B	15	150	10	3	2	4	1	8–9
3	15	150	10	7	2	0	1	8–15
4	15	150	10	6	2	0	2	8–15
5	15	150	10	5	1	2	2	15
5A	15	150	10	4	1	3	2	10–14
5B	15	150	10	3	1	4	2	8–9

There are two types of uplink dedicated physical channels; those that include TFCI (e.g. for several simultaneous services) and those that do not include TFCI (e.g. for fixed-rate services). These types are reflected by the duplicated rows of Table 4.5. It is the UTRAN that determines whether a TFCI should be transmitted and it is mandatory for all UEs to support the use of TFCI in the uplink. The mapping of TFCI bits onto slots is described in [1]. In compressed mode, DPCCH slot formats with TFCI fields are changed. There are two possible compressed slot formats for each normal slot format. They are labelled A and B and the selection between them is dependent on the number of slots that are transmitted in each frame in compressed mode.

Table 4.6 Pilot bit patterns for uplink DPCCH with N_{pilot} =3, 4, 5 and 6

Slot	$N_{pilot} = 3$			$N_{pilot} = 4$				$N_{pilot} = 5$					$N_{pilot} = 6$					
Bit	*0*	*1*	2	0	*1*	2	*3*	*0*	*1*	2	*3*	*4*	0	*1*	2	*3*	*4*	*5*
0	1	1	1	1	1	1	1	1	1	1	1	0	1	1	1	1	1	0
1	0	0	1	1	0	0	1	0	0	1	1	0	1	0	0	1	1	0
2	0	1	1	1	0	1	1	0	1	1	0	1	1	0	1	1	0	1
3	0	0	1	1	0	0	1	0	0	1	0	0	1	0	0	1	0	0
4	1	0	1	1	1	0	1	1	0	1	0	1	1	1	0	1	0	1
5	1	1	1	1	1	1	1	1	1	1	1	0	1	1	1	1	1	0
6	1	1	1	1	1	1	1	1	1	1	0	0	1	1	1	1	0	0
7	1	0	1	1	1	0	1	1	0	1	0	0	1	1	0	1	0	0
8	0	1	1	1	0	1	1	0	1	1	1	0	1	0	1	1	1	0
9	1	1	1	1	1	1	1	1	1	1	1	1	1	1	1	1	1	1
10	0	1	1	1	0	1	1	0	1	1	0	1	1	0	1	1	0	1
11	1	0	1	1	1	0	1	1	0	1	1	1	1	1	0	1	1	1
12	1	0	1	1	1	0	1	1	0	1	0	0	1	1	0	1	0	0
13	0	0	1	1	0	0	1	0	0	1	1	1	1	0	0	1	1	1
14	0	0	1	1	0	0	1	0	0	1	1	1	1	0	0	1	1	1

Table 4.7 Pilot bit patterns for uplink DPCCH with $N_{pilot} = 7$ and 8

Slot	$N_{pilot} = 7$							$N_{pilot} = 8$							
Bit	*0*	*1*	*2*	*3*	*4*	*5*	*6*	*0*	*1*	*2*	*3*	*4*	*5*	*6*	*7*
0	1	1	1	1	1	0	1	1	1	1	1	1	1	1	0
1	1	0	0	1	1	0	1	1	0	1	0	1	1	1	0
2	1	0	1	1	0	1	1	1	0	1	1	1	0	1	1
3	1	0	0	1	0	0	1	1	0	1	0	1	0	1	0
4	1	1	0	1	0	1	1	1	1	1	0	1	0	1	1
5	1	1	1	1	1	0	1	1	1	1	1	1	1	1	0
6	1	1	1	1	0	0	1	1	1	1	1	1	0	1	0
7	1	1	0	1	0	0	1	1	1	1	0	1	0	1	0
8	1	0	1	1	1	0	1	1	0	1	1	1	1	1	0
9	1	1	1	1	1	1	1	1	1	1	1	1	1	1	1
10	1	0	1	1	0	1	1	1	0	1	1	1	0	1	1
11	1	1	0	1	1	1	1	1	1	1	0	1	1	1	1
12	1	1	0	1	0	0	1	1	1	1	0	1	0	1	0
13	1	0	0	1	1	1	1	1	0	1	0	1	1	1	1
14	1	0	0	1	1	1	1	1	0	1	0	1	1	1	1

Table 4.8 TPC bit pattern

TPC bit pattern		Transmitter power control command
$N_{TPC} = 1$	$N_{TPC} = 2$	
1	11	1
0	00	0

The pilot bit patterns are described in Tables 4.6 and 4.7. The shadowed column part of pilot bit pattern is defined as FSW, which can be used to confirm frame synchronisation. (The value of the pilot bit pattern other than FSWs shall be '1'.)

Table 4.8 presents the relationship between the TPC bit pattern and transmitter power control command.

While there is only DPCCH per radio link, several parallel DPDCHs using different channelisation codes [2] can be transmitted for the multi-code operation in the uplink dedicated physical channels.

4.3.2.1 Spreading DPCCH/DPDCH

In the uplink spreading principle of DPCCH and DPDCHs real-valued sequences of +1 and −1 represent the binary values '0' and '1', respectively. We spread the DPCCH to the chip rate by the channelisation code c_c, and the *n*th DPDCH (or DPDCH$_{n}$) to the chip rate by the channelisation code $c_{d,n}$. As illustrated in Figure 4.4, we can transmit one DPCCH and up to six parallel DPDCHs simultaneously, i.e. $1 \leq n \leq 6$ [4].

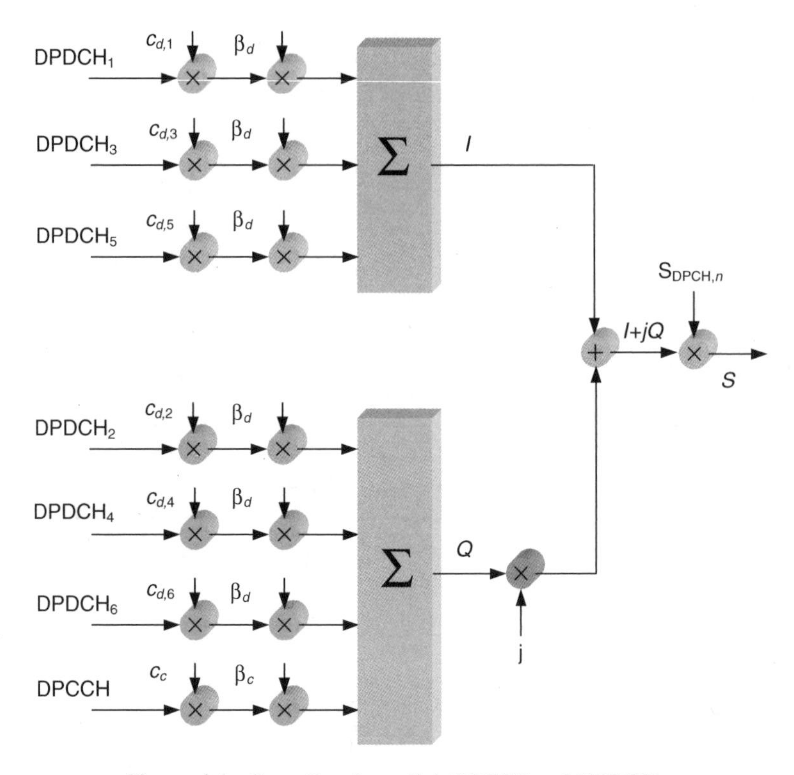

Figure 4.4 Spreading for uplink DPCCH and DPDCHs.

After channelisation, gain factors β_c for DPCCH and β_d for all DPDCHs weigh the real-valued spread signals, where at every instant in time, at least one of the values β_c and β_d have the amplitude 1.0. Likewise after the weighing, we sum the stream of real-valued chips on the I and Q branches and then treat them as a complex-valued stream of chips. After we scramble these streams by the complex-valued scrambling code $S_{\mathrm{dpch},n}$, the scrambling code application aligns with the radio frames, i.e. the first scrambling chip corresponds to the beginning of a radio frame.

Table 4.9 illustrates quantisation steps of the β values quantised into 4 bit words. After the weighing, we sum the stream of real-valued chips on the I and Q branches and then treat them as a complex-valued stream of chips. After that, we scramble these streams by the complex-valued scrambling code $S_{\mathrm{dpch},n}$. The scrambling code application aligns with the radio frames, i.e. the first scrambling chip corresponds to the beginning of a radio frame.

4.3.3 Common Uplink Physical Channels

4.3.3.1 Physical Random-Access Channel–PRACH

The PRACH carries the Random-Access Channel (RACH).

Table 4.9 The quantisation of the gain parameters

Signalling values for β_c and β_d	Quantised amplitude ratios β_c and β_d
15	1.0
14	0.9333
13	0.8666
12	0.8000
11	0.7333
10	0.6667
9	0.6000
8	0.5333
7	0.4667
6	0.4000
5	0.3333
4	0.2667
3	0.2000
2	0.1333
1	0.0667
0	Switch off

4.3.3.1.1 The Random-Access Transmission Structure

The random-access transmission uses a slotted ALOHA technique with fast acquisition indication. The UE can start the random-access transmission at the beginning of a number of well-defined time intervals, denoted as *access slots* as illustrated in Figure 4.5. There are 15 access slots per two frames and they are spaced 5120 chips apart. The timing of the access slots and the acquisition indication is described in Section 7.3. The information

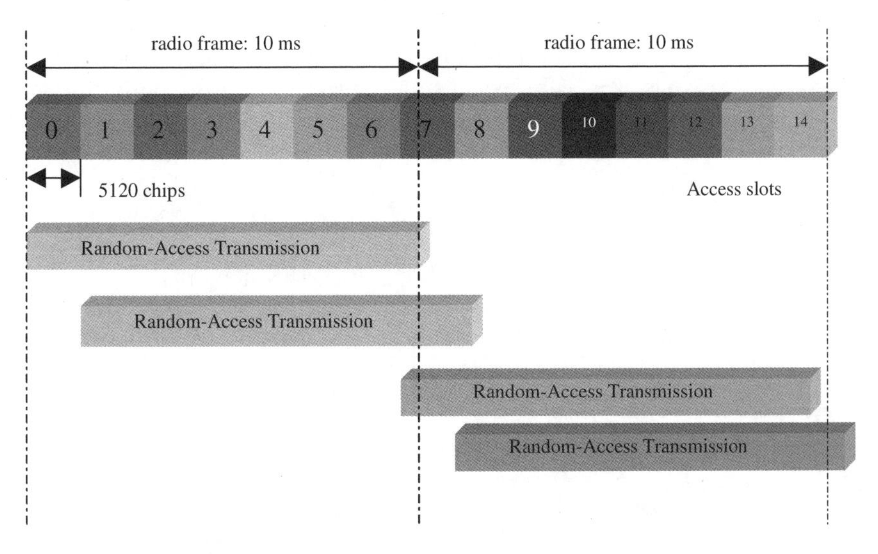

Figure 4.5 RACH access slot numbers and spacing.

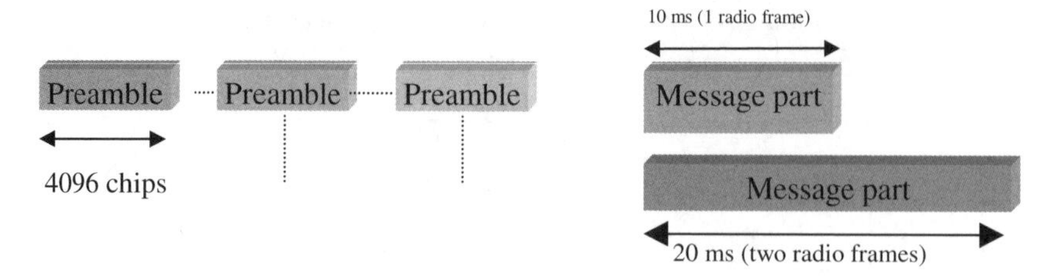

Figure 4.6 Structure of the random-access transmission.

about the type of access slots available for random-access transmission comes from the upper layers.

Figure 4.6 illustrates the random-access transmission structure, where the transmission consists of one or several *preambles* of length 4096 chips and a *message* of length 10 or 20 ms. Each preamble has 256 repetitions of 16 chips signature. Thus, there is a maximum of 16 available signatures, see [4] for more details.

4.3.3.1.2 The RACH Message Part

Figure 4.6 illustrates the random-access message part radio frame structure, where the 10 ms message part radio frame is split into 15 slots, each having a length $T_{slot} = 2560$ chips. Furthermore, each slot consists of two parts, i.e. a data part to which the RACH transport channel is mapped and a control part that carries layer 1 control information; they are transmitted in parallel.

A 10 ms message part consists of one message part radio frame, while a 20 ms message part consists of two consecutive 10 ms message part radio frames. The message part length can be determined from the used signature and/or access slot, as configured by higher layers. Table 4.10 illustrates data and control fields of the random-access message.

The data part consists of 10×2^k bits, where $k = 0,1,2,3$. This corresponds to a spreading factor of 256, 128, 64 and 32 for the message data part, respectively.

The control part consists of 8 known pilot bits to support channel estimation for coherent detection and 2 TFCI bits. This corresponds to a spreading factor of 256 for the message control part. The pilot bit pattern is described in Table 4.11. The total number of TFCI bits in the random-access message is $15 \times 2 = 30$.

Table 4.10 Random-access message data and control fields

Slot format (i)	Channel bit rate (kbps)	Channel symbol rate (ksps)	SF	Bits/frame	Bits/slot	N_{pilot}	N_{data}
0	15	15	256	150	10		10
1	30	30	128	300	20		20
2	60	60	64	600	40		40
3	120	120	32	1200	80		80
Control fields							N_{TFCI}
0	15	15	256	150	10	8	
2							

Table 4.11 Pilot bit patterns for RACH message part with $N_{pilot} = 8$

Slot	$N_{pilot} = 8$							
Bit	0	1	2	3	4	5	6	7
0	1	1	1	1	1	1	1	0
1	1	0	1	0	1	1	1	0
2	1	0	1	1	1	0	1	1
3	1	0	1	0	1	0	1	0
4	1	1	1	0	1	0	1	1
5	1	1	1	1	1	1	1	0
6	1	1	1	1	1	0	1	0
7	1	1	1	0	1	0	1	0
8	1	0	1	1	1	1	1	0
9	1	1	1	1	1	1	1	1
10	1	0	1	1	1	0	1	1
11	1	1	1	0	1	1	1	1
12	1	1	1	0	1	0	1	0
13	1	0	1	0	1	1	1	1
14	1	0	1	0	1	1	1	1

The TFCI of a radio frame indicates the transport format of the RACH transport channel mapped to the simultaneously transmitted message part radio frame. In the case of a 20 ms PRACH message part, the TFCI is repeated in the second radio frame.

4.3.3.2 Physical Common Packet Channel (PCPCH)

The PCPCH carries the CPCH. The CPCH transmission is based on the Collision Detection-Digital Sense Multiple Access (CD-DSMA) technique with fast acquisition indication. The UE can start transmission at the beginning of a number of well-defined time intervals, relative to the frame boundary of the received BCH of the current cell. The access slot timing and structure are identical to those defined for the RACH (Figure 4.7). Figure 4.8 illustrates the structure of the CPCH access transmission. The PCPCH access transmission consists of one or several Access Preambles (AP) of length 4096 chips, one Collision Detection Preamble (CDP) of length 4096 chips, a DPCCH Power Control Preamble (PCP) which is either 0 or 8 slots in length and a message of variable length $N \times 10$ ms.

4.3.3.2.1 CPCH Access–Power Control and Detection Preamble Parts

- Like in the RACH, the access CPCH preamble uses signature sequences, but the number of sequences can be lower. The scrambling codes may differ from the gold codes segment used in the RACH or could be the same scrambling code.
- Table 4.12 defines the DPCCH fields form the CPCH PC-P part. The power control preamble length parameter takes the values 0 or 8 slots, as set by the higher layers. When

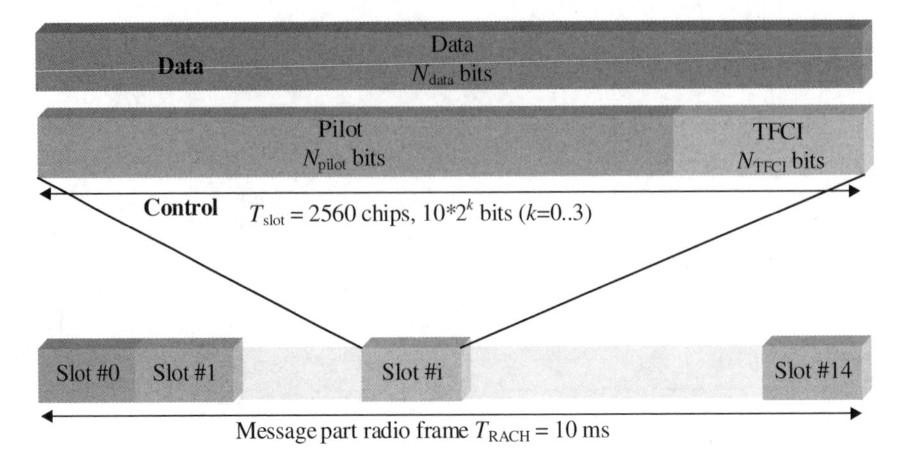

Figure 4.7 Random-access message part radio frame structure.

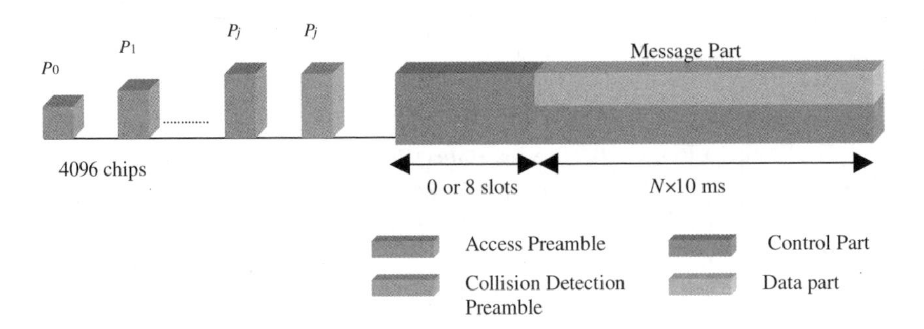

Figure 4.8 Structure of the CPCH access transmission.

the power control preamble length is set to 8 slots, pilot bit patterns from slot 0 to slot 7 defined in Table 4.7 shall be used for CPCH PCP.

- Also like in the RACH, the detection CPCH preamble uses signature sequences. However, the scrambling code set differs from the gold code segment used to form the RACH scrambling code.

Table 4.12 DPCCH fields for CPCH power control preamble segment

Slot format (i)	Channel bit rate (kbps)	Channel symbol rate (ksps)	SF	Bits/frame	Bits/slot	N_{pilot}	N_{TPC}	N_{TFCI}	N_{FBI}
0	15	15	256	150	10	6	2	2	0
1	15	15	256	150	10	8	2	0	0
2	15	15	256	150	10	5	2	2	1
3	15	15	256	150	10	7	2	0	1
4	15	15	256	150	10	6	2	0	2
5	15	15	256	150	10	5	1	2	2

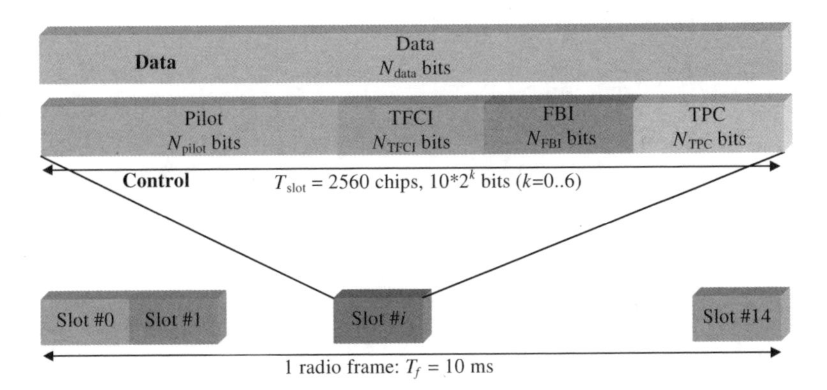

Figure 4.9 Frame structure for uplink data and control parts associated with PCPCH.

4.3.3.2.2 CPCH Message Part

With similar message part structure of the RASH, each CPCH message part consists of up to N_Max_frames[1] 10 ms frames, with a 10 ms frame split into 15 slots, each having $T_{slot} = 2560$ chips length. In addition, every slot consists of a data part that carries higher layer information and a control part that carries layer 1 control information. The data and control parts are transmitted in parallel.

The DPDCH field entries defined in Table 4.4 apply also to the data part of the CPCH message part. The control part of the CPCH message part has a spreading factor of 256, and it uses the same slot format as the control part of the CPCH PCP. The pilot bit patterns defined in Tables 4.6 and 4.7 apply also to the pilot bit patterns of the CPCH message part.

Figure 4.9 illustrates the uplink common packet physical channel frame structure. Each frame of length 10 ms is split into 15 slots having $T_{slot} = 2560$ chips length corresponding to one power-control period.

The data part consists of 10×2^k bits, where $k = 0, 1, 2, 3, 4, 5, 6$, corresponding to spreading factors of 256, 128, 64, 32, 16, 8, 4, respectively.

4.3.3.3 Spreading Common Uplink Physical Channels

4.3.3.3.1 PRACH

The PRACH preamble part consists of a complex-valued code and the message part includes the data and control parts, Figure 4.10 illustrates its spreading principle. In the message part, real-value sequences represent the binary control and data parts, i.e. the binary value '0' maps to the real value +1, while the binary value '1' maps to the real value −1. The channelisation code c_c spreads the control part, while channelisation code c_d spreads the data part.

After channelisation, gain factor β_c for the control part and β_d for the data part weigh the real-valued spread signals, where at least every instant in time one of the values β_c and β_d have the amplitude 1.0. Table 4.9 illustrates quantisation steps of the β values quantised into 4-bit words.

[1]*N_Max_frames* is a higher layer parameter.

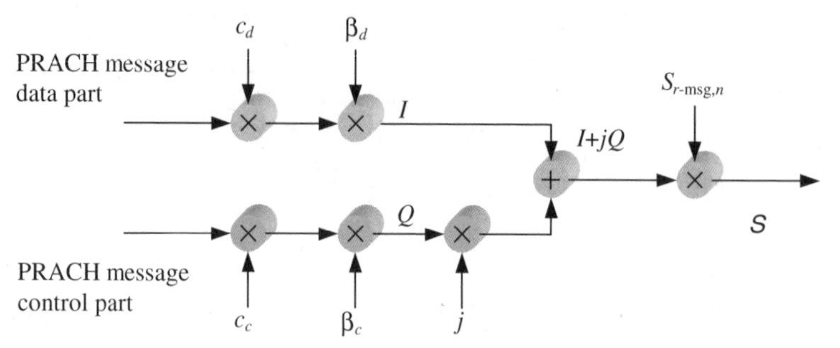

Figure 4.10 Spreading of PRACH message part.

Once the weighing takes place, we treat the stream of real-valued chips on the I and Q branches as a complex-valued stream of chips. Then the complex-valued scrambling code $S_{r\text{-msg},n}$ scrambles this complex-valued signal. The 10 ms scrambling code application aligns with the 10 ms message part radio frames, i.e. the first scrambling chip corresponds to the beginning of a message part radio frame [4].

4.3.3.3.2 PCPCH

As in the PRACH, the PCPCH preamble part consists of a complex-valued code, and the PCPCH message part includes data and control parts, Figure 4.11 illustrates its spreading principle.

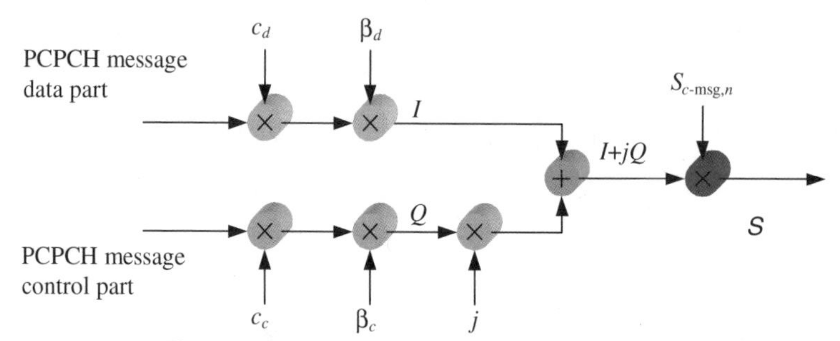

Figure 4.11 Spreading of PCPCH message part.

In the message part, real-value sequences represent the binary control and data parts, i.e. the binary value ‘0’ maps to the real value +1, while the binary value ‘1’ maps to the real value −1. The channelisation code c_c spreads the control part, while channelisation code c_d spreads the data part. Channelisation and weighing follows the same pattern as in the PRACH.

4.3.4 Uplink Channelisation codes

The Orthogonal Variable Spreading Factor (OVSF) channelisation codes preserve orthogonality between a user’s different physical channels. The tree illustrated in Figure 4.12 defines these codes.

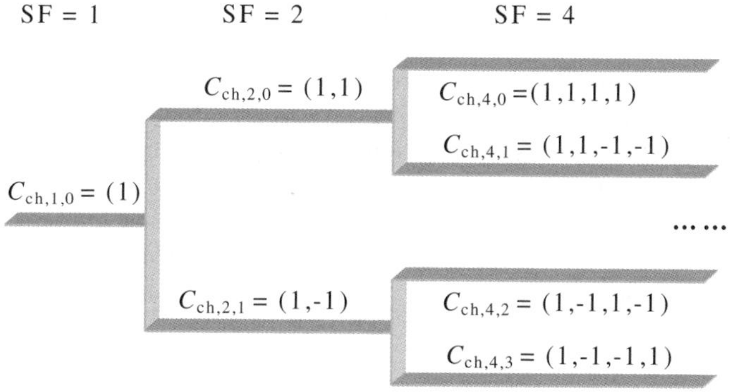

SF = 1 SF = 2 SF = 4

$C_{ch,2,0} = (1,1)$

$C_{ch,4,0} = (1,1,1,1)$

$C_{ch,4,1} = (1,1,-1,-1)$

$C_{ch,1,0} = (1)$

......

$C_{ch,2,1} = (1,-1)$

$C_{ch,4,2} = (1,-1,1,-1)$

$C_{ch,4,3} = (1,-1,-1,1)$

Figure 4.12 Orthogonal Variable Spreading Factor (OVSF) code-tree generation.

The channelisation codes in the OVSF tree have a unique description as $C_{ch,SF,k}$, where SF is the spreading factor of the code and k is the code number, $0 \leq k \leq SF - 1$. Each level in the code tree defines channelisation codes of length SF, corresponding to a spreading factor of SF. From [4] the generation method for the channelisation code is defined as

$$C_{ch,1,0} = 1 \tag{4.1}$$

$$\begin{bmatrix} C_{ch,2,0} \\ C_{ch,2,1} \end{bmatrix} = \begin{bmatrix} C_{ch,1,0} & C_{ch,1,0} \\ C_{ch,1,0} & -C_{ch,1,0} \end{bmatrix} = \begin{bmatrix} 1 & 1 \\ 1 & -1 \end{bmatrix} \tag{4.2}$$

$$\begin{bmatrix} C_{ch,2^{(n+1)},0} \\ C_{ch,2^{(n+1)},1} \\ C_{ch,2^{(n+1)},2} \\ C_{ch,2^{(n+1)},3} \\ \vdots \\ C_{ch,2^{(n+1)},2^{(n+1)}-2} \\ C_{ch,2^{(n+1)},2^{(n+1)}-1} \end{bmatrix} = \begin{bmatrix} C_{ch,2^n,0} & C_{ch,2^n,0} \\ C_{ch,2^n,0} & -C_{ch,2^n,0} \\ C_{ch,2^n,1} & C_{ch,2^n,1} \\ C_{ch,2^n,1} & -C_{ch,2^n,1} \\ \vdots & \vdots \\ C_{ch,2^n,2^n-1} & C_{ch,2^n,2^n-1} \\ C_{ch,2^n,2^n-1} & -C_{ch,2^n,2^n-1} \end{bmatrix} \tag{4.3}$$

The leftmost value in each channelisation code word corresponds to the chip transmitted first in time.

4.3.4.1 *DPCCH/DPDCH Code Allocation*

According to [4] for the DPCCH and DPDCHs the following applies: the DPCCH is always a code $c_c = C_{ch,256,0}$ as spread; and when we transmit only one DPDCH, the DPDCH$_1$ has code $c_{d,1} = C_{ch,SF,k}$ as spread, where SF is the spreading factor of DPDCH$_1$ and $k = SF/4$. However, when we transmit more than one DPDCH, all DPDCHs have spreading factors equal to 4. The DPDCH$_n$ is spread by the code $c_{d,n} = C_{ch,4,k}$, where $k = 1$ if $n \in \{1, 2\}, k = 3$ if $n \in \{3, 4\}$ and $k = 2$ if $n \in \{5, 6\}$.

4.3.4.2 PRACH Message Part Code Allocation

The preamble signature $s, 0 \leq s \leq 15$, points to one of the 16 nodes in the code tree that corresponds to channelisation codes of length 16. To spread the message part we use the sub-tree below a specified node, while to spread the control part we use the channelisation code c_c with $\mathrm{SF} = 256$ in the lowest branch of the sub-tree, i.e. $c_c = C_{\mathrm{ch},256,m}$ where $m = 16 \times s + 15$. The data part uses any of the channelisation codes from spreading factor 32 to 256 in the upper-most branch of the sub-tree. More exactly, we spread the data part by channelisation code $c_d = C_{\mathrm{ch,SF},m}$, SF is the data part spreading factor and $m = \mathrm{SF} \times s/16$ [4].

4.3.4.3 PCPCH Message Part Code Allocation

For the control part and data part the following applies: the control part has always code $c_c = C_{\mathrm{ch},256,0}$ as spread; and the data part has code $c_d = C_{\mathrm{ch,SF},k}$ as spread, where SF is the spreading factor of the data part and $k = \mathrm{SF}/4$. The data part may use the code from spreading factor 4 to 256, and a UE can increase SF during a message transmission on frame-by-frame basis [4].

Finally, the same channelisation code of the message control part applies to the PCPCH power control preamble.

4.3.5 Uplink Scrambling Codes

All uplink physical channels use a complex-valued scrambling code. While either long or short scrambling codes apply to the DPCCH/DPDCH, to the PRACH and PCPCH message parts only long scrambling codes apply. Higher layers assign the 2^{24} long and 2^{24} short uplink scrambling codes.

4.3.5.1 Long Scrambling Sequence

The long scrambling sequences $c_{\mathrm{long},1,n}$ and $c_{\mathrm{long},2,n}$ result from the position-wise mod 2 sum of 38 400 chip segments and two binary m sequences generated by means of two generator polynomials of degree 25. The first m sequences, i.e. x comes from the primitive (over GF (2)) polynomial $X^{25} + X^3 + 1$; while the second m sequences, i.e. y, come from the polynomial $X^{25} + X^3 + X^2 + X + 1$. The resulting sequences constitute a segment set of gold sequences, where the sequence $c_{\mathrm{long},2,n}$ is a 16 777 232 chip shifted version of the sequence $c_{\mathrm{long},1,n}$ [4]. Figure 4.13 illustrates a configuration of long uplink scrambling sequence generator.

For completeness in the following we include an extract of the long scrambling sequence definition from [4] where $n_{23} \ldots n_0 = 24$ bit binary representation of the scrambling sequence number n with n_0 as the least significant bit, x sequence which depends on the chosen scrambling sequence number n is denoted by x_n and $x_n(i)$ and $y(i)$ denote the ith symbol of the sequences x_n and y, respectively. Then m sequences x_n and y can be

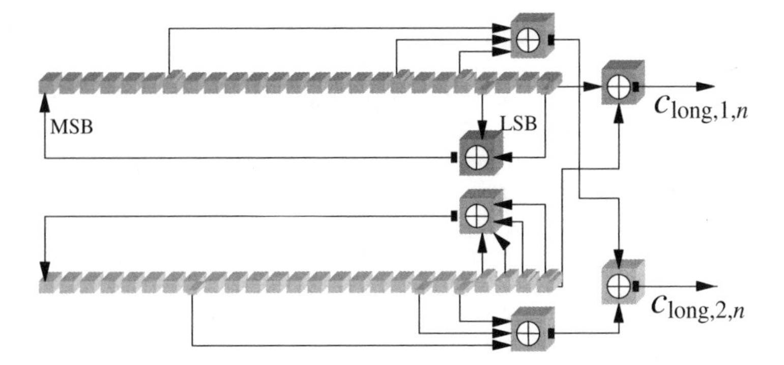

Figure 4.13 Configuration of the uplink long scrambling sequence generator.

defined as

$$x_n(0) = n_0, x_n(1) = n_1, \ldots, x_n(22) = n_{22}, x_n(23) = n_{23}, x_n(24) = 1 \tag{4.4}$$

$$y(0) = y(1) = \cdots = y(23) = y(24) = 1 \tag{4.5}$$

where $x_n(0)$ and $y(0)$ are the initial conditions.

The recursive definition of subsequent symbols include

$$x_n(i+25) = x_n(i+3) + x_n(i) \bmod 2, \quad i = 0, \ldots, 2^{25} - 27 \tag{4.6}$$

$$y(i+25) = y(i+3) + y(i+2) + y(i+1) + y(i) \bmod 2, \quad i = 0, \ldots, 2^{25} - 27 \tag{4.7}$$

The binary gold sequence z_n can be defined as

$$z_n(i) = x_n(i) + y(i) \bmod 2, \quad i = 0, 1, 2, \ldots, 2^{25} - 2 \tag{4.8}$$

then the real-valued gold sequence Z_n is defined by

$$Z_n(i) = \begin{cases} +1 & \text{if } z_n(i) = 0 \\ -1 & \text{if } z_n(i) = 1 \end{cases} \quad \text{for} \quad i = 0, 1, \ldots, 2^{25} - 2 \tag{4.9}$$

Now, the real-valued long scrambling sequences $c_{\text{long},1,n}$ and $c_{\text{long},2,n}$ are defined as

$$c_{\text{long},1,n}(i) = Z_n(i), \quad i = 0, 1, 2, \ldots, 2^{25} - 2 \tag{4.10}$$

and

$$c_{\text{long},2,n}(i) = Z_n((i + 16\ 777\ 232) \bmod (2^{25} - 1)), \quad i = 0, 1, 2, \ldots, 2^{25} - 2 \tag{4.11}$$

Finally, we define the complex-valued long scrambling sequence $C_{\text{long},n}$, as

$$C_{\text{long},n}(i) = c_{\text{long},1,n}(i)(1 + j(-1)^i c_{\text{long},2,n}(2\lfloor i/2 \rfloor)) \tag{4.12}$$

where $i = 0, 1, \ldots, 2^{25} - 2$ and $\lfloor\rfloor$ denotes rounding to the nearest lower integer.

4.3.5.2 Short Scrambling Sequence

The short scrambling sequences $c_{\text{short},1,n}(i)$ and $c_{\text{short},2,n}(i)$ originate from a family sequence of periodically extended S(2) codes, where $n_{23}n_{22}\ldots n_0 =$ 24 bit binary representation of the code number n. We obtain the nth quaternary S(2) sequence $z_n(i), 0 \leq n \leq 1\ 677\ 721$ by mod 4 addition of three sequences, a quaternary sequence $a(i)$ and two binary sequences $b(i)$ and $d(i)$, where the initial loading of the three sequences comes from the code number n. The sequence $z_n(i)$ of length 255 results from the following relation:

$$z_n(i) = a(i) + 2b(i) + 2d(i) \bmod 4, \quad i = 0, 1, \ldots, 254 \tag{4.13}$$

where we obtain the quaternary sequence $a(i)$ recursively through the polynomial $g_0(x) = x^8 + x^5 + 3x^3 + x^2 + 2x + 1$ as

$$a(0) = 2n_0 + 1 \text{ mod } 4 \tag{4.14}$$

$$a(i) = 2n_i \text{ mod } 4, \quad i = 1, 2, \ldots, 7 \tag{4.15}$$

$$a(i) = 3a(i-3) + a(i-5) + 3a(i-6) + 2a(i-7) + 3a(i-8) \bmod 4, \quad i = 8, 9, \ldots, 254r \tag{4.16}$$

and the binary sequence $b(i)$ comes also recursively from the polynomial $g_1(x) = x^8 + x^7 + x^5 + x + 1$ as

$$b(i) = n_{8+i} \text{ mod } 2, \quad i = 0, 1, \ldots, 7 \tag{4.17}$$

$$b(i) = b(i-1) + b(i-3) + b(i-7) + b(i-8) \bmod 2, \quad i = 8, 9, \ldots, 254 \tag{4.18}$$

and the binary sequence $d(i)$ is again generated recursively by the polynomial $g_2(x) = x^8 + x^7 + x^5 + x^4 + 1$ as

$$d(i) = n_{16+i} \text{ mod } 2, \quad i = 0, 1, \ldots, 7 \tag{4.19}$$

$$d(i) = d(i-1) + d(i-3) + d(i-4) + d(i-8) \text{ mod } 2, \quad i = 8, 9, \ldots, 254 \tag{4.20}$$

We extend the sequence $z_n(i)$ to length 256 chips by setting $z_n(255) = z_n(0)$.

Table 4.13 defines the mapping from $z_n(i)$ to the real-valued binary sequences $c_{\text{short},1,\text{n}}(i)$ and $c_{\text{short},2,n}(i),\ i = 0, 1, \ldots, 255$.

Table 4.13 Mapping from $z_n(i)$ to $c_{\text{short},1,n}(i)$ and $c_{\text{short},2,n}(i), i = 0, 1, \ldots, 255$

$z_n(i)$	$c_{\text{short},1,n}(i)$	$c_{\text{short},2,n}(i)$
0	+1	+1
1	−1	+1
2	−1	−1
3	+1	−1

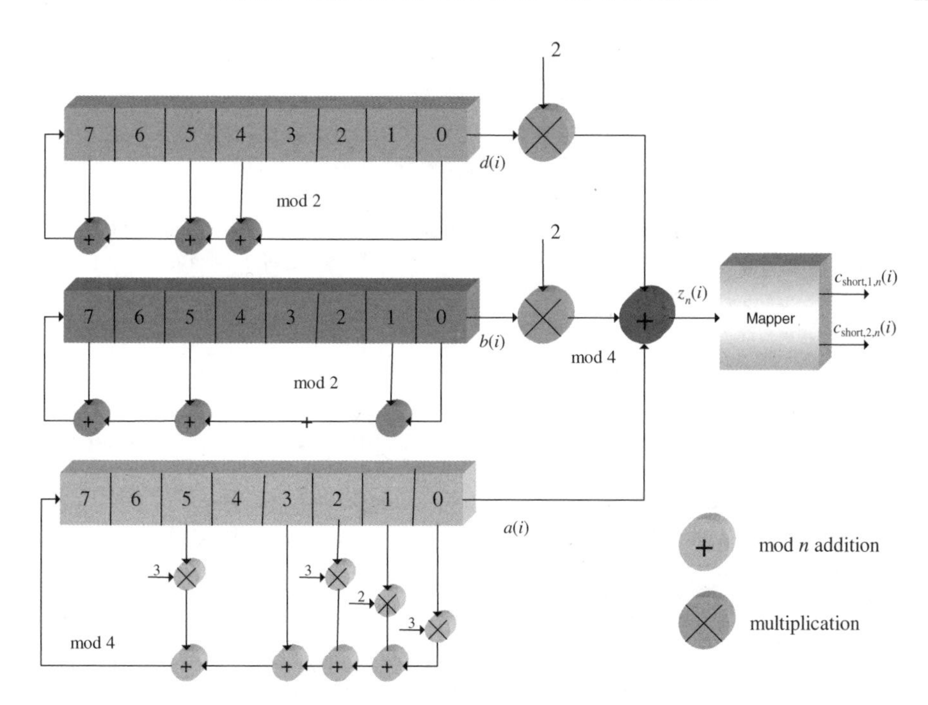

Figure 4.14 255 Chip sequence uplink short scrambling sequence generator.

Finally, we define the complex-valued short scrambling sequence $c_{\text{short},n}$ as

$$C_{\text{short},n}(i) = c_{\text{short},1,n}(i \bmod 256)(1 + j(-1)^{i} c_{\text{short},2,n}(2\lfloor(i \bmod 256)/2\rfloor)) \qquad (4.21)$$

Figure 4.14 illustrates an implementation of the short scrambling sequence generator for the 255 chip sequence extension by one chip.

4.3.5.3 *Scrambling Codes in Uplink Dedicated Physical Channels*

The uplink DPCCH/DPDCH may use either long or short scrambling codes with different constituent codes in each case.

From [4], when using long scrambling codes we define the nth uplink DPCCH/DPDCH scrambling code denoted by $S_{\text{dpch},n}$, as

$$S_{\text{dpch},n}(i) = C_{\text{long},n}(i), \quad i = 0, 1, \ldots, 38\,399 \qquad (4.22)$$

where the lowest index corresponds to the chip transmitted first in time and Section 4.3.5.1 defines $C_{\text{long},n}$. Likewise, when using short scrambling codes we define the nth uplink DPCCH/DPDCH scrambling code denoted by $S_{\text{dpch},n}$, as

$$S_{\text{dpch},n}(i) = C_{\text{short},n}(i), \quad i = 0, 1, \ldots, 38\,399 \qquad (4.23)$$

where the lowest index corresponds to the chip transmitted first in time and Section 4.3.5.2 defines $c_{\text{short},n}$.

4.3.5.4 PRACH and PCPCH Message Part Scrambling Code

The PRACH message part uses 10 ms long scrambling code, and there are 8192 possible PRACH scrambling codes. From [4] we define the nth PRACH message part scrambling code, denoted by $S_{\text{r-msg},n}$, where $n = 0, 1, \ldots, 8191$, based on the long scrambling sequence as

$$S_{\text{r-msg},n}(i) = C_{\text{long},n}(i + 4096), \quad i = 0, 1, \ldots, 38\,399 \tag{4.24}$$

where the lowest index corresponds to the chip transmitted first in time and Section 4.3.5.1 defines $C_{\text{long},n}$.

The message part scrambling code has a one-to-one correspondence to the scrambling code utilised in the preamble part. For one PRACH, we use the same code number in both scrambling codes, i.e. if the PRACH preamble scrambling code uses $S_{\text{r-pre},m}$ then the PRACH message part scrambling code uses $S_{\text{r-msg},m}$, where the number m is the same for both codes [4].

As in PRACH, PCPCH uses 10 ms long scrambling codes in the message part. They are cell specific and each scrambling code has a one-to-one correspondence to the signature sequence and the access sub-channel utilised by the access preamble part. Both long and short scrambling codes may scramble the PCPCH message part. We define up to 64 uplink-scrambling codes per cell and up to 32 768 different PCPCH scrambling codes in the system. For the long scrambling sequence we define the nth PCPCH message part scrambling code $(S_{\text{c-msg},n}, n = 8192, 8193, \ldots, 40\,959)$ as

$$S_{\text{c-msg},n}(i) = C_{\text{long},n}(i), \quad i = 0, 1, \ldots, 38\,399 \tag{4.25}$$

where the lowest index corresponds to the chip transmitted first in time and Section 4.3.5.1 defines $C_{\text{long},n}$. For the short scrambling codes we have

$$S_{\text{c-msg},n}(i) = C_{\text{short},n}(i), \quad i = 0, 1, \ldots, 38\,399 \tag{4.26}$$

A total of 512 groups each containing 64 codes comprise the 32 768 PCPCH scrambling codes. The group of PCPCH preamble scrambling codes in a cell and the primary scrambling code used in the downlink of the cell match one-to-one. $S_{\text{c-msg},n}$ as defined in the preceding paragraphs with $n = 64 \times m + k + 8176$, is the kth PCPCH scrambling code within the cell with downlink primary scrambling code m, where $k = 16, 17, \ldots, 79$ and $m = 0, 1, 2, \ldots, 511$ [4].

4.3.5.5 Scrambling Code in the PCPCH Power Control Preamble

The PCPCH power control preamble uses the same scrambling code as the PCPCH message part (Section 4.3.2.1), where the phase of the scrambling code is such that the end of the code aligns with the frame boundary at the end of the power control preamble.

4.3.5.6 PRACH Preamble Codes

Complex valued sequence constitutes the random-access preamble code $C_{\text{pre},n,}$. It originates from a preamble scrambling code $S_{\text{r-pre},n}$ and a preamble signature $C_{\text{sig},s}$ as

$$C_{\text{pre},n,s}(k) = S_{\text{r-pre},n}(k) \times C_{\text{sig},s}(k) \times \exp\left[j\left(\frac{\pi}{4} + \frac{\pi}{2}k\right)\right], \quad k = 0, 1, 2, 3, \ldots, 4095 \quad (4.27)$$

where $k = 0$ corresponds to the chip transmitted first in time and we define $S_{\text{r-pre},n}$ and $C_{\text{sig},s}$ next. A total of 8192 PRACH preamble part scrambling codes result from the long scrambling sequences. We define the nth preamble scrambling code, $n = 0, 1, \ldots, 8191$, as

$$S_{\text{r-pre},n}(i) = c_{\text{long},1,n}(i), \quad i = 0, 1, \ldots, 4095 \quad (4.28)$$

where Section 4.3.5.1 defines the sequence $c_{\text{long},1,n}$.

As for the PCPCH, we divide the 8192 PRACH preamble scrambling codes in 512 groups with 16 codes in each. And again as in the earlier scrambling codes, we have one-to-one correspondence between the group of PRACH preamble scrambling codes in a cell and the primary scrambling code used in the downlink of the cell. $S_{\text{r-pre},n}(i)$ as defined in eqn (4.28) with $n = 16 \times m + k$, represents the kth PRACH preamble scrambling code within the cell with downlink primary scrambling code m, $k = 0, 1, 2, \ldots, 15$ and $m = 0, 1, 2, \ldots, 511$.

The preamble signature s has 256 repetitions of the signature $P_s(n)$ from the set of 16 Hadamard codes of length 16 (Table 4.14), where $n = 0, \ldots, 15$. The specifications in [4] define it as

$$C_{\text{sig},s}(i) = P_s(i \bmod 16), \quad i = 0, 1, \ldots, 4095 \quad (4.29)$$

Table 4.14 Preamble signatures

Preamble	Value of n															
signature	0	1	2	3	4	5	6	7	8	9	10	11	12	13	14	15
$P_0(n)$	1	−1	−1	−1	−1	−1	−1	−1	−1	−1	−1	−1	−1	−1	−1	−1
$P_1(n)$	−1	−1	−1	−1	−1	−1	−1	−1	−1	−1	−1	−1	−1	−1	−1	−1
$P_2(n)$	−1	−1	−1	−1	−1	−1	−1	−1	−1	−1	−1	−1	−1	−1	−1	−1
$P_3(n)$	−1	−1	−1	−1	−1	−1	−1	−1	−1	−1	−1	−1	−1	−1	−1	−1
$P_4(n)$	−1	−1	−1	−1	−1	−1	−1	−1	−1	−1	−1	−1	−1	−1	−1	−1
$P_5(n)$	−1	−1	−1	−1	−1	−1	−1	−1	−1	−1	−1	−1	−1	−1	−1	−1
$P_6(n)$	−1	−1	−1	−1	−1	−1	−1	−1	−1	−1	−1	−1	−1	−1	−1	−1
$P_7(n)$	−1	−1	−1	−1	−1	−1	−1	−1	−1	−1	−1	−1	−1	−1	−1	−1
$P_8(n)$	−1	−1	−1	−1	−1	−1	−1	−1	−1	−1	−1	−1	−1	−1	−1	−1
$P_9(n)$	−1	−1	−1	−1	−1	−1	−1	−1	−1	−1	−1	−1	−1	−1	−1	−1
$P_{10}(n)$	−1	−1	−1	−1	−1	−1	−1	−1	−1	−1	−1	−1	−1	−1	−1	−1
$P_{11}(n)$	−1	−1	−1	−1	−1	−1	−1	−1	−1	−1	−1	−1	−1	−1	−1	−1
$P_{12}(n)$	−1	−1	−1	−1	−1	−1	−1	−1	−1	−1	−1	−1	−1	−1	−1	−1
$P_{13}(n)$	−1	−1	−1	−1	−1	−1	−1	−1	−1	−1	−1	−1	−1	−1	−1	−1
$P_{14}(n)$	−1	−1	−1	−1	−1	−1	−1	−1	−1	−1	−1	−1	−1	−1	−1	−1
$P_{15}(n)$	−1	−1	−1	−1	−1	−1	−1	−1	−1	−1	−1	−1	−1	−1	−1	−1

4.3.5.7 PCPCH Preamble Codes

Like in PRACH, PCPCH access preamble codes $C_{\text{c-acc},n,s}$ have complex-valued sequences. We define them from the preamble scrambling codes $S_{\text{c-acc},n}$ and a preamble signature $C_{\text{sig},s}$ as

$$C_{\text{c-acc},n,s}(k) = S_{\text{c-acc},n}(k) \times C_{\text{sig},s}(k) \times \exp\left[j\left(\frac{\pi}{4} + \frac{\pi}{2}k\right)\right], \quad k = 0, 1, 2, 3, \ldots, 4095 \quad (4.30)$$

where $S_{\text{c-acc},n}$ and $C_{\text{sig},s}$ are defined in the sequel.

Code generation takes place as in PRACH, resulting in 32 768 PCPCH scrambling codes in total. We define nth PCPCH access preamble scrambling code, where $n = 8192$, $8193, \ldots, 40\,959$, as

$$S_{\text{c-acc},n}(i) = c_{\text{long},1,n}(i), \quad i = 0, 1, \ldots, 4095 \quad (4.31)$$

where the sequence Section 4.3.5.1 defines $c_{\text{long},1,n}$.

When PRACH and PCPCH share access resources, the scrambling codes applied in PRACH preamble apply also to PCPCH preamble; and as in the PRACH part we divide the 32,768 PCPCH preamble scrambling codes into 512 groups with 64 codes in each group. There exists a one-to-one correspondence between the group of PCPCH access preamble scrambling codes in a cell and the primary scrambling code used in the downlink of the cell. The kth PCPCH scrambling code within the cell with downlink primary scrambling code m, $k = 16, 17, \ldots, 79$ and $m = 0, 1, 2, \ldots, 511$, corresponds to $S_{\text{c-acc},n}$ as defined in Section 4.3.5.7 with $n = 64 \times m + k + 8176$.

When PCPCH and PRACH share scrambling code resources and the index k is less than 16, the corresponding PRACH formulae apply. Otherwise, if the index k is greater than or equal to 16, the formula in this section applies. The CPCH-access burst preamble part carries one of the 16 different orthogonal complex signatures identical to the ones used by the preamble part of the random-access burst [4].

4.3.5.7.1 Collision Detection (CD) Preamble

As in PRACH, the PCPCH CD preamble codes $C_{\text{c-cd},n,s}$ have complex-valued sequences. We define these preamble codes from the preamble scrambling codes $S_{\text{c-cd},n}$ and a preamble signature $C_{\text{sig},s}$ as

$$C_{\text{c-cd},n,s}(k) = S_{\text{c-cd},n}(k) \times C_{\text{sig},s}(k) \times \exp\left[j\left(\frac{\pi}{4} + \frac{\pi}{2}k\right)\right], \quad k = 0, 1, 2, 3, \ldots, 4095 \quad (4.32)$$

where we define $S_{\text{c-cd},n}$ in the sequel and $C_{\text{sig},s}$ in Section 4.3.5.6.

The 32 768 PCPCH CD preamble-scrambling code originates from the same scrambling code utilised in the CPCH access preamble. We define the nth PCPCH CD access preamble scrambling code, where $n = 8192, 8193, \ldots, 40\,959$, as

$$S_{c\text{-}cd,n}(i) = c_{\text{long},1,n}(i), \quad i = 0, 1, \ldots, 4095 \quad (4.33)$$

where Section 4.3.5.1 defines the sequence $c_{\text{long},1,n}$.

When RACH and CPCH share scrambling code resources, RACH preamble scrambling codes will also apply to the CPCH CD preamble. As in the cases above, we divide the 32 768 PCPCH scrambling codes into 512 groups with 64 codes each. There exists also a one-to-one correspondence between the group of PCPCH CD preamble scrambling codes in a cell and the primary scrambling code used in the downlink of the cell. The kth PCPCH scrambling code within the cell with downlink primary scrambling code m, $k = 16, 17, \ldots, 79$ and $m = 0, 1, 2, \ldots, 511$, corresponds to $S_{\text{c-cd},n}$ as defined in eqn (4.33) with $n = 64 \times m + k + 8176$.

When PCPCH and PRACH share scrambling code resources and the index k is less than 16, the corresponding PRACH formula applies. Otherwise, when the index k is greater than or equal to 16, the preceding formulae apply. The CD preamble part of the CPCH access burst carries one of 16 different orthogonal complex signatures identical to the ones utilised by the preamble part of the random-access burst [4].

4.3.6 Uplink Power Control Procedure

The FDD mode has unique procedures compared to the TDD. These include fast power control and soft handover procedures. Other procedures are synchronisation and random access.

4.3.6.1 PRACH and DPCCH/DPDCH Power Control

The uplink PRACH message part applies gain factors to manage the control/data part of relative power similar to the uplink dedicated physical channels. Thus, power control steps in the dedicated physical channels apply also to the RACH message part, with the differences that [5]:

- β_c is the gain factor for the control part (similar to DPCCH);
- β_d is the gain factor for the data part (similar to DPDCH);
- no inner or fast loop power control is performed, but open-loop power control.

Before the uplink power control procedure simultaneously controls the power of a DPCCH and its corresponding DPDCHs when present, high layers set the initial uplink DPCCH transmit power. The network determines this relative transmit power offset between DPCCH and DPDCHs using the gain factors signalled to the UE. The inner or fast power control loop operation adjusts the power of the DPCCH and DPDCHs in steps of 1 dB or multiples of one and smaller steps through emulation at 1500 Hz command rate. The DPCCH uplink transmit power takes place immediately before the start of its pilot field. This change occurs with respect to its previous value derived by the UE, i.e. Δ_{DPCCH} (in dB). The previous DPCCH power value corresponds to the one used in the previous slot, except in the event of an interruption in transmission due to the use of compressed mode. In the latter case, the previous value corresponds to the one used in the last slot before the transmission gap. While in power control, the UE transmit power will not exceed a maximum allowed value, i.e. the lowest out of the terminal maximum output power and the one set by higher layer signalling. If the UE transmit power falls below the required minimum output power and the derived value of $\Delta_{DPCCH} < 0$, the UE may reduce the Δ_{DPCCH} magnitude [5].

4.3.6.1.1 The Transmit Power Control Function

The uplink inner-loop power control adjusts the UE transmit power to keep the received uplink Signal-to-Interference Ratio (SIR) at a given target, i.e. SIR_{target}. The serving cells in the active set estimate Signal-to-Interference Ratio (SIR_{est}) of the received uplink DPCH. Then they generate TPC commands and transmit them once per slot according to the following rules: if $SIR_{est} > SIR_{target}$ then the TPC command enables transmission of '0', otherwise if $SIR_{est} < SIR_{target}$ then the TPC command enables transmission of '1'. Upon receipt of one or more TPC commands in a slot, the UE derives a single TPC command, TPC_cmd, for each slot, i.e. it combines multiple TPC commands if more than one is received in a slot. The UTRAN uses two algorithms supported by the UE[2] to realise a TPC_cmd.

Algorithm 1

1. *UE not in soft handover.* If each slot receives only one TPC command, then:
   - if the received TPC_cmd = 0 then TPC_cmd for that slot = −1,
   - if the received TPC_cmd = 1, then TPC_cmd for that slot = 1.
2. *UE is in soft handover.* Each slot may receive multiple TPC commands from different cells in the active set. In receiver diversity (i.e. softer handover), the UTRAN transmits the same command in all the serving cells the UE is in softer handover with, and the TPC commands known to be the same get combined into one TPC command (see more details in [5]).

Algorithm 2[3]

1. *UE is not in soft handover.* Each slot receives only one TPC command and the UE processes received TPC commands on a five-slot cycle. The non-overlapping sets of five slots align to the frame boundaries. The TPC_cmd logic is as follows:
   - The first four slots of a set have TPC_cmd = 0.
   - In the fifth slot of a set, the UE uses hard decisions on each of the five received TPC commands as follows [5]:
     If all five hard decisions within a set are 1 then TPC_cmd = 1 in the fifth slot.
     If all five hard decisions within a set are 0 then TPC_cmd = −1 in the fifth slot.
     Otherwise, TPC_cmd = 0 in the fifth slot.
2. *UE is in soft handover.* Each slot receives multiple TPC commands from different cells in the active set. When UE is in soft handover, then each slot may receive multiple TPC

[2]The step size Δ_{TPC} is a UE specific parameter, under UTRAN control, which can have values 1 or 2 dB.

[3]This allows emulation of smaller step sizes than the minimum power control step or to turn off uplink power.

commands from different cells in the active set. In receiver diversity (i.e. softer handover), the UTRAN transmits the same command in all the serving cells the UE is in softer handover, and the TPC commands known to be the same get combined into one TPC command (see more details in [5]).

After deriving the combined TPC command TPC_cmd using one of the two supported algorithms, the UE adjusts the transmit power of the uplink DPCCH with a step of Δ_{DPCCH} (in dB), which is given by $\Delta_{DPCCH} = \Delta_{TPC} \times$ TPC_cmd.

Out of sync Handling. The UE shuts its transmitter off when the UE estimates the DPCCH quality over the last 200 ms period to be worse than a threshold Q_{out}. This criterion never occurs during the first 200 ms of the dedicated channel's existence. The UE can turn its transmitter on when the UE estimates the DPCCH quality over the last 200 ms period to be better than a threshold Q_{in}. This criterion always occurs during the first 200 ms of the dedicated channel's existence. At the transmission resumption the power of the DPCCH shall remain the same as when the UE transmitter went off [6].

4.3.6.2 Compressed Mode Power Control

The compressed mode, which has compressed frames containing transmission gaps, uses the same transmit power control function outlined in the preceding section, but with additional features aiming for fastest recovery of the Signal-to-Interference Ratio (SIR) close to the target SIR after each transmission gap. In this mode, compressed frames may exist either in the uplink or the downlink or both. In the first case, the DPDCH(s) and DPCCH uplink transmissions stop during the gaps. In the second case, if the gaps cause the absence of downlink TPC commands, the corresponding TPC_cmd derived by the UE goes to zero.

A transmit power change of the uplink DPCCH compensates the variation in the total pilot energy in both compressed and non-compressed frames in the uplink DPCCH due to the different number of pilot bits per slot. Thus, at the start of each slot the UE derives a power offset Δ_{pilot} value. The compensation uses the value in the most recently transmitted slot; Δ_{pilot} (in dB) follows:

$$\Delta_{PILOT} = 10 \log_{10}(N_{pilot,prev}/N_{pilot,curr}) \tag{4.34}$$

where $N_{pilot,prev}$ is the number of pilot bits in the most recently transmitted slot, and $N_{pilot,curr}$ is the number of pilot bits in the current slot. If no compensation takes place during transmission gaps in the downlink, Δ_{pilot} = zero. Furthermore, during compressed mode the UE will adjust the transmit power of the uplink DPCCH with a step of Δ_{DPCCH} (in dB) as follows:

$$\Delta_{DPCHH} = \Delta_{TPC} \times \text{TPC\_cmd} + \Delta_{PILOT} \tag{4.35}$$

The latter may not occur if otherwise specified. After an uplink transmission gap, the UE applies a change in the transmit power of the uplink DPCCH by an amount Δ_{DPCCH} (in dB) at the beginning of the first slot, with respect to the uplink DPCCH power in the most recently transmitted uplink slot, where

$$\Delta_{DPCCH} = \Delta_{resume} + \Delta_{pilot} \tag{4.36}$$

Table 4.15 Initial Transmit Power (ITP) and Recovery Period Power (RPP) control modes

ITP	*Description*
0	$\Delta_{resume} = \Delta_{TPC} \times TPC\_cmd_{gap}$
1	$\Delta_{resume} = \delta_{last}$
RPP	
0	Transmit power control applies using the algorithm determined by the value of PCA with step size Δ_{TPC}
1	Transmit power control applies using algorithm 1 with step size Δ_{RP-TPC} during RPL slots after each transmission gap

The UE determines the Δ_{resume} value (in dB) according to the Initial Transmit Power (ITP) mode, which is a UE specific parameter signalled by the network with other compressed mode parameters. Table 4.15 summarises the ITP mode. If a downlink TPC command is transmitted in the first slot of a downlink transmission gap, then $\delta_{last} = \delta_i$ is computed in the first slot of the downlink transmission gap. Otherwise $\delta_{last} = \delta_i$ is computed in the last slot before the downlink transmission gap. δ_i will be updated according to the following recursive relation:

$$\begin{aligned} \delta_i &= 0.9375\,\delta_{i-1} - 0.96875\,TPC\_cmd_i \Delta_{TPC} \\ \delta_{i-1} &= \delta_i \end{aligned} \tag{4.37}$$

δ_{i-1} is the value of δ_i computed for the previous slot. $\delta_{i-1} = 0$ when we activate the uplink DPCCH, and also at the end of the first slot after each downlink transmission gap. This relation gets executed in all slots with simultaneous uplink and downlink DPCCH transmission, and in the first slot of a downlink transmission gap if a downlink TPC command is transmitted in that slot. TPC_cmd_i is the most recent power control command derived by the UE [5].

After a transmission gap in either the uplink or the downlink, there exists a recovery period following resumption of simultaneous uplink and downlink DPCCH transmission. This period has a Recovery Period Length (RPL) and is expressed as the number of slots. The RPL is equal to the minimum value out of the transmission gap length of 7 slots. Table 4.15 illustrates the two recovery period modes for the power control algorithm, where the Recovery Period Power (RPP) control mode is signalled with the other compressed mode parameters [7]. For RPP mode 0, the step size does not change during the recovery period and ordinary transmit power control applies, i.e. the algorithm for processing TPC commands determined by the value of PCA. For RPP mode 1, during RPL slots after each transmission gap, power control algorithm 1 applies with a step size $\Delta_{RP\text{-}TPC}$ instead of Δ_{TPC}, regardless of the value of PCA. We define the uplink DPCCH transmit power change, which does not apply to the first slot after the transmission gap, as

$$\Delta_{DPCCH} = \Delta_{RP\text{-}TPC} \times TPC\_cmd + \Delta_{PILOT} \tag{4.38}$$

where $\Delta_{RP\text{-}TPC}$ denotes the recovery power control step size and is expressed in dB. When the PCA has the value $1, \ldots, \Delta_{RP\text{-}TPC}$, it is equal to the minimum value of 3 dB and 2 Δ_{TPC}, and when the PCA has the value $2, \ldots, \Delta_{RP\text{-}TPC}$, it is equal to 1 dB. After the recovery period, normal transmit power control function resumes using the algorithm specified by the value of PCA and with step size Δ_{TPC}. When the PCA has 2 as value, slot sets over which the TPC commands are processed remain aligned to the frame boundaries in the compressed frame. For both RPP modes 0 and 1, if the transmission gap or the recovery period results in any incomplete set of TPC commands, TPC_cmd = zero for those incomplete slots sets [5].

4.3.6.3 DPCCH Power-Control Preamble

DCHs can use power-control preamble for initialisation, and both UL and DL DPCCHs get transmitted during the uplink power-control preamble. However, the UL DPDCH does not start before the end of the power-control preamble. The network signals the power-control preamble length as a UE-specific parameter, where values can take 0 or 8 slots. When the preamble length is greater than zero, power-control details used during the power-control preamble differ from the ordinary power-control used afterwards. The uplink DPCCH change transmit power after the first slot of the power-control preamble can be defined as

$$\Delta_{DPCCH} = \Delta_{TPC\text{-}init} \times \text{TPC\_cmd} \tag{4.39}$$

When PCA = 1, then $\Delta_{TPC\text{-}init}$ is equal to the minimum value out of 3 dB and 2 Δ_{TPC}; and when PCA = 2, then $\Delta_{TPC\text{-}init} = 2$ dB. TPC_cmd is derived according to algorithm 1 regardless of the value of PCA. Normal transmit power-control with the power-control algorithm determined by the value of PCA and step size $\Delta_{TPC,}$ applies as soon as the sign of TPC_cmd reverses for the first time, or at the end of the power-control preamble if the power control preamble ends first. The specifications in [5] describe the setting of the uplink DPCCH/DPDCH power difference.

4.3.6.4 Power Control in the PCPCH Message Part and Preamble

Message part. The uplink inner or fast loop power control adjusts the UE transmit power to keep the received uplink SIR at a given SIR_{target} set by upper layer outer loop.

The network estimates the SIR_{est} of the received PCPCH, then it generates TPC commands and transmits the commands once per slot according to the following rule:

- if $\text{SIR}_{est} > \text{SIR}_{target}$ then the TPC command to transmit is 0
- if $\text{SIR}_{est} < \text{SIR}_{target}$ then the TPC command to transmit is 1

The UE derives a TPC_cmd for each slot. The UE will support the UTRAN controlled algorithms 1 and 2 to derive a TPC_cmd. These come in step size Δ_{TPC} and can have values of 1 or 2 dB. After deriving the TPC command TPC_cmd using one of the two supported algorithms, the UE adjusts the transmit power of the uplink PCPCH in steps of Δ_{TPC} dB according to the TPC command. If TPC_cmd = 1, then the transmit power of the uplink

PCPCH increases by Δ_{TPC} dB. If TPC_cmd = −1, then the transmit power of the uplink PCPCH decreases by Δ_{TPC} dB. If TPC_cmd = 0, then the transmit power of the uplink PCPCH remains unchanged. Any power increase or decrease takes place immediately before the start of the pilot field on the PCPCH control channel [5].

4.3.6.5 Power Control in the Power-Control Preamble Part

The UE begins the power-control preamble using the same power level applied for the CD preamble. The initial power-control step size utilised in the power control preamble differs from the one applied in the message part as follows:

- when inner loop power control algorithm 1 applies to the message part, then the initial step size in the power-control preamble = $\Delta_{TPC\text{-init}}$, where $\Delta_{TPC\text{-init}}$ is the minimum value out of 3 dB and $2\,\Delta_{TPC}$, where Δ_{TPC} is the power-control step size used for the message part.
- When inner loop power control algorithm 2 applies to the message part, then we use initially the inner loop power-control algorithm 1 in the power-control preamble, with a step size of 2 dB.
- In either one of the cases, the power-control algorithm and step size revert to the ones used for the message part as soon as the sign of the TPC commands reverses for the first time [5].

4.3.7 Downlink Physical Channels

4.3.7.1 Downlink Transmit Diversity

Table 4.16 outlines possible applications of open and closed loop transmit diversity modes on different types of downlink physical channel. Simultaneous use of STTD and closed loop

Table 4.16 Application of Tx diversity modes on downlink physical channel types

	Open loop mode[a]		Closed loop[a]
Physical channel type	TSTD	STTD	Mode
P-CCPCH	×	✓	×
SCH	✓	×	×
S-CCPCH	×	✓	×
DPCH	×	✓	✓
PICH	×	✓	×
PDSCH	×	✓	✓
AICH	×	✓	×
CSICH	×	✓	×

[a] ✓, may apply; ×, does not apply.

modes on the same physical channel is not possible. Furthermore, when Tx diversity applies to any of the downlink physical channels, it also applies to the P-CCPCH and SCH.

In addition, the PDSCH and the DPCH associated with this PDSCH shall use the same transmit diversity mode. A transmit diversity mode (open loop or closed loop) on the associated DPCH may not change during the duration of the PDSCH frame, and within the slot prior to the PDSCH frame. Nevertheless, changing from closed loop mode 1 to mode 2 or vice versa, is possible.

4.3.7.2 Open-Loop Transmit Diversity

- *Space Time Block Coding Based Transmit Antenna Diversity (STTD)* employs a space time block coding based transmit diversity. It is optional in UTRAN but mandatory in the UE. STTD encoding works on blocks of four consecutive channel bits. Figure 4.15 illustrates a block diagram of a generic STTD encoder for channel bits b_0, b_1, b_2 and b_3. Channel coding, rate matching and inter-leaving occurs as in the non-diversity mode. The bit b_i has real valued $\{0\}$ for DTX bits and $\{1, -1\}$ for all other channel bits.
- *Time Switched Transmit Diversity for SCH (TSTD)* can apply to the SCH; like STTD, it is optional in UTRAN but mandatory in the UE.

4.3.7.3 Closed Loop Transmit Diversity

Closed loop transmit diversity is described in [5] (Figure 4.15).

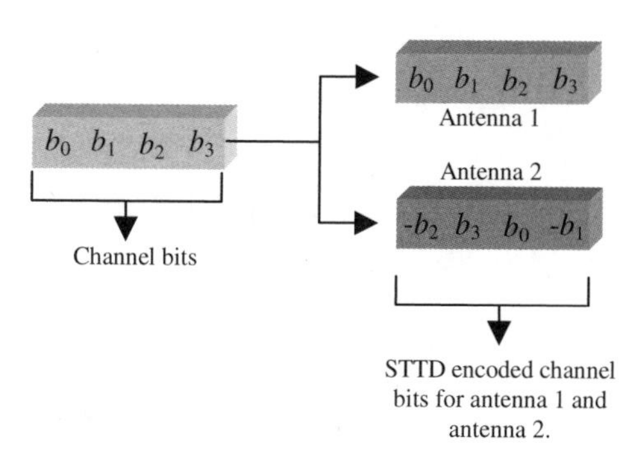

Figure 4.15 The STTD encoder–block diagram example.

4.3.8 Dedicated Downlink Physical Channels

The only downlink Dedicated Physical Channel (downlink DPCH) transmits dedicated data generated at layer 2 and above, i.e. the Dedicated transport Channel (DCH), in time multiplex with control information generated at layer 1 (known as pilot bits, TPC commands and an optional TFCI). The downlink DPCH is therefore a time multiplex of a downlink DPDCH and a downlink DPCCH.

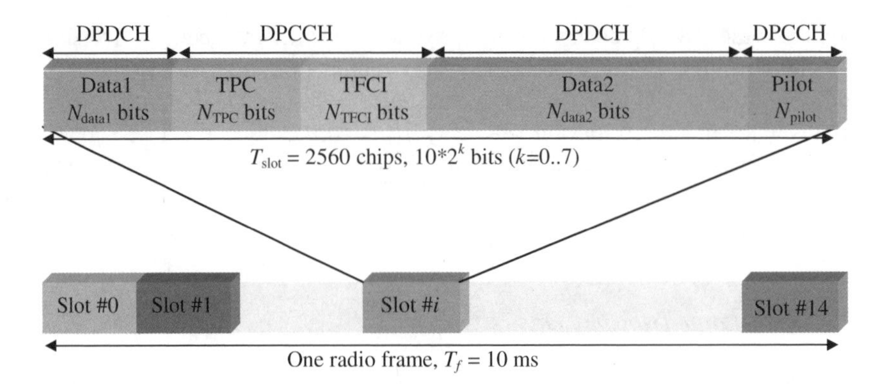

Figure 4.16 Downlink DPCH frame structure.

Figure 4.16 illustrates the downlink DPCH frame structure, where each frame has 10 ms length split into 15 slots with $T_{slot} = 2560$ chip length, corresponding to one power-control period.

As in earlier frame structures, the parameter k in Figure 4.16 determines the total number of bits per downlink DPCH slot. It is related to the spreading factor SF of the physical channel as $\text{SF} = 512/2k$. The spreading factor may thus range from 512 down to 4.

Appendix A defines the exact number of bits of the different downlink DPCH fields, i.e. N_{pilot}, N_{TPC}, N_{TFCI}, $N_{data\,1}$ and $N_{data\,2}$. Upper layers configure and reconfigure the slot format usage.

The two basic types of downlink DPCH reflected in Appendix A are the ones that contain TFCI (e.g. various simultaneous services) and those that do not contain TFCI (e.g. fixed-rate services). While support of downlink TFCI inclusion in the network may be optional, it is mandatory in all UEs. The UTRAN determines if a TFCI should be transmitted or not.

4.3.8.1 The Compressed Mode

There are two compressed slot formats, i.e. A and B. Format B is possible by the spreading factor reduction and format A by all other transmission time reduction methods.

Table 4.17 shows the DPCCH pilot bit patterns (order left to right), where shadowed columns define the Frame Synchronisation Word (FSW) part, which can be used to confirm frame synchronisation. All other non-FSW pilot bit pattern columns contain '11' for each slot.

The downlink compressed mode through 'spreading factor reduction' has double the number of bits in the TPC and pilot fields, where symbol repetition fills up the fields when necessary. In the normal mode we denote the bits in one of these fields by $x_1, x_2, x_3, \ldots, x_X$, while in the compressed mode we denote the corresponding field as $x_1, x_2, x_1, x_2, x_3, x_4, x_3, x_4, \ldots, x_X$.

- For the other slot formats, symbol repetition shall be applied to the pilot bit pattern with the half size.

Table 4.17 Pilot bit patterns for downlink DPCCH with $N_{pilot} = 2, 4, 8$ and 16

Slot	$N_{pilot} = 2$	$N_{pilot} = 4^a$		$N_{pilot} = 8^b$				$N_{pilot} = 16^c$							
Symbol[b]	0	0	1	0	1	2	3	0	1	2	3	4	5	6	7
0	11	11	11	11	11	11	10	11	11	11	10	11	11	11	10
1	00	11	00	11	00	11	10	11	00	11	10	11	11	11	00
2	01	11	01	11	01	11	01	11	01	11	01	11	10	11	00
3	00	11	00	11	00	11	00	11	00	11	00	11	01	11	10
4	10	11	10	11	10	11	01	11	10	11	01	11	11	11	11
5	11	11	11	11	11	11	10	11	11	11	10	11	01	11	01
6	11	11	11	11	11	11	00	11	11	11	00	11	10	11	11
7	10	11	10	11	10	11	00	11	10	11	00	11	10	11	00
8	01	11	01	11	01	11	10	11	01	11	10	11	00	11	11
9	11	11	11	11	11	11	11	11	11	11	11	11	00	11	11
10	01	11	01	11	01	11	01	11	01	11	01	11	11	11	10
11	10	11	10	11	10	11	11	11	10	11	11	11	00	11	10
12	10	11	10	11	10	11	00	11	10	11	00	11	01	11	01
13	00	11	00	11	00	11	11	11	00	11	11	11	00	11	00
14	00	11	00	11	00	11	11	11	00	11	11	11	10	11	01

[a]Pattern does not apply to slot formats 2B and 3B.
[b]Pattern does not apply to slot formats 0B, 1B, 4B, 5B, 8B and 9B.
[c]Pattern does not apply to slot formats 6B, 7B, 10B, 11B, 12B and 13B.

- In compressed mode through spreading factor reduction, symbol repetition is applied to the symbol patterns described in Table 4.12.

Table 4.18 illustrates the relationship between the TPC symbol and the transmitter power control command.

4.3.8.2 *Multi-Code Transmission*

Multi-code transmission in the downlink, i.e. the Coded Composite Transport Channel (CCTrCH) is mapped onto several parallel downlink DPCHs using the same spreading factor. As illustrated in Figure 4.17, the layer 1 control information is transmitted only on the first downlink DPCH. DTX bits are transmitted during the corresponding time period for the additional downlink DPCHs.

When there are several CCTrCHs mapped to different DPCHs transmitted to the same UE, different spreading factors can be used on the DPCHs. However, even in this case, layer 1

Table 4.18 TPC bit pattern

TPC bit pattern			Transmitter power control command
$N_{TPC} = 2$	$N_{TPC} = 4$	$N_{TPC} = 8$	
11	1111	11111111	1
00	0000	00000000	0

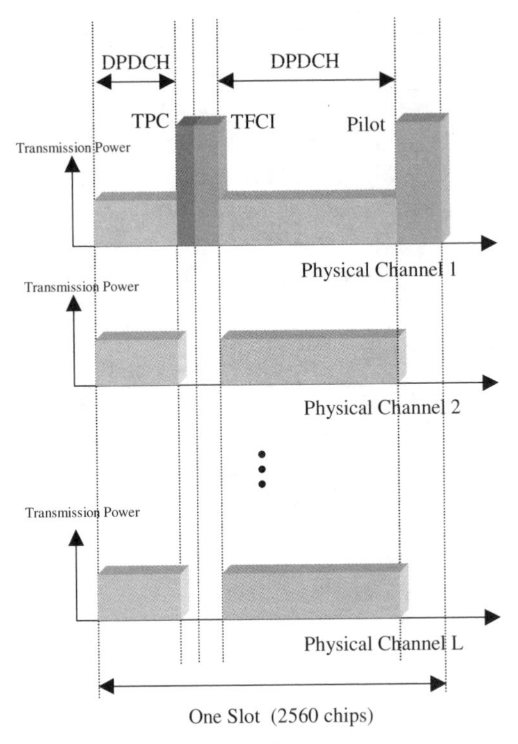

Figure 4.17 Downlink slot format in case of multi-code transmission.

control information is transmitted only on the first DPCH, while DTX bits are transmitted during the corresponding time period on the additional DPCHs.

4.3.8.3 STTD in the DPCH

In the following we describe how antenna diversity occurs using the pilot bit pattern for the DPCH channel transmitted on antenna 2 illustrated in Table 4.19:

- $N_{pilot} = 2$ diversity antenna pilot pattern results from STTD encoding the two pilot bits defined in Table 4.17 with the last two bits (data or DTX) of the second data field (data 2) of the slot. Hence, for $N_{pilot} = 2$ the last two bits of the second data field (data 2) after STTD encoding follow the diversity antenna pilot bits defined in Table 4.19.
- $N_{pilot} = 4$ diversity antenna pilot bit pattern results from STTD encoding both the shadowed and non-shadowed pilot bits in Table 4.17.
- $N_{pilot} = 8, 16$ diversity antenna pilot bit pattern shown in Table 4.19 results from STTD encoding the corresponding (shadowed) bits in Table 4.17. The non-shadowed pilot bit pattern is orthogonal to the corresponding (non-shadowed) pilot bit pattern in Table 4.17.

Table 4.19 Pilot Bit Patterns of Downlink DPCCH for Antenna 2 using STTD

	N_{pilot} =2[a]	$N_{pilot}=4$[b]			$N_{pilot}=8$[c]				$N_{pilot}=16$[d]							
Sym	0	0	1	1[e]	0	1	2	3	0	1	2	3	4	5	6	7
Slot#0	01	01	10	10	11	00	00	10	11	00	00	10	11	00	00	10
1	10	10	10	01	11	00	00	01	11	00	00	01	11	10	00	10
2	11	11	10	00	11	11	00	00	11	11	00	00	11	10	00	11
3	10	10	10	01	11	10	00	01	11	10	00	01	11	00	00	00
4	00	00	10	11	11	11	00	11	11	11	00	11	11	01	00	10
5	01	01	10	10	11	00	00	10	11	00	00	10	11	11	00	00
6	01	01	10	10	11	10	00	10	11	10	00	10	11	01	00	11
7	00	00	10	11	11	10	00	11	11	10	00	11	11	10	00	11
8	11	11	10	00	11	00	00	00	11	00	00	00	11	01	00	01
9	01	01	10	10	11	01	00	10	11	01	00	10	11	01	00	01
10	11	11	10	00	11	11	00	00	11	11	00	00	11	00	00	10
11	00	00	10	11	11	01	00	11	11	01	00	11	11	00	00	01
12	00	00	10	11	11	10	00	11	11	10	00	11	11	11	00	00
13	10	10	10	01	11	01	00	01	11	01	00	01	11	10	00	01
14	10	10	10	01	11	01	00	01	11	01	00	01	11	11	00	11

[a] The pilot bits precede the last two bits of the data 2 field.
[b] Bit pattern does not apply to compressed slot formats 2B and 3B.
[c] Bit pattern does not apply to compressed slot formats 0B, 1B, 4B, 5B, 8B, and 9B.
[d] Bit pattern does not apply to compressed slot formats 6B, 7B, 10B, 11B, 12B, and 13B.
[e] Bit pattern applies to compressed slot formats 2B and 3B.
[1] In other slot formats we apply symbol repetition to the pilot bit pattern with the half size.
[2] Appendix B illustrates the bit pattern for compressed mode with spread reduction and $N_{pilot}=4$.

- Pilot bit patterns $N_{pilot} > 4$ in compressed mode with the spreading factor reduction method will get symbol repetition for the pilot bit patterns of Table 4.19.

STTD encoding for the DPDCH, TPC and TFCI fields follows the definition in Section 4.3.7.2. For DPCH with SF = 512 the first two bits in each slot, i.e. TPC bits, are not STTD encoded. These bits are transmitted with equal power from the two antennas; however, the remaining four bits are STTD encoded.

4.3.8.4 Closed-Loop Mode Transmit Diversity and Dedicated Channel Pilots

Closed-loop mode 1 uses orthogonal pilot patterns between the transmit antennas. Table 4.17 defines the pilot patterns used on antenna 1 and Table 4.19 defines pilot patterns used on antenna 2. Figure 4.18 illustrates the two antenna slot structures with the pilot pattern bits shaded in grey.

In closed loop mode 2, the same pilot pattern is used on both antennas with bit pattern defined in Table 4.17.

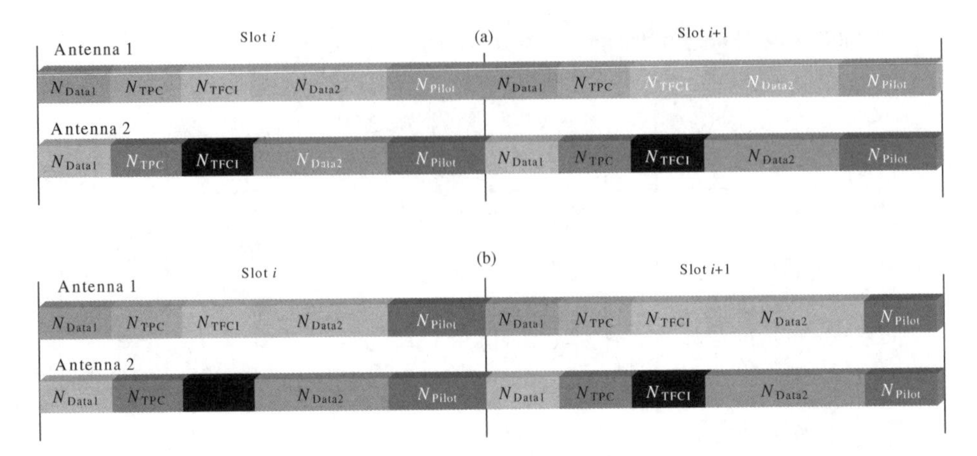

Figure 4.18 Slot structures for downlink dedicated physical channel diversity transmission. Closed-loop mode 1 uses structure (a), closed-loop mode 2 uses structure (b) and the grey shading indicates the bit pattern orthogonality.

4.3.8.5 *The Downlink DPCCH for CPCH*

Downlink DPCCH for CPCH is a special case of downlink dedicated physical channel of 'slot format 0' illustrated in the table of Appendix A, where the spreading factor for the DL-DPCCH is 512. Figure 4.19 illustrates the CPCH downlink DPCCH frame structure.

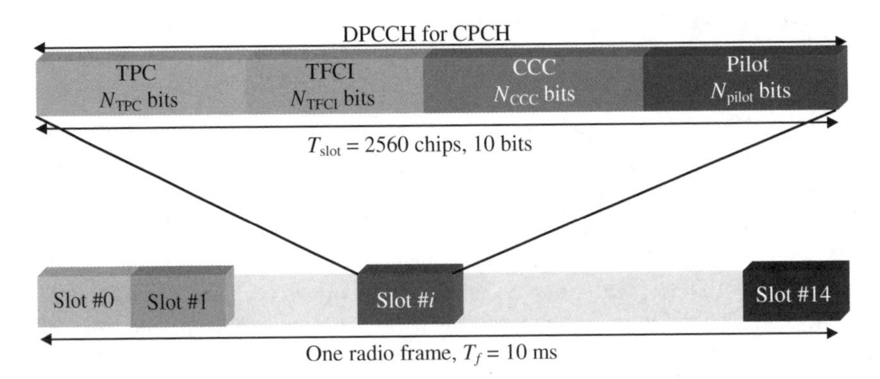

Figure 4.19 Frame structure for downlink DPCCH for CPCH.

The CPCH downlink DPCCH incorporates known pilot bits, TFCI, TPC commands and CPCH Control Commands (CCC). CPCH control commands support CPCH signalling. These commands include layer 1 control command such as start of message indicator, and higher layer control command such as emergency stop command. The exact number of bits of DL DPCCH fields (N_{pilot}, N_{TFCI}, N_{CCC} and N_{TPC}) is determined in Table 4.20. Table 4.17 defines the pilot bit pattern for $N_{pilot} = 4$ used for DPCCH for CPCH.

Table 4.20 DPCCH fields for CPCH message transmission

Slot format (i)	Channel bit rate (kbps)	Channel symbol rate (ksps)	SF	Bits/slot	DPCCH bits/slot				Transmitted slots/frame
					N_{TPC}	N_{TFCI}	N_{CCC}	N_{pilot}	N_{Tr}
0	15	7.5	512	10	2	0	4	4	15

The transmission of the CPCH control command uses the CCC field in Figure 4.19 upon request from higher layers, where a given pattern is mapped onto the CCC field. If no requests exist, nothing is transmitted in the CCC field; thus, there is one-to-one mapping between the CPCH control command and the pattern. 'CPCH emergency stop transmission' maps the [1111] pattern onto the CCC field. This stop command cannot be transmitted during the first $N_{Start\_Message}$ frames of DL DPCCH after power control preamble. The CCC field gets the [1010] pattern for the start of message indicator during the first $N_{Start\_Message}$ frames of DL DPCCH after power control preamble.

4.3.9 Common Downlink Physical Channels

4.3.9.1 Common Pilot Channel (CPICH)

Figure 4.20 illustrates the frame structure of the CPICH, which is a fixed rate (30 kbps, SF = 256) downlink physical channel that carries a pre-defined bit/symbol sequence.

During *transmit diversity* on any downlink cell channel, either with open or closed-loop power control, the CPICH shall be transmitted from both antennas using the same channelisation and scrambling code (in this case, as illustrated in Figure 4.21).

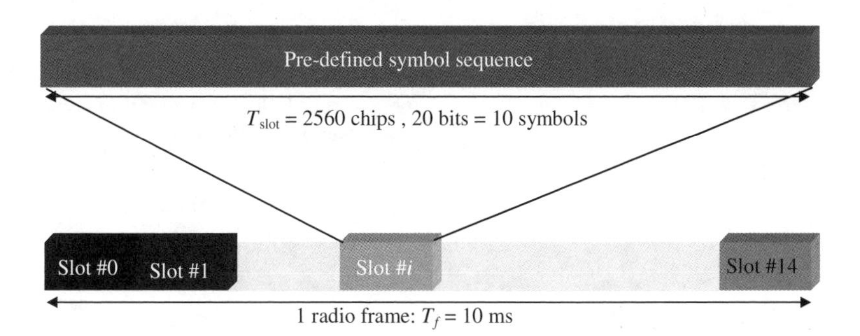

Figure 4.20 Frame structure for the common pilot channel.

Antennas 1 and 2 do have different CPICH pre-defined symbol sequences. In the absence of *transmit diversity*, the symbol sequence of antenna 1 applies.

In the sequel we describe the two types of common pilot channels, i.e. primary and secondary CPICHs, which differ in their use and the limitations placed on their physical features.

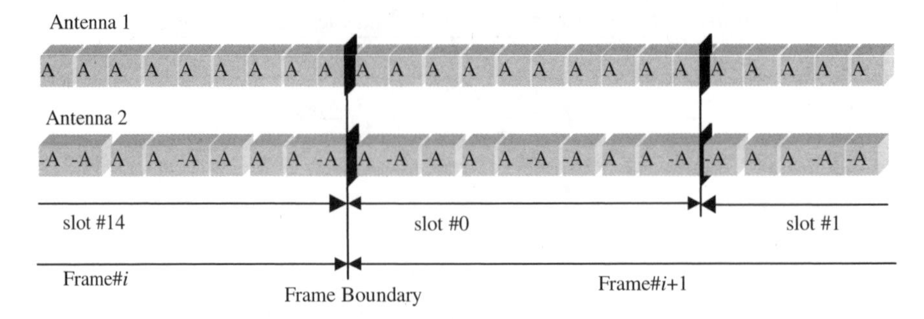

Figure 4.21 Common pilot channel (with $A = 1 + j$) modulation pattern.

4.3.9.1.1 Primary and Secondary Common Pilot Channels (P-CPICH and S-CPICH)
Table 4.21 illustrates the main characteristics of the CPICHs. The coding and spreading section presents more details on scrambling issues.

4.3.9.2 Primary Common Control Physical Channel (P-CCPCH)

The primary CCPCH, as a downlink physical channel with a fixed rate (30 kbps, SF = 256), carries the BCH transport channel. Figure 4.22 shows the frame structure of the primary

Table 4.21 Characteristics of primary and secondary CPICHs

Primary CPICH characteristics	Secondary CPICH characteristics
The P-CPICH uses always the same channelisation code	The S-CPICH uses an arbitrary channelisation code of SF = 256
The primary scrambling code scrambles the P-CPICH	Either the primary or a secondary scrambling code scrambles a S-CPICH
Each cell has only one P-CPICH	A cell may contain zero, one or several S-CPICHsl
The P-CPICH is broadcast over the entire cell	A S-CPICH may be transmitted over the entire cell or only over a part of the cell
P-CPICH serves as phase reference for DL: SCH, primary CCPCH, AICH and PICH	A secondary CPICH may be the reference for secondary CCPCH and DL-DPCH
It is also the *default* phase reference for all downlink physical channels	Upper layers inform the UE when a secondary CPICH is used as reference

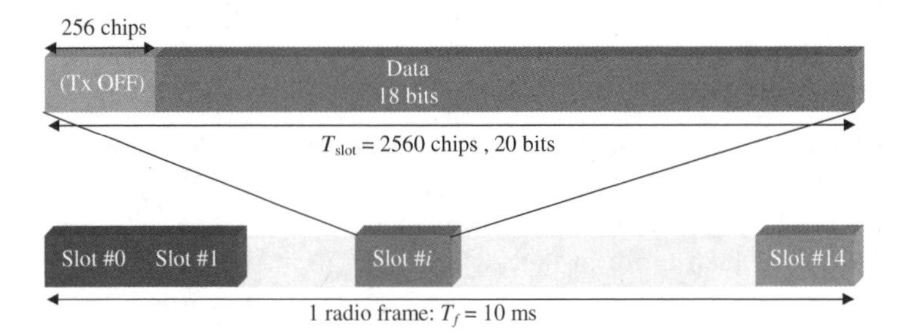

Figure 4.22 Primary CCPCH frame structure.

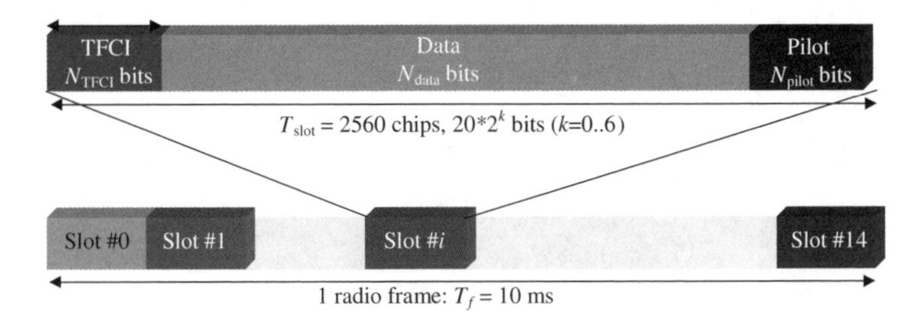

Figure 4.23 Frame structure for the secondary common control physical channel.

CCPCH, which differs from the downlink DPCH in that it does not transmit TPC commands, TFCI or pilot bits. In addition, it is not transmitted during the first 256 chips of each slot. The technical specification in [8] describes the primary CCPCH structure with STTD encoding.

4.3.9.3 Secondary Common Control Physical Channel (S-CCPCH)

The two CCPCH types, i.e. those that include TFCI and those that do not, carry the FACH and PCH. Since it is the UTRAN which determines when a TFCI shall be transmitted, it is mandatory that all UEs support the use of TFCI. Possible secondary CCPCH rates are the same as for the downlink DPCH. Figure 4.23 illustrates a secondary CCPCH frame structure, where the parameter k determines the total number of bits per downlink secondary CCPCH slot. It is related to the spreading factor $\mathrm{SF}(= 256/2^k)$ of the physical channel with spreading range of 256 down to 4.

Table 4.22 presents the number of bits per field, as well as the channel bit and symbol rates before spreading for the secondary CCPCH. Table 4.23 illustrates the pilot patterns.

We can map the FACH and PCH to the same or to separate secondary CCPCHs. If the first case occurs, one frame can serve both. Key characteristics and differences are:

- a CCPCH does not have inner-loop power control as a downlink dedicated physical channel does,
- while a transport channel mapped to the primary CCPCH (BCH) can support only a fixed pre-defined transport format combination, a secondary CCPCH can support multiple transport format combinations using TFCI,
- we can transmit a primary CCPCH over the entire cell,
- a secondary CCPCH may be transmitted in a narrow lobe in the same way as a dedicated physical channel, only when carrying the FACH.

We can use the shadowed part of the pilot symbol pattern illustrated in Table 4.23 as frame synchronisation words. Symbol patterns other than the frame synchronisation word contain ‘11’. The transmission order of the two-bit pair representing an *I*/*Q* pair of QPSK modulation illustrated in Table 4.23 goes from left to right.

Table 4.22 Secondary-CCPCH fields

Slot format (i)	Channel bit rate (kbps)	Channel symbol rate (ksps)	SF	Bits/frame	Bits/slot	N_{data}	N_{pilot}	N_{TFCI}
0	30	15	256	300	20	20	0	0
1	30	15	256	300	20	12	8	0
2	30	15	256	300	20	18	0	2
3	30	15	256	300	20	10	8	2
4	60	30	128	600	40	40	0	0
5	60	30	128	600	40	32	8	0
6	60	30	128	600	40	38	0	2
7	60	30	128	600	40	30	8	2
8	120	60	64	1200	80	72	0	8[a]
9	120	60	64	1200	80	64	8	8[a]
10	240	120	32	2400	160	152	0	8[a]
11	240	120	32	2400	160	144	8	8[a]
12	480	240	16	4800	320	312	0	8[a]
13	480	240	16	4800	320	296	16	8[a]
14	960	480	8	9600	640	632	0	8[a]
15	960	480	8	9600	640	616	16	8[a]
16	1920	960	4	19 200	1280	1272	0	8[a]
17	1920	960	4	19 200	1280	1256	16	8[a]

[a]If TFCI bits are not used, then DTX shall be used in TFCI field.

Table 4.23 Pilot symbol pattern

Slot	$N_{pilot} = 8$				$N_{pilot} = 16$							
Symbol	0	1	2	3	0	1	2	3	4	5	6	7
0	11	11	11	10	11	11	11	10	11	11	11	10
1	11	00	11	10	11	00	11	10	11	11	11	00
2	11	01	11	01	11	01	11	01	11	10	11	00
3	11	00	11	00	11	00	11	00	11	01	11	10
4	11	10	11	01	11	10	11	01	11	11	11	11
5	11	11	11	10	11	11	11	10	11	01	11	01
6	11	11	11	00	11	11	11	00	11	10	11	11
7	11	10	11	00	11	10	11	00	11	10	11	00
8	11	01	11	10	11	01	11	10	11	00	11	11
9	11	11	11	11	11	11	11	11	11	00	11	11
10	11	01	11	01	11	01	11	01	11	11	11	10
11	11	10	11	11	11	10	11	11	11	00	11	10
12	11	10	11	00	11	10	11	00	11	01	11	01
13	11	00	11	11	11	00	11	11	11	00	11	00
14	11	00	11	11	11	00	11	11	11	10	11	01

Table 4.24 Antenna 2 pilot symbol pattern for STTD encoding in the S-CCPCH

Slot	$N_{pilot} = 8$				$N_{pilot} = 16$							
Symbol	0	1	2	3	0	1	2	3	4	5	6	7
0	11	00	00	10	11	00	00	10	11	00	00	10
1	11	00	00	01	11	00	00	01	11	10	00	10
2	11	11	00	00	11	11	00	00	11	10	00	11
3	11	10	00	01	11	10	00	01	11	00	00	00
4	11	11	00	11	11	11	00	11	11	01	00	10
5	11	00	00	10	11	00	00	10	11	11	00	00
6	11	10	00	10	11	10	00	10	11	01	00	11
7	11	10	00	11	11	10	00	11	11	10	00	11
8	11	00	00	00	11	11	00	00	11	01	00	01
9	11	01	00	10	11	01	00	10	11	01	00	01
10	11	11	00	00	11	11	00	00	11	00	00	10
11	11	01	00	11	11	01	00	11	11	00	00	01
12	11	10	00	11	11	10	00	11	11	11	00	00
13	11	01	00	01	11	01	00	01	11	10	00	01
14	11	01	00	01	11	01	00	01	11	11	00	11

When a slot format uses TFCI bits, its value in each radio frame corresponds to a certain transport format combination of the FACHs and/or PCHs in actual use, which is (re)negotiated at each FACH/PCH addition/removal.

4.3.9.3.1 STTD Encoding the S-CCPCH Structure

If the UTRAN perceives antenna diversity and we transmit S-CCPCH using open-loop transmit diversity, STTD encoding applies to the S-CCPCH data symbols. Table 4.24 shows the pilot symbol pattern for antenna 2.

4.3.9.4 Synchronisation Channel (SCH)

The SCH as a downlink signal for cell search consists of two sub-channels, i.e. the primary and secondary SCH. These channels shown in Figure 4.24 have 10 ms radio frames divided into 15 slots, each having 2560 chip length.

The primary SCH consists of a modulated code of length 256 chips, the Primary Synchronisation Code (PSC) denoted as c_p in Figure 4.24, transmitted once every slot. The PSC is the same for every cell in the system.

The secondary SCH consists of repeatedly transmitting a length-15 sequence of modulated codes of length 256 chips, the Secondary Synchronisation Codes (SSC), transmitted in parallel with the primary SCH. The SSC is denoted as $c_s^{i,k}$ in Figure 4.24, where $i = 0, 1, \ldots, 63$ is the number of the scrambling code group and $k = 0, 1, \ldots, 14$ is the slot number. Each SSC is chosen from a set of 16 different codes of length 256. This sequence on the secondary SCH indicates which of the code groups the cell's downlink scrambling code belongs to.

Symbol a modulates the primary and secondary synchronisation codes as illustrated in Figure 4.24. Table 4.25 shows the presence/absence of STTD encoding on the P-CCPCH.

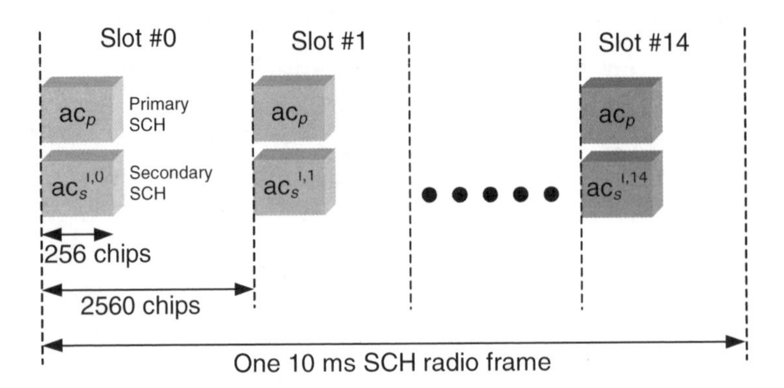

Figure 4.24 Structure of synchronisation channel.

Table 4.25 STTD encoding on the P-CCPCH

P-CCPCH STTD encoded	$a = +1$
P-CCPCH not STTD encoded	$a = -1$

4.3.9.4.1 SCH Transmitted by TSTD

Figure 4.25 shows the SCH's structure transmitted by the TSTD scheme, where we transmit on antenna 1 both PSC and SSC in even numbered slots, and on antenna 2 both PSC and SSC in odd numbered slots.

4.3.9.5 Physical Downlink Shared Channel (PDSCH)

Users share the PDSCH (carrying the transport Downlink Shared Channel (DSCH)), based on code multiplexing. Since the DSCH associates itself always with one or several DCHs,

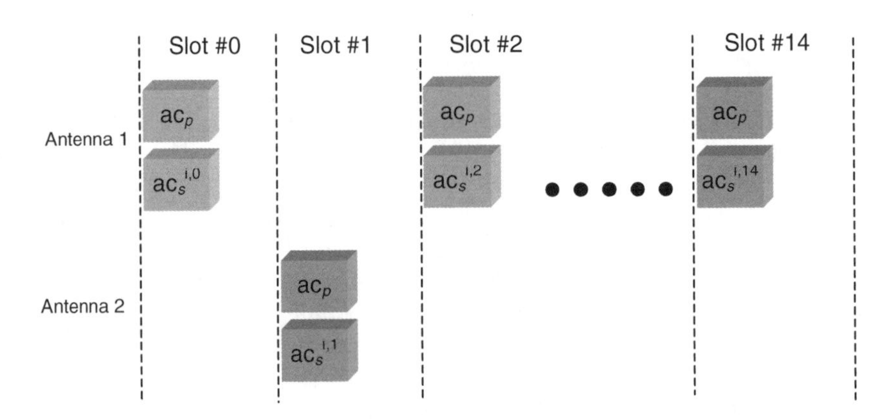

Figure 4.25 Structure of SCH transmitted by TSTD scheme.

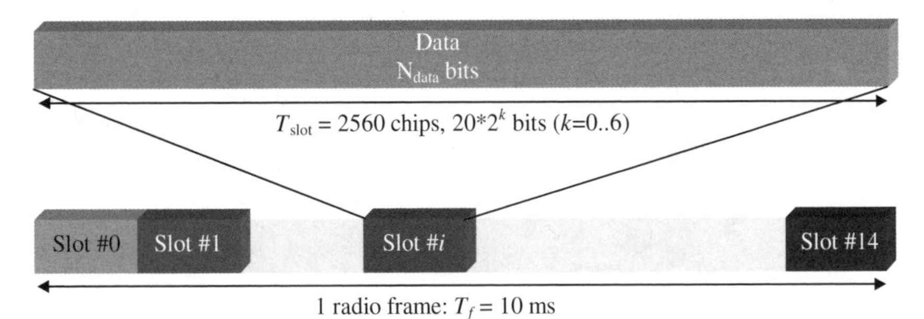

Figure 4.26 Frame structure for the PDSCH.

we also associate the PDSCH with one or several downlink DPCHs. More precisely, we associate each PDSCH radio frame with one downlink DPCH. Figure 4.26 illustrates the PDSCH frame and slot structure.

Two signalling methods indicate whether the UE has data to decode on the DSCH, i.e. through the TFCI field or higher layer signalling. For example, when the spreading factor and other physical layer parameters vary on a frame-by-frame basis, the TFCI informs the UE of PDSCH instantaneous parameters including the channelisation code from the PDSCH OVSF code tree.

Although the PDSCH and DPCH do not necessarily have the same spreading factors, and the PDSCH spreading factor may vary from frame to frame, a PDSCH transmission with associated DPCH is a special case of multi-code transmission. Thus, when mapping a DSCH to multiple parallel PDSCHs the spreading factor of all PDSCH codes will be the same. The PDSCH does not carry physical layer info, but all relevant L1 control information. Table 4.26 illustrates PDSCH T bit rates and symbol rates, where spreading factors may vary from 256 to 4.

4.3.9.6 Acquisition Indicator Channel (AICH)

As a physical channel, having the primary CPICH for phase reference, the AICH carries Acquisition Indicators (AIs) corresponding to signature s on the PRACH. Figure 4.27

Table 4.26 PDSCH[a] fields

Slot format (i)	Channel bit rate (kbps)	Channel symbol rate (ksps)	SF	Bits/frame	Bits/slot	N_{data}
0	30	15	256	300	20	20
1	60	30	128	600	40	40
2	120	60	64	1200	80	80
3	240	120	32	2400	160	160
4	480	240	16	4800	320	320
5	960	480	8	9600	640	640
6	1920	960	4	19 200	1280	1280

[a] When open-loop transmit diversity is employed for the PDSCH, STTD encoding is used on the data bits.

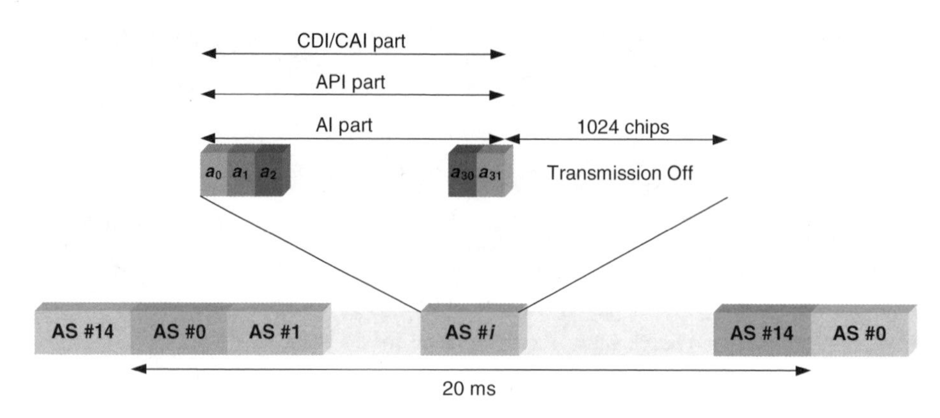

Figure 4.27 Structure of the AICH, AP-AICH and CD/CA-ICH.

illustrates its structure consisting of a repeated sequence of 15 consecutive *Access Slots* (AS), each of length 40 bit intervals. In turn, every access slot consists of two parts: an AI part containing 32 real-valued symbols $a_0, \ldots, a_{31}$ and a part of 1024 chips duration without transmission.

Equation (4.40a) defines the real-valued symbols $a_0, a_1, \ldots, a_{31}$ seen in Figure 4.27. Table 20 in [8] provides the acquisition indicator $\mathrm{AI_s}$ with values $+1$, -1 and 0 corresponding to signature s and the sequence $b_{s,0}, \ldots, b_{s,31}$. When the AICH has STTD-based open-loop transmit diversity, STTD encoding applies to each $b_{s,0}, b_{s,1}, \ldots, b_{s,31}$ sequence separately before these sequences combined into AICH symbols $a_0, \ldots, a_{31}$.

$$a_j = \sum_{s=0}^{15} \mathrm{AI}_s b_{s,j} \tag{4.40a}$$

$$a_j = \sum_{s=0}^{15} \mathrm{API}_s \times b_{s,j} \tag{4.40b}$$

4.3.9.7 CPCH Access Preamble Acquisition Indicator Channel (AP-AICH)

The physical AP-AICH carries CPCH Access Preamble Acquisition Indicators (API). The AP acquisition indicator API corresponds to the AP signature s transmitted by UE. The AP-AICH, with the primary CPICH as reference, and the AICH may use the same or different channelisation codes. Figure 4.27 with the corresponding 'API part' label illustrates the structure of AP-AICH. The AP-AICH has a part of 4096 chips duration to transmit the API, followed by a part of 1024 chips duration with no transmission.

Equation (4.40b) defines the real-valued symbols $a_0, a_1, \ldots, a_{31}$ seen in Figure 4.27, where $\mathrm{API_s}$, taking the values $+1$, -1 and 0, are the AP acquisition indicators corresponding to the access preamble signature s transmitted by UE and Table 20 in [8] provides the sequence $b_{s,0}, \ldots, b_{s,31}$. As in the AICH, when the AP-AICH has STTD-based open-loop transmit diversity, STTD encoding applies to each $b_{s,0}, b_{s,1}, \ldots, b_{s,31}$ sequence separately before these sequences combine into AICH symbols $a_0, \ldots, a_{31}$.

4.3.9.8 CPCH Collision Detection/Channel Assignment-Indicator Channel

As a physical channel, the Collision Detection/Channel Assignment-Indicator Channel (CD/CA-ICH) carries the CD Indicator (CDI) only if the CA is not active, or CD Indicator/CA Indicator (CDI/CAI) at the same time if the CA is active. Figure 4.27 illustrates the CD/CA-ICH structure with the corresponding label, where the CD/CA-ICH transmits 4096 chips duration, followed by 1024 chips duration without transmission. The same or different channelisation codes may apply to the CD/CA-ICH and AP-AICH.

As in the preceding indicator channels, when the CD/CA-ICH has STTD-based open-loop transmit diversity, STTD encoding applies to each $b_{s,0}, b_{s,1}, \ldots, b_{s,31}$ sequence separately before these sequences combined into AICH symbols $a_0, \ldots, a_{31}$.

Equation (4.41a) defines the real-valued symbols $a_0, a_1, \ldots, a_{31}$ shown in Figure 4.27 for non-active CA, where CDI_s, with values +1 and 0, is the CD indicator corresponding to CD preamble signature s transmitted by the UE. Table 20 in [8] lists the sequence $b_{s,0}, \ldots, b_{s,31}$.

$$a_j = \sum_{s=0}^{15} \mathrm{CDI}_s \times b_{s,j} \tag{4.41a}$$

$$a_j = \sum_{i=0}^{15} \mathrm{CDI}_i \times b_{s_i,j} + \sum_{k=0}^{15} \mathrm{CAI}_k \times b_{s_k,j} \tag{4.41b}$$

Equation (4.41b) defines the real-valued symbols $a_0, a_1, \ldots, a_{31}$ when CA is active, where the subscript s_i, s_k depend on the indices i, k according to Table 4.27, respectively, and indicate the signature number s in Table 20 [8]. The sequence $b_{s,0}, \ldots, b_{s,31}$ is also given in Table 20 [8]. $\mathrm{CDI_i}$, with a value of +1/0 or −1/0, is the CD indicator corresponding to the

Table 4.27 Generation of $\mathrm{CDI}_i/\mathrm{CAI}_k$

<table>
<tr><th>UE transmitted CD preamble i</th><th>CDI_i</th><th>Signature s_i</th><th>Channel assignment index k</th><th>CAI_k</th><th>Signature s_k</th></tr>
<tr><td>0</td><td>+1/0</td><td rowspan="2">1</td><td>0</td><td>+1/0</td><td rowspan="2">0</td></tr>
<tr><td>1</td><td>−1/0</td><td>1</td><td>−1/0</td></tr>
<tr><td>2</td><td>+1/0</td><td rowspan="2">3</td><td>2</td><td>+1/0</td><td rowspan="2">8</td></tr>
<tr><td>3</td><td>−1/0</td><td>3</td><td>−1/0</td></tr>
<tr><td>4</td><td>+1/0</td><td rowspan="2">5</td><td>4</td><td>+1/0</td><td rowspan="2">4</td></tr>
<tr><td>5</td><td>−1/0</td><td>5</td><td>−1/0</td></tr>
<tr><td>6</td><td>+1/0</td><td rowspan="2">7</td><td>6</td><td>+1/0</td><td rowspan="2">12</td></tr>
<tr><td>7</td><td>−1/0</td><td>7</td><td>−1/0</td></tr>
<tr><td>8</td><td>+1/0</td><td rowspan="2">9</td><td>8</td><td>+1/0</td><td rowspan="2">2</td></tr>
<tr><td>9</td><td>−1/0</td><td>9</td><td>−1/0</td></tr>
<tr><td>10</td><td>+1/0</td><td rowspan="2">11</td><td>10</td><td>+1/0</td><td rowspan="2">6</td></tr>
<tr><td>11</td><td>−1/0</td><td>11</td><td>−1/0</td></tr>
<tr><td>12</td><td>+1/0</td><td rowspan="2">13</td><td>12</td><td>+1/0</td><td rowspan="2">10</td></tr>
<tr><td>13</td><td>−1/0</td><td>13</td><td>−1/0</td></tr>
<tr><td>14</td><td>+1/0</td><td rowspan="2">15</td><td>14</td><td>+1/0</td><td rowspan="2">14</td></tr>
<tr><td>15</td><td>−1/0</td><td>15</td><td>−1/0</td></tr>
</table>

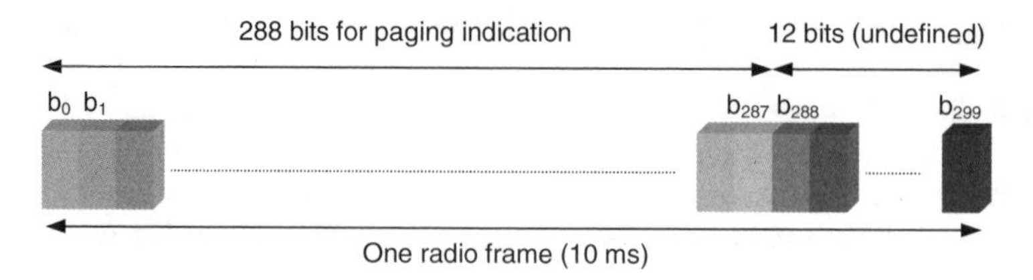

Figure 4.28 Structure of Paging Indicator Channel (PICH).

CD preamble i transmitted by the UE, and CAI_k, with a value of $+1/0$ or $-1/0$, is the CA indicator corresponding to the assigned channel index k as given in Table 4.27.

4.3.9.9 Paging Indicator Channel (PICH)

As a physical channel the PICH has a fixed rate (SF = 256) and carries the Paging Indicators (PIs). It is always associated with a S-CCPCH to which a PCH transport channel is mapped.

Figure 4.28 illustrates PICH frame structure, where one PICH radio frame of 10 ms length consists of 300 bits $(b_0, b_1, \ldots, b_{299})$. Of these, 288 bits $(b_0, b_1, \ldots, b_{287})$ are used to carry paging indicators. The remaining 12 bits $(b_{288}, b_{289}, \ldots, b_{299})$ are undefined. Each PICH frame transmits N paging indicators $\{\text{PI}_0, \ldots, \text{PI}_{N-1}\}$, where N = 18, 36, 72 or 144.

Higher layers calculate the PI mapped to the paging indicator PI_p (where p is computed as a function of the PI), the SFN of the P-CCPCH radio frame during which the start of the PICH radio frame occurs and the number of paging indicators per frame (N), see eqn (4.3). Table 4.28 shows the mapping from $\{\text{PI}_0, \ldots, \text{PI}_{N-1}\}$ to the PICH bits $\{b_0, \ldots, b_{287}\}$.

$$p = \left(\text{PI} + \left\lfloor ((18 \times (\text{SFN} + \lfloor \text{SFN}/8 \rfloor + \lfloor \text{SFN}/64 \rfloor + \lfloor \text{SFN}/512 \rfloor)) \bmod 144) \times \frac{N}{144} \right\rfloor \right) \bmod N \quad (4.42)$$

If a paging indicator in a certain frame has the value '1', it indicates that UEs associated with this paging indicator should read the corresponding frame of the associated S-CCPCH. In the event of transmit diversity for the PICH, STTD encoding applies on the PICH bits.

Table 4.28 Mapping of Paging Indicators (PIs) to PICH bits

Number of PI per frame (N)	$\text{PI}_p = 1$	$\text{PI}_p = 0$
$N = 18$	$\{b_{16p}, \ldots, b_{16p+15}\} = \{1, 1, \ldots, 1\}$	$\{b_{16p}, \ldots, b_{16p+15}\} = \{0, 0, \ldots, 0\}$
$N = 36$	$\{b_{8p}, \ldots, b_{8p+7}\} = \{1, 1, \ldots, 1\}$	$\{b_{8p}, \ldots, b_{8p+7}\} = \{0, 0, \ldots, 0\}$
$N = 72$	$\{b_{4p}, \ldots, b_{4p+3}\} = \{1, 1, \ldots, 1\}$	$\{b_{4p}, \ldots, b_{4p+3}\} = \{0, 0, \ldots, 0\}$
$N = 144$	$\{b_{2p}, b_{2p+1}\} = \{1, 1\}$	$\{b_{2p}, b_{2p+1}\} = \{0, 0\}$

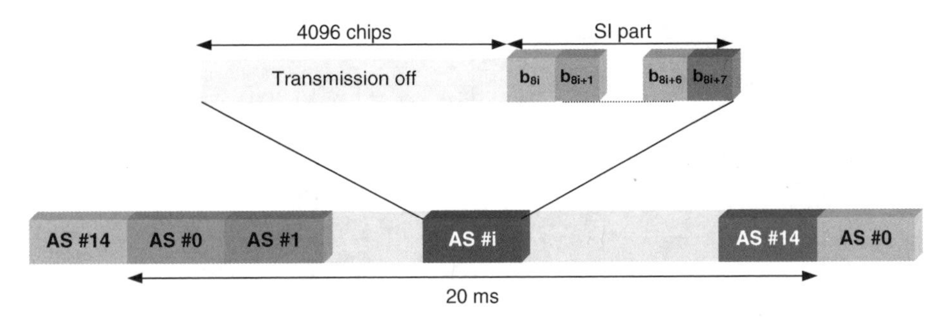

Figure 4.29 Structure of CPCH Status Indicator Channel (CSICH).

4.3.9.10 CPCH Status Indicator Channel (CSICH)

The CSICH is also a fixed rate (SF = 256) physical channel carrying CPCH status information. It has always an association with a physical channel used for transmitting a CPCH AP-AICH and uses the same channelisation and scrambling codes. Figure 4.29 illustrates the CSICH frame structure. It consists of 15 consecutive Access Slots (AS) each 40 bits long and having two parts, i.e. one of duration 4096 chips without transmission, and a Status Indicator (SI) part consisting of 8 bits $b_{8i}, \ldots, b_{8i+7}$, where i is the access slot number. The CSICH uses the same modulation of the PICH, and has the primary CPICH phase as reference.

Each CSICH frame transmits N status indicators $\{SI_0, \ldots, SI_{N-1}\}$, i.e. all the access slots of the CSICH frame transmit status indicator even if some signatures and/or access slots are shared between CPCH and RACH. The mapping from $\{SI_0, \ldots, SI_{N-1}\}$ to the CSICH bits $\{b_0, \ldots, b_{119}\}$ follows Table 4.29. In the event of transmit diversity for the CSICH, STTD encoding applies on the CSICH bits.

Higher layers set the status indicator values for UTRAN. Thus, the higher layers provide layer 1 with the mapping between the values of the status indicators and the availability of CPCH resources. At the UE the number of status indicators per frame is also a higher layer parameter.

Table 4.29 Mapping of Status Indicators (SIs) to CSICH bits

Number of SI per frame (N)	$SI_n = 1$	$SI_n = 0$
$N = 1$	$\{b_0, \ldots, b_{119}\} = \{1, 1, \ldots, 1\}$	$\{b_0, \ldots, b_{119}\} = \{0, 0, \ldots, 0\}$
$N = 3$	$\{b_{40n}, \ldots, b_{40n+39}\} = \{1, 1, \ldots, 1\}$	$\{b_{40n}, \ldots, b_{40n+39}\} = \{0, 0, \ldots, 0\}$
$N = 5$	$\{b_{24n}, \ldots, b_{24n+23}\} = \{1, 1, \ldots, 1\}$	$\{b_{24n}, \ldots, b_{24n+23}\} = \{0, 0, \ldots, 0\}$
$N = 15$	$\{b_{8n}, \ldots, b_{8n+7}\} = \{1, 1, \ldots, 1\}$	$\{b_{8n}, \ldots, b_{8n+7}\} = \{0, 0, \ldots, 0\}$
$N = 30$	$\{b_{4n}, \ldots, b_{4n+3}\} = \{1, 1, 1, 1\}$	$\{b_{4n}, \ldots, b_{4n+3}\} = \{0, 0, 0, 0\}$
$N = 60$	$\{b_{2n}, b_{2n+1}\} = \{1, 1\}$	$\{b_{2n}, b_{2n+1}\} = \{0, 0\}$

Table 4.30 Mapping transport channels to physical channels

Transport channels	Physical channels
Dedicated Channel (DCH)	Dedicated Physical Data Channel (DPDCH)
	Dedicated Physical Control Channel (DPCCH)
Random Access Channel (RACH)	Physical Random Access Channel (PRACH)
Common Packet Channel (CPCH)	Physical Common Packet Channel (PCPCH)
	Common Pilot Channel (CPICH)
Broadcast Channel (BCH)	Primary Common Control Physical Channel (P-CCPCH)
Forward Access Channel (FACH)	Secondary Common Control Physical Channel (S-CCPCH)
Paging Channel (PCH)	Synchronisation Channel (SCH)
Downlink Shared Channel (DSCH)	Physical Downlink Shared Channel (PDSCH)
	Acquisition Indicator Channel (AICH)
	Access Preamble Acquisition Indicator Channel (AP-AICH)
	Paging Indicator Channel (PICH)
	CPCH Status Indicator Channel (CSICH)
	Collision-Detection/Channel-Assignment Indicator Channel (CD/CA-ICH)

4.3.10 Mapping Transport Channels onto Physical Channels

Table 4.30 summarises the mapping of transport channels onto physical channels. The DCHs are coded and multiplexed as described in [1], and the resulting data stream is mapped sequentially (first-in-first-mapped) directly to the physical channel(s). The mapping of BCH and FACH/PCH is equally straightforward, where the data stream after coding and interleaving is mapped sequentially to the primary and secondary CCPCH, respectively. Also for the RACH, the coded and interleaved bits are sequentially mapped to the physical channel, in this case the message part of the PRACH.

4.3.11 Timing Relationship Between Physical Channels

The P-CCPCH, which carries a cell's SFN, serves as timing reference for all the physical channels, directly for downlink and indirectly for uplink. Figure 4.30 describes the frame timing of the downlink physical channels, for the AICH it includes the access slot timing. Uplink physical channels get their transmission timing from the received timing of the downlink physical channels. In general the following applies from [8]:

- SCH (primary and secondary), CPICH (primary and secondary), P-CCPCH and PDSCH have identical frame timings,
- the S-CCPCH timing may vary for different S-CCPCHs, but the offset from the P-CCPCH frame timing is a multiple of 256 chips, i.e. $\tau_{\text{S-CCPCH},k} = T_k \times 256$ chips, $T_k \in \{0, 1, \ldots, 149\}$,

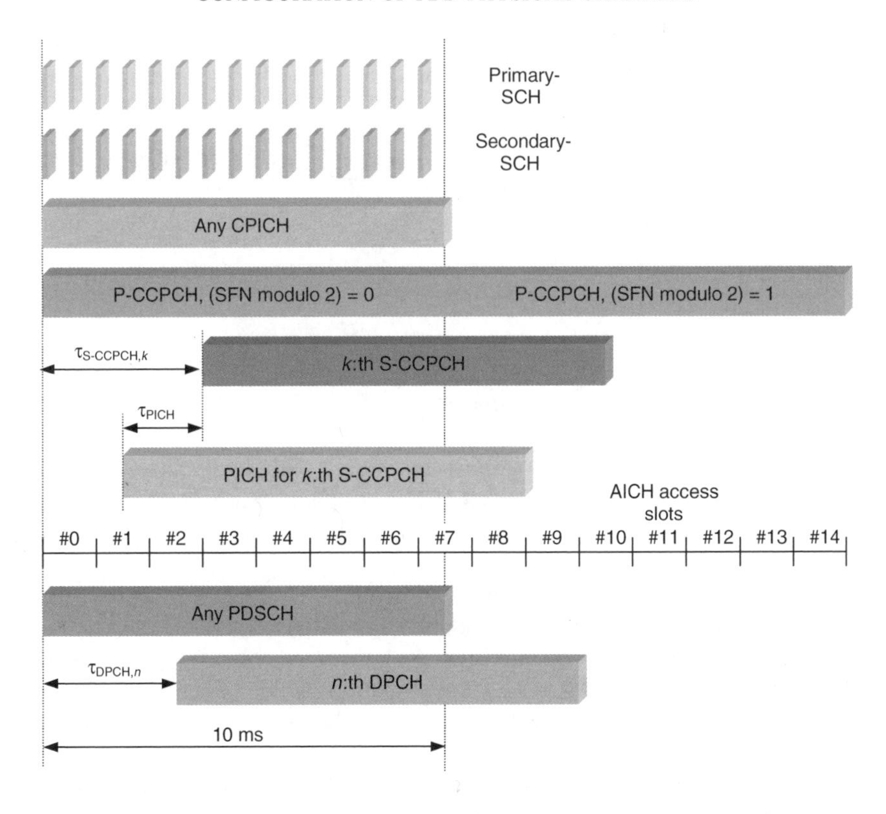

Figure 4.30 Frame timing and access slot timing of downlink physical channels.

- the PICH timing is $\tau_{\text{PICH}} = 7680$ chips prior to its corresponding S-CCPCH frame timing, i.e. the timing of the S-CCPCH carrying the PCH transport channel with the corresponding paging information,
- AICH access slots 0 start at the same time as P-CCPCH frames with SFN $\bmod 2 = 0$,
- any DPCH frame is associated to one PDSCH frame through the relation 460 80 chips $\leq \text{T}_{\text{PDSCH}} - \text{T}_{\text{DPCH}} < 84\,480$ chips,
- the DPCH timing may be different for different DPCHs, but the offset from the P-CCPCH frame timing is a multiple of 256 chips, i.e. $\tau_{\text{DPCH},n} = T_n \times 256$ chips, $T_n \in \{0, 1, \ldots, 149\}$.

4.3.11.1 PICH/S-CCPCH Timing Relation

Figure 4.31 illustrates the timing between a PICH frame and its associated S-CCPCH frame, i.e. the S-CCPCH frame carrying paging information related to the paging indicators in the PICH frame. A paging indicator set in a PICH frame means that the paging message is transmitted on the PCH in the S-CCPCH frame starting τ_{PICH} chips after the transmitted PICH frame.

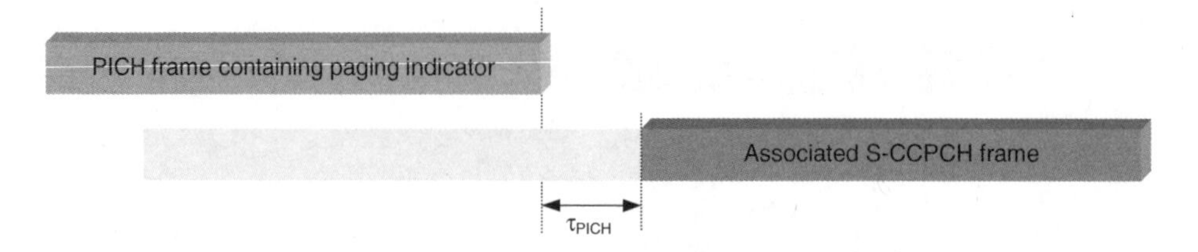

Figure 4.31 Timing relation between PICH frame and associated S-CCPCH frame.

4.3.11.2 PRACH/AICH Timing Relation

The downlink AICH has two downlink access slots, each with 5120 chips length and time aligned with the P-CCPCH. The uplink PRACH has uplink access slots, each with 5120 chips length. The UE transmits uplink access slot number n, $\tau_{\text{p-a}}$ chips prior to the reception of downlink access slot number $n, n = 0, 1, \ldots, 14$.

Downlink acquisition indicators may only start at the beginning of a downlink access slot. Likewise, transmission of uplink RACH preambles and RACH message parts may only start at the beginning of an uplink access slot.

Figure 4.32 illustrates the PRACH/AICH timing relation, where the preamble-to-preamble distance $\tau_{\text{p-p}}$ shall be larger than or equal to the minimum preamble-to-preamble distance $\tau_{\text{p-p,min}}$, i.e. $\tau_{\text{p-p}} \geq \tau_{\text{p-p,min}}$.

In addition to $\tau_{\text{p-p,min}}$, [8] defines the *preamble-to-AI distance* $\tau_{\text{p-a}}$ and *preamble-to-message distance* $\tau_{\text{p-m}}$ as follows (higher layers signal the parameter AICH_Transmission_Timing):

1. when AICH_transmission_timing = 0, then $\tau_{\text{p-p,min}} = 15\,360$ chips (three access slots), $\tau_{\text{p-a}} = 7680$ chips and $\tau_{\text{p-m}} = 15\,360$ chips (three access slots),
2. when AICH_transmission_timing = 1, then $\tau_{\text{p-p,min}} = 20\,480$ chips (four access slots), $\tau_{\text{p-a}} = 12\,800$ chips and $\tau_{\text{p-m}} = 20\,480$ chips (four access slots).

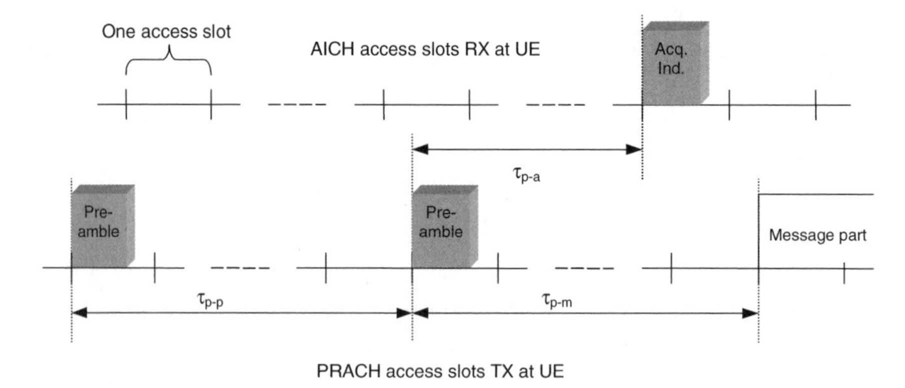

Figure 4.32 Timing relation between PRACH and AICH as seen in the UE.

4.3.11.3 PCPCH/AICH Timing Relation

AICH, the message and the PRACH/AICH have an identical timing relationship between preambles, where the collision resolution preambles follow the access preambles of the latter. The RACH preamble and AICH have the same timing relationship of the CD preamble and CD-ICH. Likewise, AICH to message in RACH and CD-ICH power-control preamble in CPCH, have identical timing relationships. Finally, the PRACH/AICH transmission timing parameter and the T_{CPCH} timing parameter are identical. See that a_1 corresponds to AP-AICH and a_2 corresponds to CD-ICH.

When $T_{CPCH} = 0$ or 1, the following PCPCH/AICH timing values apply:

1. $\tau_{p\text{-}p}$ is the time to next available access slot, between access preambles. Minimum time = 15 360 chips + 5120 chips × T_{CPCH}. Maximum time = 5120 chips × 12 = 61 440 chips. Actual time is given by time to next slot (which meets minimum time criterion) in allocated access slot sub-channel group.
2. $\tau_{p\text{-}a1}$ is the time between access preamble and AP-AICH, and has two alternative values 7680 or 12 800 chips depending on T_{CPCH}.
3. $\tau_{a1\text{-}CDP}$ is the time between receipt of AP-AICH and CD preamble $\tau_{a1\text{-}cdp}$ transmission, and has a minimum value of $\tau_{a1\text{-}CDP,\min} = 7680$ chips.
4. $\tau_{p\text{-}CDP}$ is the time between the last AP and CD preamble. It has a minimum value of $\tau_{p\text{-}cdp\text{-}min} = 3$ or 4 access slots, depending on T_{CPCH}.
5. $\tau_{CDP\text{-}a2}$ is the time between the CD preamble and the CD-ICH, and has two alternative values 7680 chips or 12 800 chips, depending on T_{CPCH}.
6. $\tau_{CDP\text{-}PCP}$ is the time between CD preamble and the start of the power control preamble, and is either 3 or 4 access slots, depending on T_{CPCH}.

The message transmission starts at 0 or 8 slots after the start of the power-control preamble depending on the length of the power-control preamble. Figure 4.33 illustrates the PCPCH/

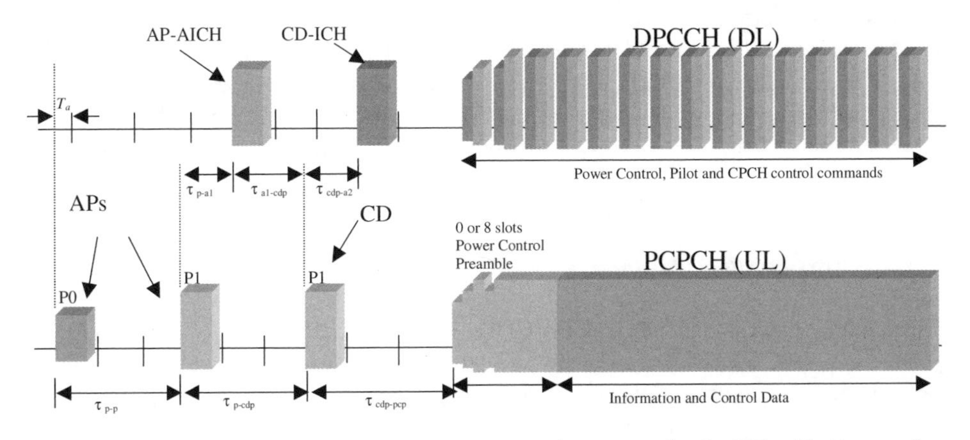

Figure 4.33 Timing of PCPCH and AICH transmission as seen by the UE, with $T_{CPCH} = 0$.

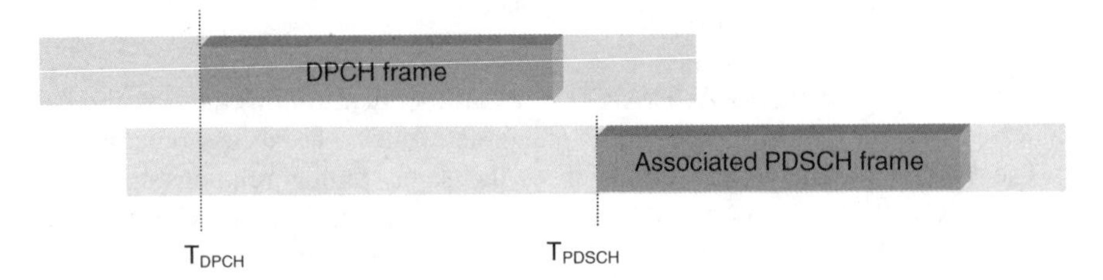

Figure 4.34 Timing relation between DPCH frame and associated PDSCH frame.

AICH timing relationship when $T_{\mathrm{CPCH}} = 0$ and all access slot sub-channels are available for the PCPCH.

4.3.11.4 DPCH/PDSCH Timing

Figure 4.34 illustrates relative timing between a DPCH frame and the associated PDSCH frame, where the start of a DPCH and of an associated PDSCH frame are denoted by T_{DPCH} and T_{PDSCH}, respectively. Any DPCH frame associates itself to one PDSCH frame through the relation 46 080 chips $\leq T_{\mathrm{PDSCH}} - T_{\mathrm{DPCH}} <$ 84 480 chips, i.e. the associated PDSCH frame starts anywhere between three slots after the end of the DPCH frame up to 18 slots after the end of the DPCH frame [8].

4.3.11.5 DPCCH/DPDCH Timing Relations

In the uplink the DPCCH and all the DPDCHs transmitted from one UE have the same frame timing. Likewise, in the downlink the DPCCH and all the DPDCHs carrying CCTrCHs of dedicated type to one UE have the same frame timing.

4.3.11.6 UE Uplink/Downlink Timing

At the UE, the uplink DPCCH/DPDCH frame transmission takes place approximately T_0 chips after the reception of the first significant path of the corresponding downlink DPCCH/DPDCH frame, where T_0 is a constant defined as 1024 chips. Other timing relations for initialisation of channels are in [8] and [3].

4.3.12 Downlink Spreading

All downlink physical channels (i.e. P-CCPCH, S-CCPCH, CPICH, AICH, PICH, PDSCH and downlink DPCH), but excluding the SCH, follow the spreading operation represented in Figure 4.35. The non-spread physical channel consists of a sequence of three (i.e. $+1$, -1 and 0) real-valued symbols, where 0 indicates DTX. However, for the AICH, the symbol values depend on the exact combination of acquisition indicators to be transmitted.

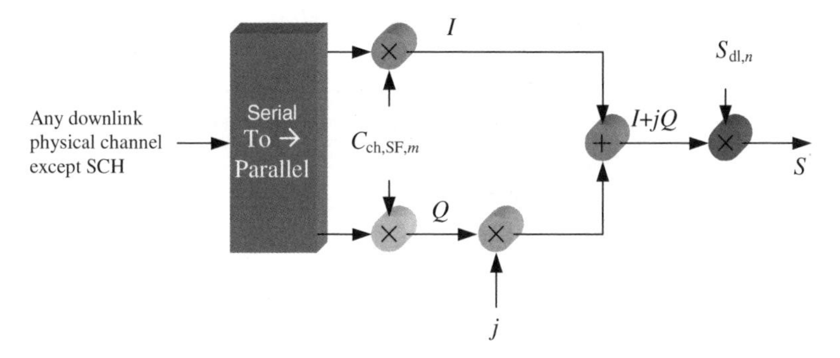

Figure 4.35 Spreading for all downlink physical channels, excluding the SCH.

In the first step each pair of two consecutive symbols pass from serial to parallel and get mapped to an I and Q branch, where even and odd numbered symbols are mapped to the I and Q branch, respectively. In all channels excluding AICH, we define symbol 0 as the first symbol in each frame. In the AICH, we define symbol 0 as the first symbol in each access slot. Then we spread I and Q branches at the chip rate by the same real-valued channelisation code $C_{ch,SF,m}$. Afterwards, we treat the sequences of real-valued chips on the I and Q branches as a single complex-valued sequence of chips. This sequence of chips is scrambled (complex chip-wise multiplication) by a complex-valued scrambling code $S_{dl,n}$. For the P-CCPCH, we apply the scrambling code aligned with the P-CCPCH frame boundary, i.e. we multiply the first complex chip of the spread P-CCPCH frame with chip 0 of the scrambling code. For other downlink channels, we apply the scrambling code aligned with the scrambling code directed to the P-CCPCH. In this case, the scrambling code application does not align with the frame boundary of the physical channel under scrambling.

4.3.12.1 Downlink Channelisation Codes

The same uplink channelisation codes presented in Section 4.3.4 apply to the downlink, i.e. Orthogonal Variable Spreading Factor (OVSF) codes that preserve the orthogonality between downlink channels of different rates and spreading factors. The specifications fix the channelisation code for the primary CPICH to $C_{CH,256,0}$ and the channelisation code for the primary CCPCH to $C_{CH,256,1}$. The UTRAN assigns the channelisation codes for all other physical channels.

Spreading Factor (SF) 512 has specific restriction in its application. For example, if we use the code word $C_{CH,512,n}$, with $n = 0, 2, 4, \ldots, 510$ in soft handover, then we do not allocate the code word $C_{CH,512,n+1}$ in the Node B, because we need the usage of timing adjustment. Likewise, if we use $C_{CH,512,n}$, with $n = 1, 3, 5, \ldots, 511$, then we do not allocate the code word $C_{CH,512,n-1}$ in the Node B, again because we need time adjustment usage. However, this restriction does not apply to softer handover operation or when the UTRAN synchronises to such a level that timing adjustments in soft handover are not applied in conjunction with SF 512.

If we do implement compressed mode by reducing the spreading factor by 2, the OVSF code used for compressed frames is $C_{CH,SF/2,\lfloor n/2 \rfloor}$ when applying ordinary scrambling code and $C_{\text{CH,SF/2},n \bmod \text{SF/2}}$ when applying alternative scrambling, where $C_{\text{CH,SF},n}$ corresponds to the channelisation code used for non-compressed frames.

If the OVSF code on the PDSCH varies from frame to frame, the OVSF codes allocation will occur in a way that the OVSF code(s) below the smallest spreading factor will be from the branch of the code tree pointed by the smallest spreading factor used for the connection. This implies that all the connecting UE-PDSCH codes can originate according to the OVSF code generation principle from smallest spreading factor code used by the UE on PDSCH.

When mapping the DSCH to multiple parallel PDSCHs, the same rule applies. However, all of the branches identified by the multiple codes, corresponding to the smallest spreading factor, may be used for higher spreading factor allocation [4].

4.3.12.2 Downlink Scrambling Codes

In principle, we can generate about $2^{18} - 1 = 262\,143$ scrambling codes, numbered $0, \ldots, 262\,142$. However, we do not use all the scrambling codes. The specifications divide these scrambling codes into 512 sets each of a primary scrambling code and 15 secondary scrambling codes. The primary scrambling codes consist of $n = 16 \times i$, where $i = 0, \ldots, 511$. The ith set of secondary scrambling codes consists of $16 \times (i + k)$, where $k = 1, \ldots, 15$. There exists a one-to-one mapping between each primary scrambling code and 15 secondary scrambling codes in a set such that the ith primary scrambling code corresponds to the ith set of secondary scrambling codes.

Thus, based on the principles above, we can use $k = 0, 1, \ldots, 8191$ scrambling codes, where for compressed frames, each of these codes are associated with a left alternative scrambling code and a right alternative scrambling code. The left alternative scrambling code corresponding to scrambling code k has scrambling code number $k + 8192$, while the right alternative scrambling code corresponding to scrambling code k has scrambling code number $k + 16\,384$. When we use alternative scrambling codes for compressed frames, left alternative scrambling code applies if $n < \text{SF}/2$ and the right alternative scrambling code applies if $n \geq \text{SF}/2$. Upper layers signal the usage of alternative scrambling code in compressed frames for each physical channel. We use channelisation code $c_{\text{CH,SF},n}$ for non-compressed frames.

The set of primary scrambling codes gets further divided into 64 scrambling code groups, each consisting of eight primary scrambling codes. The jth scrambling code group consists of primary scrambling codes $16 \times 8 \times j + 16 \times k$, where $j = 0, \ldots, 63$ and $k = 0, \ldots, 7$. *Each cell receives one and only one primary scrambling code allocation.* The primary CCPCH and primary CPICH always use the primary scrambling code to transmit. On the other hand, the other downlink physical channels can use either the primary scrambling code or a secondary scrambling code from the set associated with the primary scrambling code of the cell to transmit. The combination of primary and secondary scrambling codes in one CCTrCH may be possible. However, if the CCTrCH has a DSCH type then all the PDSCH channelisation codes that one UE may receive must be under a single scrambling code, i.e. either the primary or a secondary scrambling code.

The scrambling code sequences result from combining two real sequences into a complex sequence. We build each of the two real sequences as the position-wise mod 2 sum of 38 400 chip segments of two binary m sequences generated by means of two generator polynomials of degree 18. The resulting sequences thus constitute segments of a set of gold sequences. The scrambling codes are repeated for every 10 ms radio frame.

If we assume that x and y are two sequences, then x sequence originates from the primitive (over GF(2)) polynomial $1 + X^7 + X^{18}$, and y sequence results from using the polynomial $1 + X^5 + X^7 + X^{10} + X^{18}$. The sequence, which depends on the chosen scrambling code number n, is denoted z_n as in the sequel. In addition, we let $x(i)$, $y(i)$ and $z_n(i)$ denote the ith symbol of the sequence x, y and z_n, respectively. Then from the specifications in [4] we summarise the m sequences x and y as

Initial conditions:

x is constructed with $x(0) = 1, x(1) = x(2) = \cdots = x(16) = x(17) = 0$

$$y(0) = y(1) = \cdots = y(16) = y(17) = 1$$

Recursive definition of subsequent symbols:

$$x(i + 18) = x(i + 7) + x(i) \bmod 2, \quad i = 0, \ldots, 2^{18} - 20$$
$$y(i + 18) = y(i + 10) + y(i + 7) + y(i + 5) + y(i) \bmod 2, \quad i = 0, \ldots, 2^{18} - 20$$

The nth gold code sequence z_n, $n = 0, 1, 2, \ldots, 2^{18} - 2$, is then defined as

$$z_n(i) = x((i + n) \bmod (2^{18} - 1)) + y(i) \bmod 2, \quad i = 0, \ldots, 2^{18} - 2 \tag{4.43}$$

These resulting binary sequences get converted to real-valued sequences Z_n by the following transformation:

$$Z_n(i) = \begin{cases} +1 & \text{if } z_n(i) = 0 \\ -1 & \text{if } z_n(i) = 1 \end{cases} \quad \text{for} \quad i = 0, 1, \ldots, 2^{18} - 2 \tag{4.44}$$

Ultimately, we define the nth complex scrambling code sequence $S_{\text{dl},n}$ as

$$S_{\text{dl},n}(i) = Z_n(i) + j Z_n((i + 131\,072) \bmod (2^{18} - 1)), \quad i = 0, 1, \ldots, 38\,399 \tag{4.45}$$

where the pattern from phase 0 up to the phase of 38 399 gets repeated.

4.3.12.3 Synchronisation Codes

The Primary Synchronisation Code (PSC), C_{PSC}, results from a so-called generalised hierarchical Golay sequence. It gets chosen to have good aperiodic autocorrelation properties. We express it as

$$a = \langle x_1, x_2, x_3, \ldots, x_{16} \rangle = \langle 1, 1, 1, 1, 1, 1, -1, -1, 1, -1, 1, -1, 1, -1, -1, 1 \rangle \tag{4.46}$$

Thus, we generate the PSC by repeating the sequence a modulated by a Golay complementary sequence, and creating a complex-valued sequence with identical real and imaginary components. The PSC C_{PSC} is defined as

$$C_{PSC} = (1 + j) \times \langle a, a, a, -a, -a, a, -a, -a, a, a, a, -a, a, -a, a, a \rangle$$

where the leftmost chip in the sequence corresponds to the chip transmitted first in time.

The 16 Secondary Synchronisation Codes (SSCs), $\{C_{SSC,1,\ldots,}C_{SSC,16}\}$, are complex valued with identical real and imaginary components, and are constructed from position-wise multiplication of a Hadamard sequence and a sequence z, defined as

$$z = \langle b, b, b, -b, b, b, -b, -b, b, -b, b, -b, -b, -b, -b, -b \rangle$$

where

$$b = \langle x_1, x_2, x, x_4, x_5, x_6, x_7, x_8, -x_9, -x_{10}, -x_{11}, -x_{12}, -x_{13}, -x_{14}, -x_{15}, -x_{16} \rangle$$

and $x_1, x_2, \ldots, x_{15}, x_{16}$, are same as in the definition of the sequence a above.

The Hadamard sequences are obtained as the rows in a matrix H_8 constructed recursively by

$$H_0 = (1)$$

$$H_k = \begin{pmatrix} H_{k-1} & H_{k-1} \\ H_{k-1} & -H_{k-1} \end{pmatrix}, \quad k \geq 1$$

The rows are numbered from the top starting with row 0 (the all ones sequence).

Denote the nth Hadamard sequence as a row of H_8 numbered from the top, $n = 0, 1, 2, \ldots, 255$, in the sequel.

Furthermore, let $h_n(i)$ and $z(i)$ denote the ith symbol of the sequences h_n and z, respectively, where $i = 0, 1, 2, \ldots, 255$ and $i = 0$ corresponds to the leftmost symbol.

The kth SSC, C_{SSC}, k, $k = 1, 2, 3, \ldots, 16$, is then defined as

$$C_{SSC,k} = (1 + j) \times \langle h_m(0) \times z(0), h_m(1) \times z(1), h_m(2) \times z(2), \ldots, h_m(255) \times z(255) \rangle$$

where $m = 16 \times (k - 1)$ and the leftmost chip in the sequence corresponds to the chip transmitted first in time.

As direct extracts from [4] we note that the 64 secondary SCH sequences are constructed such that their cyclic shifts are unique, i.e. a non-zero cyclic shift less than 15 of any of the 64 sequences is not equivalent to some cyclic shift of any other of the 64 sequences. In addition, a non-zero cyclic shift less than 15 of any of the sequences is not equivalent to itself with any other cyclic shift less than 15. Table 4.48 in Appendix C (see page 581) illustrates the sequences of SSCs used to encode the 64 different scrambling code groups. The entries in Table 4.48 specify what SSC sequence applies to the different slots for the different scrambling code groups, e.g. the entry '7' means that SSC $C_{SSC,7}$ applies to the corresponding scrambling code group and slot.

4.3.13 Downlink Power Control Procedure

The network determines the transmit power of the downlink channels. Generally, the transmit power ratio between different downlink channels does not have specification and may change with time.

4.3.13.1 DPCCH/DPDCH Downlink Power Control

The downlink transmit power-control procedure controls simultaneously the DPCCH power and that of its corresponding DPDCHs. The power-control loop adjusts the power of the DPCCH and DPDCHs with the same relative power difference. The network determines the relative transmit power offset between DPCCH fields and DPDCHs. The TFCI, TPC and pilot fields of the DPCCH are offset relative to the DPDCHs power by PO1, PO2 and PO3 dB, respectively. The power offsets may vary in time.

4.3.13.1.1 The Downlink Power Control Function

The UE generates TPC commands to control the network[4] transmit power and send them in the TPC field of the uplink DPCCH. In the absence of UE the soft handover UE TPC command generated gets transmitted in the first available TPC field in the uplink DPCCH. In the presence of soft handover, the UE checks the downlink power control mode (DPC_MODE)[5] before generating the TPC command as follows [5]:

- if DPC_MODE = 0, then the UE sends a unique TPC command in each slot and the TPC_cmd generated gets transmitted in the first available TPC field in the uplink DPCCH;
- if DPC_MODE = 1, then the UE repeats the same TPC_cmd over three slots and we transmit the new TPC_cmd aiming to have a new command at the beginning of the frame.

The average DPDCH power symbols[6] transmitted over one time slot do not exceed the Maximum_DL_Power (dBm), neither do they fall below the Minimum_DL_Power (dBm). These two powers become power limits for one spreading code. When the UE cannot generate TPC commands due to lack of synchronisation, the transmitted TPC command gets set to ‘1’ during the period of out of synchronisation.

Power changes occur in multiples of the minimum step size $\Delta_{TPC,min}$ (dB), where it is mandatory for UTRAN to support $\Delta_{TPC,min}$ of 1 dB, but optional to support 0.5 dB. The UTRAN may further employ the following method.

When the *limited power raise* parameter value applies, the UTRAN will not increase the DL power of the Radio Link (RL) if it would exceed by more than *Power_Raise_Limit* (dB) the averaged DL power used in the last *DL_Power_Averaging_Window_Size* time slots of the same RL. The latter applies only after the first *DL_Power_Averaging_Window_Size* time slots precede the activation of this method. The *Power_Raise_Limit* and *DL_Power_Averaging_Window_Size* parameter configuration occur in the UTRAN [5].

4.3.13.2 Power Control in the Downlink Compressed Mode

Compressed mode or slotted power control in uplink or/and downlink aims to recover as fast as possible the SIR close to the target SIR after each transmission gap. Practically,

[4]As a response to the received TPC commands, UTRAN may adjust the downlink DPCCH/DPDCH power.
[5]The DPC-MODE parameter is a UE specific parameter controlled by the UTRAN.
[6]Transmitted DPDCH symbols imply complex QPSK symbols before spreading, which does not contain DTX.

compressed mode intervenes when taking measurements into other frequency ranges from single mode WCDMA systems. The same UE behaviour of the preceding section applies to the compressed mode. Since the specifications do not describe the details of the UTRAN behaviour during the compressed downlink mode, algorithms of the UL compress mode may apply. Downlink DPCCH and DPDCH(s) transmission stops during DL compressed mode or in simultaneous DL and UL compressed modes. As [9] puts it, the use of compressed mode has an impact on the link performance as studied in [10] for the uplink compressed mode and for the downlink in [11]. The largest impact occurs at the cell edge, where the difference in the uplink performance between compressed mode and non-compressed mode cases is very small until headroom is less than 4 dB.

4.3.13.3 Power Control in Site Selection Diversity Transmit (SSDT)

The UE in SSDT, the optional macro-diversity method in soft handover mode, selects one of the cells from its active set to be 'primary', all other cells are classed as 'non-primary'. In this context, there are two goals. First, to transmit on the downlink from the primary cell, minimising thereby the interference resulting from multiple transmissions in a soft handover mode. Second, to achieve fast site selection without network intervention, thereby maintaining the advantage of the soft handover. To select a primary cell, each cell gets a temporary Identification (ID) and the UE periodically informs a primary cell ID to the connecting cells via the uplink FBI field. The non-primary cells selected by the UE switch off their transmission power. Upper layer signalling carries out SSDT activation, termination and ID assignment. See [5] for details on cell identification.

4.3.13.4 Power Control in The PDSCH, AICH, PICH and S-CCPCH

- The network can select *inner (fast) loop power control* based on the power-control commands sent by the UE on the uplink DPCCH or *slow power control* to realise the PDSCH power control.
- The UE gets information about the relative transmit power of the AICH,[7] compared to the primary CPICH transmit power by the higher layers.
- The UE also gets information about the relative transmit power of the PICH,[8] compared to the primary CPICH transmit power by the higher layers.
- The TFCI and pilot fields may have time-varying offset relative to the power of the data field.

4.3.14 The Compressed Mode Procedure

In the compressed mode we do not use TGL slots from N_{first} to N_{last} for data transmission. As shown in Figure 4.36 the instantaneous compress-frame transmit power increases to keep

[7]Measured as the power transmitted acquisition indicator.

[8]Measured as the power over the transmitted paging indicators, excluding the undefined part of the PICH frame.

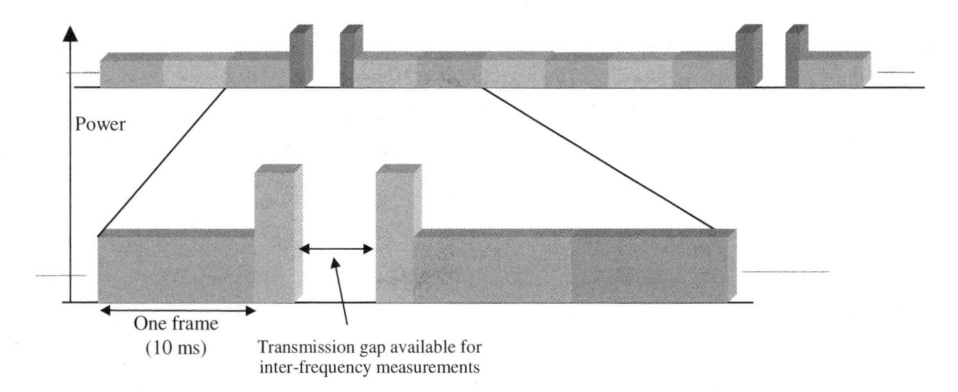

Figure 4.36 Compressed mode transmission (after [17]).

quality (e.g. BER, FER, etc.) despite reduced processing gain. The power increase depends on the transmission time reduction method under the network decision. Compressed frames may occur periodically or on demand. Compressed-frame rate and type vary and depend on the environment and the measurement requirements.

4.3.14.1 Uplink Frame Structure

Figure 4.37 illustrates the UL compressed mode structure showing the control slots and data transmission slots.

4.3.14.2 Downlink Frame Structure

We have two different types of downlink frame structures. Type A maximises the TGL and type B optimises power control.

- In type A we transmit the pilot field of the last slot in the transmission gap, and transmission gets turned off during the remaining of the transmission gap, see Figure 4.38(a).
- In type B we transmit the TPC field of the first slot and the pilot field of the last slot in the transmission gap. As in type A, transmission gets turned off during the remaining of the transmission gap, see Figure 4.38(a).

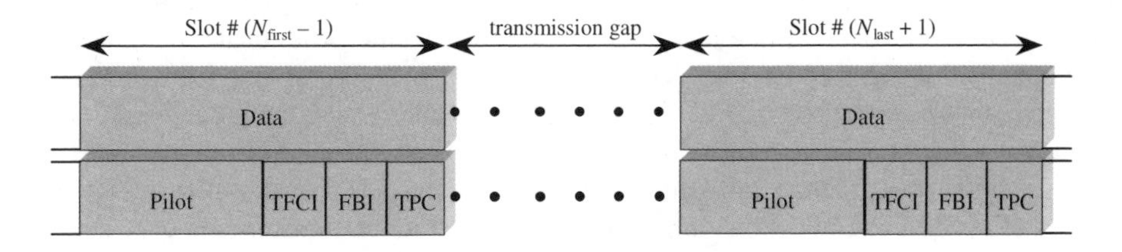

Figure 4.37 Uplink compressed frame structure.

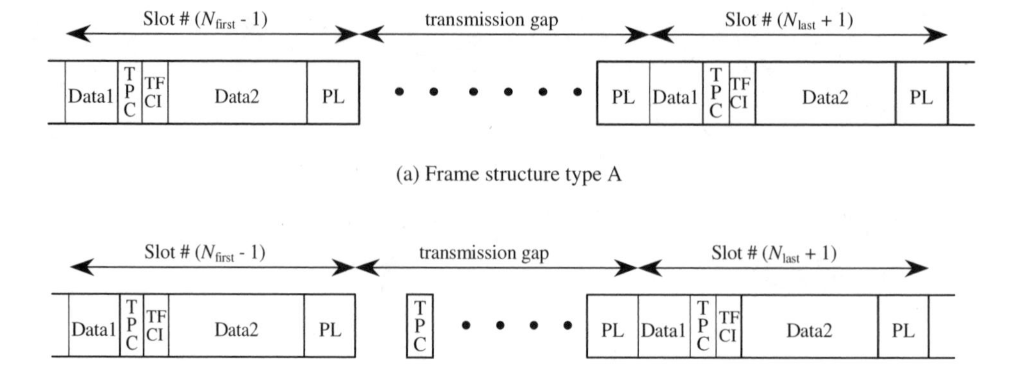

(a) Frame structure type A

(b) Frame structure type B

Figure 4.38 Frame structure types in downlink compressed transmission.

4.3.14.3 Transmission Time Reduction Method

In compressed mode we transmit in less time the information typically transferred in 10 ms frames. We achieve the latter through puncturing, halving spreading factor and higher layer scheduling. The DL compression supports all methods, while the DL excludes compression by puncturing. We define the maximum idle length as seven slots per 10 ms frame.

By Puncturing. Puncturing compression takes place through rate matching while using the rate matching or puncturing algorithm for rate matching.

By Halving the SF. In this mode we halve the SF during one radio frame to enable the transmission of the information bits in the remaining time slots of a compressed frame. In the DL the UTRAN may also command the UE to use a scrambling code different from the usual one. If the latter occurs then there exists a one-to-one mapping between the scrambling code used in normal mode and the one used in compressed mode.

Compression Through Higher Layer Scheduling. Higher layers set conditions so that only a subset of the allowed TFCs can apply in compressed mode.

4.3.14.4 Transmission Gap Position

We can place transmission gaps at different positions (Figure 4.39) for inter-frequency power measurement, acquisition of the control channel of another system/carrier and actual handover operation. Thus, when using the single frame method, we locate the transmission gap within the compressed frame depending on the Transmission Gap Length (TGL) as illustrated in Figure 4.39(a). When applying the double frame method, we locate the transmission gap at the centre of two connected frames as illustrated in Figure 4.39(b).

We calculate the parameters of the transmission gap positions as follows: TGL is the number of consecutive idle slots during the compressed mode transmission gap, i.e.

$$\mathrm{TGL} = 3, 4, 5, 7, 10, 14.$$

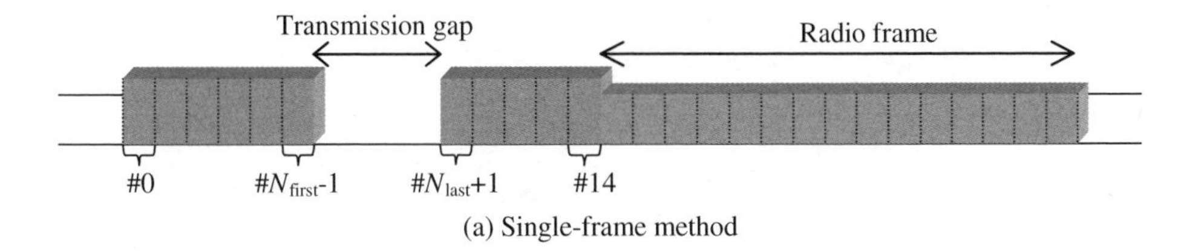

(a) Single-frame method

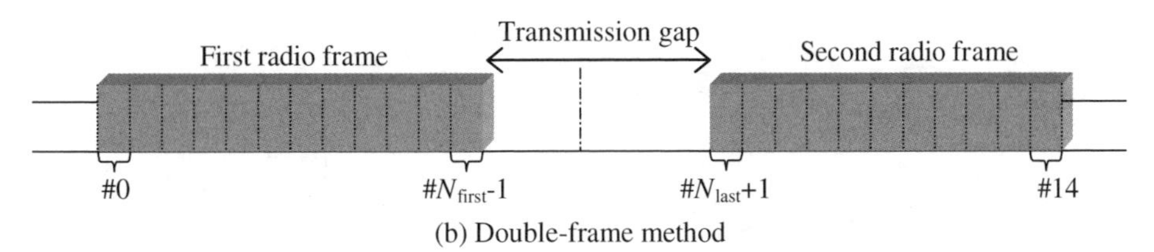

(b) Double-frame method

Figure 4.39 Transmission gap position [17].

N_{first} specifies the starting slot of the consecutive idle slots,

$$N_{\text{first}} = 0, 1, 2, 3, \ldots, 14.$$

N_{last} shows the number of the final idle slot and is calculated as follows:

- if $N_{\text{first}} + \text{TGL} \leq 15$, then $N_{\text{last}} = N_{\text{first}} + \text{TGL} - 1$ (in the same frame),
- if $N_{\text{first}} + \text{TGL} > 15$, then $N_{\text{last}} = (N_{\text{first}} + \text{TGL} - 1) \bmod 15$ (in the next frame).

When the transmission gap spans by two consecutive radio frames, we choose N_{first} and TGL so that at least eight slots in each radio frame get transmitted. Table 4.31 illustrates the detailed parameters for each transmission gap length for the different transmission time reduction methods.

4.3.15 Handover Procedures

The essential FDD mode handover types include intra-mode handover (i.e. soft, softer and hard[9] handover); inter-mode handover (i.e. handover to TDD); inter-system handover (e.g. handover to GSM).

Intra-Mode Handover. This handover depends on CPICH[10] power level measurements, which include the Received Signal Code Power (RSCP), Received Signal Strength Indicator (RSSI) and the E_c/N_o resulting from the RSCP/RSSI ratio. The other measurement involved is the relative timing information between the cells. Cells within a 10 ms window generally

[9]Hard handover may also take place as intra- or inter-frequency handover.
[10]That is, the E_c/N_o measurement performed at the common pilot channel.

Table 4.31 Parameters for compressed mode[a] [10]

TGL	Frame type	Spreading factor	Idle length (ms0)	Transmission time reduction method	Idle frame combining[b]
3	A	512–4	1.73–1.99	Puncturing Spreading factor division by 2 or Higher layer scheduling	(S) = Single frame method
	B	256–4	1.60–1.86		(D) = (1,2) or (2,1)
4	A	512–4	2.40–2.66		(S)
	B	256–4	2.27–2.53		(D) = (1,3), (2,2) or (3,1)
5	A	512–4	3.07–3.33		(S)
	B	256–4	2.94–3.20		(D) = (1,4), (2,3), (3, 2) or (4,1)
7	A	512–4	4.40–4.66		(S)
	B	256–4	4.27–4.53		(D) = (1,6), (2,5), (3,4), (4,3), (5,2) or (6,1)
10	A	512–4	6.40–6.66		(D) = (3,7), (4,6), (5,5), (6,4) or (7,3)
	B	256–4	6.27–6.53		
14	A	512–4	9.07–9.33		(D) = (7,7)
	B	256–4	8.93–9.19		

[a]Compressed mode by spreading factor reduction is not supported when SF = 4 is used in normal mode.

[b]S–single-frame method as illustrated in Figure 4.39(a); D, double-frame method as illustrated in Figure 4.39(a); x is the number of idle slots in the first frame; y is the number of idle slots in the second frame.

get relative timing from the primary scrambling code phase since the code period used is 10 ms. If large inaccuracies occur, the terminal decodes the System Frame Number (SFN) from the primary CCPCH.

Inter-Mode Handover. Multimode or dual FDD-TDD mode terminals will afford inter-mode handover. While in the FDD, an MS will measure the power level in synchronised TDD cells with useful reference midambles to execute the handover process.

Inter-System Handover. Inter-system handover for UTRA takes an important proportion of all the inter-operability and system integration tasks. It plays an important role in the evolution process of second and third generation mobile systems. In particular, UTRA needs to inter-operate seamlessly with the family of IMT-2000 networks and co-exist for some time with all types of deployment scenarios. However, complete inter-operability will not necessarily occur at the introduction of UMTS, but will follow a gradual process. Thus, in this section we mainly discuss essential handover issues with GSM. Handover with other systems such as MC-CDMA and IS-136, e.g. will not be cover at this time.

We introduce *inter-system handover* based on the handover recommendations from GSM to UMTS. Specifications in this area continue in this writing. Thus, here we look at it primarily from the requirement side. The handover may consist of the following aspects [12]: cell selection/reselection and handover features.

4.3.15.1 Cell Selection/Reselection Requirements

The MS compares GSM quality with UMTS neighbour cells to determine the most appropriate cell for cell selection/reselection. The MS in a GSM cell obtains system information of its UMTS neighbour cells through BSC broadcasting all the required information in the GSM cell. Broadcasting UMTS cell information depends on the definition of new system information messages or the adaptation of existing system information messages together with the transmission of area-based UMTS system information (e.g. UTRAN frequency/ies used in LA) [12].

4.3.15.2 Handover Requirements

The requirements for GSM to UMTS handover as an extract from [12] can be outlined as follows:

1. *Synchronisation requirements*
   - A MS will synchronise with a UTRAN cell using GSM idle frame(s).
   - A MS will be capable of blind detection; thus, it will not reject a handover command to an UMTS or GSM cell which it has not reported and to which it is not synchronised.

2. *GSM MS requirements*
   - R99 and newer MSs will support a 'blind handover' to GSM or UMTS.
   - When a handover fails, the MS will remain camped onto the original cell, and continue its measurement reporting as defined prior to the attempted handover.

3. *GSM BSC requirements*
   - The call will go to the most suitable cell for the service that the user requested.
   - The BSC needs to know the service in progress to determine suitable cells to direct the call to.
   - The BSC will provide the MS with mapping parameters, which will enable the MS to obtain the set of most suitable cells for the measurement report.
   - The BSC may provide the MS with the Q search parameters to trigger measurement of other radio-access technologies; a separate bit may be used to indicate whether the UMTS measurements are triggered when the GSM RXLEV measurement for the current cell is below or above Q search.
   - In GSM to UMTS handover, it is the working assumption that the source GSM BSC shall provide the ID of the target RNC; it is also assumed that the source RNC to target RNC transparent container will be created by the BSC.
   - In order to optimise non-synchronised ('blind') handover, the GSM BSC shall provide additional information about the target cell (e.g. scrambling code, synchronisation).

4. *UMTS RNC*
   - The RNC will broadcast the parameter M offset on its BCCH; this offset is set by the network, and is used to adjust the comparison of UMTS with GSM measurements.

5. *UMTS measurements from GSM MS*

   The requirements on the measurements made by the MS are:

   - measurement on UTRAN cells by the MS will not have a significant impact on the measurement ability and performance of the MS for support of GSM-to-GSM handover; it is assumed that the MS uses search frames for UTRAN measurements and that the UTRAN cells are only monitored during idle search frames,
   - the maximum time for detecting a new suitable UTRAN cell relates to the number of UTRAN frequencies under monitoring,
   - the time it takes to detect, confirm BSIC and report a new suitable GSM cell applies also for detecting and reporting a new suitable UTRAN cell when one UTRAN frequency remains under monitoring,
   - the UE, when in connected mode, does not need to read BCCH on the UTRAN cells under measurement,
   - the operator will be able to provide the UE with information that enables the UE to activate the actual physical measurements only when considered needed; e.g. when the quality of the GSM cell falls below a certain threshold.

6. *Measurement reports*

   The requirements on the measurement reporting are:

   - the UE will include both UTRAN and GSM measurements in the periodic measurement reports which are sent to the BSS,
   - the operator will have the capability to control the minimum number of GSM cells in the serving band, the minimum number of cells in UMTS and the minimum number of cells in other GSM bands which are included in the measurement report,
   - if there is any space remaining in the measurement report after inserting the minimum number of entries for each reporting type, the MS will use this space to insert further measurements in decreasing order of priority,
   - both the existing format measurement report and the extended format measurement report (e.g. 76/00 and 2B00-009) will be supported by the MS,
   - Release 1999 and newer MSs will support extended measurement reporting,
   - the network will inform the MS whether extended measurement reporting is supported; the default measurement reporting type is the normal measurement reporting.

7. *Signalling*

   The signalling between the GSM network and UMTS network to perform handover needs the following modifications (Table 4.32):

   - the GSM system will provide the UMTS system with the target ID of the RNC to which the call is being directed.

Table 4.32 Allowed handover[a] for GSM Release 1999 [12]

	GSM CS	GSM GPRS	GSM ECSD	GSM EGPRS nRT	GPRS	UTRA FDD	UTRA TDD	Multi-Carrier	IS-136
GSM CS	✓	×	✓	×	×	✓	✓	✓	×
GSM GPRS	×	×	×	×	×	×	×	×	×
GSM ECSD	✓	×	✓	×	×	✓	✓	✓	×
GSM EGPRS nRT	×	×	×	×	×	×	×	×	×
GSM GPRS COMPACT	×	×	×	×	×	×	×	×	×

[a]✓ stands for handover permitted; × for handover not permitted.

4.3.16 Other FDD Mode Physical Layer Procedures

Paging Channel (PCH). The MS gets a Paging Indicator (PI) belonging to a paging group once it registers in a network. The PI appears periodically on the Paging Indicator Channel (PICH) whenever paging messages exist, and the MS decodes[11] the next PCH frame transmitted on the secondary CCPCH, seeking for the messages corresponding to it.

The RACH. To cope with the power control uncertainty and near-far impacts, e.g. the following events correspond to the RACH procedures in the terminal:

1. identifying and eventual selection of available RACH sub-channels with scrambling codes and signatures through BCH decoding;
2. measurement of DL power level and setting of initial RACH power level;
3. sending of 1 ms RACH preamble with chosen signature;
4. AICH decoding to verify preamble detection[12] by the BS;
5. transmitting[13] 10 or 20 ms RACH message part at the AICH detection.

CPCH. The CPCH follows basically the same events as the RACH, differing only on L1 collision detection (see [5]). Applying fast power control on the CPCH we minimise interference due to the data transmission.

Cell search. The cell search procedure employing the synchronisation channel uses different scrambling codes with different phase shifts of the code. The events include

a. searching 256 chips primary synchronisation code;[14] since the latter is the same in every slot, the peak detected corresponds to the slot boundary;

[11] Battery life duration will increase with the lowest amount of PI detection events.
[12] In the absence of AICH the terminal resends preamble with higher power in the next available slot.
[13] When the RACH transmits data, the SF and thereby the data are fluctuate.
[14] Identical to all cells.

b. on the peak detection of the primary synchronisation code, the MS will look at the largest peak from the secondary SCH code word, i.e. from among the 64 options;

c. at the detection of the secondary SCH code word the frame timing is known.

4.4 CONFIGURATION OF TDD PHYSICAL CHANNELS

TDD physical channels illustrated in Figure 4.40 have three-layer structure with respect to Time Slots (TSs), radio frames and System Frame Number (SFN). The radio frame configurations or time slots vary according to the resource allocation. All physical channels use guard symbols in every time slot. The latter serve as the TDMA component to separate different user signals in time and code domains.

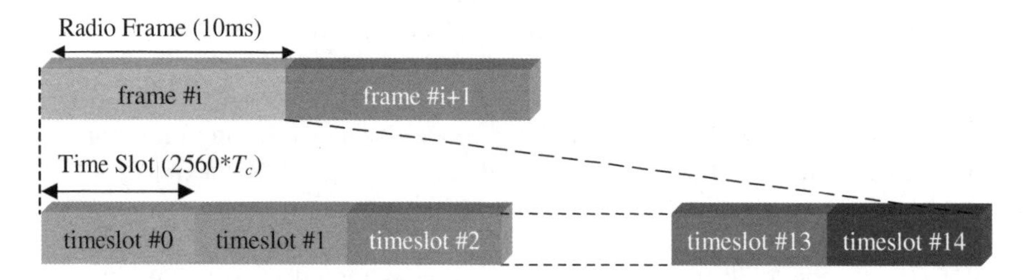

Figure 4.40 Physical channel signal format.

A TDD physical channel is burst (i.e. a combination of a data part, a midamble and a guard period), and transmitted in a particular time slot within allocated radio frames. A burst lasts one time slot and its allocation can be either continuous (i.e. in every frame) or discontinuous (i.e. only one in a subset of radio frames). Several bursts can be transmitted at the same time from one transmitter. In this case, the data part must use different OVSF channelisation codes, but the same scrambling code. The midamble part has to use the same basic midamble code, but can use different midambles [13].

The data part of the burst has a combined spread of channelisation and scrambling codes. The OVSF channelisation code can have a spreading factor of 1, 2, 4, 8 or 16, where the data rate of the physical channel will depend on the spreading factor used.

The midamble part of the burst may contain two different types of midambles, i.e. a short one with a length of 256 chips or a long one with 512 chips. The midamble size also has an impact on the data rate of the physical channel.

Thus, we define a TDD physical channel by frequency, time slot, channelisation code, burst type and radio frame allocation. Scrambling and basic midamble codes' broadcast may be constant within a cell. After the physical channel establishment a frame start event occurs with infinite or limited duration.

4.4.1 Frame Structure

As in the FDD mode, a TDMA frame in the TDD mode has a duration of 10 ms with a subdivision into 15 TSs of $2560 \times T_c$ duration each. Hence, a TS corresponds to 2560 chips,

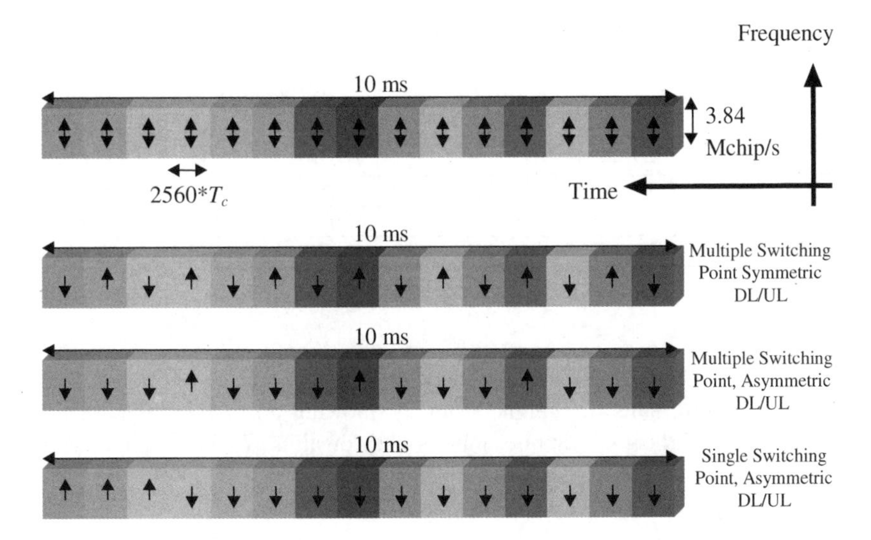

Figure 4.41 The TDD frame structure.

each allocated to either the uplink or the downlink as illustrated in Figure 4.41. This flexibility allows the TDD mode to adapt itself to different environments and deployment scenarios. Nonetheless, in any configuration there must be at least one time slot in the downlink and at least one time slot in the uplink.

4.4.2 Dedicated Physical Channel (DPCH)

4.4.2.1 Downlink and Uplink Physical Channel Spreading

We map DCH onto the dedicated physical channel. The two-step spreading of the data part in the physical channels include channelisation and scrambling operations. The first operation transforms every data symbol into a number of chips, thereby increasing the bandwidth of the signal. We call this number of chips per data symbol the Spreading Factor (SF). The second operation scrambles the spread signal through a scrambling code [14].

Downlink physical channels use SF = 16 as described in [14], and operation with a single code with SF = 1 can apply to downlink physical channels. To support higher data rates we can utilise multiple parallel physical channel transmission using different channelisation codes.

Uplink physical channels have SF ranging from 16 down to 1. In multi-code transmission a UE simultaneously uses two physical channels per time slot maximum. These parallel physical channels transmit using different channelisation codes [14].

4.4.3 Burst Types

The two types of dedicated physical channel bursts, i.e. Burst1 (B1) and Burst2 (B2), consist of two data symbol fields, a midamble and a guard period. Burst1, because of its longer

Table 4.33 TDD burst characteristics

	Symbols per data field		Chip number		Field length		Field content (symbols)	
SF	B1	B2	B1	B2	B1	B2	B1	B2
1	976	1104	0–975	0–1103	976	1104	Data	Data
2	488	552	976–1487	1104–1359	512	256	Mid	Mid
4	244	276	1488–2463	1360–2463	976	1104	Data	Data
8	122	138	2464–2559	2464–2559	96	96	GP	GP
16	61	69						

midamble of 512 chips, suits the uplink better by allowing up to 16 channel estimations. Burst2, which has only 256 chips, applies more to the downlink. We can summarise the burst use as follows:

	Burst1	Burst2
Uplink	Independent of the number of active users in one time slot	When bursts within a time slot are allocated to less than four users
Downlink	Independent of the number of active users in one time slot	Independent of the number of active users in one time slot

Burst1 has data fields of 976 chips, whereas Burst2 has data fields of 1104 chips. The corresponding number of symbols depends on the spreading factor, as seen in Table 4.33. Both bursts have 96 chip long Guard Periods (GPs).

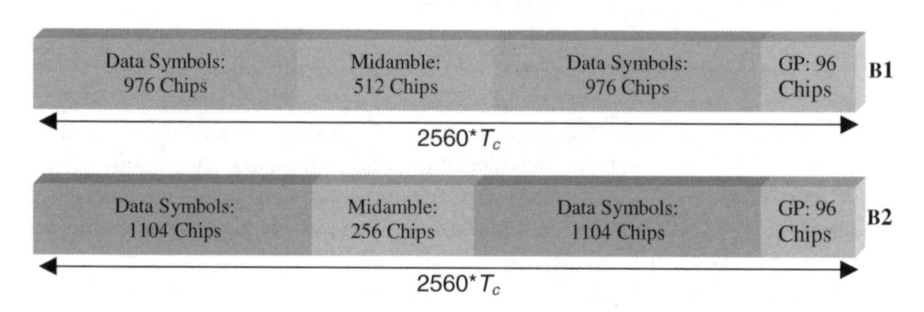

Figure 4.42 TDD burst types.

The two different bursts illustrated in Figure 4.42 and the characteristics noted in Table 4.33 can support a different set of applications and also allows optimisation for particular operational environments within the unlicensed frequency range.

4.4.3.1 Transmission of TFCI

Both bursts (B1 and B2) afford uplink and downlink TFCI transmission. This transmission is negotiated at call setup and re-negotiation may occur during a call. Upper layer signalling indicates TFCI formats for each CCTrCH with information in the presence or absence of TFCI. When a time slot contains a TFCI, transmission takes place using the first allocated channelisation code in the time slot.

The data parts of a corresponding physical channel realise the TFCI transmission following the same spreading procedures outlined in [14], and while keeping the midamble structure illustrated in Figure 4.42. We transmit the TFCI information directly adjacent to the midamble, and after the TPC in the presence of power control commands. Figure 4.43 illustrates the two cases.

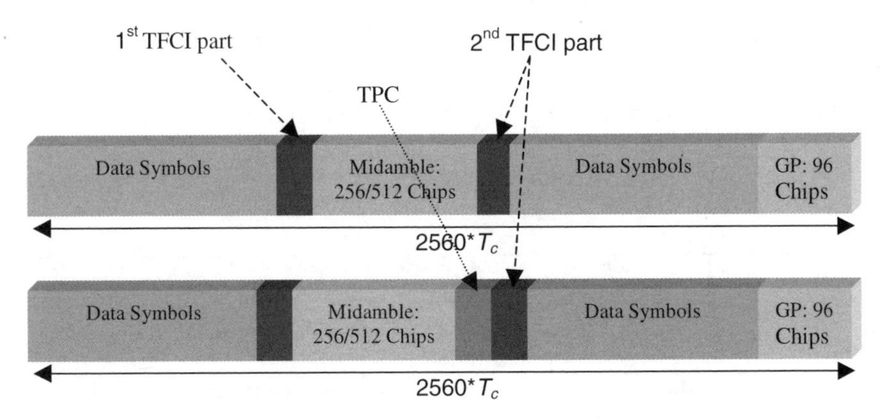

Figure 4.43 TFCI information positions in traffic bursts.

Both burst types 1 and 2 for dedicated channels provide the possibility for transmission of TPC in uplink.

The transmission of TPC is negotiated at call set-up and can be re-negotiated during the call. If applied, transmission of TPC is done in the data parts of the traffic burst. Hence the midamble structure and length is not changed. The TPC information is to be transmitted directly after the midamble. Figure 4.42 shows the position of the TPC in a traffic burst.

For every user the TPC information is to be transmitted once per frame. If the TPC is applied, then it is always transmitted using the first allocated channelisation code and the first allocated time slot, according to the order in the higher layer allocation message. The TPC is spread with the same SF and spreading code as the data parts of the respective physical channel. Specifications in [13] cover time slot formats and training sequences for spread bursts.

4.4.3.2 Midamble Transmit Power and Beamforming/Transmit Diversity

When all one-time slot downlink users have a common midamble, this common mid-amble's transmit power has no power offset between the data part and the midamble part of the transmit signal within the given slot. Likewise, transmit power of users with specific midambles does not have power offset between the data parts and the midamble part. In the event of DL beamforming or Tx diversity, the user who has beamforming/Tx diversity and a dedicated channel, will get one individual midamble.

4.4.4 Common Physical Channels

4.4.4.1 Primary and Secondary Common Control Physical Channels

We map the BCH onto the Primary Common Control Physical Channel (P-CCPCH), and obtain the position (time slot/code) of the P-CCPCH from the Physical Synchronisation Channel (PSCH).

We also map PCH and FACH onto one or more Secondary Common Control Physical Channels (S-CCPCH). Through the PCH, the FACH adapts itself to different requirements. Table 4.34 summarises the key P-CCPCH and S-CCPCH features.

4.4.5 The Physical Random Access Channel (PRACH)

The RACH maps onto one or more uplink PRACHs affording thereby flexible and scaleable capacity to the RACH.

	PRACH
Spreading	The uplink PRACH uses either spreading factor SF = 16 or SF = 8, where the BCH broadcasts the set of admissible spreading codes and the associated spreading factors.
Burst types	Mobiles send uplink access bursts randomly through the PRACH, which has a burst consisting of two data symbol fields, a midamble and a guard period as seen in Figure 4.44. The second data symbol field has only 880 chips instead 976 to allow larger guard time [13]. The collision groups depend on the selected RACH configuration. For example, with SF = 8, the first data field has 122 symbols and the second 110. Likewise, with SF = 16, field1 has 61 and field2 has 55 symbols, respectively. Table 7 in [13] illustrates the PRACH burst field content.
Training Sequences (TS)	Different active users in the same time slot have time-shifted TS or midamble versions of single periodic basic code described in [13]. We choose time shifts from *all* $k = 1, 2, 3, \ldots, K'$ (for cells with small radius) or *uneven* $k = 1, 3, 5, \ldots, \leq K'$ (for cells with large radius).
TS and channelisation code association	We base the generic rule to define this association on the channelisation codes $c_Q^{(k)}$ order given by k and midambles $m_j^{(k)}$ order given by k, first, and j, second, with the constraint that the midamble for a spreading SF factor is the same as in the upper branch for the spreading factor 2SF. The index $j = 1$ or 2 indicates whether we use the original basic midamble sequence ($j = 1$) or the time–inverted basic midamble sequence ($j = 2$) [13].

4.4.6 The Synchronisation Channel (SCH)

The synchronisation channel provides the code group of a cell. To prevent uplink/downlink asymmetry limitations we map the SCH on one or two downlink slots per frame only. The two cases of SCH and P-CCPCH allocation include: first, SCH and P-CCPCH allocated in

Table 4.34 P-CCPCH features

	P-CCPCH	S-CCPCH
Spreading	It uses fixed spreading with a spreading factor SF = 16 and always channelisation code $c_{Q=16}^{(k=1)}$	It uses fixed spreading with a spreading factor SF = 16
Burst types	Burst1 type with no TFCI applies	Utilises bursts type 1 and 2 with TFCI
Training sequences, i.e. midambles	Time slots carrying P-CCPCH transmission use midambles $m^{(1)}$, $m^{(2)}$, $m^{(9)}$ and $m^{(10)}$ in order to support block STTD antenna diversity and the beacon function; see the description in [13]	The training sequences described in [13] apply to the S-CCPCH
Block STTD antenna diversity	Its support is mandatory for the UE as follows. If no antenna diversity, it uses $m^{(1)}$ while $m^{(2)}$ remains unused. If in block STTD antenna diversity exists, the first antenna uses $m^{(1)}$ and the diversity antenna uses $m^{(2)}$	

TS#k, $k = 0, \ldots, 14$; and second SCH allocated in two TS (TS#k and TS#$k+8$, $k = 0, \ldots, 6$; P-CCPCH allocated in TS#k). The position of SCH (value of k) in the frame can change in the long term in either of the two cases and allow knowledge of the position of P-CCPCH from the SCH. Specifications in [13] and [14] give more details for the SCH.

4.4.7 Physical Uplink/Downlink Shared Channels

The Physical Uplink Shared Channel (PUSCH), which provides uplink TFCI transmission possibilities, uses the DPC burst structure, where user specific physical layer parameters, e.g. power control, timing advance or directive antenna settings come from the associated channel (i.e. FACH or DCH).

The Physical Downlink Shared Channel (PDSCH), which provides downlink TFCI transmission possibilities, uses the DPCH burst structure. As in the PUSCH, specific L1 parameters, e.g. power control or directive antenna settings, come from the associated channel (FACH or DCH).

The DSCH utilises three signalling methods to inform the UE that it has data to decode:

a. using the TFCI field of the associated channel or PDSCH;

b. using the DSCH user specific midamble derived from the set of midambles used for that cell;

c. using higher layer signalling.

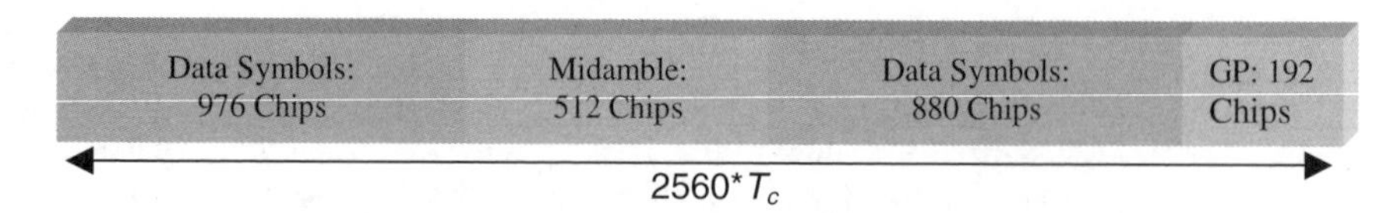

Figure 4.44 PRACH burst configuration.

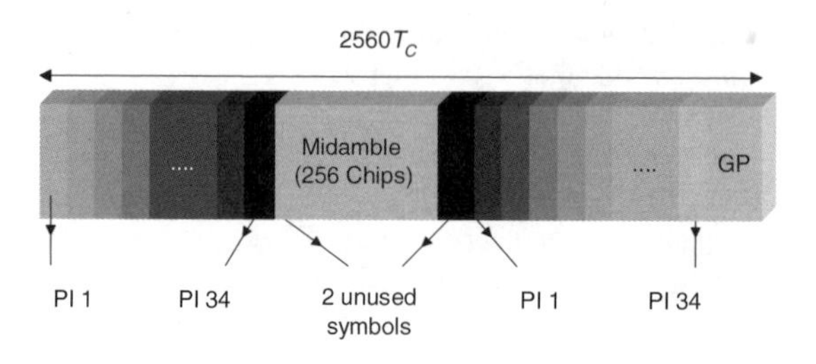

Figure 4.45 Example of PI transmission in PICH bursts ($L_{PI} = 4$).

In the last method, the UE decodes the PDSCH if the PDSCH was transmitted with the midamble assigned to the UE by UTRAN.

4.4.8 The Page Indicator Channel (PICH)

The Page Indicator Channel[15] (PICH) carries the Page Indicators (PIs), which indicate a paging message for one or more UEs associated with it, and is always transmitted at the same reference power level as the P-CCPCH. The PICH substitutes one or more paging sub-channels mapped on a S-CCPCH.

Figure 4.45 illustrates normal bursts that carry PIs of length $L_{PI} = 2$, $L_{PI} = 4$ or $L_{PI} = 8$ symbols, and Table 4.35 illustrates the number of page indicators (N_{PI}) per time slot given by the number L_{PI} symbols for the page indicators and the burst type.

The same burst type is used for the PICH in every cell. As illustrated in Figure 4.45 when $L_{PI} = 4$ or $L_{PI} = 8$, we leave behind one symbol in each data part adjacent to the midamble and fill it by dummy bits transmitted with the same power as the PI [13].

Table 4.35 Number N_{PI} of PI per time slot for the different burst types and PI lengths L_{PI}

	$L_{PI} = 2$	$L_{PI} = 4$	$L_{PI} = 8$
Burst type 1	61	30	15
Burst type 2	69	34	17

[15]Physical channel.

4.4.9 Beacon Function in Physical Channels

Depending on its allocation case, the SCH determines the location of the physical channels with beacon function for the purpose of measurements. In case 1, all physical channels with channelisation code $c_{Q=16}^{(k=1)}$ and in TS#k, $k = 0, \ldots, 14$, allocation provides the beacon function. In case 2 all physical channels with channelisation code $c_{Q=16}^{(k=1)}$ and in TS#k and TS#$k+8$, $k = 0, \ldots, 6$, allocation also provides the beacon function. Thereby, the P-CCPCH always provide the beacon function.

The physical channels providing the beacon function transmit with reference power and without beamforming, use type 1 burst employing midamble $m^{(1)}$ and $m^{(2)}$ exclusively, while midambles $m^{(9)}$ and $m^{(10)}$ remain unused in this time slot when the cell allows 16 midambles.

The reference power equals the sum of the power allocated to both midambles $m^{(1)}$ and $m^{(2)}$. According to [13] two options are as follows:

- In the absence of block STTD antenna diversity application to the P-CCPCH, all the reference power of any physical channel providing the beacon function goes to $m^{(1)}$.
- When block STTD antenna diversity applies to the P-CCPCH, physical channels providing beacon function midambles $m^{(1)}$ and $m^{(2)}$ share the reference power, i.e. midamble $m^{(1)}$ applies to the first antenna and $m^{(2)}$ applies to the diversity antenna. The data in P-CCPCH uses block STTD encoding [15]. For all other physical channels, both antennas transmit identical data sequences.

4.4.10 Allocating Midamble to Physical Channels

Generally high layers configure DL physical channels with midambles. Otherwise, they allocate default midambles by fixed association between midambles and channelisation codes. Different associations apply for different burst types and cell configurations with respect to the maximum number of midambles. Physical channels providing the beacon function shall always use the reserved midambles. For all other DL physical channels the midamble allocation is signalled or given by default.

In the UL, if the physical channel has a midamble as part of its configuration, we assign an individual midamble to all UEs in one time slot. Otherwise, when higher layers do not allocate midambles, the UE will derive the midamble from the assigned channelisation code as for DL physical channels. If the UE changes the SF according to the data rate, it shall always vary the channelisation code along the lower branch of the OVSF tree. See more midamble details in [13].

4.4.11 Mapping Transport Channels onto Physical Channels

Table 4.36 summarises the mapping of the transport channels onto the physical channels.

4.4.11.1 Dedicated Transport Channels

We map a dedicated transport channel onto one or more physical channels, where an inter-leaving period association occurs with each allocation. The frame is sub-divided into slots that are available for uplink and downlink information transfer.

Table 4.36 Mapping of transport channels to physical channels

Transport channels	Map to	Physical channels
DCH	→	Dedicated Physical Channel (DPCH)
BCH	→	Primary Common Control Physical Channel (P-CCPCH)
FACH	→	Secondary Common Control Physical Channel (S-CCPCH)
PCH	→	
RACH	→	Physical Random Access Channel (PRACH)
USCH	→	Physical Uplink Shared Channel (PUSCH)
DSCH	→	Physical Downlink Shared Channel (PDSCH)
	→	Page Indicator Channel (PICH)
	→	Synchronisation Channel (SCH)

For NRT packet data services, shared channels (USCH and DSCH) can be used to allow efficient allocations for a short period of time.

4.4.12 Mapping Common Transport Channels

4.4.12.1 The Broadcast Channel (BCH)

We map the BCH onto the P-CCPCH, where the secondary SCH indicates in which time slot a mobile can find the P-CCPCH containing a BCH. For additional resources the BCH in P-CCPCH will comprise a pointer to additional FACH S-CCPCH resources in which this additional broadcast information will occur.

4.4.12.2 The Paging Channel (PCH)

We map the PCH onto one or several S-CCPCHs while matching capacity to requirements, indicate its location on the BCH and always transmit it at a reference power level. To allow an efficient DRX, the PCH is divided into several paging sub-channels within the allocated multi-frame structure. See examples of multi-frame structures in [13]. Each paging sub-channel comes mapped onto two consecutive frames allocated to the PCH on the same S-CCPCH. Layer 3 information to a particular paging group arrives through the associated paging sub-channel. UE assignment to paging groups occurs independent of the assignment of UEs to paging indicators.

4.4.12.3 The Forward Channel (FACH)

We map the FACH onto one or several S-CCPCHs. Indication of FACH location comes on the BCC, where both capacity and location can be changed when necessary. The FACH may or may not have power control.

Table 4.37 Basic modulation parameters [14]

Chip rate	Same as FDD basic chip rate: 3.84 Mchip/s	Low chip rate: 1.28 Mchip/s
Data modulation	QPSK	QPSK
Spreading characteristics	Orthogonal Q chips/symbol, where $Q = 2^p, 0 \leq p \leq 4$	Orthogonal Q chips/symbol, where $Q = 2^p, 0 \leq p \leq 4$

4.4.12.4 The Random-Access Channel (RACH)

The RACH, which we map onto the PRACH, has intra-slot inter-leaving only. One or more cells may use the same slots for PRACH. However, more than one slot per frame may be administered for the PRACH. The BCH broadcasts the location of slots allocated to PRACH. The latter uses open-loop power control with algorithms, which may differ from the ones used on other channels. Multiple transmissions using different spreading codes may be received in parallel [13].

4.4.12.5 Shared Channels

We map the Uplink Shared Channel (USCH) on one or several PUSCHs. Likewise, we map the Downlink Shared Channel (DSCH) on one or several PDSCHs.

4.5 SPREADING AND MODULATION IN TDD

4.5.1 Modulation and Symbol Rate

Table 4.37 illustrates the TDD basic modulation parameters. Notice that it has a low chip rate option at 1.28 Mchip/s. The complex-valued chip sequence is QPSK modulated as illustrated in Figure 4.46.

In this section we use Q for the spreading, while SF denotes spreading in the FDD mode.

The symbol duration T_S depends on the spreading factor Q and the chip duration T_c: $T_s = Q \times T_c$, where $T_c = 1/\text{chip rate}$.

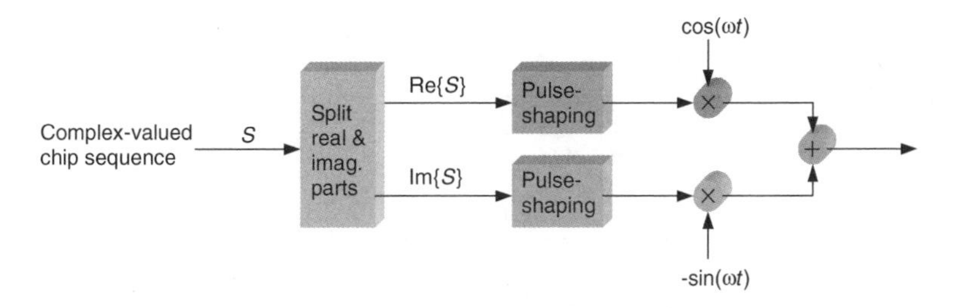

Figure 4.46 Modulation of complex-valued chip sequences.

4.5.2 Mapping of Bits onto Signal Point Constellation

4.5.2.1 Mapping for Burst Type 1 and 2

We perform data modulation on the bits from the output of the physical channel mapping procedure in [16] and combine always two consecutive binary bits to a complex valued data symbol. Each user burst has two data carrying parts, termed data blocks

$$\underline{\mathbf{d}}^{(k,i)} = (\underline{d}_1^{(k,i)}, \underline{d}_2^{(k,i)}, \ldots, \underline{d}_{N_k}^{(k,i)})^{\mathrm{T}}, \quad i = 1,2;\ k = 1, \ldots, K \tag{4.47}$$

N_k corresponds to the number of symbols per data field for the user k. We link this number to the spreading factor Q_k as described in Table 1 of [13].

Data block $\underline{\mathbf{d}}^{(k,1)}$ gets transmitted before the midamble and data block $\underline{\mathbf{d}}^{(k,2)}$ after the midamble. Each of the N_k data symbols $\underline{d}_n^{(k,i)}$; $i = 1,2$; $k = 1, \ldots, K$; $n = 1, \ldots, N_k$; of eqn (4.47) has the symbol duration $T_s^{(k)} = Q_k T_c$ as already given.

The data modulation is QPSK, thus the data symbols $\underline{d}_n^{(k,i)}$ are generated from two consecutive data bits from the output of the physical channel mapping procedure in [16]

$$b_{l,n}^{(k,i)} \in \{0,1\}, \quad l = 1,2; \quad k = 1, \ldots, K; \quad n = 1, \ldots, N_k; \quad i = 1,2 \tag{4.48}$$

using the following the mapping to complex symbols illustrated in Table 4.38.

The mapping corresponds to a QPSK modulation of the interleaved and encoded data bits $b_{l,n}^{(k,i)}$ of eqn (4.48).

4.5.2.2 Mapping for PRACH Burst Type

When mapping the PRACH burst type the preceding logic applies with a modified number of symbols in the second data block. Thus, for the PRACH burst type, the number of symbols in the second data block $\underline{\mathbf{d}}^{(k,2)}$ is decreased by $96/Q_k$ symbols.

4.5.3 Spreading Parameters and Channelisation Codes

Data spreading includes two steps, i.e. channelisation and scrambling. First, each complex valued data symbol $\underline{d}_n^{(k,i)}$ of eqn (4.47) gets spread with a real-valued channelisation code $\mathbf{c}^{(k)}$

Table 4.38 Mapping complex symbols

Consecutive binary bit pattern	Complex symbol
$b_{l,n}^{(k,i)}\ b_{2,n}^{(k,i)}$	$\underline{d}_n^{(k,i)}$
00	$+j$
01	$+1$
10	-1
11	$-j$

of length $Q_k \in \{1, 2, 4, 8, 16\}$. We then scramble the resulting sequence by a complex sequence $\mathbf{v}$ of length 16.

The elements $c_q^{(k)}; k = 1, \ldots, K; q = 1, \ldots, Q_k$; of the real-valued channelisation codes, i.e.

$$c^{(k)} = (c_1^{(k)}, c_2^{(k)}, \ldots c_{Q_k}^{(k)}), \quad k = 1, \ldots, K \tag{4.49}$$

will be taken from the set

$$V_c = \{1, -1\} \tag{4.50}$$

The $c_{Q_k}^{(k)}$ belongs to Orthogonal Variable Spreading Factor (OVSF) codes, which allow mixing in the same time slot channels with different spreading factors while preserving the orthogonality. We define the OVSF codes using the code tree illustrated in Figure 4.47.

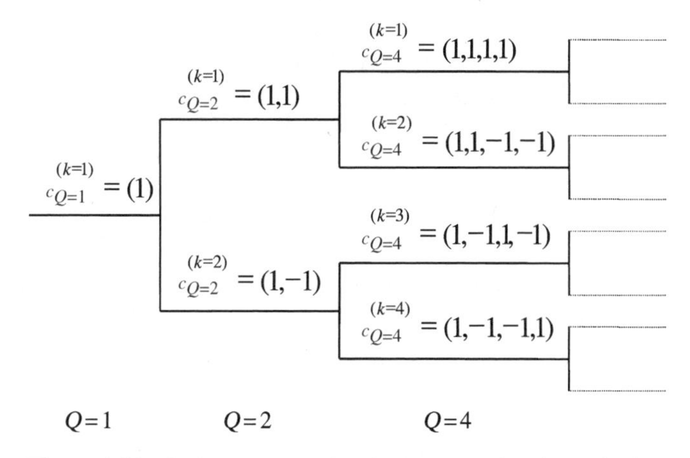

Figure 4.47 Code tree generating OVSF codes for channelisation.

Each level in the code tree defines a SF indicated by the value of Q in Figure 4.47. We may not use all codes within the code tree simultaneously in a given time slot. We can use a code in a time slot if and only if no other code on the path from the specific code to the root of the tree, or in the sub-tree below the specific code is used in this time slot. This implies that the number of available codes in a slot depends on the rate and spreading factor of each physical channel. The SF goes up to $Q_{\max} = 16$ [14].

4.5.4 Scrambling Codes

Data spreading by a real-valued channelisation code $c^{(k)}$ of length Q_k gets followed by a cell-specific complex scrambling sequence $\underline{\mathbf{v}} = (\underline{v}_1, \underline{v}_2, \ldots, \underline{v}_{16})$. The elements $\underline{v}_i$, $i = 1, \ldots, 16$, of the complex-valued scrambling codes originates from the complex set

$$\underline{V}_{\underline{v}} = \{1, j, -1, -j\} \tag{4.51}$$

where j denotes the imaginary unit.

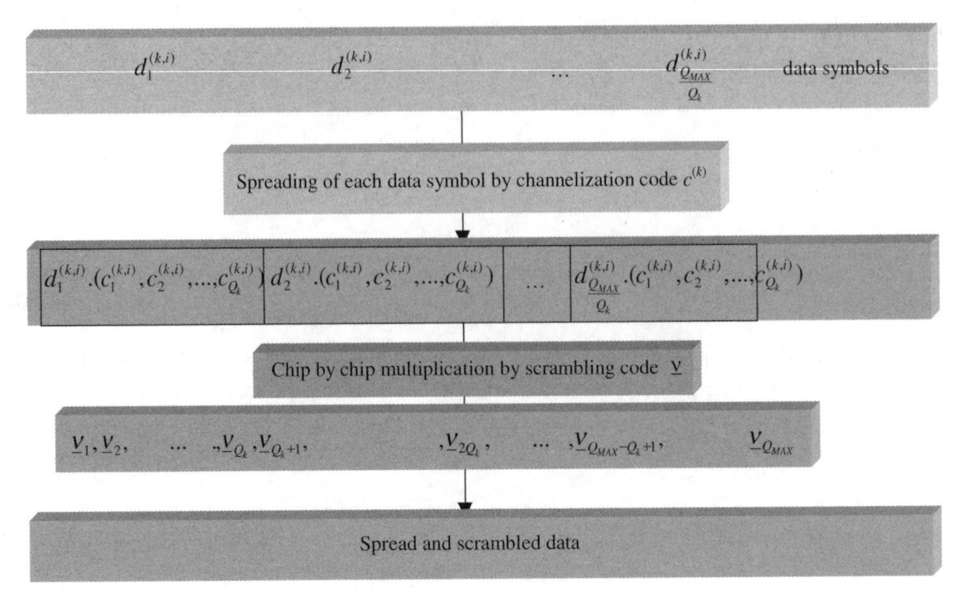

Figure 4.48 Spreading of data symbols.

We generate a complex scrambling code $\mathbf{v}$ from the binary scrambling codes $\mathbf{v} = (v_1, v_2, \ldots, v_{16})$ of length 16 described in the Annex of [14]. The relation between the elements $\underline{\mathbf{v}}$ and $\mathbf{v}$ is given by

$$\underline{v}_i = (j)^i \cdot v_i, \quad v_i \in \{1, -1\}, \quad i = 1, \ldots, 16 \tag{4.52}$$

Thus, the elements $\underline{v}_i$ of the complex scrambling code $\underline{\mathbf{v}}$ have alternating real and imaginary values.

We obtain length matching by concatenating $Q_{\max}/Q_k$ spread words before the scrambling event as illustrated in Figure 4.48.

4.5.5 Spreading Data Symbols and Data Blocks

We can see the combination of the user-specific channelisation and cell-specific scrambling codes as a user- and cell-specific spreading code $\mathbf{s}^{(k)} = (s_p^{(k)})$ with

$$s_p^{(k)} = c^{(k)}_{1+[(p-1) \bmod Q_k]} \cdot \underline{v}_{1+[(p-1) \bmod Q_{\max}]}, \quad k = 1, \ldots, K, \quad p = 1, \ldots N_k Q_k \tag{4.53}$$

With the root raised cosine chip impulse filter $Cr_0(t)$ the transferred signal belonging to the data block $\underline{\mathbf{d}}^{(k,1)}$ of eqn (4.47) transmitted before the midamble gets expressed as

$$\underline{\mathbf{d}}^{(k,1)}(t) = \sum_{n=1}^{N_k} \underline{d}_n^{(k,1)} \sum_{q=1}^{Q_k} s^{(k)}_{(n-1)Q_k+q} \cdot Cr_0(t - (q-1)T_c - (n-1)Q_k T_c) \tag{4.54}$$

and for the data block $\underline{\mathbf{d}}^{(k,2)}$ of eqn (1) transmitted after the midamble

$$\underline{\mathbf{d}}^{(k,2)}(t) = \sum_{n=1}^{N_k} \underline{d}_n^{(k,2)} \sum_{q=1}^{Q_k} s_{(n-1)Q_k+q}^{(k)} \cdot Cr_0(t-(q-1)T_C-(n-1)Q_kT_c-N_kQ_kT_c-L_mT_c) \tag{4.55}$$

where L_m is the number of midamble chips.

4.5.6 Synchronisation Codes

The primary code sequence, C_p, results from a generalised hierarchical Golay sequence. The Primary Synchronisation Channel (PSCH), in addition, has good aperiodic auto-correlation properties.

We define

$$a = \langle x_1, x_2, x_3, \ldots, x_{16}\rangle = \langle 1,1,1,1,1,1,-1,-1,1,-1,1,-1,1,-1,-1,1\rangle$$

The PSC code word gets generated by repeating the sequence 'a' modulated by a Golay complementary sequence and creating a complex-valued sequence with identical real and imaginary components.

Then we define the PSC code word C_p as

$$C_p = \langle y(0), y(1), y(2), \ldots, y(255)\rangle$$

where

$$y = (1+j) \times \langle a,a,a,-a,-a,a,-a,-a,a,a,a,-a,a,-a,a,a,\rangle$$

and the leftmost index corresponds to the chip transmitted first in each time slot. The 16 secondary synchronisation code words, $\{C_0, \ldots, C_{15}\}$, constitute complex valued with identical real and imaginary components, and they originate from the position-wise multiplication of a Hadamard sequence and a sequence z, defined as

$$z = \langle b,b,b,-b,b,b,-b,-b,b,-b,b,-b,-b,-b,-b,-b\rangle \tag{4.56}$$

where

$$b = \langle x_1, \ldots, x_8, -x_9, \ldots, -x_{16}\rangle = \langle 1,1,1,1,1,1,-1,-1,-1,1,-1,1,-1,1,1,-1\rangle$$

We build the Hadamard sequences as the rows in a matrix H_8 constructed recursively by

$$\begin{aligned} H_0 &= (1) \\ H_k &= \begin{pmatrix} H_{k-1} & H_{k-1} \\ H_{k-1} & H_{k-1} \end{pmatrix}, \quad k \geq 1 \end{aligned} \tag{4.57}$$

The rows are numbered from the top starting with row 0 (the all zeroes sequence).

We denote the nth Hadamard sequence as a row of H_8 numbered from the top, $n = 0, 1, 2, \ldots, 255$, in the sequel.

In addition, we let $h_m(i)$ and $z(i)$ and denote the ith symbol of the sequence h_m and z, respectively, where $i = 0, 1, 2, \ldots, 255$ and $i = 0$ corresponds to the leftmost symbol.

The ith SCH code word, $C_{\text{SCH},i}$, $i = 0, \ldots, 15$, is then defined as

$$C_{\text{SCH},i} = (1 + j) \times \langle h_m(0) \times z(0), h_m(1) \times z(1), h_m(2) \times z(2), \ldots, h_m(255) \times z(255) \rangle \tag{4.58}$$

where $m = (16 \times i)$ and the leftmost chip in the sequence corresponds to the chip transmitted first in time.

This code word gets selected from every 16th row of the matrix H_8, which yields 16 possible code words. We define the secondary SCH code words in terms of $C_{\text{SCH},i}$ and the definition of $\{C_0, \ldots, C_{15}\}$ now follows as:

$$C_i = C_{\text{SCH},i}, \quad i = 0, \ldots, 15 \tag{4.59}$$

Finally, more details and code allocations and evaluation of synchronisation codes can be found in [14].

4.6 MULTIPLEXING AND CHANNEL CODING

We encode/decode information from/to upper layers to afford transport services over the air interface. Thus, through channel coding we protect data flow by combining error detection and correction and adapting to transmission needs by means of rate matching, inter-leaving and mapping of transport channels to physical channels.

Multiplexing and channel coding techniques in the UTRA physical layer specifications apply to both FDD and TDD modes in most aspects. Thus, for completeness here we will introduce primarily the description for the FDD because it has more cases (e.g. description for uplink and downlink), and indicate where they differ with the TDD. Specifications in [17] and [18] provide all the details for each mode.

In UTRA data arrives at the coding/multiplexing unit in transport block sets once every transmission time interval. The transmission time interval depends on the transport channel from the set {10, 20, 40, 80 ms}. The main steps valid for both FDD and TDD from [17] are:

- add CRC to each transport block,
- transport block concatenation and code block segmentation,
- channel coding,
- rate matching,
- insertion of Discontinuous Transmission (DTX) indication bits,
- inter-leaving,
- radio frame segmentation,

- multiplexing of transport channels,
- physical channel segmentation,
- mapping to physical channels.

Figure 4.49 illustrates the coding/multiplexing steps for FDD uplink and downlink. Clearly, the uplink also applies to the TDD mode. Hence the uplink description will basically cover the needs for the TDD. However, as mentioned above we will highlight where differences exist. We should also note here that in this section for consistency we keep the structure and nomenclature of the technical specifications by incorporating direct extracts and using the same type of equations.

N.B.: In the downlink we denoted *Coded Composite Transport Channel* (CCTrCH) the single output data stream from the TrCH multiplexing, including DTX indication. This CCTrCH can get mapped to one or several physical channels.

4.6.1 Error Detection and CRC Calculations

In the sequel we cover the DL and UL in an integrated manner by indicating differences where appropriate.

Cyclic Redundancy Check (CRC) affords error detection on transport blocks. Higher layers signal what CRC (24, 16, 12, 8 or 0) bit length shall be used for each TrCH. We use the entire transport block to calculate the CRC parity bits for each transport block. The following cyclic generator polynomials generate these parity bits:

$$g_{\mathrm{CRC24}}(D) = D^{24} + D^{23} + D^6 + D^5 + D + 1 \tag{4.60}$$

$$g_{\mathrm{CRC16}}(D) = D^{16} + D^{12} + D^5 + 1 \tag{4.61}$$

$$g_{\mathrm{CRC12}}(D) = D^{12} + D^{11} + D^3 + D^2 + D + 1 \tag{4.62}$$

$$g_{\mathrm{CRC8}}(D) = D^8 + D^7 + D^4 + D^3 + D + 1 \tag{4.63}$$

In relation with Figure 4.49 and [17], we denote the bits in a transport block delivered to layer 1 by

$$a_{im1}, a_{im2}, a_{im3}, \ldots, a_{imA_i}$$

and the parity bits by

$$p_{im1}, p_{im2}, p_{im3}, \ldots, p_{imL_i}$$

where A_i is the length of a transport block of TrCH i, m is the transport block number and L_i is either 24, 16, 12, 8 or 0 depending on what the upper layers signal. Then encoding follows systematically; which means that in GF(2) we express the above polynomials as:

$$a_{im1}D^{A_i+23} + a_{im2}D^{A_i+22} + \cdots + a_{imA_i}D^{24} + p_{im1}D^{23} + p_{im2}D^{22} + \cdots + p_{im23}D^1 + p_{im24} \tag{4.64}$$

$$a_{im1}D^{A_i+15} + a_{im2}D^{A_i+14} + \cdots + a_{imA_i}D^{16} + p_{im1}D^{15} + p_{im2}D^{14} + \cdots + p_{im15}D^1 + p_{im16} \tag{4.65}$$

$$a_{im1}D^{A_i+11} + a_{im2}D^{A_i+10} + \cdots + a_{imA_i}D^{12} + p_{im1}D^{11} + p_{im2}D^{10} + \cdots + p_{im11}D^1 + p_{im12} \tag{4.66}$$

$$a_{im1}D^{A_i+7} + a_{im2}D^{A_i+6} + \cdots + a_{imA_i}D^{8} + p_{im1}D^{7} + p_{im2}D^{6} + \cdots + p_{im7}D^1 + p_{im8} \tag{4.67}$$

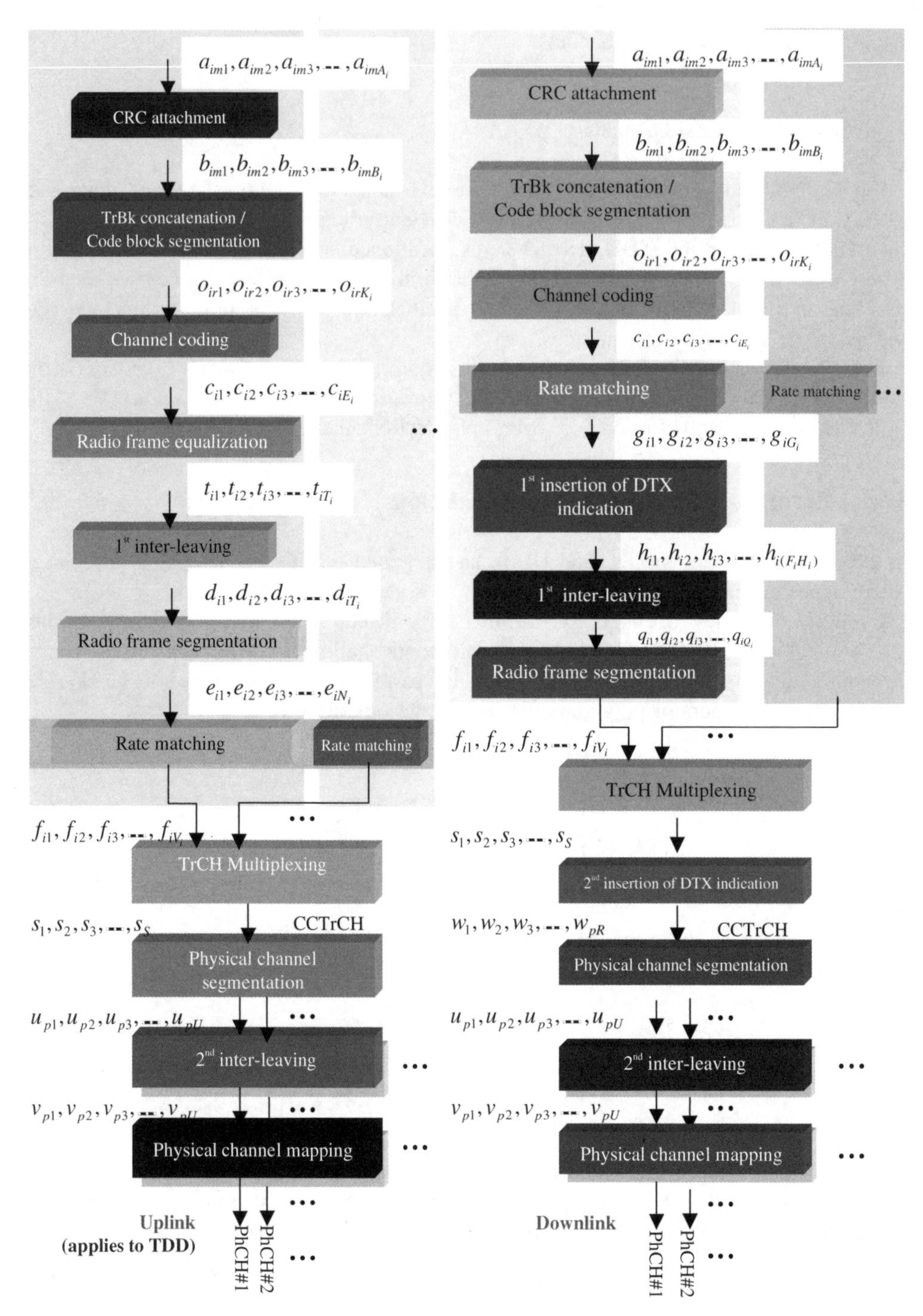

Figure 4.49 FDD uplink and downlink transport channel multiplexing structure.

The preceding polynomials yield a remainder equal to 0 when divided by $g_{CRC24}(D)$, $g_{CRC16}(D)$, $g_{CRC12}(D)$ and $g_{CRC8}(D)$, respectively. In the absence of transport block inputs to the CRC calculation ($M_i = 0$), CRC attachment does not occur. However, if transport block inputs exist in the CRC calculation (i.e. $M_i \neq 0$) and the size of a transport block equals zero ($A_i = 0$), CRC attachment occurs, i.e. all parity bits equal to zero. Denoting the bits after CRC attachment by

$$b_{im1}, b_{im2}, b_{im3}, \ldots, b_{imB_i}$$

where $B_i = A_i + L_I$; the relation between a_{imk} and b_{imk} can be defined as

$$b_{imk} = a_{imk} \quad \text{where } k = 1, 2, 3, \ldots, A_i \tag{4.68}$$

$$b_{imk} = p_{im(L_i+1-(k-A_i))} \quad \text{where } k = A_i + 1, A_i + 2, A_i + 3, \ldots, A_i + L_i \tag{4.69}$$

4.6.2 Transport Block Concatenation and Code Block Segmentation

All transport blocks in a Transmission Time Interval (TTI) have serial concatenation. When the number of bits in a TTI is larger than Z, i.e. the maximum size of a code block in question, then code block segmentation takes place after the concatenation of the transport blocks. The maximum size of the code blocks depends on whether convolutional coding, turbo coding or no coding occurs. We denote by

$$b_{im1}, b_{im2}, b_{im3}, \ldots, b_{imB_i}$$

the bits input to the transport block concatenation, where i is the TrCH number, m is the transport block number and B_i is the number of bits in each block (including CRC). M_i represents the number of transport blocks on TrCH i.

$$x_{i1}, x_{i2}, x_{i3}, \ldots, x_{iX_i}$$

denotes the bits after concatenation, where i is the TrCH number and $X_i = M_i B_i$. Then, the following relations apply:

$$x_{ik} = b_{i1k}, \quad k = 1, 2, \ldots, B_i \tag{4.70}$$

$$x_{ik} = b_{i,2,(k-B_i)}, \quad k = B_i + 1, B_i + 2, \ldots, 2B_i \tag{4.71}$$

$$x_{ik} = b_{i,3,(k-2B_i)}, \quad k = 2B_i + 1, 2B_i + 2, \ldots, 3B_i \tag{4.72}$$

$$\ldots$$

$$x_{ik} = b_{i,M_i,(k-(M_i-1)B_i)}, \quad k = (M_i - 1)B_i + 1, (M_i - 1)B_i + 2, \ldots, M_i B_i \tag{4.73}$$

Segmentation of the bit sequence from transport block concatenation transpires when $X_i > Z$, where the segmented blocks have the same size. If the number of bits input to the segmentation (i.e. X_i) is not a multiple of C_i (the number of code blocks on TrCHi), we add filler bits (0s) to the beginning of the first block. The specifications in [17] define maximum code block sizes as:

- convolutional coding: $Z = 504$,
- turbo coding: $Z = 5114$,
- no channel coding: $Z =$ *unlimited*.

From Figure 4.48 and [17] we denote by

$$o_{ir1}, o_{ir2}, o_{ir3}, \ldots, o_{irK_i}$$

the bit output from code block segmentation, where i is the TrCH number r is the code block number and K_i is the number of bits. Then number of code blocks: $C_i = \lceil X_i/Z \rceil$, and for the number of bits in each code block and filler bits, the following logic applies:

Number of bits in each code block
if $X_i < 40$ and turbo coding is used, then
$K_i = 40$
else
$K_i = \lceil X_i/C_i \rceil$
end if

Number of filler bits: $Y_i = C_iK_i - X_i$
if $X_i \leq Z$, then
$o_{i1k} = 0, \quad k = 1, 2, \ldots, Y_i$
$o_{i1k} = x_{i,(k-Y_i)}, \quad k = Y_i + 1, Y_i + 2, \ldots, K_i$
end if

If $X_i > Z$, then

$$\begin{aligned}
&o_{i1k} = 0, \quad k = 1, 2, \ldots, Y_i \\
&o_{i1k} = x_{i,(k-Y_i)}, \quad k = Y_i + 1, Y_i + 2, \ldots, K_i \\
&o_{i2k} = x_{i,(k+K_i-Y_i)}, \quad k = 1, 2, \ldots, K_i \\
&o_{i3k} = x_{i,(k+2K_i-Y_i)}, \quad k = 1, 2, \ldots, K_i \\
&\ldots \\
&o_{iC_ik} = x_{i,(k+(C_{i-1})K_i-Y_i)}, \quad k = 1, 2, \ldots, K_i
\end{aligned}$$

end if

4.6.3 Channel Coding

The concatenation or segmentation process delivers code blocks

$$o_{ir1}, o_{ir2}, o_{ir3}, \ldots, o_{irK_i}$$

to the channel coding block, where i is the TrCH number, r is the code block number and K_i is the number of bits in each code block. We denote by C_i the number of code blocks on TrCH i, and the encoded bits

$$y_{ir1}, y_{ir2}, y_{ir3}, \ldots, y_{irY_i}$$

Table 4.39 Usage of channel coding scheme and coding rate

Type of TrCH	Coding scheme	Coding rate
BCH	Convolutional coding	1/2
PCH		
RACH		1/3, 1/2
CPCH, DCH, DSCH,	Turbo coding	1/3
FACH + USCH (tdd)	No coding	

where Y_i is the number of encoded bits. The relation between O_{irk} and y_{irk} and between K_i and Y_i depends on the following channel coding scheme: convolutional coding, turbo coding and no coding. Table 4.39 illustrates the usage of these schemes, and the values of Y_i in connection with each coding scheme are:

- convolutional coding with rate 1/2:$Y_i = 2\,K_i + 16$; rate 1/3:$Y_i = 3\,K_i + 24$;
- turbo coding with rate 1/3:$Y_i = 3\,K_i + 12$;
- no coding: $Y_i = K_i$.

4.6.3.1 Convolutional Coding

In UTRA we define convolutional codes with constraint length 9 and coding rates 1/3 and 1/2. Figure 4.50 illustrates the convolutional coder configuration. Output from the rate 1/3 convolutional coder follows the order output0, output1, output2, output0, output1, output2, output0,. . .,output2; while output from the rate 1/2 convolutional coder follows the order

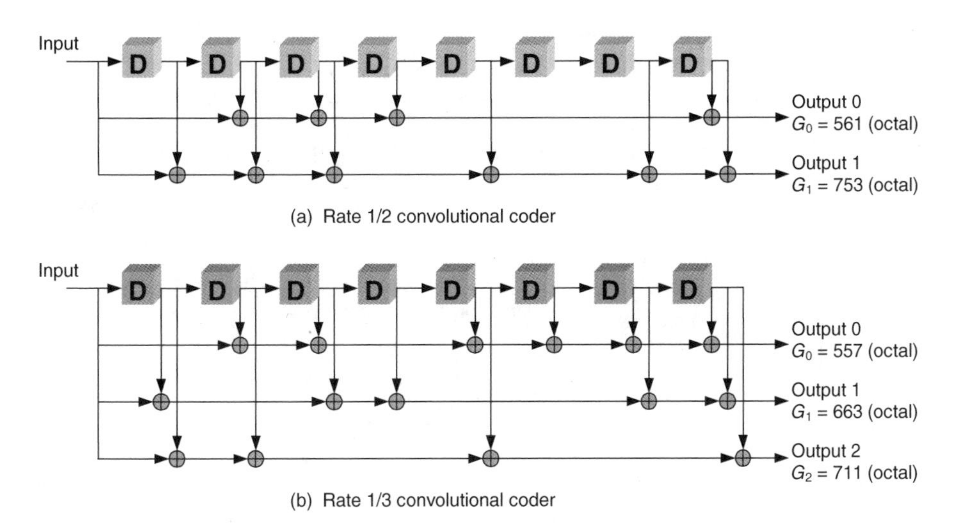

Figure 4.50 Rate 1/2 and rate 1/3 convolutional coders.

output0, output1, output0, output1, output0,..., output1. We add 8 tail bits with binary value 0 at the end of the code block before encoding, and when starting to encode the initial value of the shift register of the coder is 'all 0'.

4.6.3.2 Turbo Coding

From the two types of code concatenation, i.e. serial and parallel, the latter suits well for high quality of services in third generation systems. This can provide very low maximum bit error ratio, e.g. 10^{-6} has the lowest S/N. In parallel code concatenation we feed the information stream into a second encoder and encode data stream generated by multiplexing (and puncture) the encoded sequences resulting from both encoding processes [19].

Coder. In UTRA we apply the Parallel Concatenated Convolutional Code (PCCC) scheme of the turbo coder with two eight-state constituent encoders and one turbo code internal interleaver [17]. The turbo coder structure illustrated in Figure 4.51 has a coding rate of 1/3. We express the transfer function of the eight-state constituent code for PCCC as

$$G(D) = \left[1, \frac{g_1(D)}{g_0(D)}\right] \tag{4.74}$$

where $g_0(D) = 1 + D^2 + D^3$ and $g_1(D) = 1 + D + D^3$.

The initial value of the shift registers in the eight-state constituent encoders is all zeros when starting to encode the input bits, and the output from the turbo coder is

$$x_1, z_1, z'_1, x_2, z_2, z'_2, \ldots, x_K, z_K, z'_K \tag{4.75}$$

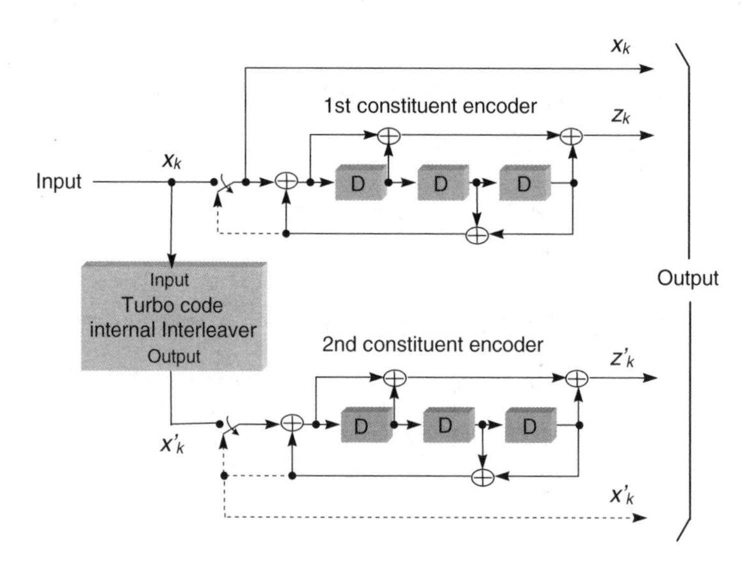

Figure 4.51 Turbo coder structure of rate 1/3 (dotted lines indicate termination only).

where $x_1, x_2, \ldots, x_K$ are the bits input to the turbo coder, i.e. both first eight-state constituent encoder and turbo code internal interleaver; K is the number of bits and $z_1, z_2, \ldots, z_K$ and $z'_1, z'_2, \ldots, z'_K$ are the bits outputs from first and second eight-state constituent encoders, respectively. The bits output from turbo code internal interleaver is denoted by $x'_1, x'_2, \ldots, x'_K$, and these bits are to be input to the second eight-state constituent encoder [17].

Other details such as trellis termination and internal interleaver of the turbo coder can be found in [17].

4.6.4 Radio Frame Size Equalisation

Radio frame size equalisation implies padding the input bit sequence in order to ensure that the output can be segmented in F_i data segments of same size, e.g. rate matching. Radio frame size equalisation occurs only in the UL (DL rate matching output block length is always an integer multiple of F_i). We denoted by

$$c_{i1}, c_{i2}, c_{i3}, \ldots, c_{iE_i}$$

the input bit sequence to the radio frame size equalisation, where i is TrCH number and E_i the number of bits. We denote the output bit sequence by

$$t_{i1}, t_{i2}, t_{i3}, \ldots, t_{iT_i}$$

where T_i is the number of bits. Then the output bit sequence follows as

$$t_{ik} = c_{ik}, \quad \text{for} \quad k = 1, \ldots, E_i \quad \text{and} \quad t_{ik} = \{0, 1\} \quad \text{for} \quad k = E_i + 1, \ldots, T_i, \quad \text{if } E_i < T_i \tag{4.76}$$

where $T_i = F_i \times N_i$; and $N_i = \lceil E_i/F_i \rceil$ is the number of bits per segment after size equalisation.

4.6.5 First Inter-leaving

In a compressed mode through puncturing, bits marked with a fourth value on top of $\{0,1,\delta\}$ and noted as p, get introduced in radio frames to be compressed at positions corresponding to the first bits of the radio frames. They will be removed in a later stage of the multiplexing chain to create the actual gap. We perform additional puncturing in the rate matching step, over the Transmission Time Interval (TTI) containing the compressed radio frame, to create room for these p bits. Specifications in [17] provide this and other first inter-leaving details.

4.6.6 Radio Frame Segmentation

When the TTI is longer than 10 ms, the input bit sequence gets segmented and mapped onto consecutive F_i radio frames. Following rate matching in the DL and radio frame size equalisation in the UL, we warrant that the input bit sequence length is an integer multiple of F_i. We denote this input bit sequence by

$$x_{i1}, x_{i2}, x_{i3}, \ldots, x_{iX_i}$$

where i is the TrCH number and X_i is the number of bits. Likewise, we denote the F_i output bit sequences per TTI by

$$y_{i,n_i1}, y_{i,n_i2}, y_{i,n_i3}, \ldots, y_{i,n_iY_i}$$

where n_i is the radio frame number in the current TTI and Y_i is the number of bits per radio frame for TrCH i. Then we define the output sequence as

$$y_{i,n_i,k} = x_{i,((n_i-1)\cdot Y_i)+k}, \quad n_i = 1, \ldots, F_i, \quad k = 1, \ldots, Y_i \tag{4.77}$$

where $Y_i(X_i/F_i)$ is the number of bits per segment. The n_i^{th} segment is mapped to the n_ith radio frame of the transmission time interval.

4.6.6.1 Input-Output Relationship of the Radio Frame Segmentation Block in Uplink

We denote the input bit sequence to the radio frame segmentation by

$$d_{i1}, d_{i2}, d_{i3}, \ldots, d_{iT_i}$$

where i is the TrCH number and T_i the number of bits. Thus, $x_{ik} = d_{ik}$ and $X_i = T_i$. Likewise, we denote the output bit sequence corresponding to radio frame n_i by

$$e_{i1}, e_{i2}, e_{i3}, \ldots, e_{iN_i}$$

where i is the TrCH number and N_i is the number of bits. Thus, $e_{i,k} = y_{i,n_ik}$ and $N_i = Y_i$.

4.6.6.2 Input-Output Relationship of Radio Frame Segmentation Block in Downlink

As in the preceding section, we denote the bits input to the radio frame segmentation by

$$q_{i1}, q_{i2}, q_{i3}, \ldots, q_{iQ_i}$$

where i is the TrCH number and Q_i the number of bits. Hence, $x_{ik} = q_{ik}$ and $X_i = Q_i$. Again, we denote the output bit sequence corresponding to radio frame n_i by

$$f_{i1}, f_{i2}, f_{i3}, \ldots, f_{iV_i}$$

where i is the TrCH number and V_i is the number of bits. Then, $f_{i,k} = y_{i,n_ik}$ and $V_i = Y_i$.

4.6.7 Rate Matching

By rate matching we mean the repetition or puncturing of bits on a transport channel based on attributes assigned by higher layers. An attribute is semi-static and can only get changed through higher layer signalling. The rate matching attribute assignment occurs after the calculation of the number of bits to be repeated or punctured.

The number of bits on a transport channel can vary between different TTIs. In the DL the transmission gets interrupted if the number of bits is lower than maximum. When the number of bits between different uplink TTIs changes, bits get repeated or punctured to ensure that the total bit rate after TrCH multiplexing is identical to the total channel bit rate of the allocated dedicated physical channels. If the rate matching event does not get input bits for all TrCHs within a CCTrCH, the rate matching does not output bits for all TrCHs within the CCTrCH and no uplink DPDCH will mean no selection of uplink rate matching. See the detailed description of rate matching characteristics, such as determination of rate matching in uplink/downlink, as well as separation and collection in uplink/downlink in [17].

4.6.8 TrCH Multiplexing

The TrCH delivers one radio frame every 10 ms to the TrCH multiplexing, which are serially multiplexed into a Coded Composite Transport Channel (CCTrCH).

We denote by

$$f_{i1}, f_{i2}, f_{i3}, \ldots, f_{iV_i}$$

the input bits going to the TrCH multiplexing, where i is the TrCH number and V_i is the number of bits in the radio frame of TrCH i. Likewise, we denote by I the number of TrCHs, and by $s_1, s_2, s_3, \ldots, s_S$ the output bits from TrCH multiplexing, where S is the number of bits, see [17].

$$S = \sum_i V_i a \tag{4.78}$$

The TrCH multiplexing is defined by the following relations

$$s_k = f_{1k}, \quad k = 1, 2, \ldots, V_i \tag{4.79}$$

$$s_k = f_{2,(k-V_1)}, \quad k = V_1 + 1, V_1 + 2, \ldots, V_1 + V_2 \tag{4.80}$$

$$s_k = f_{3,(k-(V_1+V_2))}, \quad k = (V_1 + V_2) + 1, (V_1 + V_2) + 2, \ldots, (V_1 + V_2) + V_3 \tag{4.81}$$

$$\ldots$$

$$s_k = f_{1,(k-(V_1+V_2+\cdots+V_{i-1}))} \tag{4.82}$$

$$k = (V_1 + V_2 + \cdots + V_{I-1}) + 1, (V_1 + V_2 + \cdots + V_{I-1}) + 2, \ldots, (V_1 + V_2 + \cdots + V_{I-1}) + V_i \tag{4.83}$$

4.6.9 Discontinuous Transmission (DTX) Bits Insertion

We use DL DTX to fill up the radio frame with bits, where the insertion point of these bits can have either fixed or flexible positions of the TrCHs in the radio frame. It depends on the UTRAN to decide for each CCTrCH whether it will have fixed or flexible positions during the connection. DTX indication bits communicate only when the transmission will get turned off, i.e. they are not transmitted themselves.

4.6.9.1 First Insertion of DTX Indication Bits

First DTX indication bits' insertion occurs only if the positions of the TrCHs in the radio frame are fixed. In the fixed position scheme, we reserve a fixed number of bits for each TrCH in the radio frame. We denote the bits from rate matching by

$$g_{i1}, g_{i2}, g_{i3}, \ldots, g_{iG_i}$$

where G_i is the number of bits in one TTI of TrCH i. Likewise, we denote the number of bits in one radio frame of TrCH i by H_i. Finally, we also denote by D_i the number of bits output of the first DTX insertion block.

In normal or compressed mode using spreading factor reduction, H_i is constant and corresponds to the maximum number of bits from TrCH i in one radio frame for any transport format of TrCH I, and $D_i = F_i \times H_i$.

Within compressed mode using puncturing techniques, additional puncturing occurs in the rate matching block. The empty positions resulting from the additional puncturing gets p bits inserted in the first inter-leaving block, the DTX insertion is thus limited to allow later insertion of p bits. Consequently, DTX bits get inserted until the total number of bits is D_i, where

$$D_i = F_i \times H_i, * + \Delta N_{\mathrm{cm},i,\max}^{\mathrm{TTI}} \quad \text{and} \quad H_i = N_{i,*} + \Delta N_i \tag{4.84}$$

We denote the output bits from the DTX insertion by $h_{i1}, h_{i2}, h_{i3}, \ldots, h_{iDi}$, where these three valued bits can be expressed by the following relations

$$h_{ik} = g_{ik}, \quad k = 1, 2, 3, \ldots, G_i \tag{4.85}$$

$$h_{ik} = \delta, \quad k = G_i + 1, G_i + 2, G_i + 3, \ldots, D_i \tag{4.86}$$

where we denote DTX indication bits by δ. Here $g_{ik} \in \{0, 1\}$ and $\delta \notin \{0, 1\}$.

4.6.9.2 Second Insertion of DTX Indication Bits

The DTX indication bits in the second insertion get placed at the end of the radio frame, and the DTX will be distributed over all slots after second inter-leaving. The input bits to the DTX insertion block get denoted by $s_1, s_2, s_3, \ldots, s_S$, where S is the number of bits from

TrCH multiplexing. We denote by P the number of PhCHs and the number of bits in one radio frame, including DTX indication bits, for each PhCH by R. In a normal mode

$$R = \frac{N_{\text{data},*}}{P} = 15\,N_{\text{data}\,1} + 15\,N_{\text{data}\,2} \tag{4.87}$$

where $N_{\text{data}\,1}$ and $N_{\text{data}\,2}$ are defined in the first part of this chapter and in [8]. For compressed mode, $N'_{\text{data},*}$ is defined as

$$N'_{\text{data},*} = P(15\,N'_{\text{data}\,1} + 15\,N'_{\text{data}\,2}) \tag{4.88}$$

where $N'_{\text{data}\,1}$ and $N'_{\text{data}\,2}$ are the number of bits in the data fields of the slot format used for the current compressed mode, i.e. slot format A or B as defined in [8] corresponding to the spreading factor and the number of transmitted slots in use [17].

When compressed mode by puncturing and fixed positions occurs, DTX gets inserted up to $N'_{\text{data},*}$ bits, because the exact room for the gap is already reserved, thanks to the earlier insertion of the p bits. Thus, R is defined as

$$R = N'_{\text{data},*}/P \tag{4.89}$$

If compressed mode by SF reduction and by higher layer scheduling occurs, additional DTX gets inserted when the transmission time reduction method does not exactly create a transmission gap of the desired Transmission Gap Length (TGL). The number of bits available to the CCTrCH in one radio frame of this compressed mode is denoted by

$$N^{\text{cm}}_{\text{data},*} \quad \text{and} \quad R = \frac{N^{\text{cm}}_{\text{data},*}}{P}$$

The exact value of $N^{\text{cm}}_{\text{data},*}$ is dependent on the TGL and the transmission time reduction method signalled from higher layers.

For transmission time reduction by SF/2 method in compressed mode

$$N^{\text{cm}}_{\text{data},*} = \frac{N'_{\text{data},*}}{2}$$

and for other methods it can be calculated as

$$N^{\text{cm}}_{\text{data},*} = N'_{\text{data},*} - N_{\text{TGL}}$$

For every transmission time reduction method

$$N'_{\text{data},*} = P(15\,N'_{\text{data}\,1} + 15\,N'_{\text{data}\,2})$$

where $N'_{\text{data}\,1}$ and $N'_{\text{data}\,2}$ are the number of bits in the data fields of a slot for slot format A or B as defined in [8].

N_{TGL} is the number of bits that are located within the transmission gap and defined as

$$N_{\text{TGL}} = \begin{cases} \frac{\text{TGL}}{15} N'_{\text{data},*} & \text{if } N_{\text{first}} + \text{TGL} \leq 15 \\ \frac{15 - N_{\text{first}}}{15} N'_{\text{data},*} & \text{in first frame if } N \text{ first} + \text{TGL} > 15 \\ \frac{\text{TGL} - (15 - N_{\text{first}})}{15} N'_{\text{data},*} & \text{in second frame if } N \text{ first} + \text{TGL} > 15 \end{cases}$$

N_{first} and TGL are part of the description of the compressed mode section.

Furthermore, notice that in compressed mode by SF/2 method, we also add DTX in the physical channel mapping stage. During the second DTX insertion the number of CCTrCH bits remains the same as in the normal mode. We denote the bits output from the DTX insertion block by $w_1, w_2, w_3, \ldots, w_{(PR)}$. Notice also that these bits have four values in case of compressed mode by puncturing, and three otherwise. We can define them by the following relations

$$w_k = s_k, \quad k = 1, 2, 3, \ldots, S \qquad \text{and} \qquad w_k = \delta, \quad k = S+1, S+2, S+3, \ldots, PR \tag{4.90}$$

where DTX indication bits are denoted by δ. Here $s_k \in \{0, 1, p\}$ and $\delta \notin \{0, 1\}$ [17].

4.6.10 Physical Channel Segmentation

When using more than one PhCH, the physical channel segmentation event divides the bits among the different PhCHs. The bits input to the physical channel segmentation are $x_1, x_2, x_3, \ldots, x_Y$, where Y is the number of bits input to the physical channel segmentation block. P denotes the number of PhCHs.

The bits after the physical channel segmentation are denoted by $u_{p1}, u_{p2}, u_{p3}, \ldots, u_{pU}$, where p is PhCH number and U is the number of bits in one radio frame for each PhCH, i.e. $U = (Y - N_{\text{TGL}})/P$ for compressed mode by puncturing, and $U = Y/P$ otherwise.

For all modes, we map some bits of the input flow to each code until the number of bits on the code reaches V. For modes other than *compressed mode by puncturing*, we take all bits of the input flow for mapping to the codes. For compressed mode by puncturing, only the bits of the input flow not corresponding to bits p are taken for mapping to the codes, and each bit p is removed to ensure creation of the gap required by the compressed mode, as described next.

- Bits on the first PhCH after physical channel segmentation: $u_{1,k} = x_{if(k)}$, $k = 1, 2, \ldots, U$.
- Bits on the second PhCH after physical channel segmentation: $u_{2,k} = x_{i,f(k+U)}$, $k = 1, 2, \ldots, U$.
- Bits on the Pth PhCH after physical channel segmentation: $u_{P,k} = x_{if(k+(P-1)U)}$, $k = 1, 2, \ldots, U$.

Here f is such that in modes other than compressed mode by puncturing, $x_{if(k)} = x_{ik}$, i.e. $f(k) = k$, for all k. In the compressed mode by puncturing, bit $u_{1,1}$ corresponds to the bit $x_{i,k}$ with smallest index k when the bits p are not counted, bit $u_{1,2}$ corresponds to the bit

$x_{i,k}$ with second smallest index k when the bits p are not counted, and so on for bits $u_{1,3}, \ldots, u_{1,V}, u_{2,1}, u_{2,2}, \ldots, u_{2,V}, \ldots, u_{P,1}, u_{P,2}, \ldots, u_{P,V}$.

We denote the bits input to the physical segmentation by $s_1, s_2, s_3, \ldots, s_S$. Hence, $x_k = s_k$ and $Y = S$. We denote the bits input to the physical segmentation by $w_1, w_2, w_3, \ldots, w_{(PU)}$. Hence, $x_k = w_k$ and $Y = PU$ [17].

4.6.11 Second Inter-leaving

The second inter-leaving consists of a block interleaver with inter-column permutations. We denote the input bits to the second interleaver by $u_{p1}, u_{p2}, u_{p3}, \ldots, u_{pU}$, where p is PhCH number and U is the number of bits in one radio frame for one PhCH. The matrix configuration can be as follows:

- Set the number of columns $C_2 = 30$. Number columns $0, 1, 2, \ldots, C_2 - 1$ from left to right.
- Determine the number of rows R_2 by finding minimum integer R_2 such that $U \leq R_2 C_2$.
- The bits input to the second inter-leaving are written into the $R_2 \times C_2$ rectangular matrix row by row.

$$\begin{bmatrix} u_{p1} & u_{p2} & u_{p3} & \cdots & u_{p30} \\ u_{p31} & u_{p32} & u_{p33} & \cdots & u_{p60} \\ \vdots & \vdots & \vdots & \cdots & \vdots \\ u_{p,((R_2-1)30+1)} & u_{p,((R_2-1)30+2)} & u_{p,((R_2-1)30+3)} & \cdots & u_{p,(R_2 30)} \end{bmatrix} \tag{4.91}$$

- Perform the inter-column permutation based on the pattern $\{P_2(j)\}(j = 0, 1, \ldots, C_2 - 1)$ which Table 4.39 illustrates, and where $P_2(j)$ is the original column position of the jth permutated column. After permutation of the columns, the bits are denoted by y_{pk} [17].

$$\begin{bmatrix} y_{p1} & y_{p,(R_2+1)} & y_{p,(2R_2+1)} & \cdots & y_{p,(29R_2+1)} \\ y_{p2} & y_{p,(R_2+2)} & y_{p,(2R_2+2)} & \cdots & y_{p,(29R_2+2)} \\ \vdots & \vdots & \vdots & \cdots & \vdots \\ y_{pR_2} & y_{p,(2R_2)} & y_{p,(3R_2)} & \cdots & y_{p,(30R_2)} \end{bmatrix} \tag{4.92}$$

The output of the second inter-leaving corresponds to the bit sequence read out column by column from the inter-column permuted $R_2 \times C_2$ matrix. We prune the output by deleting bits that were not present in the input bit sequence, i.e. bits y_{pk} that correspond to bits u_{pk} with $k > U$. We denote the bits after second inter-leaving by $v_{p1}, v_{p2}, \ldots, v_{pU}$, where v_{p1} corresponds to the bit y_{pk} with smallest index k after pruning, v_{p2} to the bit y_{pk} with second smallest index k after pruning and so on (Table 4.40).

Table 4.40 Inter-column permutations [17]

Number of columns, C_2	Inter-column permutation pattern
30	{0, 20, 10, 5, 15, 25, 3, 13, 23, 8, 18, 28, 1, 11, 21, 6, 16, 26, 4, 14, 24, 19, 9, 29, 12, 2, 7, 22, 27, 17}

4.6.12 Physical Channel Mapping

Specifications in [8] and the earlier sections in this chapter define the PhCH for both uplink and downlink. We denote the input bits to the physical channel mapping by $v_{p1}, v_{p2}, \ldots, v_{pU}$, where p is the PhCH number and U is the number of bits in one radio frame for one PhCH. We map the bits v_{pk} to the PhCHs so that the bits for each PhCH are transmitted over the air in ascending order with respect to k.

In the compressed mode, no bit gets mapped to certain slots of the PhCH(s). Likewise, if $N_{\text{first}} + \text{TGL} \leq 15$, no bit gets mapped to slots N_{first}–N_{last}. If $N_{\text{first}} + \text{TGL} > 15$, i.e. the transmission gap spans two consecutive radio frames, the mapping is as follows:

- in the first radio frame, no bit is mapped to slots $N_{\text{first}}, N_{\text{first}} + 1, N_{\text{first}} + 2, \ldots, 14$,
- in the second radio frame, no bit is mapped to the slots $0, 1, 2, \ldots, N_{\text{last}}$.

We describe TGL, N_{first} and N_{last} while presenting the compressed mode section.

4.6.12.1 *Uplink and Downlink*

Uplink. PhCHs used during a radio frame can go either full of bits transmitted over the air or not used at all. However, with UE in compressed mode the transmission gets turned off during consecutive slots of the radio frame.

Downlink. PhCHs do not need to be transmitted full of bits over the air, e.g. bits $v_{pk} \notin \{0, 1\}$ do not get transmitted. During compressed mode when reducing the SF by 2, no bit gets mapped to the DPDCH field. See the logic of this event and more details in the DL physical channel mapping in [17].

The preceding sections complete the functional description of the multiplexing structure illustrated in Figure 4.49. Additional details on the presentation of each block can be found in [17] and [18].

4.6.13 Detection of the Transport Format

When the transport format set of a TrCH i contains more than one transport format, we can detect them according to one of the following schemes [17]:

- *Transport Format Combination Indicator (TFCI) based detection*: this scheme applies when the transport format combination signals using the TFCI field.

- *Explicit blind detection*: consists of detecting the TF of TrCH i by means of channel decoding and CRC check.
- *Guided detection*: applies when there exists at least one other TrCH i', hereafter called guiding TrCH,

  the guiding TrCH has the same TTI duration as the TrCH under consideration, i.e. $F_{i'} = F_i$;

  different TFs of the TrCH under consideration correspond to different TFs of the guiding TrCH,

  we can use explicit blind detection on the guiding TrCH.

If the *transport format set* for a TrCH i contains one transport format only, we do not need a transport format detection event for this TrCH. In the uplink, the blind transport format detection corresponds to network controlled option. In the downlink, the UE will perform blind transport format detection, when given conditions on the configured transport channels comply. For a DPCH associated with a PDSCH, the DPCCH includes TFCI.

4.6.13.1 Blind Transport Format Detection

In the absence of TFCI explicit blind detection or guided detection takes place on all TrCHs within the CCTrCH, which have more than one transport format. However, according to [17], the UE will support blind transport format detection only if all of the following conditions apply:

1. the number of CCTrCH bits received per radio frame ≤ 600;
2. the number of transport format combinations of the CCTrCH ≤ 64;
3. the CCTrCH under detection uses fixed positions of the transport channels;
4. all explicitly detected TrCHs use convolutional coding;
5. we append CRC to all transport blocks on all explicitly detected TrCHs;
6. the number of explicitly detected TrCHs ≤ 3;
7. for all explicitly detected TrCHs i, the number of code blocks in one TTI (C_i) does not exceed 1;
8. the sum of the transport format sets[16] sizes of all explicitly detected TrCHs, ≤ 16;
9. there is at least one usable TrCH in guiding a transport channel for all transport channels using guided detection. See examples in [17].

4.6.13.2 Transport Format Detection Based on TFCI

When a TFCI exists, TFCI based detection applies to all TrCHs within the CCTrCH, where the TFCI informs the receiver about the transport format combination of the CCTrCHs.

[16]The transport format set size is defined as the number of transport formats within the transport format set.

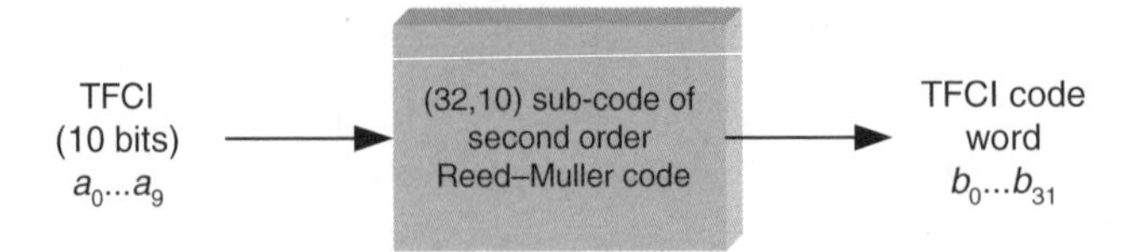

Figure 4.52 Channel coding of TFCI bits.

Right after the TFCI detection we know the transport format combination as well as the transport formats of the individual transport channels.

TFCI coding. As illustrated in Figure 4.52, we encode TFCI bits using a (32,10) sub-code of the second-order Reed–Muller code.

- If the TFCI < 10 bits, we pad it with 0–10 bits, by setting the most significant bits to zero. The length of the TFCI code word = 32 bits. The code words of the (32,10) sub-code of second order Reed–Muller code has a linear combination of ten basis sequences.
- If we define the TFCI information bits as $a_0, a_1, a_2, a_3, a_4, a_5, a_6, a_7, a_8, a_9$, where $a_0 = \text{LSB}$ and $a_9 = \text{MSB}$, the TFCI information corresponds to the TFC defined by the RRC layer to reference the TFC of the CCTrCH in the associated DPCH radio frame. The output code word bits b_i are then given by

$$b_i = \sum_{n=0}^{9} (a_n \times M_{i,n}) \bmod 2 \tag{4.93}$$

where $i = 0, \ldots, 31$, and the output bits are denoted by b_k, $k = 0, 1, 2, \ldots, 31$.

In the downlink if the SF < 128, the encoded TFCI code words get repeated yielding 8 encoded TFCI bits per slot in normal mode and 16 encoded TFCI bits per slot in compressed mode.

4.6.13.3 *TFCI Operation in the Split Mode*

If one of the DCHs has association with a DSCH, the TFCI code word gets split in such a way that the code word relevant for TFCI activity indication is not transmitted from every cell. Higher layer signalling indicates usage of this latter functionality (Table 4.41).

In this case we encode the TFCI bits using a (16,5) bi-orthogonal (or first-order Reed–Muller) code as illustrated in Figure 4.53. Table 4.42 illustrates the code words of the (16,5) bi-orthogonal code, which are linear combinations of five basic sequences.

If we define the first set of TFCI information bits as $a_{1,0}, a_{1,1}, a_{1,2}, a_{1,3}, a_{1,4}$, where $a_{1,0} = \text{LSB}$ and $a_{1,4} = \text{MSB}$, we can assume that this set of TFCI information bits will correspond to the TFC index[17] defined by the RRC layer to reference the TFC of the DCH CCTrCH in the associated DPCH radio frame. Likewise, if we define the second set of TFCI

[17]Expressed in unsigned binary.

Table 4.41 Basis sequences for (32,10) TFCI code [17]

i	$M_{i,0}$	$M_{i,1}$	$M_{i,2}$	$M_{i,3}$	$M_{i,4}$	$M_{i,5}$	$M_{i,6}$	$M_{i,7}$	$M_{i,8}$	$M_{i,9}$
0	1	0	0	0	0	1	0	0	0	0
1	0	1	0	0	0	1	1	0	0	0
2	1	1	0	0	0	1	0	0	0	1
3	0	0	1	0	0	1	1	0	1	1
4	1	0	1	0	0	1	0	0	0	1
5	0	1	1	0	0	1	0	0	1	0
6	1	1	1	0	0	1	0	1	0	0
7	0	0	0	1	0	1	0	1	1	0
8	1	0	0	1	0	1	1	1	1	0
9	0	1	0	1	0	1	1	0	1	1
10	1	1	0	1	0	1	0	0	1	1
11	0	0	1	1	0	1	0	1	1	0
12	1	0	1	1	0	1	0	1	0	1
13	0	1	1	1	0	1	1	0	0	1
14	1	1	1	1	0	1	1	1	1	1
15	1	0	0	0	1	1	1	1	0	0
16	0	1	0	0	1	1	1	1	0	1
17	1	1	0	0	1	1	1	0	1	0
18	0	0	1	0	1	1	0	1	1	1
19	1	0	1	0	1	1	0	1	0	1
20	0	1	1	0	1	1	0	0	1	1
21	1	1	1	0	1	1	0	1	1	1
22	0	0	0	1	1	1	0	1	0	0
23	1	0	0	1	1	1	1	1	0	1
24	0	1	0	1	1	1	1	0	1	0
25	1	1	0	1	1	1	1	0	0	1
26	0	0	1	1	1	1	0	0	1	0
27	1	0	1	1	1	1	1	1	0	0
28	0	1	1	1	1	1	1	1	1	0
29	1	1	1	1	1	1	1	1	1	1
30	0	0	0	0	0	1	0	0	0	0
31	0	0	0	0	1	1	1	0	0	0

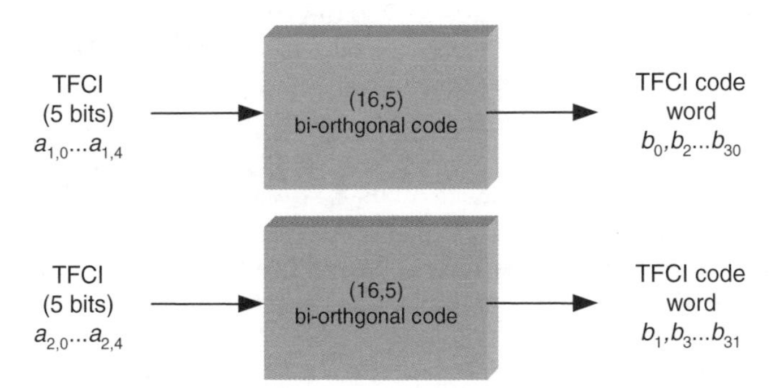

Figure 4.53 Channel coding of split mode TFCI bits.

Table 4.42 Basis sequences for (16,5) TFCI code [17]

i	$M_{i,0}$	$M_{i,1}$	$M_{i,2}$	$M_{i,3}$	$M_{i,4}$
0	1	0	0	0	1
1	0	1	0	0	1
2	1	1	0	0	1
3	0	0	1	0	1
4	1	0	1	0	1
5	0	1	1	0	1
6	1	1	1	0	1
7	0	0	0	1	1
8	1	0	0	1	1
9	0	1	0	1	1
10	1	1	0	1	1
11	0	0	1	1	1
12	1	0	1	1	1
13	0	1	1	1	1
14	1	1	1	1	1
15	0	0	0	0	1

information bits as $a_{2,0}, a_{2,1}, a_{2,2}, a_{2,3}, a_{2,4}$, where $a_{2,0} = \text{LSB}$ and $a_{2,4} = \text{MSB}$, we can assume that this set of TFCI information bits will correspond to the TFC index defined by the RRC layer to reference the TFC of the associated DSCH CCTrCH in the corresponding PDSCH radio frame. Then, the output code word bits b_k are given [17] by

$$b_{2i} = \sum_{n=0}^{4} (a_{1,n} \times M_{i,n}) \bmod 2, \quad b_{2i+1} = \sum_{n=0}^{4} (a_{2,n} \times M_{i,n}) \bmod 2 \tag{4.94}$$

where $i = 0, \ldots, 15$, $j = 0, 1$ and the output bits are b_k, $k = 0, 1, 2, \ldots, 31$.

4.6.14 Mapping of TFCI Words

In non-compressed mode, we map cord word bits directly to the slots of the radio frame, the bit with lower index gets transmitted before the bit with higher index. The coded bits b_k, get mapped to the transmitted TFCI bits d_k as follows: $d_k = b_{k \bmod 32}$.

For the UL physical channels, despite the SF and DL physical channels, if the SF ≥ 128, $k = 0, 1, 2, \ldots, 29$.[18] In a like manner, for the DL physical channels whose SF < 128, $k = 0, 1, 2, \ldots, 119$. The latter implies that bits b_0–b_{23} get transmitted four times and bits b_{24}–b_{31} get transmitted three times [17].

In the uplink compressed mode, we map TFCI bits differently for downlink with SF ≥ 128 and downlink with SF < 128. The slot format gets changed so that we do not lose TFCI bits.

[18]This implies that bits b_{30} and b_{31} do not transmitted.

We repeat TFCI bits because the different slot formats in compressed mode do not match the exact number of TFCI bits for all possible TGLs.

Denoting the number of bits available in the TFCI fields of one compressed radio frame by D and the number of bits in the TFCI field in a slot by N_{TFCI}, we obtain the first bit to get repeated, $E = N_{\mathrm{first}} N_{\mathrm{TFCI}}$. When $N_{\mathrm{last}} \neq 14$, then E corresponds to the number of the first TFCI bits in the slot directly after the TG. The following expressions define the uplink mapping

$$d_k = b_{k \bmod 32} \quad \text{where } k = 0, 1, 2, \ldots, \min(31, D-1) \tag{4.95}$$

If $D > 32$, the remaining positions get filled in reverse order by repetition:

$$d_{D-k-1} = b_{(E+k) \bmod 32} \quad \text{where } k = 0, \ldots, D-33 \tag{4.96}$$

In the downlink compressed mode we change the slot format to prevent TFCI bit losses. When the slot formats do not match the exact number of TFCI bits for all possible TGLs and the number of TFCI fields exceeds the number of TFCI bits, we use DTX. The block of fields, where we use DTX, starts on the first field after the gap. If fewer TFCI fields exist after the gap than DTX bits, the last fields before the gap can also get filled with DTX. Denoting the number of bits available in the TFCI fields of one compressed radio frame by D and the number of bits in the TFCI field in a slot by N_{TFCI}, then we can express E, the first bit to be repeated as

$$E = N_{\mathrm{first}} N_{\mathrm{TFCI}} \quad \text{if } N_{\mathrm{first}} + \mathrm{TGL} \leq 15, \quad \text{else } E = 0 \tag{4.97}$$

When the transmission gap does not extend to the end of the frame, then E corresponds to the number of the first TFCI bits in the slot directly after the TG. We denote the total number of TFCI bits to be transmitted by N_{tot}. Thus, if $\mathrm{SF} \geq 128$ then $N_{\mathrm{tot}} = 32$, else $N_{\mathrm{tot}} = 128$. Afterwards, the following relations define the mapping:

$$d_k = b_{k \bmod 32} \quad \text{where } k = 0, 1, 2, \ldots, \min(E, N_{\mathrm{tot}}) - 1 \text{ and if } E < N_{\mathrm{tot}} \tag{4.98}$$

$$d_{k+D-N_{\mathrm{tot}}} = b_{k \bmod 32} \quad \text{where } k = E, \ldots, N_{\mathrm{tot}} - 1 \tag{4.99}$$

DTX bits are sent on d_k where $k = \min(E, N_{\mathrm{tot}}), \ldots, \min(E, N_{\mathrm{tot}}) + D - N_{\mathrm{tot}} - 1$ [17].

4.6.15 Examples on Channel Coding and Multiplexing

In the sequel we illustrate channel coding and multiplexing examples from [20] following the principles outlined in [16] and [17]. The examples aim to practically show the patters and fields to code different frames in the UTRA FDD and TDD modes. Thus, the number and variables in the forthcoming figures show the number of bits in corresponding fields.

4.6.15.1 Downlink FDD BCH

The parameters for the BCH shown in Table 4.43 indicate CRC bits, coding type, TTI, the number of codes used and the Spreading Factor (SF).

Table 4.43 Downlink FDD BCH parameters

Transport block size	246
CRC	16 bits
Coding	CC, coding rate = 1/2
TTI	20 ms
The number of codes	1
SF	2561

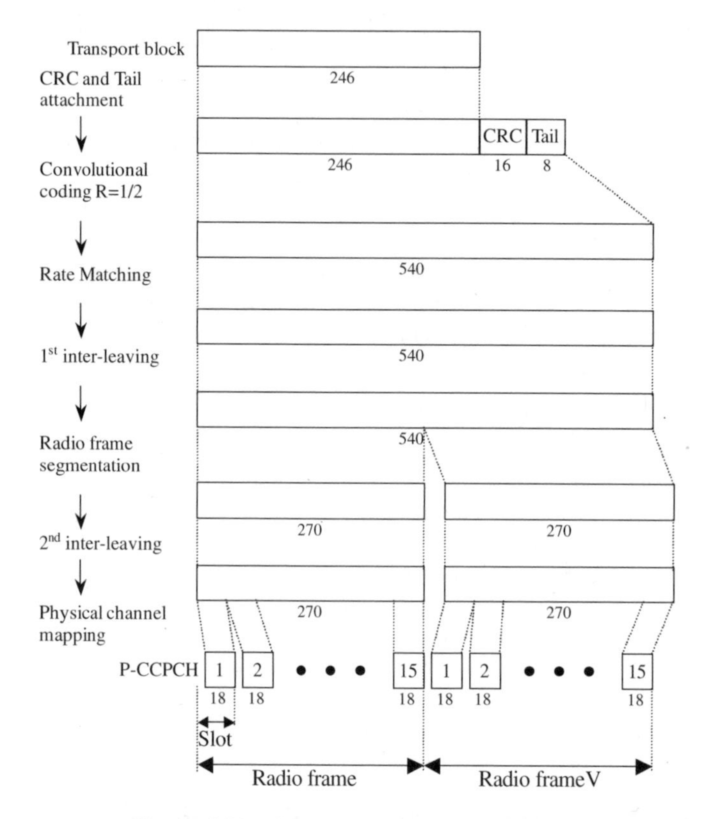

Figure 4.54 Coding the FDD downlink BCH.

Table 4.44 Parameter examples for 12.35 kbps speech information

The number of TrChs	3
Transport block size	81, 103 and 60 bits
CRC	12 bits (attached only to TrCh 1)
Coding	CC, coding rate = 1/3 for TrCh 1, 2 coding rate = 1/2 for TrCh 3
TTI	20 ms

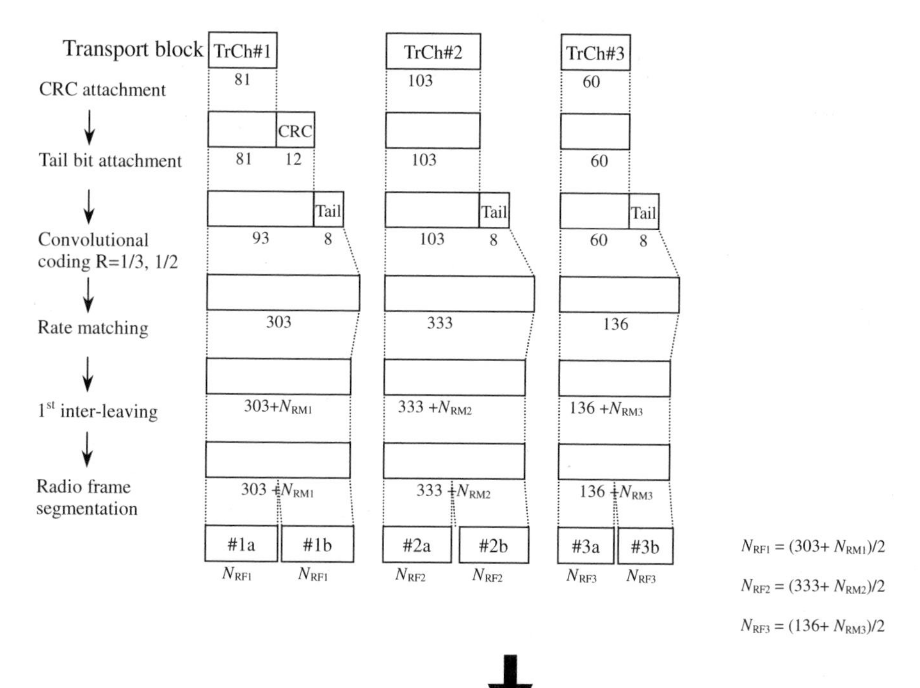

Figure 4.55 Channel coding and multiplexing 12.2 kbps speech.

Figure 4.54 illustrates the patterns of the bits in the corresponding fields of the DL FDD BCH example. Notice that we do not necessarily follow all the steps outlined in the preceding section. Thus, each particular channel will use only the corresponding steps.

4.6.15.2 Speech Channel Coding in the FDD Mode

We next illustrate the coding of transport channel for CS data or speech services. Notice how this example applies to 12.2 kbps AMR speech.

Table 4.44 and Figure 4.55 illustrate the key parameters for the 12.35 kbps AMR speech data. See the three the number of TrChs coding example when compared to the control channel in the previous example.

Table 4.45 Packet data parameters for 64/128/384 kbps services

The number of TrChs		1
Transport block size		640 bits
Transport block	64 kbps	$640 \times B$ bits $(B = 0, 1)$
Size	128 kbps	$640 \times B$ bits $(B = 0, 1, 2)$
	384 kbps	$640 \times B$ bits $(B = 0, 1, 2, \ldots, 6)$
CRC		16 bits
Coding		Turbo coding, coding rate = 1/3
TTI		10 ms

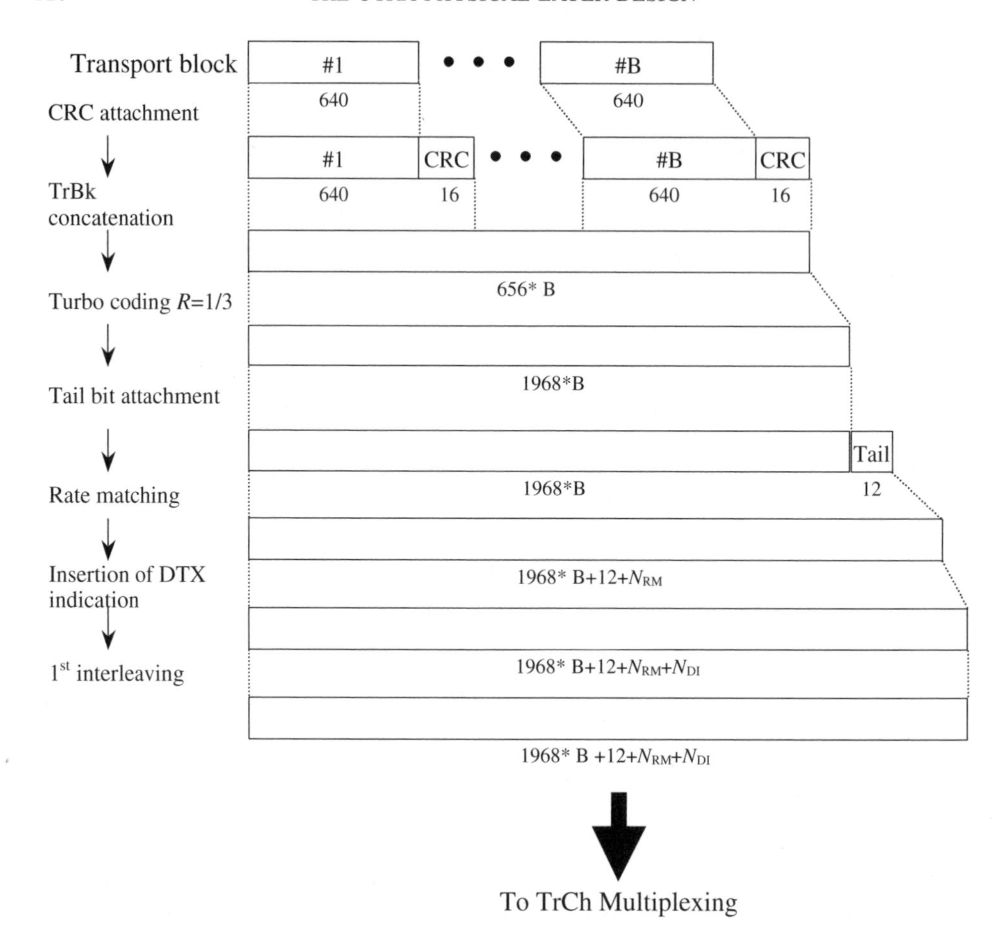

Figure 4.56 Channel coding and multiplexing 64/128/384 kbps packet data channels.

4.6.15.3 Coding FDD 64/128/384 kbps Packet Data Channels

Table 4.45 and Figure 4.56 show the key coding and multiplexing parameters for the aforementioned packet data channels. Here the number of blocks used logically depends on the transmission data rate required.

Table 4.46 Physical channel parameters to multiplex 64/128/384 kbps packet data and 4.1 kbps data

Data rate (kbps)	Symbol rate (ksps)	No. of phy. Channel: P	N_{pilot} (bits)	N_{TFCI} (bits)	N_{TPC} (bits)	$N_{\text{data 1}}$ (bits)	$N_{\text{data 2}}$ (bits)
64	120	1	8	8	4	4	56
128	240	1	16	8	8	48	240
384	240	3	16	8	8	48	240

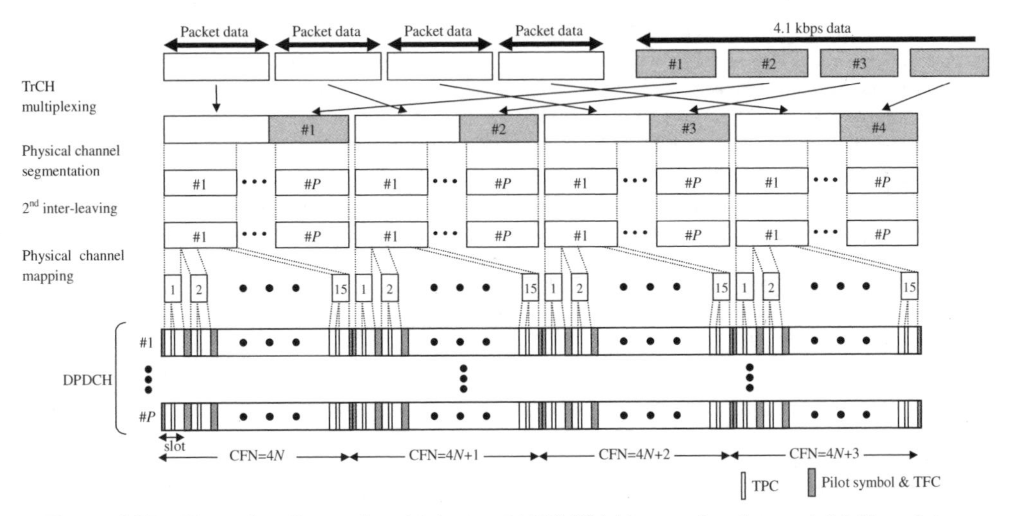

Figure 4.57 Channel coding and multiplexing 64/128/384 kbps packet data and 4.1 kbps data.

Notice also that when coding these transport channels we use turbo coding instead of convolution codes in the preceding examples.

4.6.15.4 Multiplexing of 64/128/384 kbps Packet Data and 4.1 kbps Data

This example applies to multiplexing 64/128/384 kbps packet data and DCCH. Table 4.46 and Figure 4.57 show a second view of the key physical channel parameters for multiplexing of 64/128/384 kbps packet data and 4.1 kbps data [21].

Other examples for FDD and TDD can be found in [20].

REFERENCES

1. 3GPP, Technical Specification Group (TSG) RUN WG4, UTRA (BS) FDD; Radio Transmission and Reception (3G TS 25.104, Version 3.1.0, 1999–2012).
2. 3GPP, Technical Specification Group (TSG) RUN WG4, UTRA (BS) TDD; Radio Transmission and Reception (3G TS 25.105, Version 3.1.0, 1999–2012).
3. 3GPP Technical Specification 25.214: Physical layer procedures (FDD).
4. 3G TS 25.213: Spreading and modulation (FDD).
5. 3G TS 25.214: Physical layer procedures (FDD).
6. 3G TS 25.101: UE Radio transmission and Reception (FDD).
7. 3G TS 25.215: Physical layer–Measurements (FDD).

8. 3G TS 25.211: Physical channels and mapping of transport channels onto physical channels (FDD), Version 3.2.0, 2000–2003.
9. Holma, H. and Toskaly, A., *WCDMA for UMTS, Radio Access for Third Generation Mobile Communications*. New York: Wiley, 2000.
10. Toskala, A., Lehtinen, O. and Kinnunen, P., UTRA GSM handover from physical layer perspective, Proc. ACTS Summit 1999, Sorrento, Italy, June 1999.
11. Gustafsson, M., Jamal, K. and Dahlman, E., Compressed Mode Techniques for Inter-Frequency Measurements in a Wide-band DS-CDMA System, Proc. IEEE Int. Conf. on Personal Indoor and Mobile Radio Commun., PIMRC'97, Helsinki, Finland, September 1997, Vol. 1, pp. 231–235.
12. ETSI STC SMG2 Plenary, GSM to other Systems Handover and cell selection/reselection project scheduling and open issues, GSM 10.89, Version 0.0.3, Tdoc 536, Meeting 35, Schaumburg, Illinois, USA, 3–7 April 2000.
13. 3G TS 25.221: Physical channels and mapping of transport channels onto physical channels (TDD), Version 3.2.0, 2000–2003.
14. 3G TS 25.223: Spreading and modulation (TDD).
15. 3G TS 25.224: Physical layer procedures (TDD).
16. 3G TS 25.222, Multiplexing and channel coding (TDD).
17. 3G TS 25.212: Multiplexing and channel coding (FDD).
18. 3G TS 25.222: Multiplexing and channel coding (TDD).
19. Swarts, F., *et al.*, *CDMA Techniques for Third Generation Mobile Systems*, Dordrecht: Kluwer Academic Publishers, 1999, p. 242.
20. 3G TS 25.944, Channel Coding and Multiplexing Examples, V3.0.0, 2000–2003.
21. 3GPP, Technical Specification Group (TSG) RUN WG4, UTRA (UE) TDD; Radio Transmission and Reception (3G TS 25.102, Version 3.1.0, 1999–2012).

APPENDIX A: DPDCH AND DPCCH FIELDS

Table 4.47 DPDCH and DPCCH fields

Slot format (i)	Channel bit rate (kbps)	Channel symbol rate (ksps)	SF	Bits/slot	DPDCH bits/slot		DPCCH bits/slot			Transmitted slots per radio frame, N_{Tr}
					N_{Data1}	N_{Data2}	N_{TPC}	N_{TFCI}	N_{Pilot}	
0	15	7.5	512	10	0	4	2	0	4	15
0A	15	7.5	512	10	0	4	2	0	4	8–14
0B	30	15	256	20	0	8	4	0	8	8–14
1	15	7.5	512	10	0	2	2	2	4	15
1B	30	15	256	20	0	4	4	4	8	8–14
2	30	15	256	20	2	14	2	0	2	15
2A	30	15	256	20	2	14	2	0	2	8–14
2B	60	30	128	40	4	28	4	0	4	8–14
3	30	15	256	20	2	12	2	2	2	15
3A	30	15	256	20	2	10	2	4	2	8–14
3B	60	30	128	40	4	24	4	4	4	8–14

Table 4.47 (*Continued*)

Slot format (i)	Channel bit rate (kbps)	Channel symbol rate (ksps)	SF	Bits/slot	DPDCH bits/slot		DPCCH bits/slot			Transmitted slots per radio frame, N_{Tr}
					N_{Data1}	N_{Data2}	N_{TPC}	N_{TFCI}	N_{Pilot}	
4	30	15	256	20	2	12	2	0	4	15
4A	30	15	256	20	2	12	2	0	4	8–14
4B	60	30	128	40	4	24	4	0	8	8–14
5	30	15	256	20	2	10	2	2	4	15
5A	30	15	256	20	2	8	2	4	4	8–14
5B	60	30	128	40	4	20	4	4	8	8–14
6	30	15	256	20	2	8	2	0	8	15
6A	30	15	256	20	2	8	2	0	8	8–14
6B	60	30	128	40	4	16	4	0	16	8–14
7	30	15	256	20	2	6	2	2	8	15
7A	30	15	256	20	2	4	2	4	8	8–14
7B	60	30	128	40	4	12	4	4	16	8–14
8	60	30	128	40	6	28	2	0	4	15
8A	60	30	128	40	6	28	2	0	4	8–14
8B	120	60	64	80	12	56	4	0	8	8–14
9	60	30	128	40	6	26	2	2	4	15
9A	60	30	128	40	6	24	2	4	4	8–14
9B	120	60	64	80	12	52	4	4	8	8–14
10	60	30	128	40	6	24	2	0	8	15
10A	60	30	128	40	6	24	2	0	8	8–14
10B	120	60	64	80	12	48	4	0	16	8–14
11	60	30	128	40	6	22	2	2	8	15
11A	60	30	128	40	6	20	2	4	8	8–14
11B	120	60	64	80	12	44	4	4	16	8–14
12	120	60	64	80	12	48	4	8[a]	8	15
12A	120	60	64	80	12	40	4	16[a]	8	8–14
12B	240	120	32	160	24	96	8	16[a]	16	8–14
13	240	120	32	160	28	112	4	8[a]	8	15
13A	240	120	32	160	28	104	4	16[a]	8	8–14
13B	480	240	16	320	56	224	8	16[a]	16	8–14
14	480	240	16	320	56	232	8	8[a]	16	15
14A	480	240	16	320	56	224	8	16[a]	16	8–14
14B	960	480	8	640	112	464	16	16[a]	32	8–14
15	960	480	8	640	120	488	8	8[a]	16	15
15A	960	480	8	640	120	480	8	16[a]	16	8–14
15B	1920	960	4	1280	240	976	16	16[a]	32	8–14
16	1920	960	4	1280	248	1000	8	8[a]	16	15
16A	1920	960	4	1280	248	992	8	16[a]	16	8–14

[a] If TFCI bits are not used, then DTX shall be used in TFCI field.

N.B. 1: Compressed mode is only supported through spreading factor reduction for SF = 512 with TFCI.

N.B. 2: Compressed mode by spreading factor reduction is not supported for SF = 4.

APPENDIX B: BIT PATTERNS COMPRESSED MODE AND $N_{pilot} = 4$

For slot formats 2B and 3B, i.e. compressed mode through spreading factor reduction and $N_{pilot} = 4$, the pilot bits on antenna 1 are STTD encoded. Thus, the pilot bit pattern is as shown in the rightmost set of Table 4.14.

Slot	$N_{pilot} = 4$	
Symbol	0	1
0	01	10
1	10	01
2	11	00
3	10	01
4	00	11
5	01	10
6	01	10
7	00	11
8	11	00
9	01	10
10	11	00
11	00	11
12	00	11
13	10	01
14	10	01

5

The UMTS Development Platform

5.1 ARCHITECTURE AND DEPLOYMENT SCENARIOS

The architecture at the domain and functional levels, as well as the deployment scenarios are presented based on the 3GPP (ETSI) specifications noted in [1,2]. The terminology and basic principles are kept for consistency with a simplified approach in some cases for a pragmatic representation of the subject.

5.1.1 The UMTS High Level System Architecture

5.1.1.1 The UMTS Domains

Figure 5.1 illustrates the different UMTS domains. The identified domains imply the evolution of current or existing network infrastructures, but do not exclude new ones. The Core Network (CN) domain can evolve, for example, from the GSM, N-ISDN, B-ISDN and PDN infrastructures.

The generic architecture incorporates two main domains, i.e. the *user equipment domain* and *the infrastructure domain*. The first concerns the equipment used by the user to access UMTS services having a radio interface to the infrastructure. The second consists of the physical nodes, which perform the various functions required to terminate the radio interface and to support the telecommunication services requirements of the users. The rest of the sub-domains are defined in Table 5.1.

Figure 5.26 in Appendix A illustrates the four (Application, Home, Serving and Transport) strata. It also shows the integrated UMTS functional flow, i.e. the interactions between the USIM, MT/ME, Access Network, Serving Network and Home Network domains, including interactions between TE, MT, Access Network, Serving Network, Transit Network domains and the Remote Party.

5.1.1.2 The IMT-2000 Family

The UMTS high level architecture integrates the physical aspects through the domain concept and functional aspects through the strata concept. The separation according to [1]

All IP in 3G CDMA Networks J. Castro
 ISBN: 0-470-85322-0

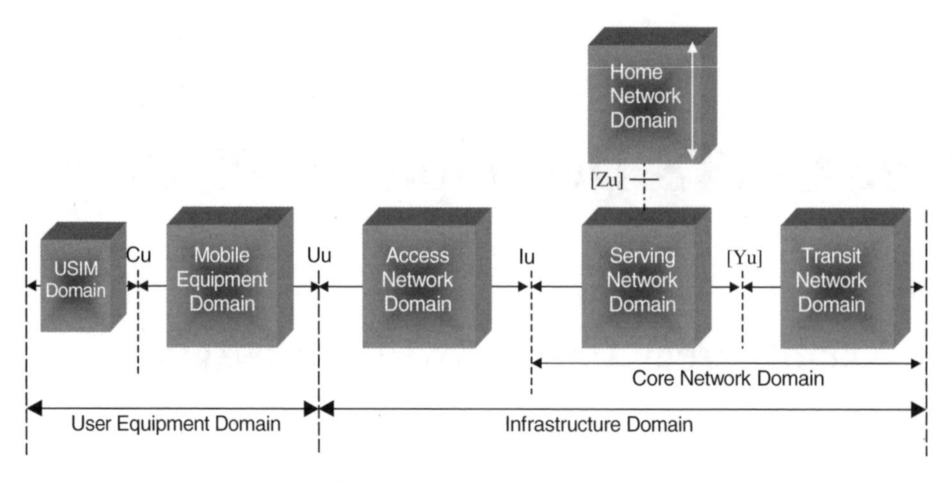

Figure 5.1 UMTS architecture domains and reference points.

allows a UMTS network to fit within the context of the IMT-2000 family of networks as illustrated in Figure 5.2.

Basically there are two CN options for the air interface of the IMT-2000 family of networks, i.e. GSM and IS-41 networks. The first one, which also includes the IP packet network, will serve the UTRA modes and the UWC-136 (packet) evolving based on EDGE. While GPRS may become an IP core network on its own, where UMTS and other air interfaces will directly connect to it, today it is part of the GSM infrastructure. IS-41 will serve primarily USA regions in the evolution of IS-136 in TDMA and IS-95 in CDMA.

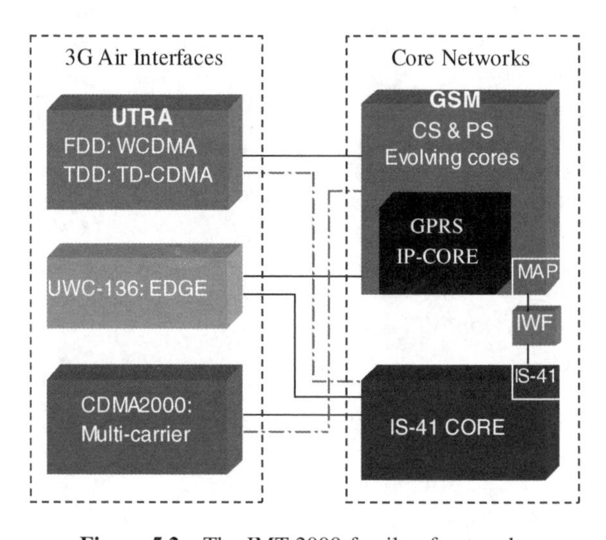

Figure 5.2 The IMT-2000 family of networks.

Table 5.1 The UMTS architecture domains

User equipment domains: dual mode and multi-mode handsets, removable smart cards, etc.	
Mobile Equipment (ME) domain	Consists of: • the Mobile Termination (MT) entity performing the radio transmission and related functions • the Terminal Equipment (TE) entity containing the end-to-end application (e.g. a laptop connected to a handset).
USIM domain	The User Services Identity Module (USIM) domain contains data and procedures to unambiguously and securely identify itself (e.g. smart card)
Infrastructure domains	
Access Network (AN) domain	Consists of the physical entities managing the access network resources and provides the users with mechanisms to access the core network
Core Network (CN) domain	Consists of the physical entities providing support for the network features and telecommunication services; e.g. management of user location information, control of network features and services, switching and transmission mechanisms for signalling and for user generated information. It includes: 1. Serving Network (SN) domain representing the core network functions local to the user's access point and thus their location changes when the user moves. 2. Home Network (HN) domain representing the core functions conducted at a permanent location regardless of the user's access point. The USIM is related by subscription to the HN. 3. Transit Network (TN) domain, which is the CN part between the SN and the remote party.

5.1.2 Co-existence of Present and Future Networks

While UMTS will bring new services and allow new access options, its deployment and introduction will be in several phases. The first no doubt will evolve within a mixed environment where co-existence with second generation systems like GSM (including GPRS) will be predominant. Figure 5.3 illustrates the main network elements of a typical GSM network incorporating the Circuit-Switched (CS) segment and the GPRS entities as part of the Packet-Switched (PS) segment. It also includes the future UMTS elements on the radio interface side. Hence while some operators or service providers will deploy completely new network infrastructures, others will use GSM architecture as the basis for UMTS or 3G systems. This means that *UMTS will complement the existing GSM system in some cases, not replace it.*

Clearly for all the elements to co-exist as illustrated in Figure 5.3 they must all contain the necessary HW/SW (including protocols) enabling features for inter-working. Today, for example, the SMG (ETSI) and 3GPP organisation have the task of making second and third generation elements inter-work seamlessly through specification and recommendations. The

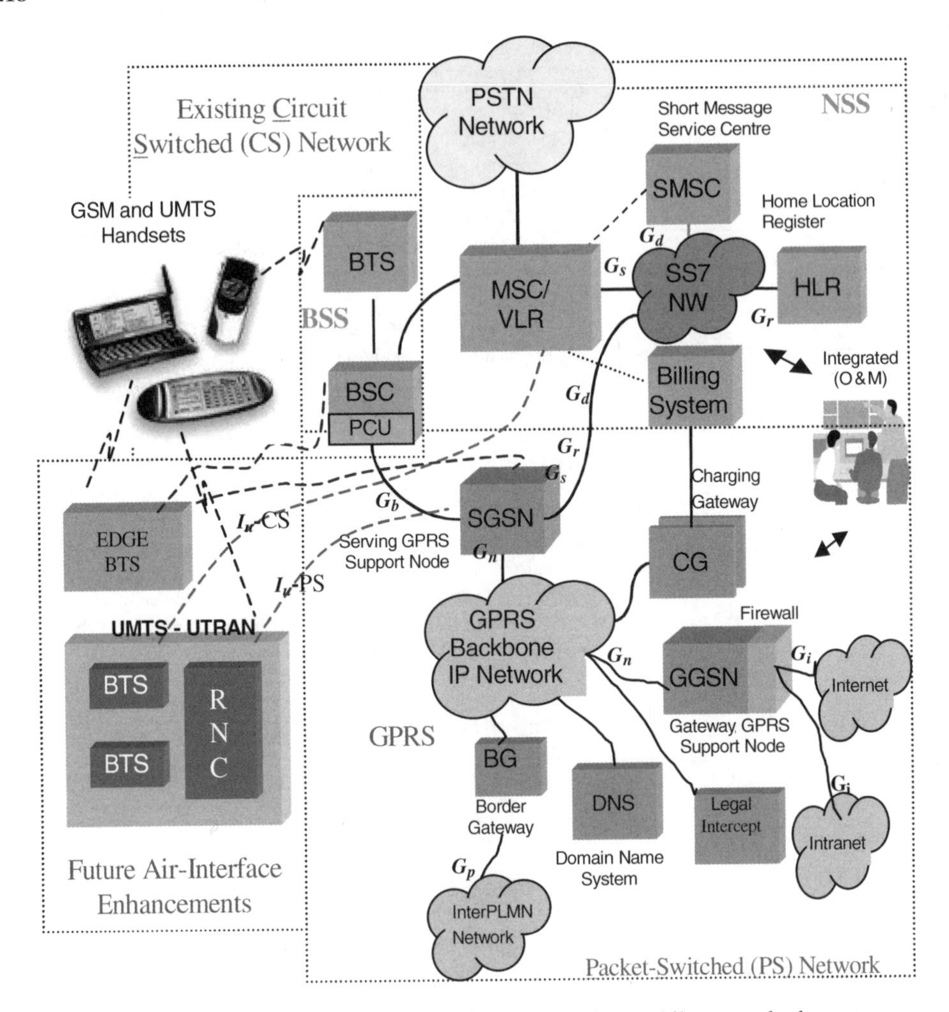

Figure 5.3 Co-existence of second and third generation mobile network elements.

technical specifications for practical reasons are issued in releases, e.g. the contents of the first edition of this book will be based on Release 1999 covering the evolution of GSM and the introduction of UMTS. However, it will cover also R4 and R5.

5.2 THE CORE NETWORK DOMAIN

5.2.1 Network Evolution Towards UMTS

Evolution here implies seamless and dynamic inter-operability of 2G (2.5G) and 3G technologies in the CN and Radio-Access Network (RAN) sides. We will thus cover these evolution implications next taking into account the integrated network elements illustrated in Figure 5.3, i.e. the PS and CS building blocks in the CN side and UTRAN and EDGE in the RAN side.

To structure the presentation following the domain concept, we first cover the core network domain. Preceding chapters have addressed the access network and mobile equipment domains. Furthermore, for completeness we also define the basic functions of the CN building blocks.

The UMTS platform as illustrated in Figure 5.3 will incorporate a number of 2G/3G[1] functional elements joined by standard interfaces. Together these network elements will route multifarious information traffic and provide:

- resource allocation;
- mobility management;
- radio link management;
- call processing;
- billing record generation;
- operational and maintenance functions;
- collection of performance statistics.

The CN comprises of circuit and packet switching systems, trunk transmission, signalling systems, the access network and service platforms.

Figure 5.4 highlights the 3G side represented in layers to point out some of the new elements incorporated to the legacy GSM network. Each layer contains a distinct network

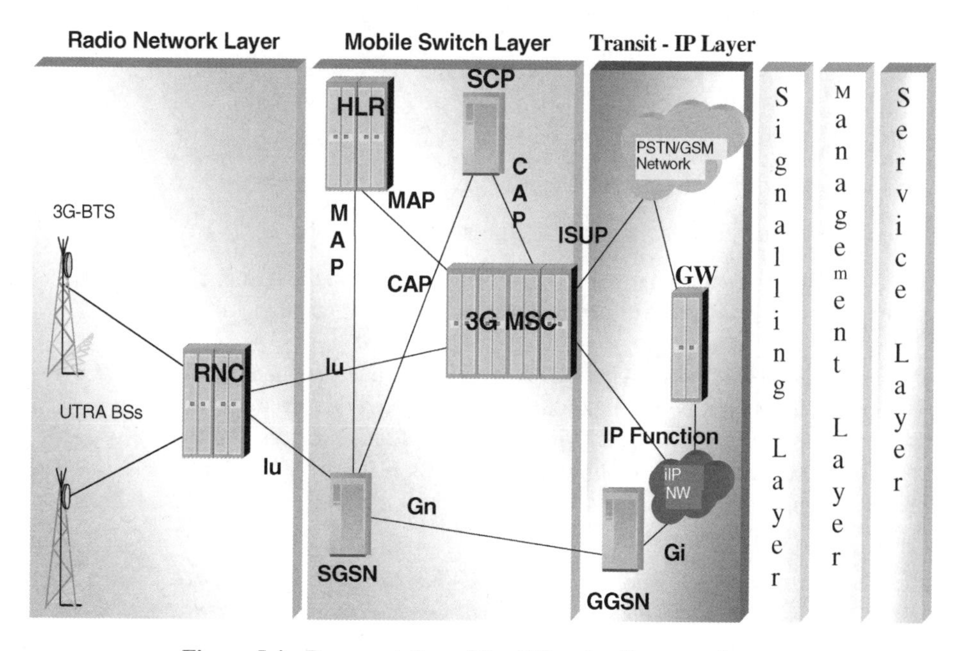

Figure 5.4 Representation of the CN and radio-access layers.

[1]GSM 1800 MHz evolving elements and new UMTS or 3G specific elements co-existing seamlessly.

element based on the CN infrastructure evolution. However, it is not restricted to the CN layers, it includes, e.g. the radio layer and others as follows:

- the radio network layer illustrating new WCDMA Base Stations (BSs) and the RNC described in Chapter 4;
- the mobile switching layer, which regroups the 3G SGSN, 3G MSCs with their upgraded associated components, such as HLR, VLR, AuC, EIR and their new SPC unit enabling IP telephony;
- the transit–IP layer, which not only serves as the backbone layer for transiting traffic between nodes, but also incorporates the IP bypass mediation device for signalling and user data between CS and PS. The GGSN may also be a part of this layer;
- the signalling layer comprising mainly of STPs connected to the other elements;
- the management layer, composed of the integrated network management systems and network mediation systems as illustrated in Figure 5.3;
- the service layer, which will comprise all value-added service platforms, such as SMC, VMS, intelligent network platform and customer care centres, ISP, billing platform, etc.

For background completeness on the CN side, the main GPRS elements and terminal connections to a GPRS network are described next from the functional level.

5.2.1.1 Main Packet-Switched Network Elements

The *Serving GPRS Support Node* (SGSN) performs the following key tasks:

- authentication and mobility management;
- protocol conversion between the IP backbone and the protocols used in the BSS and MS;
- collection of charging data and traffic statistics;
- routing data to the relevant GGSN when connection to an external network is required (all intra-network MS to MS connections must also be made via a GGSN).

The *Gateway GPRS Support Node* (GGSN) acts as the interface between the GPRS network and external networks; it is simply a router to a sub-network. When the GGSN receives data addressed to a specific user, it checks if the address is active. If it is, the GGSN forwards the data to the SGSN serving the mobile: if the address is inactive the data are discarded. The GGSN also routes mobile originated packets to the correct external network.

5.2.1.1.1 Terminal Attachment to the GPRS Network

The connection between a GPRS terminal and the network has two parts:

1. *Connection to the GSM network (GPRS Attach)* – when the GPRS terminal is switched on, it sends an ‘attach’ message to the network. The SGSN collects the user data from the HLR and authenticates the user before attaching the terminal.

2. *Connection to the IP network (PDP context)* – once the GPRS terminal is attached, it can request an IP address (e.g. 172.19.52.91) from the network. This address is used to route data to the terminal. It can be static (the user always has the same IP address) or dynamic (the network allocates the user a different IP address for each connection).

Dedicated standard (ETSI specified) interfaces assuring the inter-connection between the key network elements and enabling multi-vendor configurations include:

- Gb-interface (SGSN-BSS);
- Gn-interface (GSN-GSN);
- Gp-interface (inter-PLMN interface);
- Gi (GGSN-external IP networks);
- Gr (SGSN-HLR);
- Gs (SGSN-MSC/VLR);
- Gd (SGSN to SMS-GMSC/SMS-IWMSC).

Other GPRS elements illustrated in Figure 5.3 are the following:

1. *Domain Name Servers* – these are standard IP devices that convert IP names into IP addresses, e.g. vms.orange.ch → 172.19.52.92
2. *Firewalls* – these protect the IP network against external attack (e.g. from hackers). The firewall might reject all packets that are not part of a GPRS subscriber initiated connection.
3. *Border Gateway* – this is a router providing, e.g. a direct GPRS tunnel between different operators' GPRS networks via an inter-PLMN data network, instead via the public Internet.
4. *Charging Gateway* – GPRS charging data are collected by all the SGSNs and GGSNs in the network. The charging gateway collects all these data together, processes it and passes it to the billing system.

5.2.1.2 Open Interfaces

To practically visualise the inter-operating environment we will take as reference a GSM Network in transition towards a 3G system.

Evolving CN elements, e.g. will concurrently support interfaces (thereby signalling) for both 2G and 3G radio networks, i.e. existing elements will be enabled through field upgrades with Iu interface towards high-capacity 3G.

The 3G RAN, e.g. will connect to a GSM CN via the Iu interface. This interface provides a logical separation between CS and PS signalling giving the possibility to physically separate the interfaces, i.e.

- Iu–CS interface for circuit-switched traffic, based on the ATM transport protocol,
- Iu–PS interface for packet-switched traffic, based on IP over ATM.

The Iu interfaces above assume that the MSC can also multiplex the Iu–PS interface to the SGSN with only one physical interface from RNC to the core network, and that the MSC will get an ATM module to interact with the ATM-based RAN.

A second new interface besides the Iu in the CN concerns IP links. It is foreseen that by the time UMTS is deployed, MCSs will support IP connections. Thus the solution can be envisaged as follows:

- a new feature in the MSC will be the integrated IP function protocol between two MSCs signalling and user data between CS and PS;
- the integrated IP function will introduce a new type of trunk signalling to the MSC switching system, i.e. SS7 over the IP network;
- the transmission over the IP network will be done using the User Datagram Protocol (UDP) from the TCP/IP stack; both signalling transmission and media transmission will use the protocol;
- data, fax and compressed speech will be packetised to IP packets and transmitted to the other switch using the Real-time Transport Protocol (RTP) on the UDP.

Other key interfaces in the evolution to 3G include:

- A-Interface MSC to GSM BSS will continue as needed for applications like Radio Resource Management (RRM), Mobility Management (MM) and Link Management (LM);
- MAP performing signalling between the MSC and other NSS elements and performing critical operations between switching and database elements to support roaming;
- CCS7 – Common Channel Signalling system (7) links the MSC to a PSTN or to an ISDN using a single channel to carry the signalling of multiple speech circuits; the digital Channel Associated Signalling (CAS) used between exchanges will also continue as needed;
- in the short term, the File Transfer Access and Management (X.25 FTAM) interface will continue to communicate with billing systems as IP links to new billing centres develop;
- standard V.24 interfaces connecting O&M terminals to the MSC will probably continue, while more sophisticated WWW-type interfaces will be implemented with evolving MSC operating systems.

In conclusion, we can say that the two critical interfaces that UMTS introduces to the CN are primarily the Iu and IP. These interfaces add new dimension to the existing GSM infrastructures besides enriching the type of links a CN may have.

In the following we cover the essential transition steps in terms of the 3G architecture requirements.

5.2.2 Key Release 1999 Architectural Requirements

The general working assumptions for Release 1999 (R99), which cover the phase 1 UMTS/Release 1999 GSM standards and reflecting in part the elements illustrated in Figure 5.3, can be summarised from [3] as follows:

- a core network based on an evolved 2G MSC and an evolved SGSN;
- an optionally evolved Gs interface;
- mobile IPv4 with Foreign Agent (FA) care-of addresses to end-users over the UMTS/GPRS network, where the FA is located in the GGSN;
- class A GSM mobiles;
- transcoder location shall be according to the 'Evolution of the GSM platform towards UMTS' outlined in 3G TS 23.930;
- UMTS/IMT-2000 Phase1 (R99) network architecture and standards shall allow the operator to choose between Integrated and Separated CNs for transmission (including L2);
- the UMTS standard shall allow for both separated and combined MSC/VLR and SGSN configurations;
- the UE shall be able to handle separated or combined MSCs and SGSNs;
- there can be several user planes to these CN nodes.

The following general concepts should be followed:

- separate the layer 3 control signalling from the layer 2 transport discussion (do not optimise layer 3 for one layer 2 technology);
- MSC–MSC layer 3 call control is out of scope of standardisation in 3GPP;
- as future evolution may lead to the migration of some services from the CS domain to the PS domain without changes to the associated higher-layer protocols or functions. UMTS Release 1999 shall provide the flexibility to do this in a way that is backwards compatible with Release 1999 UEs, provided this does not introduce significant new complexity or requirements in the system.

5.2.3 Co-existence Inter-operability Issues

Although it seems that only one new interface, i.e. Iu, appears when incorporating the UMTS radio network to the 2G or 2.5G CN, inter-networking impacts spread to all the integrated network elements as shown in Figure 5.3. In particular, mobility management and call control bring in new inter-operability requirements. These requirements are summarised next before we concentrate on describing the different UMTS building blocks of the radio network in forthcoming chapters.

5.2.3.1 Iu Interface Inter-working Characteristics

The Iu principles presented in [4] apply to PS and CS networks. In this context, UTRAN supports two logically independent signalling flows via the Iu interface to combined or separated network nodes of different types like MSC and SGSN [3].

Thus, UTRAN contains *domain distribution function* routing application independent UE control signalling to a corresponding CN domain. The UE indicates the addressed application type through a protocol discriminator, for example. Then UTRAN maps this onto a correct Iu instance to forward signalling. UTRAN services, including radio-access bearers, are CN domain independent, e.g. we can get speech bearer either through the PS or CS core network. The Iu includes control and user planes.

Because only a RNC can identify the actual packet volume successfully transferred to a UE, it indicates the volume of all not transferred downlink data to the 3G-SGSN so this latter can correct its counter.

5.3 THE UMTS NETWORK CONFIGURATION BEYOND R99

As illustrated in Figure 5.3, the core extension of the GSM footprint does not appear high with the introduction of UMTS Release 1999, even though we introduced a complete radio-access network. This picture will change very much as we add Release 5 Network Elements (NE), and re-shape completely the landscape of Public Land Mobile Networks (PLMNs). In this section we will aim to update the overall PLMN architecture and highlight the new building blocks of the UMTS core following the technical specification, which continues to progress.

5.3.1 Entities Common to PS and CS Sub-domains

Here we will mainly list the new entities or the old ones with renewed features and interfaces; then, in subsequent sections or chapters we will define their logical functions.

5.3.1.1 The Home Subscriber Server (HSS)

Common to both the PS and CS domains, the HSS is the master database for a given user. It is the entity containing the subscription-related information to support the network entities actually handling calls/sessions [5].

The number of mobile subscribers in a Home Network will determine the number of HSSs, including their capabilities, type of equipment and network organisation. For example, the HSS supports the *control servers* while they complete the routing/roaming procedures by solving authentication, authorisation, naming/addressing resolution, location dependencies, etc. Thus, the HSS holds the following user-related information:

- User identification; numbering and addressing information.
- User security information: network access control information for authentication and authorisation.
- User location information at inter-system level: the HSS supports the user registration, and stores inter-system location information, etc.
- User profile information.

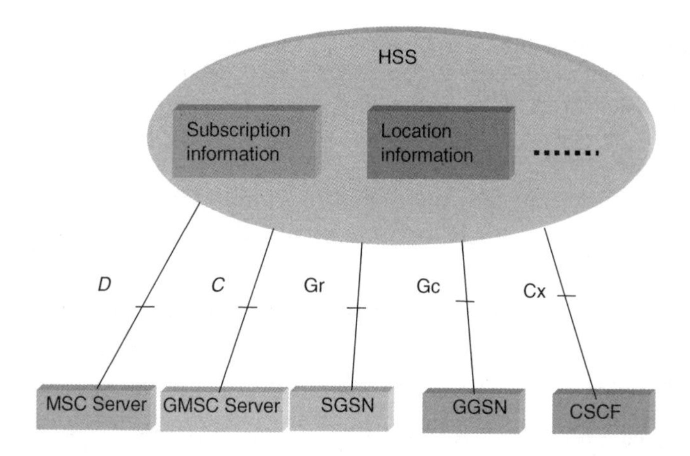

Figure 5.5 Generic HSS structure example with basic interfaces [5].

In addition, the HSS also generates *User Security Information* (USI) for mutual authentication, communication integrity check and ciphering.

Based on the USI, the HSS also supports the *call control* and *session management* entities of the different Domains and Subsystems. Figure 5.5 illustrates a HSS example.

The HSS will integrate heterogeneous information and enable CN enhanced features to be offered to the application and services domain, while hiding the heterogeneity. It includes the following functionalities:

- Access network independent IP-multimedia functionality to support IMS control functions (e.g. CSCF), which enables IMS-service subscriber usage.
- The HLR/AUC functionality subset required by the PS Domain.
- The HLR/AUC functionality subset required by the CS Domain in order to enable subscriber access to the CS Domain or to support roaming to legacy GSM/UMTS CS Domain networks.

3GPP TS 23.008 [6] outlines the organisation of the subscriber data, and indicates as well the numbers, addresses and identifiers specified in TS 23.003 [7], which the HSS needs to store [5].

5.3.1.1.1 The Home Location Register Functions

The HLR, containing features referenced in the preceding architectures up to Rel-4, today is considered as a subset of the HSS containing the following functionalities:

- enables subscriber access to the PS Domain services by supporting PS Domain entities such as the SGSN and GGSN, through the Gr and Gc interfaces;
- enables subscriber access to the CS Domain services and allows roaming in legacy GSM/UMTS CS Domain networks by supporting CS Domain entities such as the MSC/MSC server and GMSC/GMSC server through the C and D interfaces.

5.3.1.1.2 The Authentication Centre (AuC) Functions

As in the HLR, the AuC referenced in the preceding architectures up to and including Rel-4, it is also considered a subset of the HSS holding the following functionalities for the CS and PS domains:

- Associated with an HLR, the AuC stores an *identity key* for each mobile subscriber registered with the associated HLR. We use this key to generate security data for each mobile subscriber [5], i.e.:
  - International Mobile Subscriber Identity (IMSI) and network data;
  - communication integrity check over the radio path between the MS and the network;
  - communication ciphering over the radio path between the MS and the network.
- The AuC communicates only with its associated HLR over a non-standardised interface denoted H-interface. The HLR requests required data for authentication and ciphering from the AuC via the H-interface; then, it stores and delivers these data to the VLR and SGSN, which use it to perform MS security functions.

5.3.1.1.3 Logical Functions in The HSS

Figure 5.6 illustrates the main high-level HSS logical functions, these include [5]:

- *mobility management*, which supports user mobility through the CS and PS domains, as well as the IP-Multimedia Subsystem (IMS);
- *call and/or session establishment support*, e.g. traffic termination, info on which call and/or session control entity currently hosts the user;

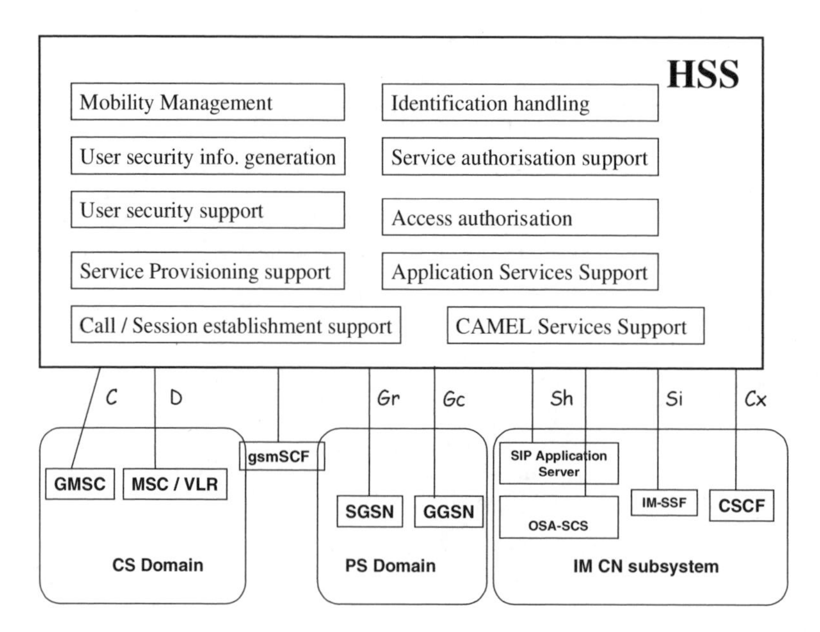

Figure 5.6 The HSS logical functions [5].

- *user security information generation*;
- *user authentication generation* integrity and ciphering data for the CS and PS Domains and IMS; and user security support. The HSS does it by generating data for authentication, integrity and ciphering and by providing these data to the appropriate entity in the CN (i.e. MSC/VLR, SGSN or CSCF);
- *user identification handling*–the HSS provides the appropriate relations among all the identifiers uniquely determining the user in the CS and PS Domains, as well as IMS (e.g. IMSI and MSISDNs for CS Domain; IMSI, MSISDNs and IP addresses for PS Domain, private identity and public identities for IMS);
- *access authorisation*–the HSS authorises the user for mobile access at the requests of the MSC/VLR, SGSN or CSCF by checking user rights for roaming in visited network;
- *service authorisation support*–the HSS provides basic authorisation for MT call/session establishment and service invocation. In addition, it updates the appropriate serving entities (i.e. MSC/VLR, SGSN, CSCF) with the relevant service information to be provided to the user, (i.e. the service profile);
- *service provisioning support*–the HSS provides access to the service profile data for use within the CS and PS Domains, IMS and CAMEL Services Support. It communicates with the SIP Application Server and the OSA–SCS to support Application Services in IMS. It also communicates with the IM–SSF to support the CAMEL Services related to the IMS, and communicates as well with the gsmSCF to support CAMEL Services in the CS and PS Domains.

Although the remaining common entities do introduce major changes, we will list them for completeness.

5.3.1.2 The Visitor Location Register (VLR)

Although, we may not find changes from earlier functions, in the following we will describe the basic VLR functions for completeness.

The VLR, which may be in charge of one or more MSC areas, controls an MS roaming in an MSC area or within a pool-area. As soon as MS enters a new location area, it starts a registration procedure. A serving MSC in that area notices this registration and transfers to a VLR the identity of the location area where the MS is. If this MS is not yet registered in the VLR, the VLR and the HLR exchange information to allow the proper handling of calls involving the MS [5].

In addition, the VLR contains required information to handle call set-ups or received calls by the MSs registered in its DB,[2] these elements include the:

- International Mobile Subscriber Identity (IMSI);
- Mobile Station International ISDN number (MSISDN);

[2]For some supplementary services the VLR may have to obtain additional information from the HLR.

- Mobile Station Roaming Number (MSRN), see TS 23.003 [7] for allocations;
- Temporary Mobile Station Identity (TMSI), if applicable;
- Local Mobile Station Identity (LMSI), if used;
- location area where the mobile station has been registered;
- identity of the SGSN where the MS has been registered. Applicable only to PLMNs supporting GPRS and which have a Gs interface between MSC/VLR and SGSN;
- last known location and the initial location of the MS.

Finally, the VLR contains also supplementary service parameters related to the mobile subscriber, which are received from the HLR [6].

5.3.1.3 The Equipment Identity Register (EIR)

As in the case of the VLR, we do not perceive major changes in the EIR functions, except that it will take into account devices dedicated to IMS. Thus, we describe the EIR as in GSM systems, where it is the logical entity, which is responsible for storing in the network the International Mobile Equipment Identities (IMEIs).

The EIR classifies the equipment in separate lists as 'white listed', 'grey listed', 'black listed' or as unknown (TS 22.016 [8] and TS 29.002 [9]). An EIR shall as a minimum contain a 'white list' (Equipment classified as 'white listed').

5.3.1.4 SMS Gateway MSC (SMS-GMSC) and SMS Inter-working MSC

The SMS-GMSC acts as an interface between the Short Message Service Centre (SMSC) and the PLMN to allow short messages delivery to MSs from SMSCs.

The SMS Inter-working MSC acts also as interface between the PLMN and a SMSC to allow short messages to be submitted to the SMSC.

The choice of the SMS-GMSC and the SMS Inter-working MSC will depend on the network configuration chosen by the operator, e.g. all MSCs or designated MCSs. In practice, a designated MSC serves the aforementioned functions.

5.3.2 Entities in the CS Sub-domain

Here again, we do not see major changes from the generic tasks seen in earlier architectures, e.g. R99. As a result, the Mobile-services Switching Centre (MSC) still serves as interface between the radio system and the fixed networks. It performs all necessary functions in order to handle the circuit-switched services to and from the MSs.

Nevertheless, there are new functions introduced with later releases, e.g. R4 or R5; thus, we will briefly recapitulate these CS entities in this section.

- Each MSC interfaces to one or more BSS(s) and/or RNS(s) providing radio coverage of a given geographical area. In addition, several MSCs may be required to cover a country.

- The MSC acts as an exchange, which performs all the switching and signalling functions for MSs located in a geographical area designated as the MSC area.
- When we apply Intra Domain Connection of RAN Nodes to Multiple CN Nodes, all the MSCs serving a pool-area share the responsibility to serve the MSs located in the pool-area; but each individual MS is served by only one out of these MSCs, as described in TS 23.236 [10]. All these MSCs interface to all the BSS(s) and/or RNS(s) forming the pool-area.

The main difference between an MSC and an exchange in a fixed network arises from the fact that the MSC has to take into account the impact of the allocation of radio resources, the mobile nature of the subscribers and has to perform in addition, at least the following location registration [11] and handover [12] procedures.

From R4 onwards the MSC can be implemented in two different entities; i.e. the MSC Server, handling only signalling and the CS-MGW, handling user's data; thus, an MSC Server and a CS-MGW make up the full functionality of an MSC after R4.

5.3.2.1 The MSC Server

The MSC Server comprises mainly of the Call Control (CC) and mobility control parts of an MSC. It:

- controls Mobile Originated (MO) and Mobile Terminated (MT) CC CS Domain calls;
- terminates the user-network signalling and translates it into the relevant network–network signalling;
- contains a VLR to hold the mobile subscriber's service and CAMEL related data;
- controls the parts of the call state that pertain to connection control for media channels in a CS–MGW.

5.3.2.2 The CS Media Gateway (CS-MGW)

Here we refer specifically to the CS domain gateway entity as CS-MGW to differentiate it from the IP-Multimedia Subsystem (IMS) gateway entity, i.e. the IM-MGW. The CS-MGW performs the following functions:

- joins the UTRAN to the core system over the Iu interface;
- serves as the PSTN/PLMN transport termination point for a defined network;
- may terminate bearer channels from a CS network and media streams from a packet network (e.g. RTP streams in an IP network);
- may support media conversion over the Iu bearer control and payload processing (e.g. codec, echo canceller, conference bridge) to enable different Iu options for CS services (i.e. AAL2/ATM based as well as RTP/UDP/IP based);

- interacts with the Media Gateway Control Function (MGCF) entity, MSC server and Gateway MSC (GMSC) server for resource control;
- owns and handles resources such as echo cancellers, and it may also use codecs;
- it is provisioned with the necessary resources to support UMTS/GSM transport media;
- it may require additional packages like H.248 [13] to support additional codecs and framing protocols, etc;
- The CS–MGW bearer control and payload processing capabilities may also need to support mobile specific functions[3] such as SRNS[4] relocation/handover and anchoring.

5.3.2.3 The GMSC–GMSC Server and IWF

The GMSC performs the routing function to the actual location of the MS, i.e. if a network delivering a call to the PLMN cannot interrogate the HLR, the call is routed to a GMSC, which interrogates the appropriate HLR and then route the call to the MSC where the mobile station is located. The HLR interrogation acceptance of the call and GMSC routing will depend on the operator's service arrangements, generally a given HLR and MSC will be assigned to these functions.

When deploying CS-domain solutions beyond R9, we can implement the GMSC in two different entities, i.e. the GMSC Server, handling only signalling and the CS-MGW as defined in the preceding section. A GMSC Server and a CS-MGW will thus make up the full functionality of a GMSC.

The GMSC server comprises mainly the call control and mobility function of the GMSC.

The Internet-Working Function (IWF), a functional entity associated with the MSC, provides the functionality necessary to allow inter-work between a PLMN and the fixed networks (ISDN, PSTN and PDNs). Its functions [14] depend on the services and the type of fixed network, e.g. it may convert the protocols used in the PLMN to those used in the appropriate fixed network.

5.3.3 PS Domain Entities

The UMTS PS domain, which in practice is GPRS[5] with additional functionalities in some of its building blocks, includes the Serving GPRS Support Node (SGSN) and the Gateway GPRS Supporting Node (GGSN) or GSNs. They constitute the interface between the radio system and the fixed networks for packet-switched services. The GSNs in releases beyond R99 or R4 and R5 still perform all necessary functions in order to handle the packet transmission to and from the mobile stations. In the forthcoming chapters, in particular when covering the IMS segment we will describe more details on the GSN functions.

[3] It is expected that current H.248 standard [13] mechanisms can be applied to enable this.
[4] Serving radio network subsystem.
[5] General packet radio service–the packet segment of GSM.

5.3.3.1 The SGSN and GGSN

Section 5.2 has already covered GPRS elements to some degree. Here we will complement briefly earlier information on mobility management based on [5] to follow the flow on the main PS domain entities.

The SGSN's location register function stores two types of subscriber data needed to handle originating and terminating packet data transfer:

1. Subscription information:
   - IMSI;
   - one or more temporary identities;
   - zero or more PDP addresses.
2. Location information:
   - the cell or the routing area where the MS is registered depending on the operating mode of the MS;
   - the VLR number of the associated VLR (when the Gs interface is implemented);
   - the GGSN address of each GGSN for which an active PDP context exists.

We find more descriptions on the procedures for information transfer between the SGSN, the GGSN, the VLR and the HLR in TS 23.016 [15] and TS 23.060 [16]. TS 23.008 [17] and [16] describe as well the organisation of the subscriber data in the SGSN.

The GGSN's location register function stores two types of subscriber data received from the HLR and the SGSN to handle originating and terminating packet data transfer, these include:

3. subscription information: IMSI and zero or more PDP addresses;
4. location information: SGSN address for the SGSN, where the MS is registered.

As for the SGSN, we find more descriptions on the procedures for information transfer between the GGSN, the SGSN and the HLR in TS 23.016 [15] and TS 23.060 [16]. TS 23.008 [17] and [16] describe as well the organisation of the subscriber data in the GGSN.

5.3.4 Other Specific Core Mobile System Entities

As [5] puts it, the other core entities presented next serve for provisioning of a given (set of) service(s), and their presence or absence in a given PLMN have only limited impact on the main architecture entities of a PLMN. Nonetheless, Table 5.2 list them here as background for the overall PLMN configuration, which takes into account R99, R4 and R5 building blocks.

Table 5.2 Other specific core mobile system entities

Entity	Sub-entities	Description
Group Call Register (GCR)		A register holding information about Voice Group Call Service (VGCS) or Voice Broadcast Service (VBS) call attributes [5], [18] and [19].
Location Communication Services (LCS) entities		
	Serving Mobile Location Centre (SMLC)	In GSM it manages the overall co-ordination and scheduling of resources required to perform positioning of MSs, and calculates the final location estimate and accuracy. In UMTS, the SMLC functionality is integrated in the SRNC. GSM supports NSS-based SMLC for Ls interface and BSS based SMLC for Lb interface [5].
	Gateway Mobile Location Centre (GMLC)	As the first external location application access node, it performs registration authorisation and requests routing information from the HLR.
	Location Measurement Unit (LMU)	It makes radio measurements to support one or more positioning methods. The two types include: Type A LMU → accessed over the normal GSM air interface. Type B LMU → accessed over the BS controller interface (Abis for GSM and Iub for UMTS.
Camel entities		
	GSM Service Control Function (gsm SCF)	Contains the CAMEL service logic to implement Operator Specific Service. It interfaces with the: gsmSSF, gsm SRF and the HLR.
	GSM Service Switching Function (gsmSSF)	Interfaces the MSC/GMSC to the gsmSCF. Based on the IN SSF, but using triggering mechanisms dedicated to mobile networks.
	GSM Specialised Resource Function (gsmSRF)	Provides various specialised resources. It interfaces with the gsmSCF and with the MSC [20,21].
	GPRS Service Switching Function (gprsSSF)	It interfaces the SGSN to the gsmSCF. Based on the IN SSF but using triggering mechanisms dedicated to mobile networks.
CBS-specific entities	The Cell Broadcast Service (CBS) is a Teleservice, which enables an Information Provider to submit short messages for broadcasting to a specified area within the PLMN [22]	

Table 5.2 (*Continued*)

Entity	Sub-entities	Description
	Cell Broadcast Centre (CBC)	It manages CBS messages and determines the CBS delivery parameters of the RNS, and it may be connected to several BSCs/RNCs. In UMTS the CN contains the CBC as a node [5].
NMP specific entities	Number Portability (NMP) allows a mobile subscriber to change the GSM subscription network within a portability cluster (e.g. a country) whilst retaining his/her original MSISDN or MSISDNs.	
	IN-based solution: Number Portability Database (NPDB)	The NPDB is the central element of the IN-based solution. The NPDB stores the table of correspondence between MSISDNs and Subscription networks. Upon request of the (gateway or visited) MSC, the NPDB retrieves from the MSISDN the Routing Number pointing out the Subscription network.
	Signalling Relay-based solution (SRF): (MNP-SRF)	The MNP-SRF obtains the routing information from a MNP database to identify the subscription network associated with a particular national MSISDN. Upon request from gateway MSC the MNP-SRF may perform one of the following actions: 1. the MNP-SRF will reply back to the GMSC with the necessary routing information to route the call; 2. the message is relayed to the HLR; 3. the message is relayed to MNP-SRF in the subscription network. For non-call related signalling (e.g. delivery of SMS), only cases 2 and 3 are applicable [5].
Signalling Gateway Function (SGW)	It performs the signalling conversion (both ways) at transport level between the SS7 based transport of signalling used in pre-Rel four networks, and the IP-based transport of signalling possibly used in post-R99 networks (i.e. between Sigtran SCTP/IP and SS7 MTP). The SGW does not interpret the application layer (e.g. MAP, CAP, BICC, ISUP) messages but may have to interpret the underlying SCCP or SCTP layer to ensure proper routing of the signalling [5].	
Global text telephony specific entities	Corresponds to the Inter-working between Cellular Text Modem (CTM) and text telephony standards (e.g. V.18) used in external networks can be supported by three methods [5,22,23]:	

Table 5.2 (*Continued*)

Entity	Sub-entities Description
	• Routing calls through a CTM Special resource function (CTM-SRF) in the CN. The CTM-SRF linksin to the call path via CAMEL procedures. Depending on operator configuration the CTM-SRF may also be linked in to the call path for Emergency calls. • A CTM/Text telephone converting function included along the speech call path selected by the network after an indication from the terminal that CTM is required. • A CTM/Text telephone converting function included in all speech call paths.
Security Gateway (SEG)	• The SEGs protect the interface between different security domains (operators borders). • The SEGs enforce the security policy of a IP security domain towards other SEGs in the destination IP security domain. • All NDS/IP traffic shall pass through a SEG before entering or leaving a security domain [5,24].

5.3.5 The IP-Multimedia Subsystem (IMS) Entities

The IMS entities will be covered in detail in the 'ALL IP' sections of Chapter 6. Because they will be the building blocks of the all IP architecture, we will describe their functions in an integrated context.

5.3.6 Access Network Entities

The main entities of the access network will be covered in Section 5.4, other details are covered in [24–27]. For completeness and to refer these elements as part of the PLMN configuration next [28], we list them in this section as follows:

- a Base Station Controller (BSC) controls one or more BTSs;
- a Base Transceiver Station (BTS) serves one or more cells depending on the antenna set-up, e.g. one BTS can serve six antennas;
- a Radio Network Controller (RNC) controls of one or more Node B's.

The Mobile Station (MS) as the physical equipment used by a PLMN subscriber comprises [27–29]:

- The Mobile Equipment (ME) and the Subscriber Identity Module (SIM), called UMTS Subscriber Identity Module (USIM) for Release 1999 and following.
- The ME comprises the Mobile Termination (MT) which, depending on the application and services, may support various combinations of Terminal Adapter (TA) and Terminal Equipment (TE) functional groups.

5.3.7 The Public Land Mobile Network (PLMN) Configuration

Now that we have described the PLMN entities in the preceding sections, we will now present its overall configuration, and show the inter-connection between these different entities. Figure 5.7 illustrates the generic view of the integrated entities connected through their corresponding interfaces.

The configuration in Figure 5.7 takes into account the position of the HSS and MSC server, which are the key new building blocks in the architecture beyond R99. The (G)MSC server and associated CS-MGW can be implemented as single node [5].

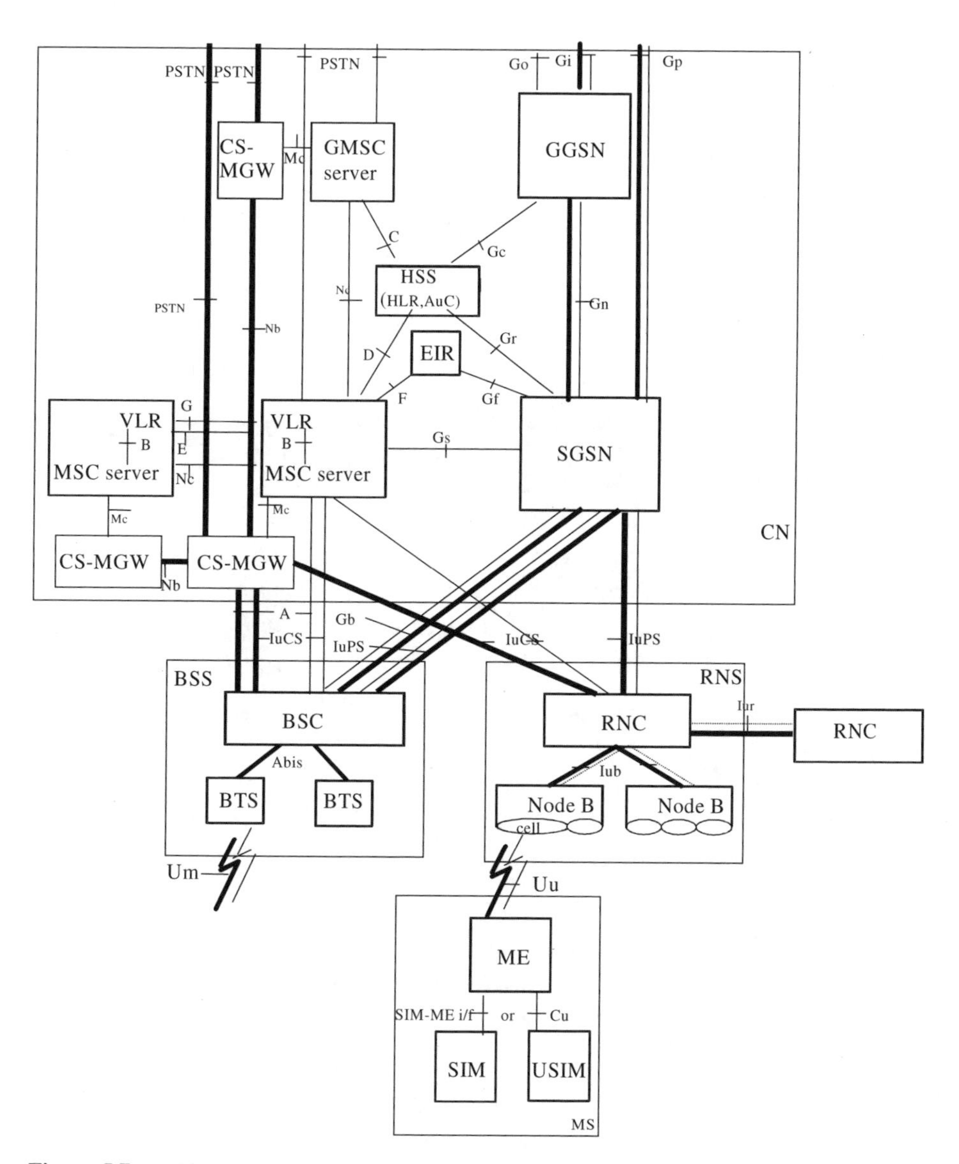

Figure 5.7 A PLMN configuration example supporting PS & CS services and interfaces [5].

Notice the difference between Figure 5.3 and Figure 5.7 with the roles of the MSC-server and the HSS. In the next chapter we will also bring in the links of IMS to the HSS. In Figure 5.7, bold lines indicate interfaces supporting traffic, while dashed lines indicate interfaces supporting signalling; where an underlying network (e.g. SS7 or IP) can provide the actual signalling links.

5.4 THE ACCESS NETWORK DOMAIN

While the access network domain incorporates elements starting the physical layer and radio protocols, here we concentrate on the UTRAN as part of the radio-access network.

5.4.1 UTRAN Architecture

The UTRAN as illustrated in Figure 5.8, contains Radio Network Subsystems (RNSs) communicating with the CN through the Iu interface. In turn a RNS contains a Radio Network Controller (RNC) and one or more Node B.[6] A Node B connects to the RNC through the Iub interface, and it can support either a FDD or TDD or combined dual mode operation.

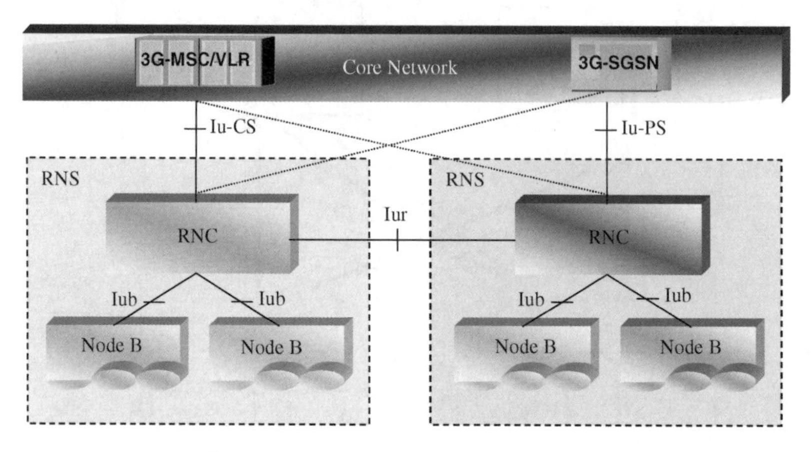

Figure 5.8 UTRAN architecture example.

The RNC takes care of handover decisions requiring signalling to the UE; it comprises a combining/splitting function to support macro-diversity between different Node B's. RNCs can inter-connect each other through the Iur logical interface. The latter can be conveyed over a direct physical connection between RNCs or through any appropriate transport network.

[6]Throughout this book we use '*Node B*' as noted by the 3GPP specifications. However, we should know that a Node B is just a 3G-BTS performing more functions than a 2G BTS in GSM, for example.

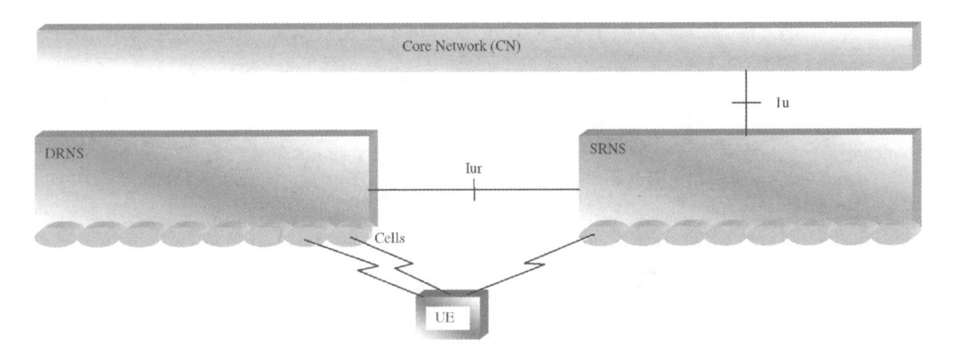

Figure 5.9 Serving and drift RNS.

All UE connections between UTRAN have a serving RNS. When service relocation demands it, a drift RNSs support the serving RNS by providing new radio resources. Figure 5.9 illustrates the role of a RNS (serving or drift) on a per connection basis between a UE and UTRAN.

5.5 UTRAN IDENTIFIERS AND FUNCTIONS

5.5.1 Identifiers

The unique RNC-IDs illustrated in Table 5.3 are defined by the operator, and set in the RNC via O&M.

5.5.2 System Access Control

Through the system access 3G subscribers connect to the UMTS network to use services and/or facilities. Subscriber system access may be initiated from either the mobile side, e.g. a mobile originated call, or the network side, e.g. a mobile terminated call. In the following we summarise key system access control functions; specifications in [30] describe additional details.

5.5.2.1 Admission and Congestion Control

Admission control admits or denies new users, new radio-access bearers or new radio links resulting from network tasks, e.g. handover events. It aims to avoid overload situations and bases its decisions on interference and resource measurements. It also serves during initial UE access, RAB assignment/reconfiguration and handover depending on the required events. Finally, its functions depend on UL interference and DL power information located in the controlling RNC. The serving RNC performs admission control towards the Iu interface.

Congestion control monitors, detects and handles situations when the system reaches near overload or an overload situation while users remain connected. Thus, when somewhere in

Table 5.3 RNC identifiers

Element	Identifiers
PLMN	The PLMN-ID is made of Mobile Country Code (MCC) and Mobile Network Code (MNC): PLMN-ID = MCC + MNC
CN domain	Identifies a CN domain edge node for relocation tasks. It is made up of the PLMN-ID and of the LAC or RAC of the first accessed cell in the target RNS. The two CN domain identifiers are CN CS Domain-ID = PLMN-ID + LAC CN PS Domain-ID = PLMN-ID + LAC+ RAC
RNC	RNC-ID together with the PLMN identifier is used to globally identify the RNC. RNC-ID or the RNC-ID together with the PLMN-ID is used as RNC identifier in UTRAN Iub, Iur and Iu interfaces. SRNC-ID is the RNC-ID of the Serving RNC. C-RNC-ID is the RNC-ID of the controlling RNC. D-RNC-ID is the RNC-ID of the Drift RNC. Global RNC-ID = PLMN-ID + RNC-ID
Service area	Used to uniquely identify an area consisting of one or more cells belonging to the same location area Such an area is called a service area and can be used for indicating the location of a UE to the CN. The Service Area Code (SAC) together with the PLMN-ID and the LAC will constitute the service area identifier. SAI = PLMN-ID + LAC + SAC
Cell	Used to uniquely identify a cell within an RNS The Cell-ID together with the identifier of the Controlling RNC (CRNC-ID) constitutes the UTRAN Cell Identity (UC-ID) UC-ID or C-ID is used to identify a cell in UTRAN Iub, Iur and Iu interfaces. UC-ID = RNC-ID + C-ID
Local cell	Used to uniquely identify the set of resources within a Node B required to support a cell (as identified by a C-ID). Also used for the initial configuration of a Node B when no C-ID is defined.
UE Radio Network Temporary Identities (RNTI) as UE identifiers	Used as in UTRAN and in signalling messages between UE and UTRAN. They include: Serving RNC RNTI (s-RNTI), Drift RNC RNTI (d-RNTI), Cell RNTI (c-RNTI), UTRAN RNTI (u-RNTI). See their use in [30]
Resource identifiers (see [30])	Radio network control plane identifiers, Transport network control plane identifiers and binding identifier.

the network, limited resources degrade service quality, congestion control brings the system back and restores stability seamlessly.

5.5.2.2 *System Information Broadcasting*

This function provides the mobile station with the access stratum and non-access stratum information used by the UE for its operation within the network.

5.5.3 Radio Channel Ciphering and Deciphering

This computation function protects radio-transmitted data against unauthorised third parties. Ciphering and deciphering usage may depend on a session key, derived through signalling and/or session dependent information.

5.5.4 Mobility Functions

5.5.4.1 Handover

Handover manages radio interface mobility based on radio measurements in order to maintain CN quality of service. It may be directed to/from another system (e.g. UMTS to GSM handover). Control for this function may originate in the network, or may come independently from the UE. Hence, it may be located in the SRNC, the UE or both.

5.5.4.2 SRNS Relocation

This function co-ordinates events when a SRNS role passes to another RNS. It manages the Iu interface connection mobility from one RNS to another. The SRNC initiates the *SRNS relocation*, which finds a home in the RNC and CN as illustrated in Figure 5.10.

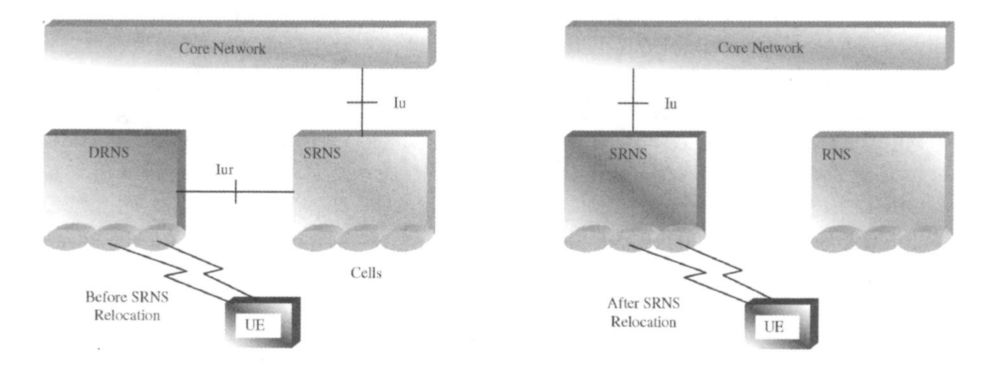

Figure 5.10 A serving RNS relocation example.

5.5.5 Radio Resource Management and Control Functions

Radio resource management concerns the allocation and maintenance of radio communication resources. In UMTS CS and PS services share these resources.

Not all functions apply to both FDD and TDD modes. For example, macro-diversity applies only to FDD while dynamic channel allocations applies only to TDD.

5.5.5.1 Radio Resource Configuration

This function configures the radio network resources, i.e. cells and common transport channels (e.g. BCH, RACH, FACH, PCH), and takes the resources into or out of operation.

5.5.5.2 Radio Environment Survey

The radio environment survey performs quality estimates and measurements on radio channels from current and surrounding cells; as in [30] these functions, located in the UE and UTRANS, include:

- received signal strengths (current and surrounding cells);
- estimated bit error ratios (current and surrounding cells);
- estimation of propagation environments (e.g. high-speed, low-speed, satellite, etc.);
- transmission range (e.g. through timing information);
- Doppler shift;
- synchronisation status;
- received interference level;
- total DL transmission power per cell.

5.5.5.3 Macro-diversity Control – FDD

In FDD, *macro-diversity control* manages duplication/replication of information streams to receive/transmit the same information through multiple physical channels (or different cells) from/towards a single mobile terminal.

This function also controls combining of information streams generated by a single source (diversity link), but conveyed via several parallel physical channels (diversity sub-links). Macro-diversity control interacts with *channel coding control* to reduce bit error ratio when combining different information streams. Depending on the physical network configuration, combining/splitting may occur at the SRNC, DRNC or Node B level.

5.5.5.4 TDD – Dynamic Channel Allocation (DCA)

The TDD mode uses fast or slow DCA. Fast DCA implies assigning resources to the radio bearers in relation to the admission control. Slow DCA implies assigning radio resources, including time slots, to different TDD cells depending on the varying cell load.

5.5.5.5 Allocation/De-allocation and Control of Radio Bearers

The allocation/de-allocation function located in the CRNC and SRNC, translates the connection element setup requests into physical radio channel allocation according to the QoS of the radio-access bearer. It gets activated, e.g. during a call when user service request varies or during macro-diversity.

Radio bearer control located both in the UE and in the RNC, manages connection element setup and release in the radio-access sub-network. It participates in the processing of the end-to-end connection set-up and release, as well as the managing and maintenance of the end-to-end connection, which is located in the radio-access sub-network.

5.5.5.6 Radio Protocols Function

This function provides user data and signalling transfer capability across the UMTS radio interface by adapting the services (according to the QoS of the radio-access bearer) to the radio transmission. This function includes:

- multiplexing of services and multiplexing of UEs on radio bearers;
- segmentation and reassembly;
- acknowledged/unacknowledged delivery according to the radio-access bearer QoS.

5.5.5.7 RF Power Control

Power control manages the transmitted power level in order to minimise interference and keep connection quality. Table 5.4 illustrates the different functions.

5.5.5.8 Radio Channel Coding and Control

This function located in both UE and UTRAN brings redundancy into the data source flow, thereby increasing its rate by adding information calculated from the data source. This allows detection or correction of signal errors introduced by the transmission medium. Channel coding algorithm(s) and redundancy level may vary in the different types of logical channels and different types of data.

Channel coding control residing in both UE and UTRAN generates control information required by the channel coding/decoding execution functions, e.g. channel coding type, code rate, etc.

5.5.5.9 Radio Channel Decoding

Channel decoding aims to reconstruct the information source using the added redundancy by the channel coding function to detect or correct possible errors in the received data flow. This function may also employ *a priori* error likelihood information generated by the demodulation function to increase the efficiency of the decoding operation. The channel decoding function, located in both the UE and UTRAN, complements the channel coding function.

5.5.5.10 Initial Random Access

This function, located in the UTRAN, detects initial MS access attempts and responds accordingly. Handling this initial access may incorporate procedures to resolve colliding events. Successful attempts will obtain the right to resource allocation request.

Table 5.4 RF power control functions

Power control	Link	Function description
Outer loop	UL	Located in the SRNC sets the target quality value for the UL inner loop power control, which is located in Node B for FDD and is located in the UE for TDD. It receives input from quality estimates of the transport channel. The UL outer loop power control is mainly used for long-term quality control of the radio channel. In FDD, if the connection involves both a SRNS and a DRNS the function UL outer loop power control (located in the SRNC) sets the target quality for the UL inner loop power control function (located in Node B).
	DL	Sets the target quality value for the DL inner loop power control. It receives input from quality estimates of the transport channel measured in the UE. It is used for a long-term quality control of the radio channel. Located mainly in the UE with some control parameters are set by the UTRAN, where the SRNC under an algorithms control sends the target down link power range based on the measurement report from UE.
Inner loop	UL	Sets the power of the uplink dedicated physical channels. *In FDD*, it is a closed loop process receiving quality target from UL outer loop power control and quality estimates of the uplink dedicated physical control channel. The UE gets power control commands on the downlink dedicated physical control channel. Resides in both the UTRAN and the UE. *In TDD* located in the UE, it is a open loop process receiving quality target from the UL outer loop power control and uses the quality target and quality estimates of downlink channels to set the transmit power.
	DL	This function located in both UTRAN and UE sets the power of the downlink dedicated physical channels while receiving the quality target from DL outer loop power control and quality estimates of the downlink dedicated physical control channel. Power control commands are sent on the uplink dedicated physical control channel to the UTRAN.
Open loop	UL	This function located in both UTRAN and UE sets the initial power of the UE at random access. It uses UE measurements and broadcasted cell/system parameters as input.
	DL	Function located in both UTRAN and UE, sets the initial power of downlink channels. It receives downlink measurement reports from the UE.

5.5.5.11 NAS Core Network Distribution Functions

Non-Access Stratum (NAS) messages in the RRC protocol have transparent transfer within the access stratum through a direct transfer procedure. A UE/SRNC distribution function handles a CN domain indicator, service descriptor and flow ID being part of the AS message to direct messages to the corresponding NAS entity, i.e. the appropriate mobility management instance in the UE domain and the corresponding CN domain.

In the downlink the SRNC provides the UE with the necessary information on the originating CN domain for the individual NAS message.

In the uplink, the UE distribution function inserts the appropriate CN domain values, domain indicator, service descriptor and flow ID IEs in the AS message. The SRNC evaluates the CN domain indicator, service descriptor and flow ID contained in the AS message and distributes the NAS message to the corresponding RANAP instance for transfer over the Iu interface.

5.5.5.12 Timing Advance in TDD

This function aligns uplink radio signals from the UE to the UTRAN. It is based on uplink burst timing measurements performed by the Node B L1, and on timing advance commands sent downlink to the UE.

5.5.5.13 NAS Service Specific Function

A UE or SRNC service specific function provides a SAP for particular services (e.g. priority levels). In the downlink direction, the SRNC may base the routing on this SAP.

5.6 MOBILITY MANAGEMENT

5.6.1 Signalling Connection

The UE may or may not have a signalling connection, and in the radio interface dedicated or common channels can be used [31].

When an established signalling connection exists over the Dedicated Control Service Access Point (DC-SAP) from the access stratum, the CN reaches the UE by a dedicated connection SAP on the CN side with a context between UTRAN and UE for the given connection. This context disappears when the connection is released and a *dedicated connection* can be initiated only from the UE.

When a dedicated connection does not exist, the CN reaches the UE through the notification SAP, where the CN message may request the UE to establish a dedicated connection. The UE is addressed with a user/terminal identity and a geographical area.

The location of the UE is known either at cell level (higher activity) or in a larger area consisting of several cells (lower activity). Knowing the location minimises the number of location update messages for moving UEs with low activity and removes paging needs for UEs with high activity.

5.6.2 Impacts of Mobility Handling

In the presence of a dedicated connection to the UE, the UTRAN handles the UE radio interface mobility, such as soft handover, and procedures for handling mobility in the RACH/PCH substrate. The radio network cell structure should not necessarily be known outside the UTRAN.

In the absence of a dedicated connection to the UE, mobility handling occurs directly between the UE and CN outside the access stratum, e.g. through registration procedures. While paging the UE, the CN indicates a geographical area which becomes the actual paged cell in UTRAN. Within a cell structure we may use location area identities or other means to identify a geographical area independently.

While a dedicated connection lasts, the UE suppresses its registrations to the CN and re-registers if required. Thus, the UTRAN does not contain any permanent location registers for the UE, but only temporary contexts for the duration of the dedicated connection. This context may typically contain location information (e.g. current cell(s) of the UE) and information about allocated radio resources and related connection references [30].

5.7 UTRAN SYNCHRONISATION AND O&M REQUIREMENTS

5.7.1 Synchronisation Model

The main synchronisation issues in UTRAN include: network, node, transport channel, radio interface and time alignment synchronisation. Figure 5.11 illustrates the nodes involved in these issues (with exception of network and node synchronisation).

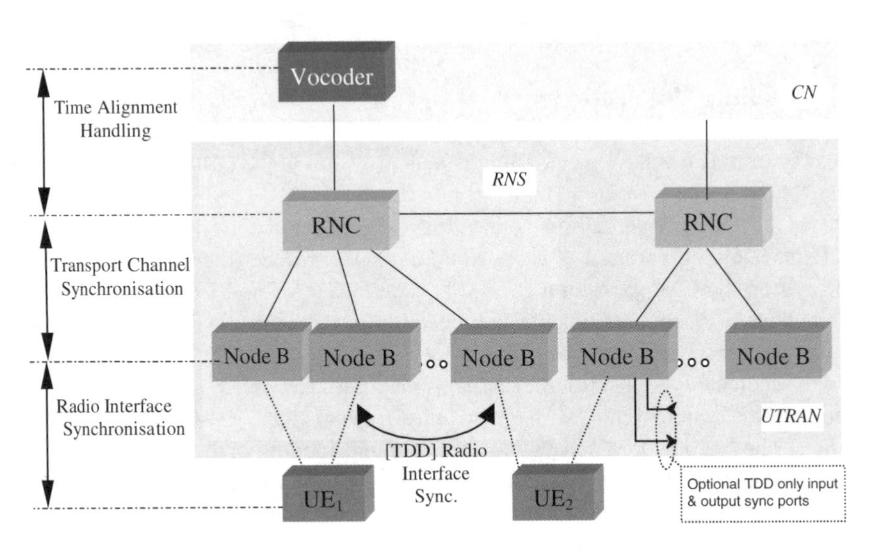

Figure 5.11 Synchronisation issues model.

5.7.2 Node B O&M

Figure 5.11 illustrates the two Node B O&M types, i.e. the *implementation specific O&M* linked to the actual implementation of Node B, and the *logical O&M* having impacts on the traffic carrying resources in Node B controlled from the RNC.

5.7.2.1 *Implementation Specific O&M*

Implementation specific O&M functions depend on both HW and SW management components of Node B, and their transport from Node B to the management system occurs via the RNC. The implementation specific O&M interface shares the same physical bearer with the Iub interface, where [32] specifies the routing function and the transport bearer.

Routing across the RNC in the UTRAN is optional, but signalling between co-located equipment and its management system is required, this may be carried over the same bearer as the implementation specific O&M.

5.7.2.2 *Logical O&M*

The logical O&M represents the signalling associated with the control of logical resources owned by the RNC but physically implemented in Node B (e.g. channels, cells, etc.). The RNC controls these logical resources. A number of O&M procedures physically implemented in Node B impact on the logical resources requiring an information exchange between RNC and Node B. All messages needed to support this information exchange are classified as logical O&M forming an integral part of NBAP [30].

Figure 5.12, with representative logical connections, shows the concept of an interface from the RNC to the management system and O&M functions within the management

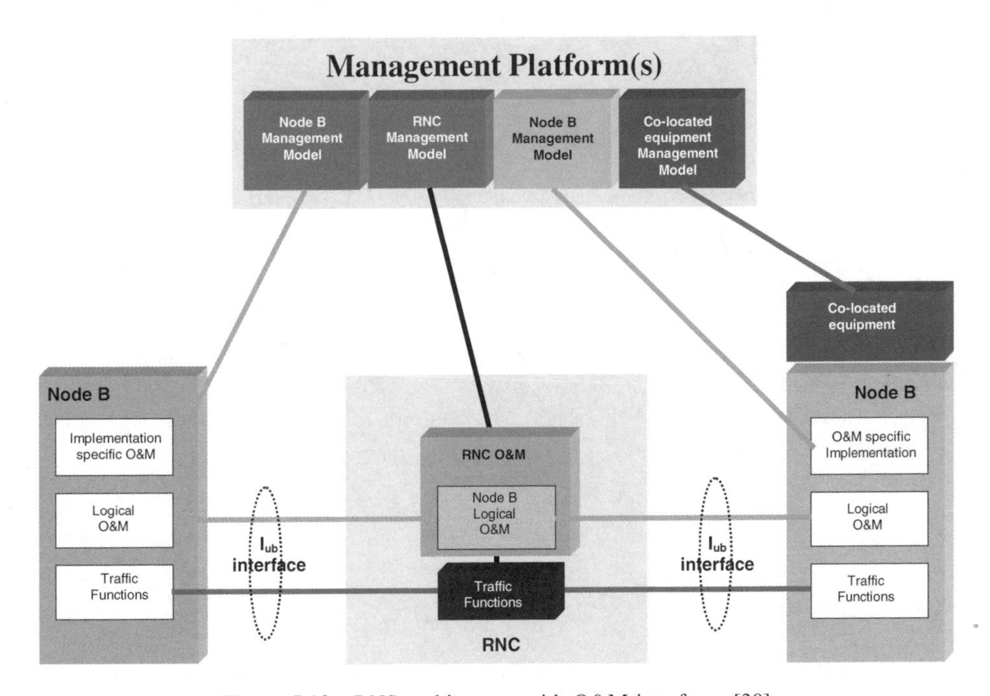

Figure 5.12 RNS architecture with O&M interfaces [30].

system for clarity only. The specifications of these functional connections do not correspond to the scope of the radio network specification.

5.8 UTRAN INTERFACES

5.8.1 General Protocol Model for UTRAN Interfaces

Figure 5.13 presents the general UTRAN interfaces protocol model. The structure assumes that layers and planes are logically independent of each other, and if needed, the whole structure may evolve later with standardisation work.

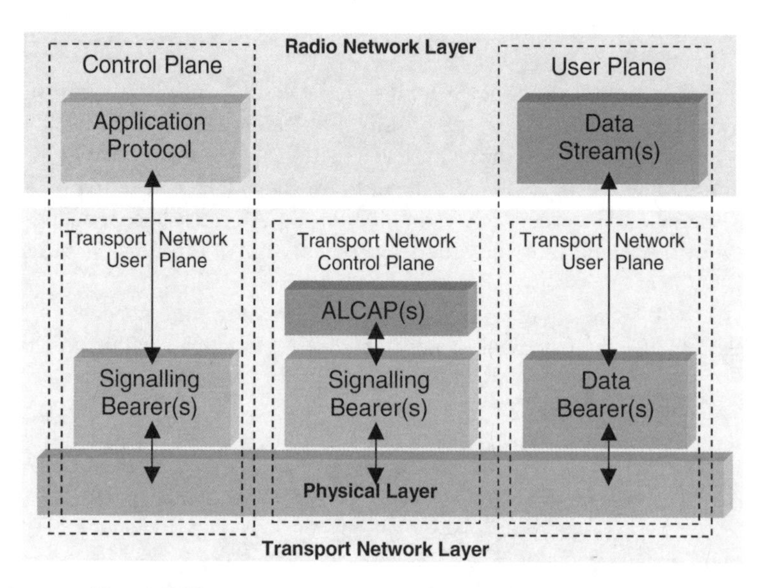

Figure 5.13 General protocol model for UTRAN interfaces.

5.8.2 Horizontal Layers

The radio network and transport network layers constitute the main components of the protocol structure. The first layer contains all visible UTRAN related issues, and the second layer represents standard UTRAN transport technology for selection without any specific requirements.

5.8.3 Vertical Planes

5.8.3.1 Control Plane

The control plane includes the application protocol, i.e. RANAP, RNSAP or NBAP, and the signalling bearer for transporting the application protocol messages.

Among other things, the application protocol is used for setting up bearers for (i.e. radio-access bearer or radio link) the radio network layer. In the three plane structure the bearer parameters in the application protocol are not directly tied to the user plane technology, but are rather general bearer parameters.

The signalling bearer for the application protocol may or may not be of the same type as the signalling protocol for the ALCAP. The signalling bearer is always set up by O&M actions.

5.8.3.1.1 Iu Control Plane

Both PS and circuit CS domains use the SCCP protocol to transport RANAP messages over the Iu interface. Likewise, both SCCP and RANAP protocols comply with ITU-T recommendations. In R99, SCCP messages in CS domain use a broadband SS7 stack comprising MTP3b on top of SAAL-NNI. In the PS domain UMTS specs allow operators to chose one out of two standardised protocol suites, i.e. broadband SS7 stack comprising MTP3b on top of SAAL-NNI or IETF/Sigtran CTP protocol suite for MTP3 users with adaptation to SCCP. Figure 5.14 illustrates the different RANAP stack options.

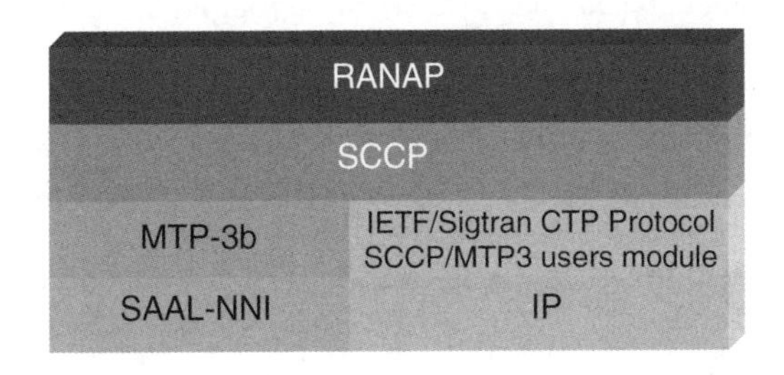

Figure 5.14 Stack options in the RANAP protocol.

5.8.3.2 User Plane

The user plane includes the data stream(s) and the data bearer(s) for the data stream(s). The data stream(s) is/are characterised by one or more frame protocols specified for that interface.

5.8.3.2.1 Iu User Plane

The user plane towards the IP domain works based on an evolved Gn interface, where we achieve tunnelling of user data packets over the Iu interface through the user plane part of GTP over UDP/IP. The tunnelling protocol corresponds to an evolution of the user plane part of the GTP protocol used in GPRS stacked on top of UDP/IP. When transport data uses ATM PVCs, the Iu IP layer provides Iu network layer services such as routing, addressing, load sharing and redundancy. This leads to an IP network configured to transfer Iu data units between RNSs and 3G-SGSNs.

We can access common layer 2 resources between UTRAN and the IP domain of a CN through one or several AAL5/ATM permanent VCs. More than one permanent AAL5/ATM VCs, for example, allows load sharing and redundancy.

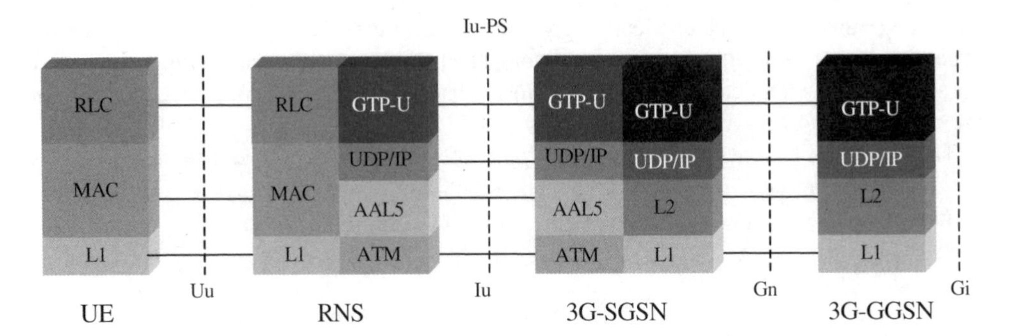

Figure 5.15 Protocol architecture for the IP domain user plane.

The UMTS user data plane in the network consists of two tunnels, i.e. a first IP/UDP/GTP tunnel between RNC and 3G SGSN on the Iu interface and a second IP/UDP/GTP tunnel between GGSN and 3G SGSN on the Gn interface.

The double tunnel architecture provides hierarchical mobility, allows direct RNC connection to the IP domain backbone, ensures traffic routing through 3G-SGSN to perform appropriate charging and legal interface functions, and it also makes room for future exploitation of Iu and Gn interfaces. The protocol stack is shown in Figure 5.15.

Specifications in [3] outline user data retrieval principles in UMTS and at GSM-UTMS handover for the PS domain. In the following we cover the radio-access domain, which in part will cover UTMS Mobility Management (UMM) and UMTS call control to complete the context of inter-operability between 2G and 3G systems, i.e. GSM and UMTS.

5.8.3.3 Transport Network Control Plane

The transport network control plane does not include any radio network layer information, and is completely in the transport layer. It includes the ALCAP protocol(s) that is/are needed to set up the transport bearers (data bearer) for the user plane. It also includes the appropriate signalling bearer(s) needed for the ALCAP protocol(s).

The transport network control plane is a plane that acts between the control plane and the user plane. The introduction of transport network control plane makes it possible for the application protocol in the radio network control plane to be completely independent of the technology selected for data bearer in the user plane.

When the transport network control plane is used, the transport bearers for the data bearer in the user plane are set up in the following fashion. First there is a signalling transaction by the application protocol in the control plane, which triggers the set-up of the data bearer by the ALCAP protocol that is specific for the user plane technology.

The independence of control plane and user plane assumes that ALCAP signalling transaction takes place. It should be noted that ALCAP might not be used for all types data bearers. If there is no ALCAP signalling transaction, the transport network control plane is not needed at all. This is the case when pre-configured data bearers are used.

It should also be noted that the ALCAP protocol(s) in the transport network control plane is/are not used for setting up the signalling bearer for the application protocol or for the ALCAP during real time operation.

The signalling bearer for the ALCAP may or may not be of the same type as the signalling bearer for the application protocol. The signalling bearer for ALCAP is always set-up by O&M actions.

5.8.3.4 Transport Network User Plane

The data bearer(s) in the user plane, and the signalling bearer(s) for application protocol, belong also to the transport network user plane. As described in the previous section, transport network user plane data bearers, are directly controlled by the transport network control plane during real time operation; but the control actions required for setting up the signalling bearer(s) for application protocol are considered O&M actions.

The following section is an informative section, which aims to provide an overall picture of how the MAC layer is distributed over Uu, Iub and Iur for the RACH, FACH and DCH [31].

5.9 INTER-WORKING OF UTRAN INTERFACES

In the preceding section we have outlined the different protocol layers and their corresponding functions. In this section will then present the practical integration of these layers through the different interfaces. However, before we describe the integration steps we shall briefly review the ATM terminology and its key characteristics, which will determine the integration patterns of the UTRAN interfaces.

5.9.1 ATM Principles

In the datagram approach, packets get treated independently because they may follow different paths to their destination. The latter may as result in out-of-sequence packets and cause demanding processing for sequencing them. Asynchronous Transfer Mode (ATM) attempts to meet these and other Packet Switching (PS) needs by combining the advantages of both circuit- and message-switching. In addition, ATM extends current PS capabilities towards the incorporation of *real-time traffic*, the most desired feature of Circuit Switching (CS).

5.9.1.1 ATM Cell Characteristics

Before starting with resource management, we will briefly describe the ATM's cell components, (more in [29]); these include the following fields:

- Generic Flow Control (GFC): 4 bit field providing UNI flow control at the user device
- Virtual Path Identifier (VPI): 8 or 12 bit field contains cell routing information
- Virtual Circuit Identifier (VCI): 16 bit field does also contain cell routing information
- Payload Type (PT): 3 bits indicating payload type

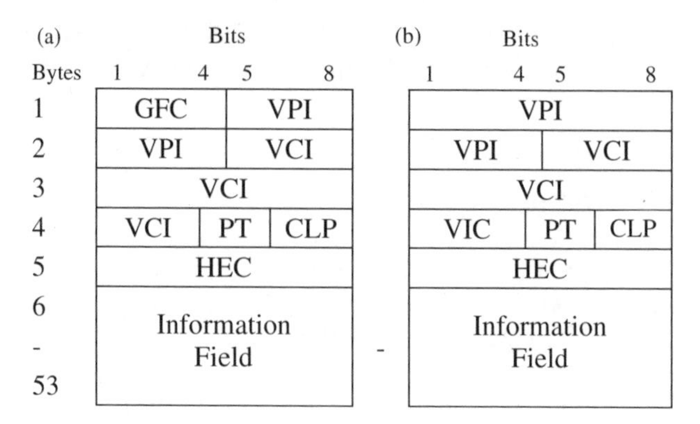

Figure 5.16 (a) UNI cell and (b) NNI cell formats [29].

- Cell Loss Priority (CLP): 1 bit field indicating cell discarding or not status
- Header Error Control (HEC): 8 bit field providing singly-bit error correction

Figure 5.16 illustrates the User Network Interface (UNI) and Network Node Interface cells. The content distribution clearly indicated the ATM cell size of 53 (5 overhead + 48 information) bytes.

5.9.1.2 Traffic and Resource Management Features

ATM allows optimum utilisation of network capacity through its Quality of Service (QoS) mechanism, which in turn affords efficient resource allocation to different traffic demands. It manages and controls traffic through the *Connection Admission Control* (CAC) and *Usage/ Network Parameter Control* (UPC/NPC) functions, besides other functions such as: feedback control, traffic shaping, priority control and Network Resource Management (NRM).

- The *CAC function* checks bandwidth and buffer resources for the desired links. We define the CAC as a group of events to take place during the link establishment period, in order to accept or reject a Virtual Channel (VC)/Virtual Path (VP) requests. Through dedicated algorithms the CAC validates the network node capabilities and the type of QoS it can offer to VC/VP demands.
- The *UPC/NPC* policies agreed end-to-end traffic contracts, e.g. peak cell rate for each link. It safeguards resources from QoS affecting misconduct by catching violations of negotiated parameters. It may use typical ATM cell level operations, e.g. cell passing, cell tagging and cell discarding.
- The *feedback control* regulates submitted traffic within a network with ATM links, e.g. it will announce presence of congestion within given nodes.
- *Traffic shaping* modifies traffic characteristics of cell streams to increase link efficiency while assuring conformance to QoS; e.g. it buffer cells to maintain traffic parameters agreed.

- *Priority Control* enforces different cell flow priorities within VCs or VPs based on the cell loss priority bit of the ATM cell header. Thus, it allows discriminated cell discarding looking at the CLPs, e.g. = 0, = 1, etc.
- The *NRM* depicts the network-resource allocation provision to split traffic according to service features.
- *Frame Discard* corresponds to one of the classical ATM congestion control mechanisms, which includes Partial Packet Discard (PPD) and Early Packet Discard (EPD):
  - *PPD* implies dropping all-but last[7] subsequent cells in upper layer protocol datagram, (e.g. a Protocol Data Unit–PDU) once a cell gets dropped from a switch buffer. PPD occurs only with buffer overflows.
  - EPD discards entire PDUs starting from the first arriving cell until buffer overflow threat does no longer exist or it drops to threshold levels.

5.9.1.3 Traffic Contract Negotiation and Settlement

Traffic contract negotiation implies primarily aligning QoS, traffic descriptors and service complying links during connection establishment. We can specify implicit (network default rules) or explicit (Switched Virtual- or Permanent Virtual-Connection) traffic parameters. Aligning QoS means defining the service performance level, which in ATM will depend from the negotiable and non-negotiable parameters as illustrated in Table 5.5.

Using the connection-oriented metrics we can also define values for the three key call control parameters [29], which allows end-to-end link performance visibility:

- Connection set-up delays in ATM switches, e.g. mean 4.5–8.3 s max 95%.
- Connection release delay, mean 0.3 s with 95% delays less than 0.85 s.
- Connection acceptance probability depends on the call attempt rate (λ) and the average call holding time ($1/\mu$).

Table 5.5 ATM QoS classes and negotiation parameters

Parameters	QoS classes			
Negotiable	Class 1	Class 2	Class 3	U Class
Cell Loss Transfer Delay (CTD)	300–400 ms	Unspecified	Unspecified	Unspecified
Cell Delay Variation (CDV)	1–130 ms	Unspecified	Unspecified	Unspecified
Cell Loss Ratio (CLR)	10^{-6}–10^{-7}	10^{-5}	Unspecified	Unspecified
Non-negotiable				
Cell Error Ratio (CER)	10^{-6}	10^{-6}	10^{-6}	Unspecified
Cell Misinsertion Rate (CMR)	1/day	1/day	1/day	Unspecified
Severely Errored Cell block Ratio SECBR	10^{-4}	10^{-4}	10^{-4}	Unspecified

[7]Indicates the beginning of the next PDU.

Table 5.6 Traffic parameter descriptors

Traffic descriptor	Definition
Peak Cell Rate (PCR)	Maximum allowable cell rate of the source
Sustained Cell Rate (SCR)	Theoretical cell average rate during a transmission
Maximum Cell Rate (MCR)	Minimum user cell rate guaranteed, applies to Available Bit Rate (ABR)
Maximum Burst Size (MBS)	Maximum cells transmitted at PCR while complying with agreed SCR
Cell Delay Variation Tolerance (CDVT)	Delay tolerance difference between delayed cells. Identifies the jitter during the cell multiplexing and applies to PCR and SCR.

Network load through the queuing delays will influence variable delay components of the CTD, while propagation conditions will influence the fixed component. Queuing effects and buffer size will also have impact on the CDV performance and thereby on the CLR. Transmission errors will come from propagation effects, which in turn will have an impact on the CMR, SECBT and CER characteristics. Exact QoS class values will depend on the applications; the ones provided in Table 5.5 are only reference.

5.9.1.4 Traffic Parameters (TP) and Compliant Links

TPs correspond to the generic list of traffic descriptors defining ATM traffic characteristics as illustrated by Table 5.6.

Figure 5.17 summarises the ingredients for the link compliance. A traffic agreement will result in link compliance when descriptor parameters do remain within given threshold values. Non-compliant links do not get guaranteed QoS support from the network.

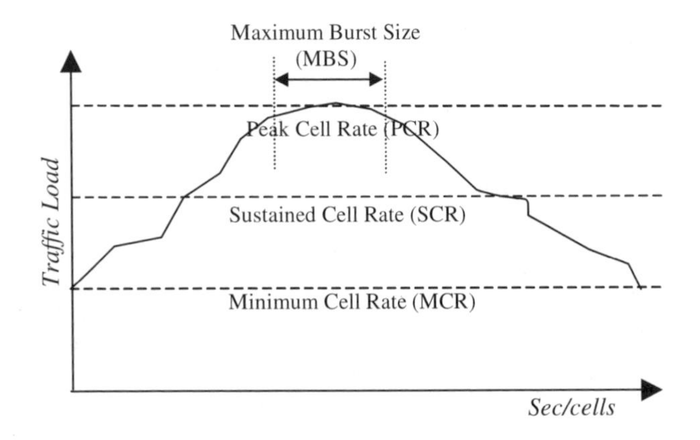

Figure 5.17 Key ATM traffic parameters.

Table 5.7 Service category attributes

Service category	Traffic descriptor	Applications suitability	QoS guarantees			Feedback control
			CLR	Cell Jitter	Bandwidth	
CBR	PCR, CDVT	Video conf. Telephony,	✓	✓	✓	×
Rt-VBR	PCR, SCR, MBS, CDVT	VBR video, compact voice	✓	✓	✓	×
Nrt-VBT	PCR, SCR, MBS, CBVT	Critical data (e.g. financing)	✓	×	✓	×
ABR	PCR, MCR, CDVT	LANs, WANs (e.g. ATM, FR)	✓	×	✓	✓
UBR	PCR, CDVT	FTP, Email, etc.	×	×	×	×

5.9.1.5 ATM Service Classes

ATM service classes enable service categorisation based on the QoS requirements; thus, services with equal QoS belong to the same category. These service classes, which use traffic parameters to define bandwidth requirements complying with QoS demands, include:

1. Constant Bit Rate (CBR) → applies to static bandwidth connections;
2. real time Variable Bit Rate (rt-VBR) → real time links varying during the link;
3. non-real time Variable Bit Rate (nrt-VBR) → applies to nrt-bursty traffic;
4. Available Bit Rate (ABR) → for nrt-apps with flexible bandwidth requirements;
5. Unspecified Bit Rate (UBR) → for nrt-apps without strict delay requirements.

Table 5.7 summarises the service category attributes. It also qualifies the application suitability. Notice that CBR and Rt-VBR require all QoS guarantees, while UBR can survive without them. The feedback control mechanism usage applies only to the ABR exploited primarily by transport networks base, e.g. in ATM, Frame Relay (FR), etc.

5.9.2 ATM-Network Resource Management

The inter-working between the Iu-b, Iu-CS, Iu-PS and Iu-r take place within the ATM environment. Thus, optimum interface configuration implies managing appropriately ATM network resources, which include:

1. ATM-interface management within the IMA[8] or terminal exchange;
2. ATM-interface access profile management;

[8]Inverse Multiplexing in ATM network.

3. VP Link termination point management (VPLpt);
4. VC Link termination point management (VCLpt).

5.9.2.1 The ATM Interface (ATM-I)

The ATM-I is a logical interface working with either (Figure 5.16) the:

- User Network Interface (UNI) → between terminal equipment and network termination (e.g. RNC & Node B), or
- Network Node Interface (NNI) → between two network nodes (e.g. RNC & MGW).

After creating an ATM-I we connect it to an IMA group, which can be either a group-of or simply a PDH exchange terminal, or a SDH exchange terminal. The object capacity to which the ATM-I gets connected shall be higher than of the ATM-I, and there can be only one ATM interface per IMA group or one ATM-I per exchange terminal.

IMA enables the combination of numerous low-bit-rate lines to be seen as one virtual ATM link. It applies, e.g. to PDH or SDH transmission mediums between the Node B and RNC or between two RNCs. Practically the IMA illustrated in Figure 5.18 works as follows:

1. ATM cell streams pass from the ATM layer to the IMA sub-layer.
2. From the IMA sub-layer cells pass evenly cell-by-cell through different physical links.
3. At the receiving end ATM cells pass back to the ATM layer from the IMA sub-layer.
4. The transmitter end aligns IMA frames to physical links to compensated jitter delays.

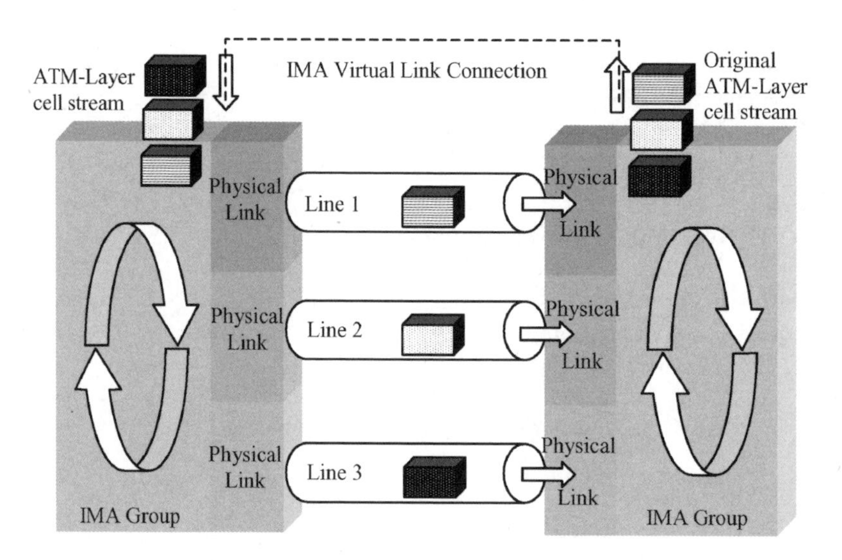

Figure 5.18 The inverse multiplexing ATM (IMA) flow example.

As illustrated in Figure 5.18, both ends of the transmission lines have IMA groups appearing as one IMA virtual link connection. These lines are identified with a PDH Exchange Terminal (PET), e.g. which have the same physical route to minimise delays.

5.9.2.2 The ATM Interface Access Profile

Each ATM interface created needs an access profile, which defines the connection structure and characteristics with the following five key components:

1. *Maximum ingress/egress transmission bandwidth → depends on the link*:
   - A STM-1 interface will provide 149 760 kbit/s, i.e. about 353 207 cells/s
   - An E1 transmission link interface will provide a bandwidth of 4528 c/s
   - A T1 transmission link interface will provide 3622 c/s.
   - For an IMA connection we can use the following formula [33]:

     $(0.99) \times (\text{NoE1s}) \times (\text{maximum E1 bandwidth})$, e.g.
     $0.99 \times 5 \times 4528\text{c/s} = 13{,}448$ c/s, i.e. five E1 links grouped.
2. *Maximum number of VPI bits*
   - They are limited to the capacity of the closest ATM switch they interact with, in order to prevent cell loss.
   - They define the available space or # of bits to specify the interface VPLpts, which determine the maximum # of Virtual Paths (VPs); e.g.
   - $\text{VPI} = 4$ bits $\rightarrow 2^4 = 16$ ATM interface VPLpts.
3. *Maximum number of VCI bits*
   - It defines the available or # of bits available to specify ATM interface VCLtp's, which in turn determines the maximum # of VCs in a VP, e.g.:
   - $\text{VCI} = 8 \rightarrow 2^8 - 1 = 255$ VCs, where $\text{VCI} = 0$ does not exist.
4. *Maximum number of Virtual Channel Connections (VCCs)*
   - It is determined by the maximum number of VCIs as illustrated above.
5. *Maximum number of VPCs*
   - It is determined by the maximum number of VPIs as also illustrated above.

5.9.2.3 Virtual Path and Virtual Channel Link Termination Points (VP/VC Lpt)

Once we determine the capacity of VPs and VCs through the VPIs and VCIs, respectively, we go on with creating the links and establishing the corresponding Link terminating points, i.e. VPLpts and VCLpts, the first must precede the second. The termination points serve:

- signalling links at VC level;
- VCC endpoints (for AAL2 type user traffic connections);

- IP over ATM connections over VCs;
- Permanent Virtual Connections (PVCs) at the VP and VC level.

Present ATM networks use Constant Bit Rate (CBR) for user and signalling traffic while Unspecified Bit Rate (UBR) applies to IP over ATM connections. Thus, VCLtp's must correspond to the appropriate service category [34].

5.9.3 Mapping ATM Layer to UTRAN Interfaces

In the preceding section we outlined the main ATM interface characteristics and resource management. We will now aim to map these principles to the different UTRAN interfaces and describe the inter-working details including the main protocol stacks.

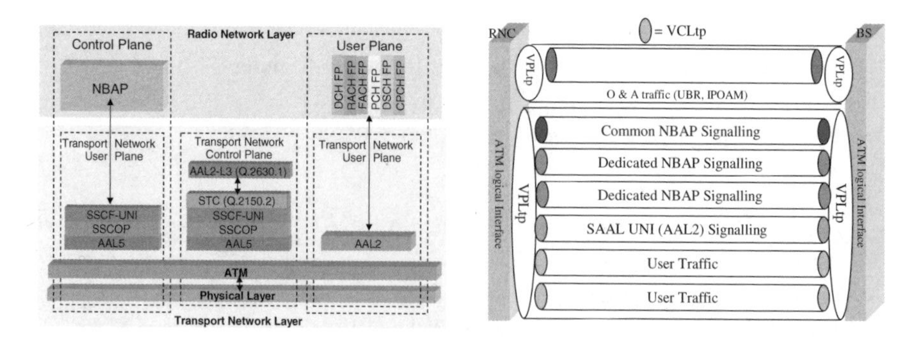

Figure 5.19 Protocol structure and ATM resource management for Iub interface.

5.9.3.1 The Iub Interface

Figure 5.19 illustrates the Iub protocol stack structure and its corresponding ATM logical interface mapping. For Iub, the Layer 3 (L3) Access Link Control Application Part (ALCAP) signalling (see the enables bearer setup) to transmit data through the User Plane. It dynamically establishes, maintains, releases and controls AAL2 connections. It does also join or links a connection control to a higher layer control protocol.

In the ATM logical interface side of Figure 5.19, the content of Table 5.8 applies.

Table 5.8

VPLtp	VCLtp	RNC interface profile	ATM logical interface
VPI	VCI	Max bandwidth	Interface ID
VPL service level, i.e. VP/VC	Service category	Max VPI/VCI bits	UNI
MTP3SL usage, e.g. AAL2UD, ALL2SL, DNBAP, CNBAP Service category: CBR/UBR Traffic and QoS parameters	Traffic and QoS parameters	Max VPC/VCC	IMAGR

To function appropriately with ITU-T Q.2150.1 specs, the protocol layer illustrated in Figure 5.19 uses a Signalling Transport Converter (STC), in addition to the following:

The ALL2 layer, which enables low-bit rate and delay-sensitive applications to share a single ATM connection. It also guarantees delay requirements and maximises the number of subscriber accommodation in fixed bandwidth.

Signalling ATM Adaptation Layer (SAAL), which serves to establish signalling virtual channels with two functional parts, i.e.:

- user equipment signalling between the network at the access → UNI cells;
- signalling between network elements within the network → NNI cells.

The Service Specific Co-ordination Function (SSCF), which provides mapping between the SSOP capabilities and signalling-protocol-module needs, e.g. NBAP. It applies differently to UNI and NNI options, where the:

- SSCF-UNI provides restricted subset of SSOP functions to the UNI signalling L3 (ITU-T recommendations Q.2130);
- the SSCF-NNI uses and extension of UNI signalling for the P-NNI interface.

The Service Specific Connection Oriented Protocol (SSCOP) offers, e.g. to user equipment and local exchange a reliable signalling protocol message carried by an ATM link often denoted as VPC (VPI = 0, VCI = 5).

5.9.3.1.1 Typical NBAP Procedures in the Iub

The Node B Application Part (NBAP) procedures enable signalling to more than one UE context already existing in a BTS, e.g. the C-NBAP defines all O&M logical procedures of a BTS, which includes configuration and fault management. Table 5.9 summarises these procedures.

For example, Radio Link Set-up (RL-set-up) occurs when the RNC decides to add a cell to the active set of dedicated RRC connection upon the request of a BTS, where one message request may serve more than one link. Once the BTS passes a RL-set-up complete message to the RNC, the latter communicates the UE the link parameters through RRC message.

5.9.3.2 Iur Interface Structure

Figure 5.20 illustrates the main components of the Iur protocol stack and the corresponding ATM interface configuration. The key elements can be summarised as follows:

- We use the Q.2150.1 stack to adapt the AAL2 signalling protocol to the MTP-3b protocol.
- The SSCP offers to other applications connection-oriented and connectionless services, while using itself MTP as a service.
  - Connection-oriented links enable virtual connections between Network Elements (NE) by providing establish and release functions.
  - Connectionless links allow non-called related communications between NEs exchanging info for small periods.

Table 5.9 Summary of Iub NBAP procedures

Radio link setup	BTS logical resource management (by RNC)	Channel handling	Fault management
Radio link addition Increasing RLs in a EU having already other RLs	*Broadcast information* RNC tells BTS to start/stop BCCH transmission	*Common Ch power control* Used by RNC to change cell size	*RNC Restart indication (to BTS)* Used to communicate, e.g. lost of AAL2 connection, Radio link, etc.
Radio link reconfiguration Radio Bearer setup renegotiation, e.g. based on new L1 conditions	*BTS configuration* *Allows RNC to update* BTS radio network parameters, e.g. resource, load, control, power control, cell config. (radius, scrambling code, etc.)	*RACH setup, release and reconfiguration*	*Traffic termination point failure alert* Used to communicate BTS ↔ RNC peer failure in traffic termination point
Radio link delition Releasing *existing RLs*	*BTS capability query* The RNC queries, e.g. *NBAP, CCH, DCH,* versions.	*FACH/PCH setup, release and reconfiguration*	*Error Indication* *The RNC or BTS* reports detected errors.
DL power drifting prevention	*Operational state query* RNC checks BTS operational status per cell	*DSCH Control Ch setup*	
Compressed mode control Tx/Rx halted for external frequency measurements.		*Common Packet Channel setup*	

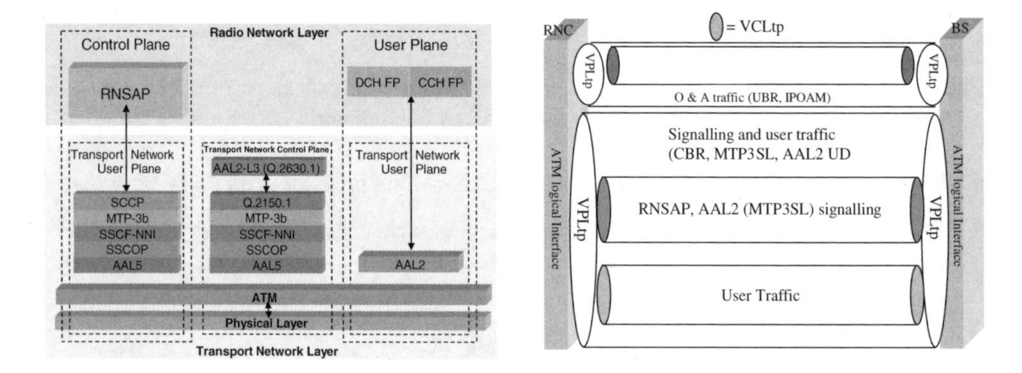

Figure 5.20 The Iur protocol stack and ATM interface mapping.

- The MTP-3b (Message Transfer Part Level 3 broadband) includes functions, which route and distribute to corresponding user part and network management the necessary procedures for optimal signalling.
- The SSCF-NNI receives L3 SS7 signalling and maps it to the SSCOP and vice versa; i.e. it performs co-ordination between higher and lower layers.
  - It offers additional services such as link error monitoring and SDU retrieval at link-failures for re-transmission over other links.

5.9.3.2.1 Iur RNSAP Procedures

The main Radio-Network Subsystem Application Part (RNSAP) includes:

- basic Mobility Management procedures → only signalling between two RNCs
  - UL/DL signalling transfer;
  - relocation execution and paging.
- DCH procedures → dedicated channel traffic between two RNCs
  - radio link management;
  - physical channel reconfiguration;
  - measurements on dedicated resources → Handover control;
  - compressed mode control;
  - DL power drifting correction.
- Common Transport Channel procedures → handling of common and shared CHs
  - where the SRNC controls the cell used for common or shared channel.
- Global procedures → they apply to all UEs
  - e.g. failure indication enabling for errors, which do not have error message definition.

5.9.3.3 The Iu–CS and Iu–PS Interface Structure

The Radio-Access Network Application Part (RANAP) as a main component in the Iu interface serves the CS and PS domains. It performs the following main functions:

- Overall Radio-Access Bearer (RAB) management, which includes the RAB's set-up, maintenance and release
- Iu connection management
- Handover procedures → initiated by the Serving RNC to:
  - move UTRAN side connection (UTRAN-CN) point to another RNC;
  - realise hard handover.
- Reset procedures
  - UTRAN and CN re-initialisation reset after a failure
- UTRAN Flow control → CCCH scheduler or processor overload at a cell

- Cipher Mode control → to select and load the user data and signalling encryption algorithm while applying the appropriate key
- Resource Check → RNC provides cell resource availability status upon CN request
- Direct Transfer → UE-CN non-interpreted signalling transfer over the Iu interface
- Transport of Non-Access Stratum (NAS) information between the UE and CN
- Tracing request → By which CN demands transaction record tracing to the RNC
- UE location exchange between RNC and CN
- Paging requests from the CN and UE
- General error handling

Figure 5.21 illustrates the CS-interface protocol stack and ATM resource management for which the RENAP procedure can be detailed further as follows:

In the Iu–PS interface case (Figure 5.22), the network control plane does not apply because setting up the GTP tunnel requires only an identifier for the tunnel and the IP

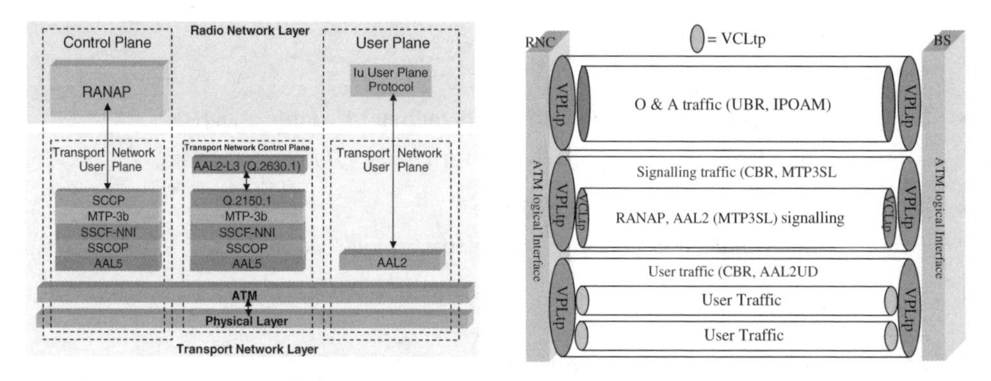

Figure 5.21 The Iu–CS-interface ATM resource management.

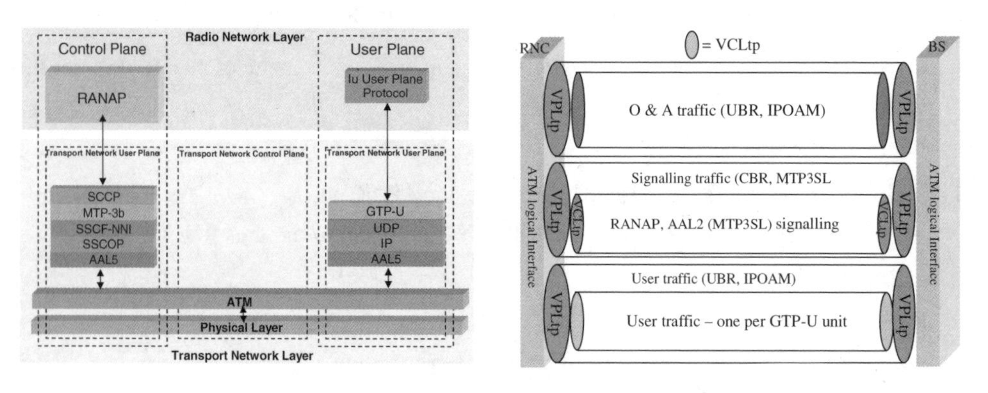

Figure 5.22 The Iu–PS-interface ATM resource management.

addresses for both directions, which are already included in the RENAP message RAB allocation.

5.10 RADIO INTERFACE PROTOCOL ARCHITECTURE

This section covers essential aspects on the radio interface protocols based on [35]. For completeness and to remain close to the technical specifications we use the same terminology and keep the approach of the proposed architecture.

5.10.1 Protocol Structure

Radio interface protocols establish, adapt and free radio bearer services in the UTRA platform. They have functions in Layers 1–3, i.e. physical (L1), link (L2) and network (L3) layers in the OSI terminology. L2 has in turn the following sub-layers: Medium Access Control (MAC), Radio Link Control (RLC), Packet Data Convergence Protocol (PDCP) and Broadcast and Multicast Control (BMC). L3 and RLC consist of Control (C) and User (U) planes. The PCDP and BMC sub-layers exist only in the U plane.

Layer 3 has sub-layers in the C-plane. The lowest one, the Radio Resource Control (RRC), interfaces with L2 and terminates in the UTRAN. The next sub-layer provides duplication avoidance functionality [36] and terminates in the CN. It remains part of the access stratum to provide access stratum services to higher layers. However, we assume that higher layer signalling such as Mobility Management (MM) and Call Control (CC) does not belong to the non-access stratum.[9]

In the architecture representation shown in Figure 5.13, each block indicates an instance of the respective protocol. At the interface between sub-layers, we mark with ovals Service Access Points (SAP) for peer-to-peer communication. The SAP between MAC and the physical layer provides the transport channels, and SAPs between RLC and the MAC sub-layer provide the logical channels. In the C-plane, the General Control (GC) defines through Notification (Nt) and Dedicated Control (DC) SAPs the interface between duplication avoidance and higher L3 sub-layers (CC, MM).

Figure 5.23 also illustrates connections between RRC and MAC as well as RRC and L1 affording local inter-layer control services. We have as well, an equivalent interface control between RRC and the RLC sub-layer, between RRC and the PDCP sub-layers and between RRC and the BMC sub-layer. These interfaces enable the RRC to control the configuration of the lower layers. Thus, separate control SAPs defined between RRC and each lower layer (PDCP, RLC, MAC and L1) exist.

The RLC sub-layer provides ARQ functionality in conjunction with the applied radio transmission technique. In this case, we do not see a difference between RLC instances in C and U planes. When the Iu connection-point remains unchanged, the CN may request the UTRAN full data protection. However, when the Iu connection point changes (e.g. SRNS relocation, streamlining, etc.), the UTRAN may not guarantee full data protection, but rely on duplication avoidance functions in the CN.

[9]Higher level signalling is not in the scope of 3GPP TSG RAN. On the other hand, the UTRA radio interface protocol architecture has similarities to the current ITU-R protocol architecture, ITU-R M.1035.

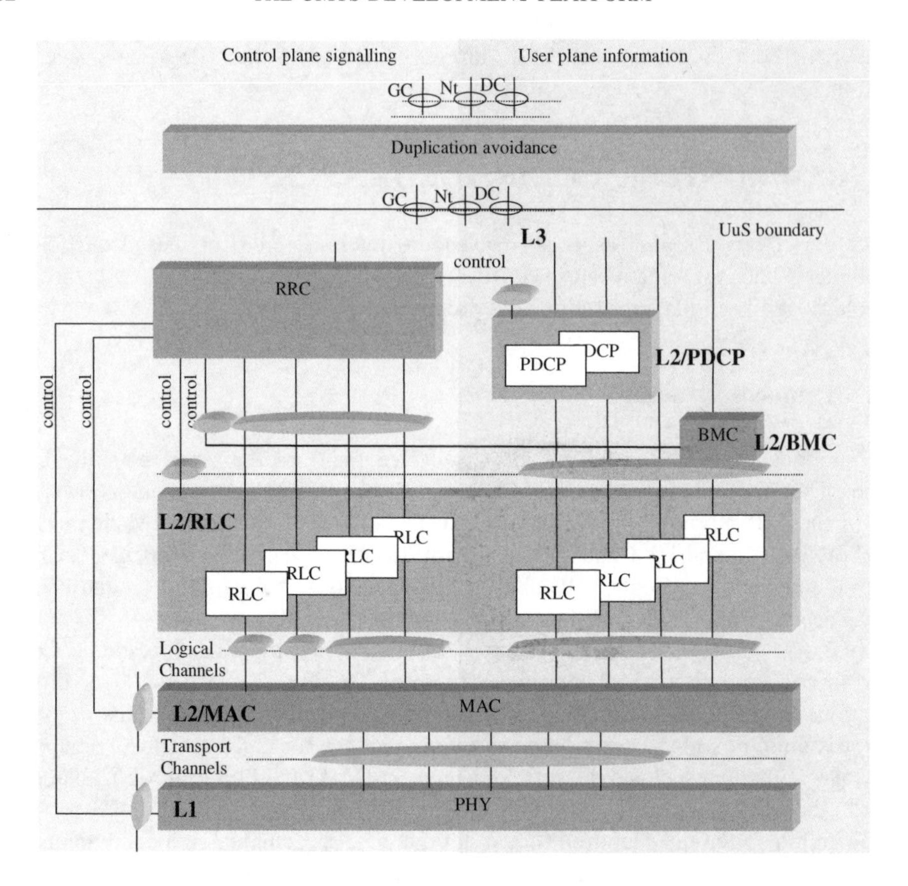

Figure 5.23 Radio interface protocol architecture (ovals are service access points) (after [35]).

5.10.1.1 Service Access Points and Service Primitives

At SAPs each layer provides services where a set of primitives or operations defines every service that a layer provides to the upper layer(s). There exists control services at Control SAPs (C-SAP) enabling the RRC layer to control lower layers locally (i.e. not requiring peer-to-peer communication).[10] See primitives in [35].

5.10.2 Services and Functions in Layer 1

L1 or the physical layer provides information transfer services to MAC and upper layers. We characterise these services by *how* and with *what* features information gets transferred over the air–interface, and we denote them *transport channels.*[11]

[10]C-SAP primitives can bypass one or more sub-layers.
[11]They transport signal and traffic information.

Table 5.10 Transport channel summary [32]

Transport channels	Description
Common	
Random-Access Channel (RACH)	A contention based uplink channel used for transmission of relatively small amounts of data, e.g. for initial access or non-real-time dedicated control or traffic data.
Common Packet Channel (CPCH)	A contention based channel used for transmission of bursty data traffic. This channel only exists in FDD mode and only in the uplink direction. The common packet channel is shared by the UEs in a cell and therefore, it is a common resource. The CPCH is fast power controlled.
Forward Access Channel (FACH)	Common downlink channel without closed-loop power control used for transmission of relatively small amount of data.
Downlink Shared Channel (DSCH)	A downlink channel shared by several UEs carrying dedicated control or traffic data.
Uplink Shared Channel (USCH)	An uplink channel shared by several UEs carrying dedicated control or traffic data, used in TDD mode only.
Broadcast Channel (BCH)	A downlink channel used for broadcast of system information into an entire cell.
Paging Channel (PCH)	A downlink channel used for broadcast of control information into an entire cell allowing efficient UE sleep mode procedures. Currently identified information types are paging and notification. Another use could be UTRAN notification of change of BCCH information.
Dedicated	
Dedicated Channel (DCH)	A channel dedicated to one UE used in uplink or downlink
Fast Uplink Signalling Channel (FAUSCH)	An uplink channel used to allocate dedicated channels in conjunction with FACH

5.10.2.1 Services

5.10.2.1.1 Transport Channels

In principle we classify transport channels in two groups, i.e. (Table 5.10) common and dedicated channels. The first group has in-band identification of UEs when addressing particular UEs. The second group has identification of UEs through the physical channel, i.e. code and frequency for FDD and code, time slot and frequency for TDD.

Each transport channel, excluding the FAUSCH,[12] gets an associated *transport format* when having a fixed or slow changing rate, or an associated *transport format set* when having a fast changing rate. We define the transport format as a combination of encoding, inter-leaving, bit rate and mapping onto physical channels [37]. We define the transport format set as a group of transport formats. In the context of the latter, e.g. variable rate DCH

[12]It only conveys a reservation request.

has a transport format set, i.e. one transport format for each rate, whereas a fixed rate DCH has a single transport format [35].

5.10.2.2 L1 Functions

Chapter 4 describes the main functions of L1; here we list a summary from [35] to complete the services and functions introduction.

- Error detection on transport channels and indication to higher layers;
- FEC encoding/decoding and inter-leaving/de-inter-leaving of transport channels;
- multiplexing of transport channels and de-multiplexing of coded composite transport channels;
- rate matching;
- modulation and spreading/de-modulation and de-spreading of physical channels;
- macro-diversity distribution/combining and soft handover execution;
- mapping of coded composite transport channels on physical channels;
- power weighing and combining of physical channels;
- frequency and time (chip, bit, slot, frame) synchronisation;
- measurements and indication to higher layers (e.g. FER, SIR, interference power, transmit power, etc.);
- closed-loop or fast power control;
- RF processing;
- support of uplink synchronisation (TDD only);
- support of timing advance on uplink channels (TDD only).

5.10.3 Services and Functions in Layer 2

5.10.3.1 Services and Functions in The MAC Sub-layer

Specification in [38] provides the details of the MAC protocol; here we simply summarise the main services and functions.

5.10.3.1.1 Services to Upper Layers

- *Data transfer* → provides unacknowledged transfer of MAC SDUs between peer MAC entities without segmentation.
- *Reallocation of radio resources and MAC parameters* → performs on request of RRC execution of radio resource reallocation and change of MAC parameters. In TDD mode, in addition, the MAC handles resource allocation autonomously.

Table 5.11 Summary of logical channels

Logical channels	Description
Control Channels (CCH)	*Broadcast Control Channel (BCCH)*
Transfer of control plane information only	A downlink channel for broadcasting system control information
	Paging Control Channel (PCCH) A downlink channel transferring paging information.
	Dedicated Control Channel (DCCH) A point-to-point bi-directional channel that transmits dedicated control information between a UE and the network
	Common Control Channel (CCCH) Bi-directional channel for transmitting control information between network and UEs
	Shared Channel Control Channel (SHCCH) Bi-directional channel that transmits control information for uplink and downlink shared channels between network and UEs
Traffic Channel (TCH)	*Dedicated Traffic Channel (DTCH)* A DTCH is a point-to-point channel, dedicated to one UE, for the transfer of user information. A DTCH can exist in both uplink and downlink.
	Common Traffic Channel (CTCH) A point-to-multipoint uni-directional channel for transfer of dedicated user information for all or a group of specified UEs.

- *Reporting of measurements* → reports local measurements, e.g. traffic volume and quality indication to the RRC.

5.10.3.1.2 Logical Channels

The MAC layer provides data transfer services on logical channels. We classify these channels in two groups, i.e. *control channels* for control-plane information transfer, and *traffic channels* for user-plane information transfer (see Table 5.11).

5.10.3.1.3 Mapping between Logical Channels and Transport Channels

Table 5.12 illustrates connections between logical and transport channels.

Table 5.12 Connections between logical and transport channels

Channel	Connected to
BCCh	BCH, may also FACH
CCCH	PCH
CCCH	RACH and FACH
SHCCH	RACH, USCH/FACH and DSCH
DTCH	Either RACH and FACH; RACH and DSCH to DCH and DCSCH, DCH and DSCH, DCH, CPCH (FDD only) or USCH (TDD only)
CTCH	FACH
DCCH	Either RACH and FACH, RACH and DSCH, DCH and DSCH, DCH, CPCH (FDD only), FAUSCH, USCH (TDD only)

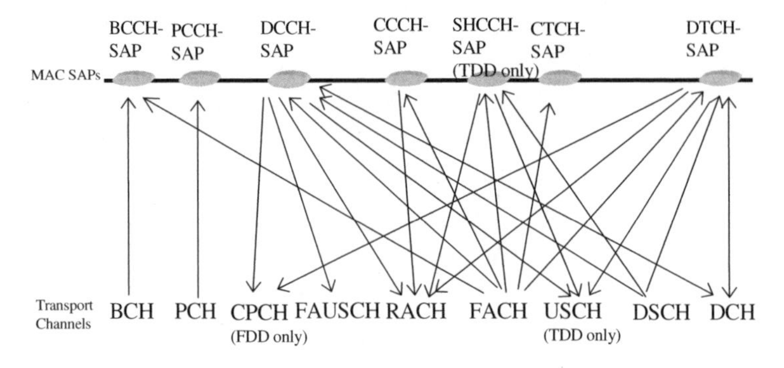

Figure 5.24 Logical channels mapped onto transport channels, seen from the UE side [35].

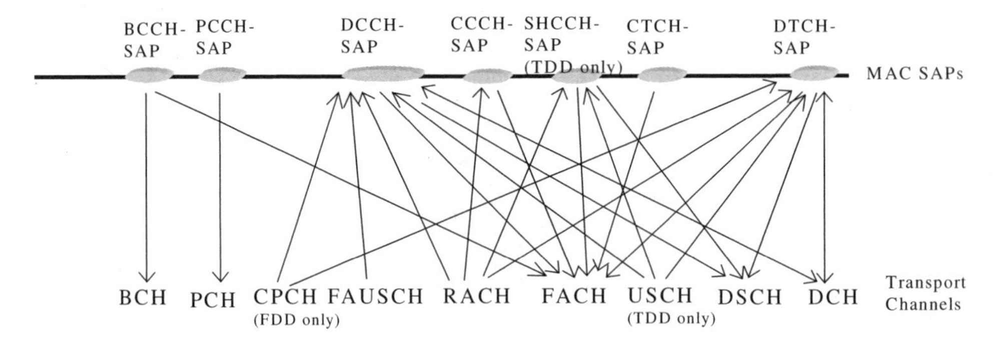

Figure 5.25 Logical channels mapped onto transport channels, seen from the UTRAN side [35].

Figures 5.24 and 5.25 illustrate the mappings as seen from the UE and UTRAN sides including both the FDD and TDD modes.

5.10.3.2 MAC Functions

The functions of MAC include:

- *Mapping between logical channels and transport channels.*
- *Selection of appropriate transport format for each transport channel depending on instantaneous source rate.*
- *Priority handling between data flows of one UE.* 'Priorities are, e.g. given by attributes of Radio Bearer services and RLC buffer status [39–44]. The priority handling is achieved by selecting a transport format combination for which high priority data is mapped onto L1 with a "high bit rate" transport format, at the same time letting lower priority data be mapped with a "low bit rate" (could be zero bit rate) Transport Format.

Transport format selection may also take into account transmit power indication from Layer 1' [35].

- *Priority handling between UEs by means of dynamic scheduling* → aiming for a dynamic scheduling for efficient spectrum utilisation.[13] The MAC realises priority handling on common and shared transport channels.
- *Identification of UEs on common transport channels.* When addressing a particular UE on a common downlink channel, or when a UE uses the RACH, we need in-band identification of the UE.
- *Multiplexing/de-multiplexing of higher layer PDUs into/from transport blocks delivered to/from the physical layer on common transport channels.* The MAC supports multiplexing for common transport channels as complement to the physical layer.
- *Multiplexing/de-multiplexing of higher layer PDUs into/from transport block sets delivered to/from the physical layer on dedicated transport channels.* The MAC allows service multiplexing for dedicated transport channels.
- *Traffic volume monitoring.* Based on the MAC's measurement of traffic volume on logical channels and reported to RRC, the latter performs transport channel switching decisions.
- *Dynamic transport channel type switching* → switching execution between common and dedicated transport channels takes place based on a switching decision derived by the RRC.
- *Ciphering.* This function prevents unauthorised acquisition of data. Ciphering occurs in the MAC layer for transparent RLC mode.
- *Access service class selection for RACH transmission.* The RACH resources (i.e. access slots and preamble signatures for FDD, timeslot and channelisation code for TDD) may be divided between different Access Service Classes (ASC) in order to provide different priorities of RACH usage. More than one ASC or all ASCs can get assigned to the same access slot/signature space. Each ASC will also have a set of back-off parameters associated with it, some or all of which may be broadcasted by the network. The MAC function applies the appropriate back-off and indicates to the PHY layer the RACH partition associated with a given MAC PDU transfer [35].

5.10.3.3 RLC Services and Functions

5.10.3.3.1 Services

- *RLC connection establishment/release.*
- *Transparent data transfer* → transmits higher layer PDUs without adding any protocol information, but may include segmentation/re-assembly functionality.
- *Unacknowledged data transfer* → transmits higher layer PDUs without guaranteeing delivery to the peer entity. The unacknowledged data transfer mode has the following

[13]In the TDD we represent transportable data in terms resource units sets.

characteristics:

  - *Detection of erroneous data*: delivering only correct SDUs to the receiving higher layer by using the sequence-number check function.
  - *Unique delivery*: delivering SDUs only once to the receiving upper layer using the duplication detection function.
  - *Immediate delivery*: delivering SDUs to the higher layer receiving entity as soon as it arrives at the receiver.

- *Acknowledged data transfer.* Transmits higher layer PDUs and guarantees delivery to the peer entity. It has the following characteristics:
  - *Error-free delivery*: ensured by means of retransmission.
  - *Unique delivery*: delivering each SDU only once to the receiving upper layer using duplication detection function.
  - *In-sequence delivery*: supports for in-order delivery of SDUs, i.e. delivering SDUs to the receiving higher layer entity in the same order as the transmitting higher layer entity submits them to the RLC sub-layer.
  - *Out-of-sequence delivery*: it shall also be possible to allow the receiving RLC entity to deliver SDUs to a higher layer in a different order than submitted to RLC sublayer at the transmitting side.
- *QoS setting.* Configurable by Layer 3 to provide different levels of QoS.
- *Notification of unrecoverable errors.* Notifying the upper layer of errors that cannot be resolved by RLC[14] itself by normal exception handling procedures, e.g. by adjusting the maximum number of retransmissions according to delay requirements.

5.10.3.4 RLC Functions

- *Segmentation and reassembly.* This function performs segmentation/reassembly of variable-length higher layer PDUs into/from smaller RLC Payload Units (PUs). The RLC PDU size is adjustable to the actual set of transport formats.
- *Concatenation*
- *Padding.* In the absence of concatenation and non-filled RLC PDUs of given size, the remainder of the data field gets filled with padding bits.
- *Transfer of user data.* Conveyance of data between users of RLC services.
- *Error correction.* Error correction by re-transmission (e.g. Selective Repeat, Go Back N or a Stop-and-Wait ARQ) in acknowledged data transfer mode.
- *In-sequence delivery of higher layer PDUs.* Preserves the order of higher layer PDUs when submitted for transfer by RLC using the acknowledged data transfer service.
- *Duplicate detection.* Detects duplicated received RLC PDUs and ensures that the resultant higher layer PDU get delivered only once to the upper layer.

[14]There is a single RLC connection per radio bearer.

- *Flow control.* Allows an RLC receiver to control the rate at which the peer RLC transmitting entity may send information.
- *Sequence number check (unacknowledged data transfer mode).* Guarantees the integrity of reassembled PDUs and provides a mechanism for the detection of corrupted RLC SDUs through checking the sequence number in RLC PDUs when they are reassembled into a RLC SDU.
- *Protocol error detection and recovery.* Detects and recovers from errors in the operation of the RLC protocol.
- *Ciphering.* Prevents unauthorised acquisition of data. Ciphering occurs in the RLC layer for non-transparent RLC mode.
- *Suspend/resume function.* Suspension and resumption of data transfer as in, e.g. LAPDm.

5.10.4 PDCP Services and Function

The Packet Data Convergence Protocol (PDCP) service provides transmission and reception of network PDUs in acknowledged/unacknowledged and transparent RLC mode. As part of its function, first it maps network PDUs from one network protocol to one RLC entity. Second it compresses in the transmitting entity and decompresses in the receiving entity redundant network PDU control information (header compression/decompression), including TCP/IP header compression and decompression when necessary. See more service and function details in [35–45].

5.10.5 Broadcast and Multicast Control–Services and Functions

The BMC provides broadcast/multicast transmission service in the user plane on the radio interface for common user data in transparent or unacknowledged mode. Its essential functions include from [45]:

- *Storage of cell broadcast messages* → stores messages received over the CBC–RNC interface for scheduled transmission.
- *Traffic volume monitoring and radio resource request for CBS* → at the UTRAN side, it calculates the required transmission rate for cell broadcast service based on the messages received over the CBC–RNC interface, and requests appropriate CTCH/FACH resources from RRC.
- *Scheduling of BMC messages* → the BMC receives scheduling information along with each cell broadcast message over the CBC–RNC-interface. Based on this UTRAN scheduling information, it generates schedule messages and schedules BMC message sequences correspondingly. At the UE side, it evaluates the scheduled messages and indicates scheduling parameters to RRC, which are used by the RRC to configure lower layers for CBS discontinuous reception.
- *Transmission of BMC messages to UE* → transmits BMC messages (scheduling and cell broadcast messages) based on a schedule.

- *Delivery of cell broadcast messages to upper layer (NAS)* → delivers correctly received cell broadcast messages to upper layer (NAS) in the UE, neglecting corrupted ones.

Specifications are given in [45], 'Data flows through Layer 2'.

5.10.6 Uu Stratum Services and Functions in Layer 3

Here we provide an overview on Layer 3 services and functions based on the Uu Stratum. Further detailed description of the RRC protocol and structured procedures involving RRC can be found in [46–48].

The main Uu stratum services include general control, notification and dedicated control. The first provides a common information broadcast service to all UEs in a certain geographical area. The second provides paging and notification broadcast services to a specific UE(s) in a certain geographical area. The third provides services for establishment/release of a connection and transfer of messages using this connection. It should also be possible to transfer a message during the establishment phase.

5.10.7 The Radio Resource Control (RRC) Functions

The RRC layer handles the control plane signalling of Layer 3 between the UEs and UTRAN. Its main functions include:

- *Broadcast of information provided by the access and non-access stratum (core network).* It performs information broadcasting from the network to all UEs. The system information is normally repeated on a regular basis.
- *Establishment, re-establishment, maintenance and release of an RRC connection between the UE and UTRAN.* Higher layers request the UE side to establish the first signalling connection for the UE. The establishment of an RRC connection includes an optional cell re-selection, an admission control and a Layer 2 signalling link establishment.
- *Establishment, reconfiguration and release of radio bearers.* Can, on request from higher layers, perform the establishment, reconfiguration and release of radio bearers in the user plane.
- *Assignment, reconfiguration and release of radio resources for the RRC connection.* It handles the assignment of radio resources (e.g. codes, CPCH channels) needed for the RRC connection including needs from both the control and user plane.
- *RRC connection mobility functions.* Performs evaluation, decision and execution related to RRC connection mobility during an established RRC connection, e.g. handover, inter-system handover preparation, cell re-selection and cell/paging area update procedures, based on, e.g. measurements done by the UE.
- *Paging/notification.* May broadcast paging information from the network to selected UEs, upon request from higher layers on the network side when necessary, or can also initiate paging during an established RRC connection.

- *Routing of higher layer PDUs.* Performs at the UE side routing of higher layer PDUs to the correct higher layer entity at the UTRAN side to the correct RANAP entity.
- *Control of requested QoS.* It ensures that the QoS requested for the radio bearers can be met, e.g. allocation of a sufficient number of radio resources.
- *UE measurement reporting and control of the reporting.* The RRC layer controls the measurements performed by the UE in terms of what to measure, when to measure and how to report, including both UMTS air interface and other systems. It also performs the reporting of the measurements from the UE to the network.
- *Outer loop power control.* The RRC layer controls setting of the target of the closed loop power control.
- *Control of ciphering.* Provides procedures for setting of ciphering (on/off) between the UE and UTRAN.
- *Slow DCA.* It applies only to the TDD mode and involves allocating preferred radio resources dynamically based on long-term decision criteria.
- *Arbitration of radio resources on uplink DCH.* Controls rapid radio resource allocations on uplink DCH using a broadcast channel to send control information to all involved users.
- *Initial cell selection and re-selection in idle mode.* Selection of the appropriate cell based on idle mode measurements and cell selection criteria.
- *Integrity protection.* Adds a Message Authentication Code (MAC-I) to sensitive and/or containing sensitive information RRC messages.
- *Initial configuration for CBS.* Performs the initial configuration of the BMC sub-layer.
- *Allocation of radio resources for CBS.* Allocates radio resources for CBS based on traffic volume requirements indicated by BMC.
- *Configuration for CBS discontinuous reception.* Configures the lower layers (L1, L2) of the UE when the latter listen to the resources allocated for CBS.
- *Timing advance control.* Controls the operation of timing advance, which is applicable only to the TDD mode [49].

REFERENCES

1. 3GPP, Technical Specification Group Services and System Aspects, General UMTS Architecture (3G TS 23.101. Version 3.01, 1999–2004).
2. 3GPP, Technical Specification Group Services and System Aspects, Evolution of the GSM platform towards UMTS (3G TS 23.920. Version 3.1, 1999–2004).
3. 3GPP, Architectural Requirements for Release 1999 (3G TS 23.121. Version 3.1.0, 1999–2010).
4. 3GPP, Technical Specification Group Services and System Aspects, Iu Principles (3G TS 23.930. Version 3.0.0, 1999–2007).
5. 3GPP TS 23.002 V5.7.0 (2002–2006), System Aspects (SA), Network Architecture Release 5.
6. 3GPP TS 23.008: Organisation of subscriber data.
7. 3GPP TS 23.003: Numbering, addressing and identification.
8. 3GPP TS 22.016: International Mobile station Equipment Identities (IMEI).
9. 3GPP TS 29.002: Mobile Application Part (MAP) specification.

10. 3GPP TS 23.236: Intra Domain Connection of RAN Nodes to Multiple CN Nodes.
11. 3GPP TS 23.012: Location Management Procedures.
12. 3GPP TS 23.009: Handover procedures.
13. ITU-T Recommendation H.248: Gateway Control Protocol.
14. 3GPP TS 29.007: General requirements on interworking between the Public Land Mobile Network (PLMN) and the Integrated Services Digital Network (ISDN) or Public Switched Telephone Network (PSTN).
15. 3GPP TS 23.016: Subscriber data management; Stage 2.
16. 3GPP TS 23.060: General Packet Radio Service (GPRS); Service description; Stage 2.
17. 3GPP TS 23.008: Organisation of subscriber data.
18. 3GPP TS 43.068: Voice Group Call Service (VGCS); Stage 2.
19. 3GPP TS 43.069: Voice Broadcast Service (VBS); Stage 2.
20. 3GPP TS 23.078: Customised Applications for Mobile network Enhanced Logic (CAMEL) Phase 3 – Stage 2.
21. ITU-T Recommendation Q.1214 (05/1995): Distributed Functional Plane for Intelligent Network CS-1.
22. 3GPP TS 23.226: Global Text Telephony (GTT); Stage 2.
23. 3GPP TS 26.226: Cellular Text Telephone Modem; General Description.
24. 3GPP TS 23.060: General Packet Radio Service (GPRS); Service description; Stage 2.
25. 3GPP TS 48.002: Base Station System - Mobile-services Switching Centre (BSS-MSC) interface; Interface principles.
26. 3GPP TS 25.410: UTRAN Iu Interface: general aspects and principles.
27. 3GPP TR 43.051: Technical Specification Group GSM/EDGE Radio Access Network; Overall description, Stage 2.
28. 3GPP TS 24.002: GSM – UMTS Public Land Mobile Network (PLMN) access reference configuration.
29. R. O. Onvural, Asynchronous Transfer Mode Networks: Performance Issues, Artech House, Inc., 1994.
30. 3GPP, Technical Specification Radio Access Network Group, UTRAN Overall Description (3G TS 25.401 Version 3.1.0, 2000–2001).
31. UMTS 23.10 UMTS Access Stratum Services and Functions.
32. TS 25.442: Implementation Specific O&M Transport.
33. ATM Forum, Inverse Multiplexing for ATM (IMA) Specification, Version 1.1.
34. ATM Forum, Traffic Management Specification, Version 4.1.
35. 3GPP, Technical Specification Radio Access Network Group, Radio Interface Protocol Architecture (3G TS 25.301 Version 3.4.0, 2000–2003).
36. 3G TS 24.007: Mobile Radio Interface Signalling Layer 3; General Aspects.
37. 3GPP, Technical Specification Radio Access Network Group, Services Provided by the Physical Layer (3G TS 25.302 Version 3.4.0, 2000–2003).
38. 3GPP, Technical Specification Radio Access Network Group, MAC Protocol Specifications (3G TS 25.321 Version 3.3.0, 2000–2003).
39. 3GPP, Technical Specification Radio Access Network Group, RLC Protocol Specifications (3G TS 25.322 Version 3.2.0, 2000–2003).
40. 3GPP, Technical Specification Radio Access Network Group, PDCP Protocol Specifications (3G TS 25.323 Version 3.1.0, 2000–2003).
41. 3GPP, Technical Specification Radio Access Network Group, BMC Protocol Specification (3G TS 25.324 Version 3.1.0, 2000–2003).
42. 3GPP, Technical Specification Radio Access Network Group, RRC Protocol Specification (3G TS 25.331 Version 3.2.0, 2000–2003).
43. 3GPP, Technical Specification Radio Access Network Group, Physical Layer Procedures (TDD) (3G TS 25.224 Version 3.4.0, 2000–2003).

44. 3G TS 24.007: Mobile Radio Interface Signalling Layer 3; General Aspects.
45. 3G TS 25.324: BMC Protocol Specification.
46. 3G TS 25.303: Interlayer Procedures in Connected Mode.
47. 3G TS 25.304: UE Procedures in Idle Mode and Procedures for Cell Reselection in Connected Mode.
48. 3G TS 25.331: RRC Protocol Specification.
49. Hentschel, T. and Fettweis, G., Software Radio Receivers, in CDMA Techniques for Third Generation Mobile Systems, Kluwer, Dordrecht, 1999, pp. 257–283.

APPENDIX A: UMTS FUNCTIONAL DOMAINS

Figure 5.16 illustrates the four (application, home, serving and transport) strata. It also shows the integrated UMTS functional flow, i.e. the interactions between the USIM, MT/ME, access network, serving network and home network domains, including interactions between TE, MT, access network, serving network, transit network domains and the remote party.

The direct flows between non-contiguous domains (i.e. non-directly inter-connected domains) are transparently transported across all the domains and interfaces located on the communication path between these end domains. The protocols may or may not be UMTS specific, as long as they can inter-work seamlessly to facilitate roaming.

When looking at the lower part of Figure 5.26, the home network domain becomes the transit network domain in the upper part. Thus, the integrated UMTS functional flow illustrated in Figure 5.16 includes the representation or notation of the remote party

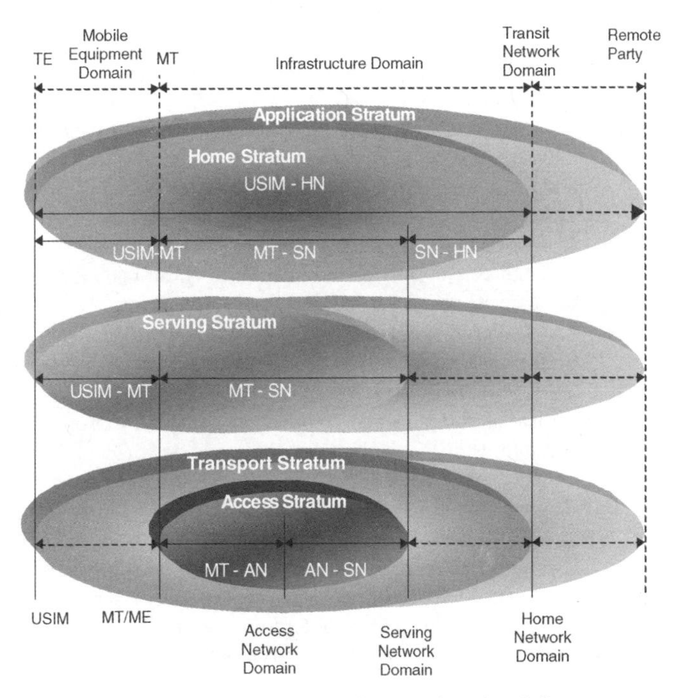

Figure 5.26 UMTS Architecture functional flow.

indicating the remote-end entity (e.g. user or machine). It shows the end-to-end character of the communication. However, the specification of the remote party is outside the scope of the UMTS specification [1].

Because of the incorporation of the remote party in Figure 5.16, the home (through the application stratum, the serving and transport strata are extended all the way to the remote party line in the representation. Hence, two diversion paths reflect the flows exchanged between serving and home domains on one side and between serving and transit on the other side. Starting the first layer the stratum levels are defined in Table 5.13.

Table 5.13 Definition of the strata layers

Stratum levels	
Transport stratum	Supports the transport of user data and network control signalling from other strata through UMTS. It includes: • considerations of the physical transmission formats used for transmission; • mechanisms for error correction and recovery; • mechanisms to encrypt data over the radio interface and in the infrastructure part if required; • mechanisms for adaptation of data to use the supported physical format (if required); and • mechanisms to transcode data to make efficient use of, e.g. the radio interface (if required); • may include resource allocation and routing local to the different interfaces (if required); • the *access stratum*, which is specific to UMTS, as the part of the transport stratum.
Access stratum	Consists of User Equipment (UE) and infrastructure parts, as well as access-technique specific protocols between these parts (i.e. specific physical media formats between the UE and the infrastructure used to carry information). • It provides services related to the transmission of data over the radio interface and the management of the radio interface to the other parts of UMTS. The access stratum includes the following protocols: • *Mobile Termination–Access Network (MT–AN)* protocol supporting transfer of detailed radio-related information to co-ordinate the use of radio resources between the MT and AN. • *Access Network–Serving Network (AN–SN)* protocol supporting the access from the SN to the resources provided by the AN. It is independent of the specific radio structure of the AN.
Serving stratum	Consists of protocols and functions to route and transmit user or network generated data/information from source to destination. The source and destination may be within the same or different networks. It contains functions related to telecommunication services and includes: • *USIM–Mobile Termination (USIM–MT)* protocol supporting access to subscriber-specific information to allow functions in the user equipment domain. • *Mobile Termination–Serving Network (MT–SN)* protocol supporting access from the MT to the services provided by the serving network domain. • *Terminal Equipment–Mobile Termination (TE–MT)* protocol supporting exchange of control information between the TE and the MT.

Table 5.13 (*Continued*)

Stratum levels	
Home stratum	• Consists of protocols and functions related to the handling and storage of subscription data and possibly home network specific services. • It also includes functions to allow domains other than the home network domain to act on behalf of the home network. • It contains functions related to subscription data management and customer care, as well as billing and charging, mobility management and authentication. The home stratum includes the following protocols: • *USIM–Home Network (USIM–HN)* protocol supporting co-ordination of subscriber-specific information between USIM & HN. • *USIM–Mobile Termination (USIM–MT)* protocol providing the MT with access to user specific data and resources necessary to perform actions on behalf of the home network. • *Mobile Termination–Serving Network (MT–SN)* protocol supporting user specific data exchanges between the MT and the SN. • *Serving Network–Home Network (SN–HN)* protocol providing the SN with access to HN data and resources necessary to perform its actions on behalf of the HN, e.g. to support the users communications, services and features (including VHE).
Application stratum	• It represents the application process itself, provided to the end-user. • It includes end-to-end protocols and functions making use of services provided by the *home*, *serving* and *transport* strata and necessary infrastructure supporting services and/or value added services. • The functions and protocols within the application stratum may adhere to GSM/UMTS standards or may be outside the scope of the UMTS standards. • End-to-end functions are applications consumed by users at the edge of/outside the overall network. • Authenticated and authorised users may access the applications by using any variety of available user equipment.

6

IP-Multimedia Subsystem (IMS)

6.1 BACKGROUND

In the preceding chapters we covered UMTS in the context of the 3GPP Release 1999 specifications. This chapter covers the forthcoming releases of UMTS, primarily Release 4 and 5, formerly Release 2000. However, before we describe the reference architecture we outline the vision of the UMTS technical specification evolution from [1].

6.1.1 UMTS Release 1999 and Medium-Term Architecture

6.1.1.1 Release 1999

Figure 6.1 illustrates the service drivers of the UMTS architecture for R99 and future releases starting with R00. The latter has now been broken into Release 4 and 5.

The service drivers for R99 based on [1] include: compatibility with GSM, access to high-speed data services and managed QoS. The CS domain provides circuit-oriented services based on nodal MSCs (an evolved GSM), while the PS domain provides IP connectivity between the mobiles and IP networks (an evolved GPRS).

6.1.1.2 Release R4 and R5

The medium-term vision (starting R4 and R5) has the added feature of IP-multimedia as illustrated in Figure 6.1. The service drivers include: compatibility with Release 1999, addition of IP-based multimedia services and an efficient support of voice-over-IP-over-radio for the multimedia service, but not necessarily fully compatible with the telephony service and its supplementary services.

- The CS domain retains and provides 100% backwards compatibility for R99 CS domain services. We can implement this domain through the evolution of MSCs, or MSC servers and a packet backbone.
- The PS domain also retains and provides IP connectivity. It gets upgraded to support QoS for IP-multimedia services.

All IP in 3G CDMA Networks J. Castro
 ISBN: 0-470-85322-0

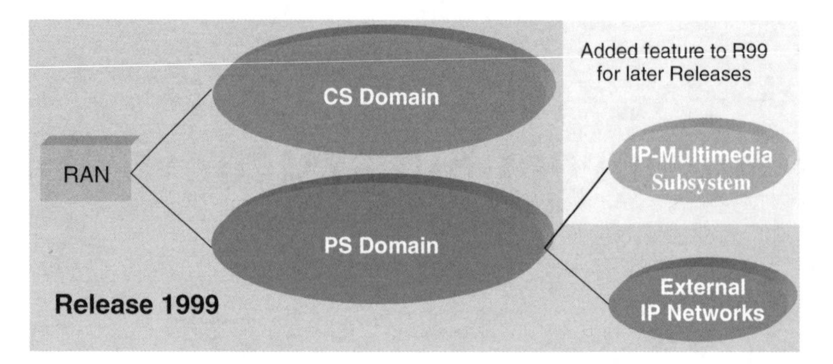

Figure 6.1 Release 1999 and medium term architecture.

The added IP-multimedia subsystem provides new IP-multimedia services that complement the services provided by the CS domain. These services will not necessarily align with the CS domain in the medium term.

6.1.2 Long-Term UMTS Architecture Vision

After the evolution of R99 culminating with R00 (R4 and R5), we aim to have an integrated platform based entirely on a packet-switched system. The service drivers for the long-term include: migration of many users to IP-multimedia services and widespread adoption of IP-multimedia outside UMTS.

By this time (Figure 6.2), we assume that the IP-multimedia subsystem has evolved to the degree that it can practically stand as a substitute for all services previously provided by the CS domain. Here we retain the PS domain but phase out the CS domain. Whether the latter can be achieved in its integrity including all security aspects remains to be seen, since it is still under standardisation or technical specification.

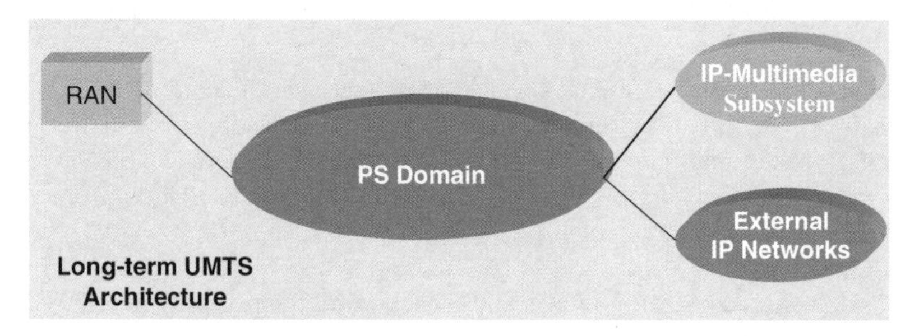

Figure 6.2 Long-term UMTS architecture.

6.1.3 All IP and Service Evolution

As noted in Chapter 1, the widespread usage of Internet and IP's ability to communicate between different networks has made IP a convergence layer to evolve from a simple data platform to larger structure for services. By aiming to reach further than the circuit switch, IP now leads mobile communications to new dimensions.

The IP protocol has opened up a whole range of wireless applications, which will allow service providers and operators to develop totally new and innovative services while enhancing their existing infrastructures. Thus, the main drivers for IP services include a full range of new multimedia applications besides IP telephony.

6.1.3.1 Transition to All IP Services

Passing to ALL IP-multimedia services will take some time; therefore both classical CS mobile services and IP-multimedia services will co-exist concurrently. As a result, networks will have to support traditional CS services and new PS services such as multimedia with the variety of terminals these services will bring in order to offer seamless roaming between evolving 2G networks and optimised 3G networks. This means that Release 2000 (now broken up into R4 and R5) will need to support service offerings while remaining independent from transport technology. The R00 platform will have to support at least the following [2]:

- hybrid architecture,
- network evolution path,
- new capabilities,
- IP-based call control,
- real-time (including voice) services over IP with end-to-end QoS,
- GERAN (support for GMS/EDGE radio-access network)
- services provided using toolkits (e.g. CAMEL, MExE, SAT, VHE/OSA),
- backwards compatibility with Release 1999 services,
- no degradation in QoS, security, authentication, privacy,
- support for inter domain roaming and service continuity.

The future UMTS releases will have new and improved enabling mechanisms to offer services without using circuit-switched network capabilities, as shown in Figure 6.3. Here, we assume that the set of services available to the user and the quality of the services offered will match those available in networks that use CS enablers.

6.1.4 Classifying Releases 4 and 5 Services

Following the suggested classification in [2], we can divide basic services into circuit teleservices [3] and bearer services [4], where both can utilise standardised supplementary services [5]. These basic services have not changed much in 2G networks like GSM. GPRS [6] provides IP-bearer services, and SMS, USSD and UUS can also be considered as a bearer service for some applications.

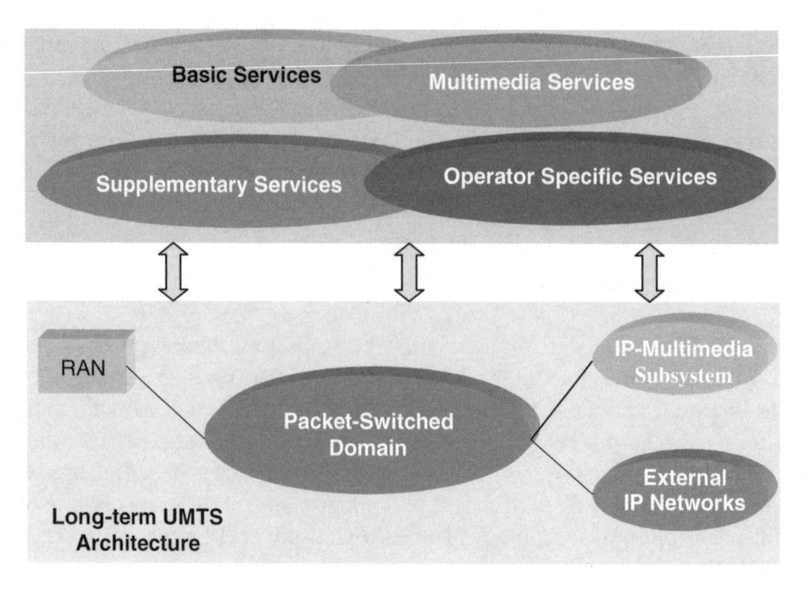

Figure 6.3 Services in the forthcoming UMTS network architecture.

IP-multimedia services (including IP telephony) using GPRS as a bearer correspond to the new services in R4 and R5. Supplementary services for IP-multimedia services do not get standardised but they can get implemented using the toolkits or at the call control level.

Value added non-call related services (not necessarily standardised) correspond to a large variety of different operator specific services. These services may use proprietary protocols or standardised protocols outside 3GPP.

To create or modify the above services (both call and non-call related services), service providers or operators may utilise standardised 3GPP toolkits (e.g. CAMEL or LCS) or external solutions (e.g. IP toolkit mechanisms). Pre-payment can serve as an example of an application created with toolkits that may apply to all of the above service categories. Additional information and details on general and IP-multimedia requirements can be found in [2].

In the following we introduce the reference architecture which will realise the type of services presented above and illustrated in Figure 6.4.

6.2 FRAMEWORK FOR THE IP-MULTIMEDIA SUBSYSTEM

3GPP groups have split R00 in R4 and R5[1] in order to achieve its specification pragmatically in phases. In the following section, for introductory purposes we will briefly make a reference to R00 as background from [7]. However, the remaining of the chapter will focus on IMS and describe the updated definition of the ALL IP building block of the UMTS core network.

[1]Recently there is also Release 6 or R6 picking up items left from R5 or new items.

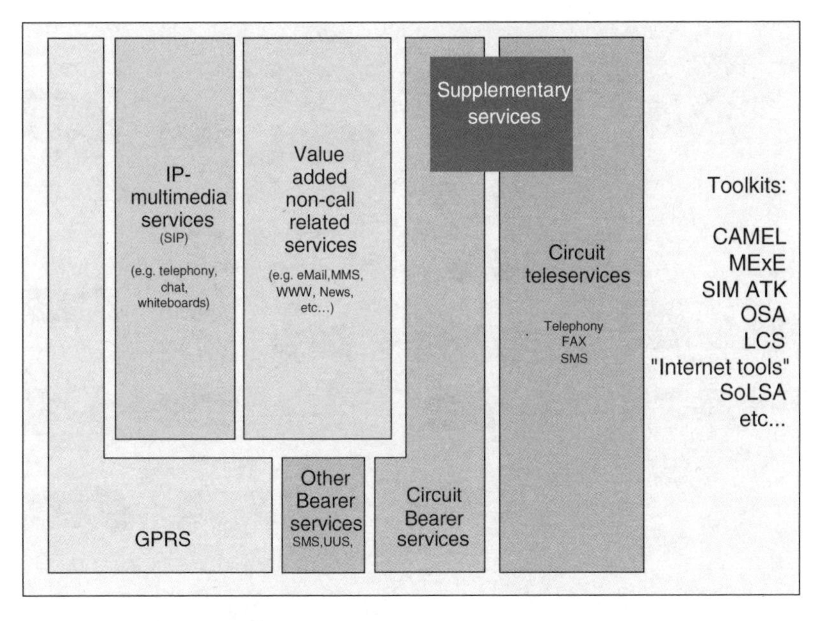

Figure 6.4 Service classification [2].

6.2.1 Overview of the IMS Release 2000 Architecture Reference

R00 served as the starting block to define IMS. Since its initiation things have evolved or represented differently, so we presented it here for historical reasons. Figure 6.5 provides a generic view of the R00 architecture with the following interfaces: E interface—between MSCs (including MSC server/MGW); G interface—between VLRs, G interface; Gn interface between SGSNs, Gm interface—between CSCF and UE; Gs interface (optional) between MSC (or MSC server) and SGSN. The forthcoming sections will describe in more detail these interfaces and any other, which Figure 6.5 does not include.

6.3 IP-MULTIMEDIA SUBSYSTEM (IMS) CONFIGURATION

IMS incorporates all CN elements to provision multimedia services. It includes the collection of signalling and bearer related network elements as defined in TS 23.002 [8]. IMS services follow session control capabilities defined by IETF, which along with multimedia bearer utilize the PS domain. By conforming to the IETF standards, it uses, for example, protocols such as SIP and Presence.

IMS will enable PLMN operators to offer multimedia services based on and built upon Internet applications, services and protocols. Therefore, these services will not need any standardisation but adoption according to market demands. IMS will facilitate convergence of and access to, voice, video, messaging, data and web-based technologies for the wireless user, and combine the growth of the Internet with the growth in mobile communications [9].

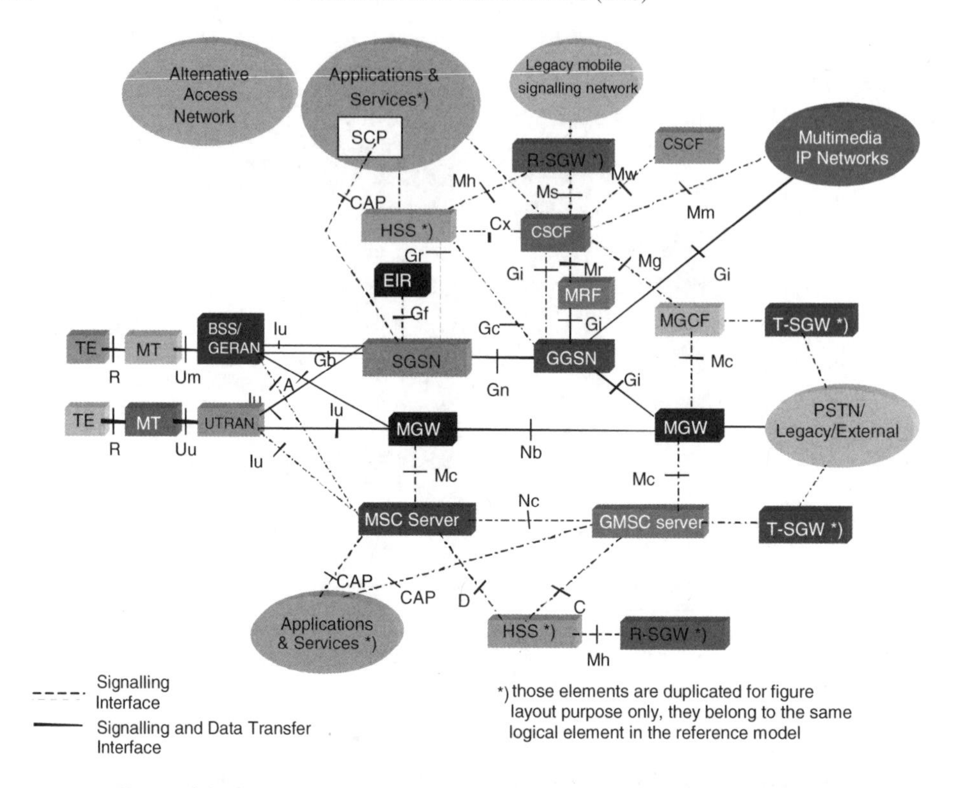

Figure 6.5 Reference Architecture for Release 2000 (R4 and R5) after [7].

IMS has been defined within principle of radio-access independence. Thus, it will apply to GERAN or UTRAN radio-access networks, GPRS evolved core network and the specific IMS functional elements described in this chapter. While IMS utilises the PS domain to transport multimedia signalling and bearer traffic, it is independent of the CS domain, i.e. the latter is not necessary to deploy IMS services.

6.3.1 Key IMS Entities

Figure 6.6 and Table 6.1 [10] illustrate the IMS entity configuration. They show how all the functions are implemented in different logical nodes. When two logical nodes use the same physical equipment, the relevant interfaces may become internal to that equipment. The illustration shows only the interfaces specifically linked to the IMS, i.e. all the SGSN, GGSN and HSS interfaces described in Figure 3.7 of the preceding chapter are still supported by these entities even though they are not shown.

6.3.2 Summary of IMS Interface and Reference Points

There are a total of about 16 interface reference points in the IMS configuration (Table 6.2).

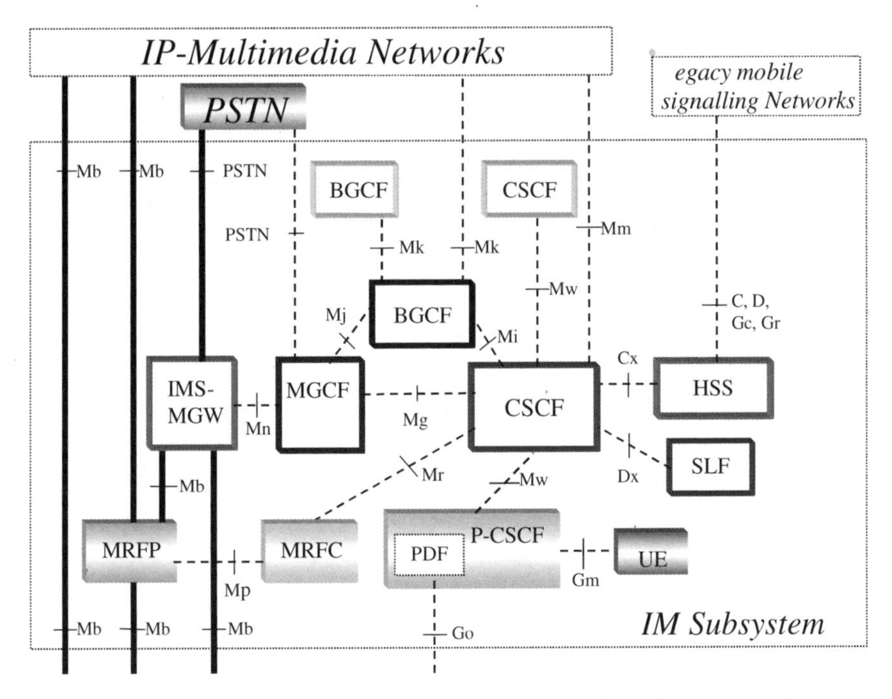

Figure 6.6 Configuration of IMS entities [10].

Table 6.1 Summary of main IMS entities

Acronym	Description	Main function
CSCF	Call Session Control Function	Can act as proxy, has 3 components
MGCF	Media Gateway Control Function	Controls call state parts for connections
IM-MGW	IP-Multimedia Media Gateway Function	Terminates bearer CHs and media streams
MRFC	Multimedia Resource Function Controller	Controls media streams
MRFP	Multimedia Resource Function Processor	Controls bearers on Mb reference point
SLF	Subscription Location Function	Queried for subscription specific data
BGCF	Breakout Gateway Control Function	Selects network for PSTN breakout
AS	Application Server	Can be, e.g.: SIP, OSA or CAMEL server
SGW	Signalling Gateway Function	Performs signalling conversion
SEG	Security Gateway	Protect assigned domains
GTTS-E	Global Text Telephony Specific Entities	Routing calls through cellular text modems

6.3.3 Cx Reference Point (HSS–CSCF)

The Cx reference point supports information transfer between CSCF and HSS, where the main procedures requiring information transfer between CSCF and HSS include:

- procedures related to serving CSCF assignment,
- procedures related to routing information retrieval from HSS to CSCF,

Table 6.2 Outline of IMS interface Ref. points

Ref.	Interfaces	Ref.	Interfaces
CAP	e.g. SGSN ↔ SCP	Mi	CSCF ↔ BGCF
Cx	HSS ↔ CSDF	Mj	BGCF ↔ MGCF
Dx	CSCF ↔ SLF	Mk	BGCF ↔ BGCF
Gi	GGSN ↔ MM[a] IP network	Mm	CSCF ↔ Multimedia IP Nets
Gm	CSCF ↔ UE	Mn	MGCF ↔ IMS-MGW
Gn	GGSN ↔ SGSN	Mp	MRFC ↔ MRFP
Go	GGSN ↔ PDF	Mr	CSCF ↔ MRF
Iu	RAN ↔ CS and PS domains	Ms	CSCF ↔ R-SGW
Mb	(IMS ↔ IPv6 Net services	Mw	CSCF ↔ CSCF
Mc	MGCF ↔ IM-MGW	ISC	CSCF ↔ AS
Mg	MGCF ↔CSCF	Sh	HSS ↔ SIP or OSA SCS
Mh	HSS ↔ R-SGW	Si	HSS ↔ CAMEL IMS-SSF

[a]Multimedia.

- procedures related to authorisation (e.g. checking of roaming agreement),
- procedures related to authentication: transfer of security parameters of the subscriber between HSS and CSCF,
- Procedures related to filter control: transfer of filter parameters of the subscriber from HSS to CSCF.

Details on these procedures can be found in TS 23.228 [9].

6.3.4 Dx Reference Point (CSCF–SLF)

We use the Dx interface (between CSCF and Subscription Locator Function—SLF) to retrieve the address of the HSS, which holds the subscription for a given user. The Dx interface is not required in a single HSS environment (e.g. a server farm architecture). See more details in 23.228 sub-clause 5.8.1 [9].

6.3.5 Gf Reference Point (SGSN–EIR)

The SGSN server supports the standard Gf interface towards the EIR server. MAP signalling is used over this interface in order to support identity (IMEI) check procedures. For more details refer to TS 23.060.

6.3.6 Gi (GGSN–Multimedia IP Network)

The GGSN supports the Gi interface. It is used for transportation of all end user IP data between the UMTS core network and external IP networks. The interface is implemented according to TS 23.060, the internet protocol according to RFC791 and RFC792 (ICMP). Finally, the IPSec is implemented or have; finally according to the following RFCs: 2401,

2402, 2403, 2404, 2405, 2406, 2410 and 2451. IP packets get transported over AAL5 according to RFC 2225 and RFC 1483.

6.3.7 Gm Reference Point (CSCF–UE)

This interface allows the UE to communicate with the CSCF, e.g. register with a CSCF, call origination and termination and supplementary services control.

The Gm reference point supports information transfer between UE and serving CSCF. The main procedures that require information transfer between UE and serving CSCF are:

- procedures related to serving CSCF registration,
- procedures related to user service requests to the serving CSCF,
- procedures related to the authentication of the application/service,
- procedures related to the CSCF's request for core network resources in the visited network.

The Gm reference point uses SIP protocol (as defined by RFC 3261 [11], other relevant RFCs and additional enhancements introduced to support 3GPP's needs) [9].

6.3.8 Gn Reference Point (GGSN–SGSN)

We use the Gn interface both for control signalling (i.e. mobility and session management) between SGSN servers and GGSN, as well as for tunnelling of end user data payload within the backbone network. The GTP-C protocol (running over UDP/IP) used for control signalling can also be included here. The interface is implemented according to TS 23.060 and TS 29.060.

6.3.9 Go Reference Point (GGSN–PDF)

The Go interface allows the Policy Decision Function (PDF) to apply policy to the bearer usage in the GGSN. The PDF is a logical entity of the P-CSCF. If the PDF is implemented in a separate physical node, the interface between the PDF and the P-CSCF is not standardised [10].

6.3.10 Iu Reference Point

The Iu remains as the reference point between UTRAN and the R00 core network. We realise this reference point by one or more of the following interfaces [7]:

- Transport of user data between UTRAN and SGSN takes place based on IP.
- Transport of signalling between UTRAN and SGSN takes place based on IP or SS#7.

- Transport of user data between UTRAN and MGW takes place based on different technologies (e.g. IP, AAL2), and includes the relevant bearer control protocol in the interface.
- Transport of signalling between UTRAN and MSC server takes place based on IP or SS#7.

When we base the Iu_CS on ATM, then we can apply R99 protocols or an evolving version, and when we base the Iu_CS on IP, we need to add new IP transport related protocols as part of the Iu protocols. On the other hand, it will be possible to have a R99 Iu interface with MSCs compliant with R99 specifications in a R5-6 network.

6.3.11 Mb Reference Point (Reference to IPv6 Network Services)

Through the Mb reference point IPv6 network services are accessed. We use these IPv6 network services for user data transport.[2]

6.3.12 Mc Reference Point (MGCF–MGW)

The Mc reference point describes the interfaces between the MGCF and MGW, between the MSC server and MGW, and between the GMSC server and MGW. It has the following features [7]:

- full compliance with the H.248 standard, baseline work of which is currently being carried out by ITU-T Study Group 16, in conjunction with IETF MEGACO WG;
- flexible connection handling which allows support of different call models and different media processing purposes not restricted to H.323 usage;
- open architecture where extensions/packages definition work on the interface may be carried out;
- dynamic sharing of MGW physical node resources; a physical MGW can be partitioned into logically separate virtual MGWs/domains consisting of a set of statically allocated terminations;
- dynamic sharing of transmission resources between the domains as the MGW controls bearers and manage resources according to the H.248 protocols.

The functionality across the Mc reference point will require to support mobile specific functions, e.g. SRNS relocation/handover and anchoring. The current H.248/IETF Megaco standard mechanisms will enable these features.

[2]The GPRS provides IPv6 network services to the UE, i.e. the GPRS Gi reference point and the IMS Mb reference point may be the same.

6.3.13 Mg Reference Point (MGCF–CSCF)

The SIP-based Mg reference point allows the transfer of session related information between the CSCF and the MGCF. We use this interface to communicate between the IP-multimedia networks and the legacy PSTN/ISDN/GSM networks.

The Mg reference point uses the SIP protocol (As defined by RFC 3261 [11], other relevant RFCs and additional enhancements introduced to support 3GPP's needs.).

6.3.14 Mh Reference Point (HSS–R-SGW)

This interface supports the exchange of mobility management and subscription data information between HSS and R99 and 2G networks. We need this interface to support Release 2000 (R4 and R5) network users who are roaming in R99 and 2G networks, and we implement it with MAP/IP using SCTP and other adaptation protocols developed by the IETF SIGTRAN working group.

6.3.15 Mi Reference Point (CSCF–BGCF)

The Mi reference point allows the Serving CSCF to forward the session signalling to the Breakout Gateway Control Function for inter-working with the PSTN networks. It is based on external specifications, i.e. SIP [11].

6.3.16 Mj Reference Point (BGCF–MGCF)

The Mj reference point allows the Breakout Gateway Control Function to forward the session signalling to the Media Gateway Control Function for inter-working with PSTN networks. It is also based on external specifications, i.e. SIP [11].

6.3.17 Mk Reference Point (BGCF–BGCF)

The Mk reference point allows the Breakout Gateway Control Function to forward the session signalling to another Breakout Gateway Control Function. It is also based on external specifications, i.e. SIP [11].

6.3.18 Mm Reference Point (CSCF–Multimedia IP Networks)

The Mm SIP based reference point stands as an IP interface between CSCF and IP networks. We use the interface, e.g. to receive a call request from another VoIP call control server or terminal. A network in principle will support SIP/SDP between the CSCF and other multimedia networks, with SIP signalling compliant with RFC 2543 and subsequent SIP releases, and with SDP compliant with RFC 2327 and also with its subsequent releases. The inter-working between SIP and other protocols, e.g. H.323, occurs at the edge of the IP-multimedia network.

6.3.19 Mn Reference Point (MGCF–IMS-MGW)

The Mn reference point describes the interfaces between the MGCF and IMS-MGW in the IMS. It includes the following properties [8]:

- Full compliance with the H.248 standard functions for IMS—PSTN/PLMN inter-working.
- Flexible connection handling, which allows support of different call models and different media processing purposes not restricted to H.323 [12] usage.
- Open architecture, where interface extensions/packages definition work may be carried out.
- Dynamic sharing of IMS-MGW physical node resources. We can partition a physical IMS-MGW into logically separate virtual MGWs/domains consisting of a set of statically allocated terminations.
- Dynamic sharing of transmission resources between the domains as the IMS- MGW controls bearers and manage resources according to the H.248 [13] protocols and functions for IMS.

6.3.20 Mp Reference Point (MRFC–MRFP)

The Mp reference point allows an MRFC to control media stream resources provided by an MRF. It has the following properties:

- full compliance with the H.248 standard [13],
- open architecture where extensions (packages) definition work on the interface may be carried out.

6.3.21 Mr Reference Point (CSCF–MRF)

The Mr affords the CSCF to control the resources within the MRF, thus allowing a network to support communication between the CSCF-MRF with either SIP or H.248 depending on the selection by standards. There is an interest in the acceptance of IETF protocols such as SIP, e.g. for Mr.

6.3.22 Ms Reference Point (CSCF–R-SGW)

The Ms corresponds to the interface between the CSCF and R-SGW. It will most likely be implemented using M3UA/SCTP.

6.3.23 Mw Reference Point (CSCF–CSCF)

This interface enables the interrogating CSCF to direct mobile terminated calls to the serving CSCF. The protocol supported is SIP according to RFC 2543. However, some additions to SIP beyond what is defined in RFC 2543bis might be required to cope, e.g. with accounting, security or supplementary services requirements.

6.3.24 Nb Reference Point (MGW-MGW)

We perform bearer control and transport over the Nb reference point. We may use RTP/UDP/IP or AAL2 to transport user data. In the R00 architecture we aim for different options to transport user data and bearer control, e.g. AAL2/Q.AAL2, STM/none, RTP/H.245.

6.3.25 Nc Reference Point (MSC Server–GMSC Server)

We perform the network–network based call control over the Nc reference point. Examples of this include ISUP or an evolution of ISUP for Bearer Independent Call Control (BICC). In the R00 architecture we aim to have different options (including IP) for signalling transport on Nc.

6.3.26 CAP-based Interfaces

This corresponds to the interfaces from the SGSN to the Signalling Control Point (SCP), from the serving CSCF (and possibly the interrogating CSCF) to the SCP, from the MSC server to the SCP and from the GMSC server to the SCP.

From [7], the interface from the SGSN to the SCP in the applications and services domain corresponds to the interface defined for UMTS GPRS to support charging application interworking.

We require the interface from the CSCF to the SCP to allow the support of existing CAMEL-based services.

The interface from the MSC server to the SCP and the GMSC server to the SCP corresponds to the standard interface defined for the CAMEL feature, which provides the mechanisms to support non-standard UMTS/GSM services of operators even when roaming outside the home PLMN.

We can implement the CAP-based interfaces by using CAP over IP, or CAP over SS7 as illustrated in Table 6.3

The above includes the interfaces from the GGSN to the HSS (i.e. Gc reference point), from the SGSN to the HSS (i.e. Gr reference point) from the GMSC server to the HSS (i.e. *C*

Table 6.3 Protocol stack for CAP [7]

<table>
<tr><td colspan="3">CAP</td></tr>
<tr><td colspan="3">TCAP</td></tr>
<tr><td colspan="3">SCCP</td></tr>
<tr><td>M3UA</td><td>MTP-3B</td><td rowspan="2">Narrow-band SS7</td></tr>
<tr><td>SCTP[a]</td><td>SAAL</td></tr>
<tr><td>IP[b]</td><td>ATM[b]</td><td>STM[b]</td></tr>
</table>

[a] In IETF work is ongoing (e.g. SCTP/UDP/IP or directly SCTP/IP).

[b] The protocols do not correspond to the same OSI layer.

reference point), and the MSC server to the HSS (i.e. D reference point). We can implement the MAP-based interfaces using MAP transported over IP, or MAP over SS7, and we can transport it on the same protocol CAP stacks as illustrated in Table 9.1.

6.3.27 IMS Service Control Interface (ISC)

As illustrated in Figure 6.7, the ISC interface stands between the Serving CSCF and the service platform(s). These platforms, i.e. the Application Servers (AS) offering value-added IM services reside either in the user's home network or in a 3rd party location. The latter could be a network or simply a stand-alone AS.

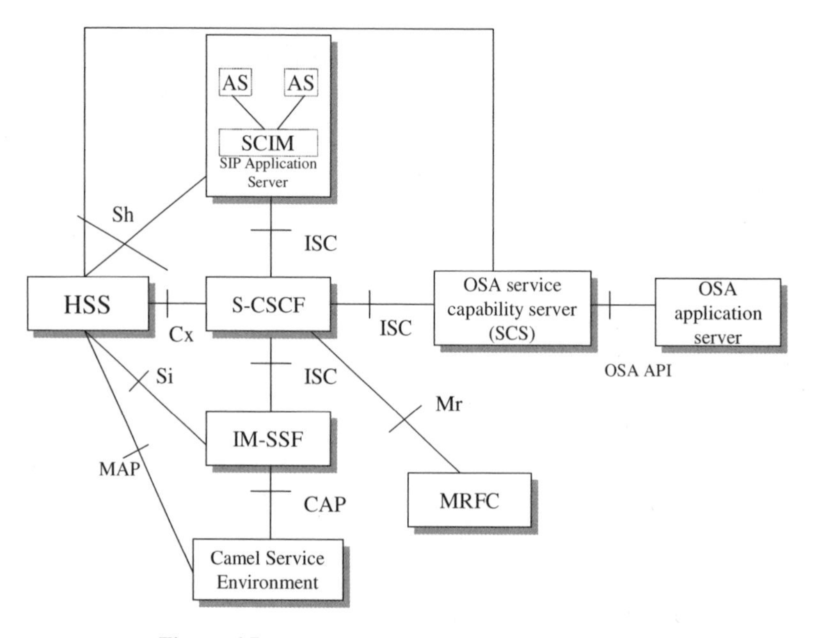

Figure 6.7 IMS service provision architecture [9].

Thus, we use the Serving-CSCF–AS interface to provide services residing in an AS for two scenarios:

- serving-CSCF to an AS in Home Network,
- serving-CSCF to an AS in External Network (e.g. 3rd Party or Visited).

In the interface process the following takes place [9]:

1. The SIP Application Server, which uses the ISC interface may host and execute services as well as influence and impact SIP sessions on behalf of the services.
2. The S-CSCF decides if there is a need of an AS to receive information related to an incoming SIP session request to ensure appropriate service handling. It bases its decision on the filtered information received from the HSS, which is stored and

conveyed on a per application server basis for each user. The type of information received from the HSS includes: name(s)/address(es), information of the application server(s).

3. The S-CSCF does not handle service interaction issues.

4. Once the IP-Multimedia Service Switching Function (IM-SSF), Open Service Architecture Service Capability Server (OSA-SCS) or SIP Application Server has been informed of an SIP session request by the S-CSCF, the IM-SSF, OSA-SCS or SIP Application Server ensures that the S-CSCF knows of any resulting activity by sending messages to the S-CSCF.

5. From the perspective of the S-CSCF, the 'SIP Application Server', 'OSA-SCS' and 'IM-SSF' will exhibit the same interface behaviour.

6. If the name/address of more than one[3] 'application server' gets transferred from the HSS, the S-CSCF will contact the application servers in the order supplied by the HSS. The response from the 1st 'application server' will serve as the input to the 2nd application server.

7. The S-CSCF does not provide authentication and security functionality for secure direct 3rd party access to the IMS. The OSA framework provides the standardised way for 3rd party secure access.

8. When an S-CSCF receives an SIP request on the ISC interface originated by an AS destined to a user served by that S-CSCF, then the S-CSCF treats the request as a terminating request to that user and provides the terminating request functionality as described above. Thus, the S-CSCF supports both registered and unregistered terminating requests.

9. The ISC interface enables conveyance of charging information as per 3GPP Technical specs 32.200[14] and 3GPP TS 32.225[15].

10. The protocol on the ISC interface allows the S-CSCF to differentiate between SIP requests on Mw, Mm and Mg interfaces and SIP requests on the ISC interface [9].

6.3.28 Call State Control Function (CSCF) and its Basic Tasks

Logically, the CSCF can be divided into three sub-components: the *serving* CSCF (S-CSCF), the *proxy* CSCF (P-CSCF) and the *interrogating* CSCF (I-CSCF).

We use the *interrogating* CSCF (I-CSCF) for Mobile Terminated (MT) communications and to determine routing for mobile terminated calls. With its function always located at the entrance to the home network, we can compare this (I-CSCF) to the GMSC in a GSM network. The I-CSCF interrogates the HSS to get information to enable calls going to the serving CSCF. The interrogating CSCF provides the Incoming Call Gateway (ICGW) and AH functionality.

[3]These multiple 'application servers' may be any combination of the SIP AS, OSA SCS or IM-SSF types.

The *proxy* CSCF, which we may compare to the visited MSC in a GSM network, manages address translation/mapping and handles call control for certain types of calls like emergency calls, legally intercepted calls, etc.

MT communications can use both *serving* CSCF and *interrogating* CSCF functionality, while MO communications do not require the *interrogating* CSCF functionality. Both *serving* CSCF and *interrogating* CSCF components may come in a single CSCF when needed. We can summarise the CSCF functions from [7] as follows:

ICGW (*Incoming Call Gateway*)

- acts as a first entry point and performs routing of incoming calls;
- incoming call service triggering (e.g. call screening/call forwarding unconditional) may need to reside for optimisation purposes;
- query address handling (implies administrative dependency with other entities);
- communicates with HSS.

CCF (*Call Control Function*)

- call set-up/termination and state/event management;
- interacts with the Multimedia Resource Functions (MRF) in order to support multi-party and other services;
- reports call events for billing, auditing, intercept or other purpose;
- receives and processes application level registration;
- query address handling (implies administrative dependency);
- can provide service trigger mechanisms (service capabilities features) towards application and services network (VHE/OSA);
- can invoke location based services relevant to the serving network;
- can check whether the requested outgoing communication is allowed given the current subscription.

SPD (*Serving Profile Database*)

- interacts with HSS in the home domain to receive profile information for the R00 all-IP network user and may store them depending on the SLA with the home domain;
- notifies the home domain of initial user's access (includes, e.g. CSCF signalling transport address, user ID, etc.; needs further study);
- may cache access related information (e.g. terminal IP address(es) where the user may be reached, etc.).

AH (*Address Handling*)

- analysis, translation, modification if required, address portability, mapping of alias addresses;
- may do temporary address handling for inter-network routing.

6.3.29 IMS Service Provision Architecture

We use the S-CSCF to support mobile originated/terminated communications. It provides the Serving Profile Database (SPD) and Address Handling (AH) functionality. The serving CSCF supports the signalling interactions with the UE through the Gm interface. The HSS sends the subscriber data to the serving CSCF for storage. It also gets updated through the latter.

The CSCF acts as the central point of the IP-multimedia control system as well as general call control (set-up, supervision and release). It triggers user controlled supplementary services and call leg handling controlled by user call control supplementary services, e.g. three-party call using Multimedia Resource Function Controller (MRFC). In addition, it handles user charging and security.

Figure 6.7 shows how the S-CSCF interacts with Application Servers (AS) through the IP-Multimedia Service Control (ISC) interface, where the ASs can be:

- *SIP Application Servers*[4] → host and execute services, as well as influence and impact the SIP session on behalf of the services.
- *The IM-SSF*—a particular server type hosting the CAMEL network features (i.e. trigger detection points, CAMEL Service Switching Finite State Machine, etc.) and CAP interface [16].
- *The OSA Service Capability Server (OSA-SCS)*—interfaces to the OSA [17] framework Application Server and provides also a standardised way for 3rd party secure access to the IM subsystem.
- *The Service Capability Interaction Manager (SCIM)*—performs the role of interaction management between other application servers.
- The MRFC interacts with the S-CSCF via the Mr interfaces in order to control Multimedia Resource Function processing.

6.3.29.1 Service Interaction

The Service Point Triggers[5] (SPTs), which set Filter Criteria in the SIP signalling (Figure 6.8), can be defined as follows [17]:

1. any initial known or unknown SIP method (e.g. REGISTER,[6] INVITE, SUBSCRIBE, MESSAGE);
2. presence or absence of any header field;
3. content of any header field or Request-URI;

[4]All the App Servers, (i.e. IM-SSF & OSA SCS) behave as SIP application servers on the ISC interface.
[5]The S-CSCF shall verify if the end user is barred before checking if any trigger applies for that end user.
[6]Which is considered part of the Mobile Origination.

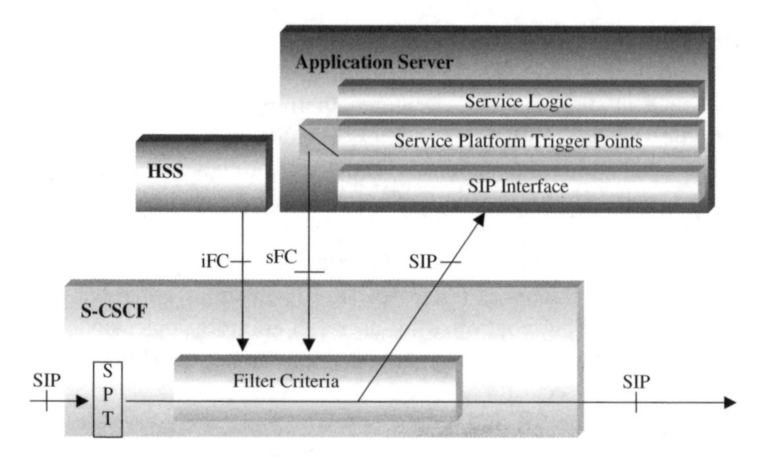

Figure 6.8 Application triggering architecture [17].

4. direction of the request is with respect to the served user, i.e. either Mobile originated (MO) or Mobile Terminated (MT) to registered user, or mobile terminated to unregistered user;
5. session description information.

The sequence of Filter Criteria (FC) can be summarised as follows [17]:

- A Filter Criteria triggers one or more SPTs to send the related request to one specific application server.
- The set of Filter Criteria stored for a service profile of a specific user, is called 'Application Server Subscription Information'.
- To handle the different Filter Criteria in the right sequence, the S-CSCF assigns a priority to each of them.
- When the S-CSCF cannot reach the AS, the S-CSCF applies the default handling associated with the trigger. This default handling implies:
    1. continuing verification if the triggers of lower priority in the list match; or
    2. abandoning verification of trigger matching of lower priority in the list and releasing the dialogue.

Thus, a Filter Criteria shall contain the following information:

1. address of the Application Server to be contacted;
2. Filter Criteria priority providing the sequence in which the criteria gets applied;
3. registered, unregistered or both trigger points, which indicated the Service Point Triggers (SPTs)[7] triggered by this Filter Criteria;

[7]The SPTs may be linked by means of logical expressions (AND, OR, NOT, etc.).

4. default handling (as described above);
5. Optional Service Information added to the message body before it is sent to the AS (e.g. IMSI for the IM-SSF).

- Each initial Filter Criteria (iFC) for a given end user will have different priority.
- The S-CSCF requests from the HSS the relevant set of iFCs, which applies to the end user (i.e. registered, unregistered or both). If the S-CSCF has a set of iFCs that is deemed valid (e.g. from a previous request), the S-CSCF does not need to request a new set.
- When the HSS sends multiple Filter Criteria to the S-CSCF, and the latter receives a message via the Mw interface, the S-CSCF checks the FC one by one according to their allotted priority, i.e. the S-CSCF:

  1. sets the list of a session[8] FC for that request according to their priority,
  2. parses the received request to find incorporated Service Point Triggers (SPTs),
  3. checks if request SPTs match FC trigger points with the next highest priority, and
     (a) if it does not match, the S-CSCF immediately proceeds with step 4;
     (b) if it does match, the S-CSCF:
        (i) adds an indication to the request, to enable the S-CSCF to identify the incoming messages, despite dialogue identification changes (e.g. due to the AS performing 3rd party call control);
        (ii) forwards the request via the ISC interface to the AS indicated in the current filter criteria. The AS then performs the service logic, may modify the request and may send the request back to the S-CSCF via the ISC interface;
        (iii) proceeds with step 4 if the request was received again from the AS via the ISC interface.
  4. repeats above the steps 2 and 3 for every FC, which was initially set up (in step 1) until the last FC has been checked;
  5. routes the request based on normal SIP routing process.

- When an Application Server terminates locally an S-CSCF request by sending back a final response to the request via the ISC interface, the S-CSCF abandons lower priority trigger verification matching. The final response includes the indicator defined in step 3 (b) (i) above, for S-CSCF message correlation.
- Each invoked Application Server/service logic may decide not to reply an invoked session and indicate so in the very first SIP transaction Record-Route/Route generated for subsequent SIP requests.
  - The non-reply implies that subsequent requests will not be routed to such Application Servers/service logic any more during the lifetime of that session.
  - Any non-replying Application Server, cannot revoke its determination by means of initial Filter Criteria (iFC).

[8]The filter criteria sequence does no change until the request leaves the S-CSCF via the Mw interface again.

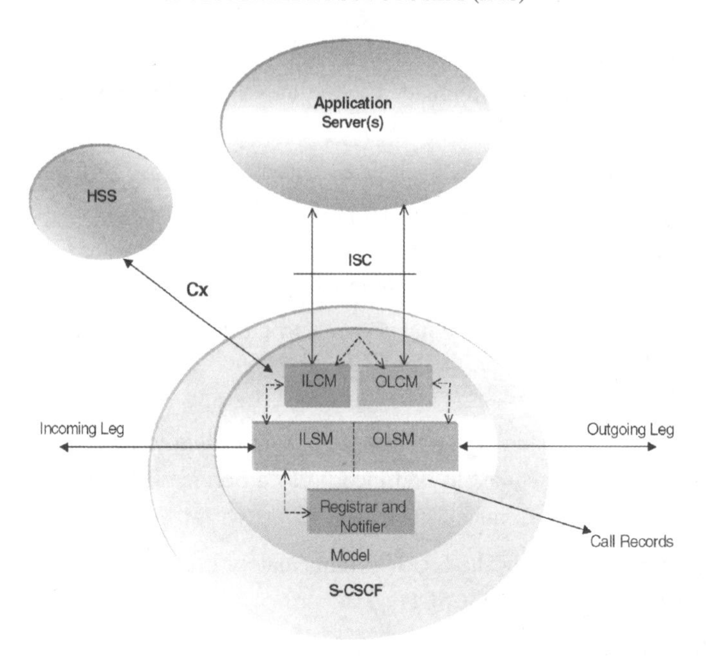

Figure 6.9 S-CSCF functional model with incoming leg control and outgoing leg control [17].

6.3.30 S-CSCF Functional Models and Operating Modes

- Figure 6.9 identifies the components of a S-CSCF functional model, which include the combined I/OLSM (Incoming/Outgoing Leg State Model), the ILCM (Incoming Leg Control Model) and OLCM (Outgoing Leg Control Model) and the Registrar and Notifier.

A single combined I/OLSM is able to store session state information. It may act on each leg independently, e.g. as a SIP Proxy, Redirect Server or User Agent dependant on the information received in the SIP request, the filter conditions specified or the state of the session [17]. Thus, we can summarise the functions as follows:

- we can split the application handling on each leg and treat each end point differently;
- a single ILCM and/or single OCLM can store transaction state information;
- the Registrar and Event Handler component handles registration and subscription to and notification of registration events;
- the Notifier handles subscription to and notification of events.

6.3.31 IP-Multimedia SIP Registration Handling

At the reception of the user initial registration request, the S-CSCF authenticates the user. Once it receives a subsequent registration request containing valid authentication credentials, the S-CSCF request the HSS to send the relevant service profile(s)[9] for the user's subscription [17,18].

The S-CSCF stores locally the initial Filter Criteria (profile subset) and verifies if the triggers match, from the highest to the lowest priority (see preceding section) [17].

Following a successfully authenticated registration, the S-CSCF downloads from the HSS all the implicitly registered public user identities associated with the registered public user identity. Then, the S-CSCF verifies, in their order of priority, if the triggers downloaded from the HSS match. If the registration request from the user matches a trigger, the S-CSCF performs a 3rd party registration to the application servers, which are interested to be informed about the user registration event of these public user identities, which in turn may trigger services to be executed by an AS.

The key information carried in the 3rd party REGISTER request includes: the public user identity, the S-CSCF address and the expiration time. It shall be possible[10] to use one of the implicitly registered public user identities as the public user identity in the 'To' header of the third party REGISTER request sent to the Application Server. In addition, the application server specific data, which is associated with the Filter Criteria and obtained from the HSS, gets added to the REGISTER request body. This data includes, e.g. the IMSI for an Application Server that supports CAMEL services or the private user identity for other Application Servers as received from the HSS. The 3rd party registration includes an expiration time equal to the expiration time sent to the UE by the S-CSCF in the 200 OK response to the incoming REGISTER request [17] (See Figure 6.10).

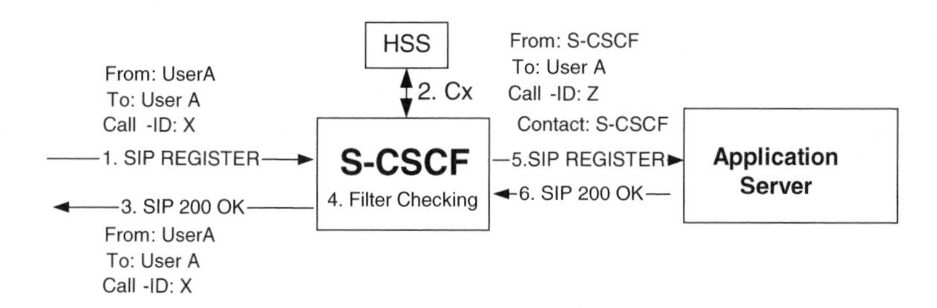

Figure 6.10 S-CSCF handling registration [17].

On receiving a failure response to one of the REGISTER requests, the S-CSCF applies the 'default handling' related with the initial Filter Criteria's trigger used.

Application Servers (AS) can in addition subscribe to the S-CSCF Registration Event Package, which allows them to discover all the implicitly registered public user identities without requiring multiple Register requests to be sent to the APs. The S-CSCF will send NOTIFY requests to the AP that has subscribed to the registration event package for the registered public user identity [18].

[9]More than one service profile may be sent, depends on config options to identify registered public user IDs.
[10]Based or service provider or operator configuration.

6.3.32 MO and MT IMS Session Handling

The S-CSCF verifies if the public user identity is barring, if so, it responds with an error code and stops further session processing.

When receiving an initial request the S-CSCF looks only for the initial Filter Criteria (iFC), which has already been downloaded from the HSS and is stored locally at the S-CSCF [9,18].

When a session request comes in, the S-CSCF checks 1st its trigger points to discriminate if it is dealing with an MO or an MT request.

6.3.32.1 Mobile Originated (MO)

When an MO request comes in the S-CSCF proceeds as follows:

1. checks if the user's highest priority matches the iFC by inspecting the service profile against the request's public user ID;
2. if the request's iFC matches, the S-CSCF forwards this request to the target Application Servers (AS), then checks for the next lower priority Filter Criteria (FC) match, and applies the FC on the SIP method received from the earlier reached AS;
3. if the request does not match the highest iFC priority, matching check passes to following FC priorities until one applies;
4. if none of the iFCs apply, the S-CSCF forwards the request downstream based on the route decision;
5. when communications with the AP fail, the S-CSCF uses 'default handling' associated with the iFC to determine call termination or call continuation based on FC information; in the absence of failure instructions, the S-CSCF allows the call to continue as default.

6.3.32.2 Mobile Terminated (MT)

As indicated above, the S-CSCF only looks for iFC when receiving an initial request, even if a terminating initial request originates from an AS via the ISC interface. Thus, if this request is a terminating request, the S-CSCF:

1. downloads the relevant subscriber profile including the initial filter criteria from the HSS if unavailable;
2. uses the iFC for the MT request to registered/unregistered user;
3. subsequent S-CSCF requirements (i.e. 3, 4, etc.) are the same as those for handling MO requests.

An originating UE and terminating UE may share the same S-CSCF and AS; hence the shared AS may interact with the S-CSCF twice in one transaction but in originating and terminating procedures, respectively [9].

6.3.33 IMS Session Release Handling

When handling session release, the S-CSCF may either proxy the release request or initiate a release request itself. In the 1st case the S-CSCF receives a release request from some entities (application server, user agent etc.) for a dialogue, it proxies the release request to the destination according to route information in that release request. In the 2nd case, the S-CSCF sends a release request to all the entities that are involved in this dialogue, e.g. AS and UE. Figure 6.11 illustrates the two release cases.

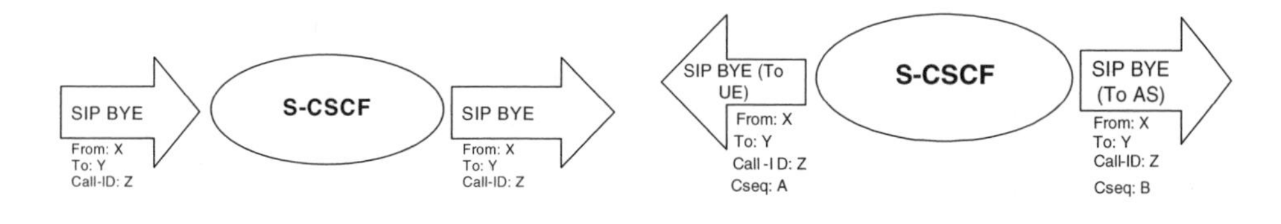

Figure 6.11 IMS session release handling [9].

6.3.34 Subscription and Notification Handling

Through the UE, P-CSCFs and APs using mechanisms specified in IETF RFC 3265 [19] the S-CSCF supports subscription to and notification of user registration events. The subscribing entity may subscribe to the registration state of individual public user identities in order to discover the implicitly registered public user identities. When notifying a subscribing entity of a change in the registration state of a subscribed to public user identity, the S-CSCF includes in the notification all the implicitly registered public user identities associated with the registered public user identity, in addition to the registered public user identity [9]. Figure 6.12 illustrates briefly the subscriber dialogue.

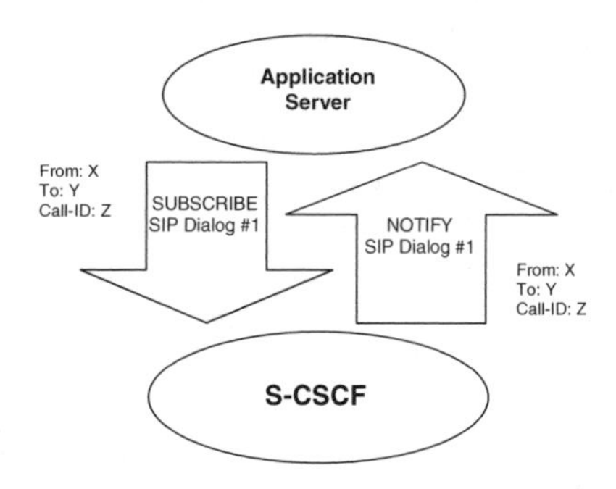

Figure 6.12 Application Server—S-CSCF subscribe notify dialogue [9].

6.3.35 IMS Charging

Before we outline the handling of IMS charging, we will briefly describe the correlation information and summarise them afterwards.

6.3.35.1 Charging Correlations

IMS generates and retrieves the following charging correlation information for use on offline and online charging [18]:

1. IMS Charging Identifier (ICID);
2. Access network information:
   a. GPRS Charging Information;
3. Inter Operator Identifier (IOI);
4. Charging function addresses:
   a. Charging Collection Function (CCF);
   b. Event Charging Function (ECF).

The P-Charging-Vector header includes the encoded charging correlation information, which contains the following parameters: ICID, access network information and IOI.

The offline and online charging function addresses are encoded in the P-Charging-Function-Addresses, which contain the CCF and ECF parameters.

6.3.35.2 Handling IMS Charging

The S-CSCF may send a 3rd party REGISTER message to an AS during registration, which may include the IMS Charging Identifier (ICID), Inter Operator Identifier (IOI) and charging function addresses. Thus, while in a session, the S-CSCF generates charging CDRs.

- In a session-originating case, the S-CSCF stores[11] the session from incoming initial request generated by the up-stream P-CSCF, which serves the originating user.
  - After processing this request the S-CSCF includes the ICID and charging function addresses received from the HSS in the outgoing message. These addresses identify on-line and off-line charging entities in the home network.
  - Implementation depends on how IMS related entities such as P-CSCF in the visited network get the local Charging Collection Function (CCF) addresses when the P-CSCF resides in the visited network.
  - Charging function addresses may be locally pre-configured addresses, where:
    - if sent outside the mobile network, the S-CSCF includes Inter Operator Identifier[12] (IOI) that identifies the home network into the message;
    - the response to the outgoing message may contain a separate IOI that identifies the home network of the called party. The S-CSCF retains either IOI in the message when contacting the Application Servers;
    - the S-CSCF receives GPRS charging[13] information from subsequent requests and responses. It stores these parameters and removes them from the outgoing message,

[11]Based on filter criteria.
[12]The IOI is globally unique identifier used fro inter-operator accounting purposes.
[13]The GPRS charging information may be sent to application servers.

if this message goes to the terminating UE's home network or the originating UE's visited network.

- In a session terminating case, the incoming initial request carries the ICID generated by the originating UE's P-CSCF while the S-CSCF stores the ICID for this session and handles this request based on Filter Criteria.
  - Once the request is processed, the S-CSCF includes the ICID and function addresses received from the HSS in the outgoing message.
  - As in the session-originated case, the addresses identify on-line and off-line charging entities in the home network.
  - The remaining steps follow as above, except that an IOI may be received from another network or is inserted by the MGCF to identify the originating PSTN/PLMN (see more details in [18]).

6.3.35.3 IMS Charging Identifier (ICID)

The ICID stands as the *session level data* shared among IMS entities including Application Servers (ASs) in both the calling and called IMSs.

The first IMS entity involved in a dialogue (session) or standalone (non-session) method will generate the ICID and include it in the icid parameter of the P-Charging-Vector (PCV) header in the SIP request to set format ICID requirements [20]. Table 6.4 illustrates the other entities, which can also generate an ICID.

Table 6.4 Entities generating ICIDs

Entity	Condition
P-CSCF	Mobile Originated (MO) calls if not ICID in initial request
	REGISTER request passed at unique instance of PCV head[a]
I-CSCF	Mobile Terminated (MT) calls if not ICID in initial request[b]
APs	When acting as an originating UA
MGCF	For PSTN/PLMN originated calls
I-CSCF/S-CSCF	For MT when received from another network

[a]This ICID is valid only for the duration of the registration and is associated with the signalling PDP context.

[b]For example, when the calling party network does not behave as an IMS.

Each entity that processes the SIP request will extract the ICID for possible later use in a CDR. Any requests that include the P-Charging-Vector header contain the icid parameter. However, the P-Charging-Vector (and ICID) is not passed to the UE. Likewise, ICID passes from the P-CSCF/PDF to the GGSN, but it does not pass to the SGSN.

6.3.35.4 Access Network Information (ANI)—GPRS Charging Info

The ANI corresponds to the media component level data shared among the IMS entities for one side of the session (either the calling or called side). It includes, e.g. GPRS charging information (i.e. GGSN identifier and PDP context information).

The GGSN, at the first opportunity after the resources allocation, provides the GPRS charging information to IMS, which is the common information used to correlate GGSN-CDRs with IMS subsystem CDRs. Thus, GPRS charging information passes from GGSN → P-CSCF/PDF → S-CSCF → AS[14] and gets updated with new information during the session as media streams is added or removed.

The GPRS charging information for the originating network is used only within that network, and similarly the GPRS charging information for the terminating network is used only within that network. Thus, the GPRS charging information are not shared between the calling and called networks. The GPRS charging information is not passed towards the external ASs from its own network. The GPRS charging information gets populated in the P-Charging-Vector using the GPRS-charging-info parameter [18].

6.3.35.5 The Inter Operator Identifier (IOI)

Operators networks/service providers/content providers share the IOI as unique globally identifier. They exchange the IOI, e.g. at the originating side, *orig-ioi*, and one for the terminating side, *term-ioi*.

The originating network S-CSCF populates the P-Charging-Vector *orig-ioi* parameter of the initial request,[15] which identifies the operator network from which the request originated. The S-CSCF in the originating network also retrieves the *term-ioi* parameter from P-Charging-Vector header response message sent to the initial request, which identifies the operator network from which the response was sent.

The terminating network S-CSCF retrieves the *orig-ioi* parameter from the P-Charging-Vector header in the initial request, which identifies the operator network from which the request originated. The S-CSCF in the terminating network populates the *term-ioi* parameter of the P-Charging-Vector header in the response to the initial request, which identifies the operator network from which the response was sent [18].

The MGCF populates the *orig-ioi* parameter for PSTN/PLMN originating call/sessions, as well as the *term-ioi* parameter for PSTN/PLMN terminating call/sessions.

IOIs do not pass along within the network, except when proxied by BGCF and I-CSCF to get to MGCF and S-CSCF. However, IOIs will be sent to the AS for accounting purposes [18].

6.3.35.6 IMS Charging Function Addresses

IMS entities have charging function in the home network for one side of the session (either the calling or called side) and provide a common location for each entity to send charging information. These functions with multiple addresses populated into the P-Charging-Function-Addresses header of the SIP request or response, with *ccf* and *ecf* parameters[16] include:

[14]Which may be needed for online pre-pay applications.
[15]The *term-ioi* parameter does not show in the initial P-Charging-Vector parameter request.
[16]Which may be redundant.

1. the Charging Collection Function (CCF) addresses used for offline billing;
2. Event Charging Function (ECF) addresses used for online billing.

We obtain the CCF and ECF addresses from an HSS via the Cx interface passing through the S-CSCF to subsequent entities. The charging function addresses pass from the S-CSCF to the IMS entities in their home network, but do not pass to the visited network or the UE. We get charging function addresses once the P-CSCF gets allocated in the visited network.

The S-CSCF provides charging function addresses to the AS through the ISC interface, where these CCF and/or ECF addresses may be allocated as locally pre-configured addresses. The AS may also retrieve the charging function address from the HSS through the Sh interface [18].

6.3.36 Transport Signalling Gateway Function (T-SGW)

This component serves as the PSTN/PLMN termination point for a defined network. Terminates, e.g. the call control signalling from GSTN mobile networks (typically ISDN) and maps the information onto IP (SIGTRAN) towards the Media Gateway Control Function (MGCF). The functionality defined within T-SGW should be consistent with existing/ ongoing industry protocols/interfaces that will satisfy the following requirements:

- maps call related signalling from/to PSTN/PLMN on an IP bearer and sends it to/from the MGCF;
- needs to provide PSTN/PLMN $\leftrightarrow$ IP transport level address mapping.

6.3.37 Roaming Signalling Gateway Function (R-SGW)

The role of the R-SGW concerns only roaming to/from 2G/R99 CS and the GPRS domain to/from the R5-6 UMTS teleservices domain and the UMTS-GPRS domain and does not involve the multimedia domain. According to [7] the main functions are:

- to ensure proper roaming, the R-SGW performs the signalling conversion at transport level (conversion: Sigtran SCTP/IP vs. SS7 MTP) between the legacy SS7 based transport of signalling and the IP-based transport of signalling. The R-SGW does not interpret the MAP/CAP messages but may have to interpret the underlying SCCP layer to ensure proper routing of the signalling;
- to support 2G/R99 CS terminals: we use R-SGW services to ensure transport inter-working between the SS7 and the IP transport of MAP_E and MAP_G signalling interfaces with a 2G/R99 MSC/VLR.

6.3.38 Media Gateway Control Function (MGCF)

The MGCF serves as the PSTN/PLMN termination point for a defined network. Its defined functionality will satisfy the standard protocols/interfaces to:

- control parts of the call state that pertain to connection control for media channels in a MGW;
- communicate with CSCF;
- select the CSCF depending on the routing number for incoming calls from legacy networks;
- perform protocol conversion between the legacy (e.g. ISUP, R1/R2, etc.) and the R00 network call control protocols;
- assume reception out-of-band information for forwarding to the CSCF/MGW.

6.3.39 Media Gateway Function (MGW)

The MGW serves as the PSTN/PLMN transport termination point for a defined network and UTRAN interfaces with the CN over Iu. It may terminate bearer channels from a switched circuit network (i.e. DSOs) and media streams from a packet network (e.g. RTP streams in an IP network). Over Iu, the MGW may support media conversion, bearer control and payload processing (e.g. codec, echo canceller, conference bridge) for support of different Iu options for CS services, AAL2/ATM based as well as RTP/UDP/IP based. The main functions include:

- interaction with MGCF, MSC server and GMSC server for resource control;
- ownership and resources handling, e.g. echo cancellers, etc.;
- ownership of codecs.

The MGW will have the necessary resources to support UMTS/GSM transport media. It will also have customised H.248 packages to support additional codecs and framing protocols, etc. from other networks besides GSM and UMTS. The MGW bearer control and payload processing capabilities will also support mobile specific functions, e.g. SRNS relocation/ handover and anchoring through H.248 protocol enabling. The following principles apply to the CS-MGW resources:

- It shall not be necessary to have the CS-MGW co-located with the MSC server.
- The CS-MGW resources need not be associated with any particular MSC server.[17]
- It shall be possible for any MSC server to request resources of any CS-MGW in the network.[1]
- It shall be possible for an RNC to connect to the CS-MGW indicated by the MSC mserver.

[17]Extensions to H.248 may be required.

6.3.40 Multimedia Resource Function (MRF)

The MRF performs:

- multiparty call and multimedia conferencing functions, i.e. would have the same functions as an MCU in an H.323 network;
- performs bearer control (with GGSN and MGW) in cases of multiparty/multimedia conferencing;
- communication with the CSCF for service validation and for multiparty/multimedia sessions.

6.3.41 MSC and Gateway MSC Server

The MSC server includes mainly the call control and mobility control parts of a GSM/ UMTS MSC. It has responsibility for the control of MO and MT 04.08CC CS domain calls. It terminates the user-network signalling (04.08 + CC + MM) and translates it into the relevant network–network signalling. The MSC server also contains a VLR to hold the mobile subscriber's service data and CAMEL-related data, controls the parts of the call state that pertain to connection control for media channels in an MGW [7].

The GMSC server comprises primarily the call control and mobility control parts of a GSM/UMTS GMSC. An MSC server and an MGW make up the full functionality of an MSC, while the Gateway MSC and a GMSC server and an MGW make up the full functionality of a GMSC.

6.4 MOBILITY MANAGEMENT

6.4.1 Address Management

We can implement a UMTS network as a number of logically separated IP networks, which contain different parts of the overall system. Here we refer to these elements as an IP Addressing Domain. In an IP addressing domain we expect to have nodes with non-overlapping IP address space and be able to route IP packets from any node in the domain to any other node in the domain by conventional IP routing. An IP addressing domain implementation can take place through a physically separate IP network or an IP VPN.

We can interconnect the IP addressing domains at various points where gateways, firewalls or NATs may be present. However, we do not guarantee that IP packets from one IP addressing domain can be directly routed to any interconnected IP addressing domain. Instead inter-domain traffic will most likely be handled via firewalls or tunnels. Therefore, different IP addressing domains can have different (and possibly overlapping) address spaces [7]. Figure 6.13 illustrates the IP addressing domains involved in PS domain and IP subsystem services.

UMTS allows usage of different IP addressing domains as shown in Figure 6.13; nonetheless, it is possible that several different IP addressing domains come under a

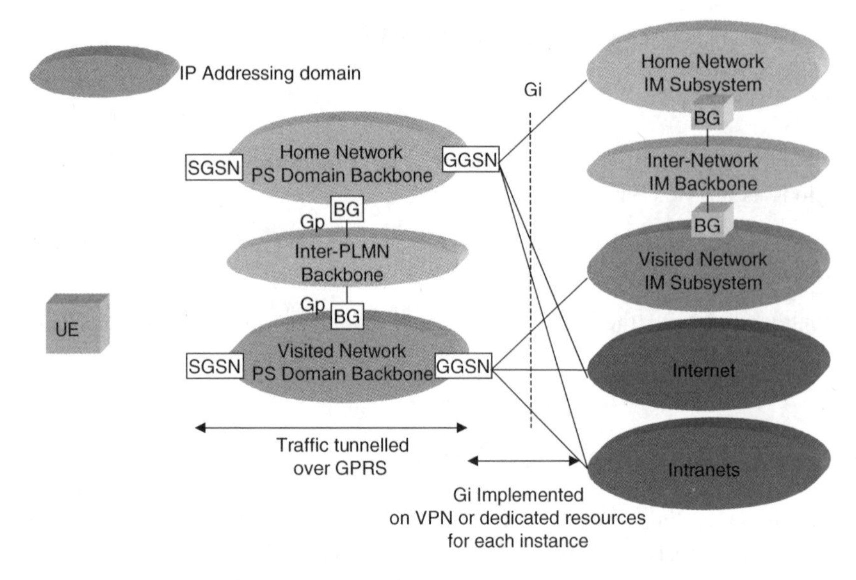

Figure 6.13 IP addressing domains involved in PS domain and IM services [7].

common management. Hence, we can physically implement the different IP addressing domains as a single domain.

6.4.2 Addressing and Routing to Access IM-Subsystem Services

When a UE gets access to IM subsystem services, an IP address is required, which is logically part of the visited network IM subsystem IP addressing domain. We established this address using an appropriate PDP context, and for routing efficiency this context gets connected though a GGSN in the visited network. Figure 6.14 illustrates the connection between the UE and the visited network IM subsystem.

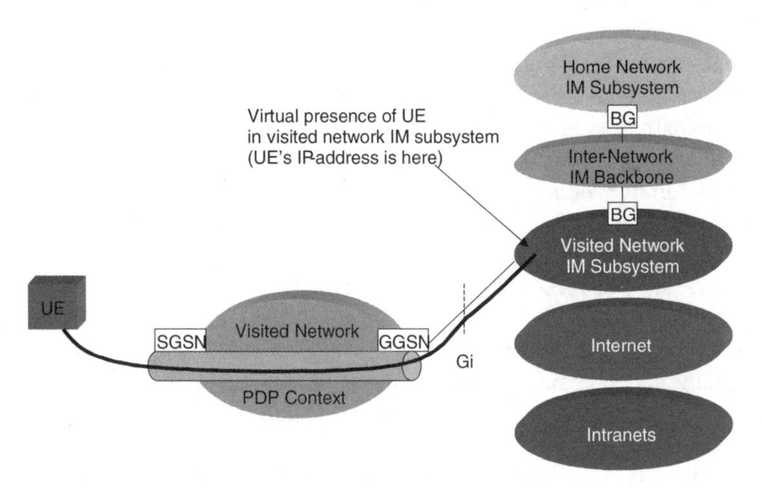

Figure 6.14 UE accessing IM subsystem services in the visited network.

6.4.3 Context Activation and Registration

An IP address allocated to a UE either by GPRS or some other means, e.g. by DHCP, can get used for (but not limited to) the following [7]:

- the exchange application level signalling (e.g. registration, CC) with the serving CSCF from the access network currently used;
- application level registration to an IP MM-CN subsystem as an address used to reach the UE;
- an address used to reach the UE for multimedia calls.

In GPRS, we associate the terminal with an IP address when we activate the primary PDP context. This IP address used for the purpose described above can be:

- the IP address obtained by the UE during the activation of a primary PDP context (e.g. if the UE does not have any existing PDP context active or desires to use a different IP address);
- the IP address of one of the already active PDP contexts.

Figure 6.15 illustrates the order in which we execute the registration procedure and how the IP address gets allocated.

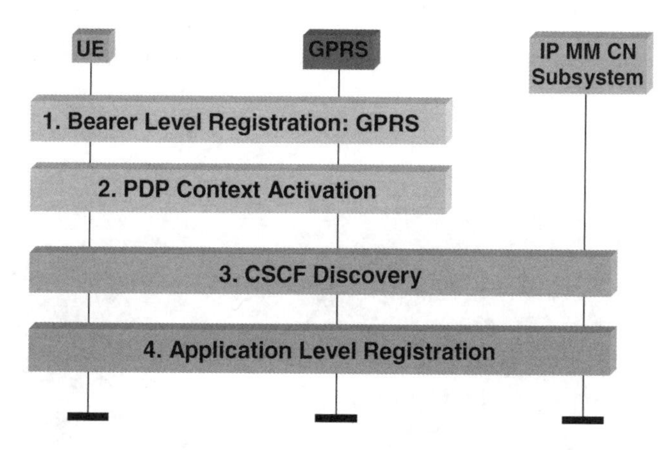

Figure 6.15 Registration of the IP address.

The steps performed include:

1. bearer level registration (e.g. after an MS gets switched on or upon explicit user demand);
2. when the PDP context gets activated, the UE has two options:
   - activate a primary PDP context and obtain a new IP address (e.g. if the UE does not have any existing PDP context active or desires to use a different IP address),

- activate a secondary PDP context and re-use the IP address of one of the already active PDP contexts.

3. UE performs the CSCF discovery procedure to select the CSCF to register with.

   The procedures can have time gaps between them, e.g. the UE may perform PDP context activation and the CSCF discovery, but not the application level registration. The UE may use the activated PDP context for other types of signalling, e.g. for CSCF discovery [7].

4. the UE performs application level registration by providing the IP address obtained at step 2 to the CSCF selected at step 3.

In the last step, the signalling IP address gets allocated in association with PDP context activation and not on an incoming call basis, then the selected CSCF becomes the serving CSCF.[18] From the point of view of the latter, the IP address provided by the UE corresponds to the address where the UE is reachable for MT call control signalling and/or any other type of MT signalling. Whether a procedure gets activated individually by the UE or automatically depends on the implementation of the terminal and on the UE's configuration. For example, a UE multimedia application may start the application level registration and steps 2–4 would need to follow in response to support the operation initiated by the application.

6.4.4 Location Management

Figure 6.16 illustrates the registration concept for a R00 subscriber roaming into a UMTS/GSM CN domain.

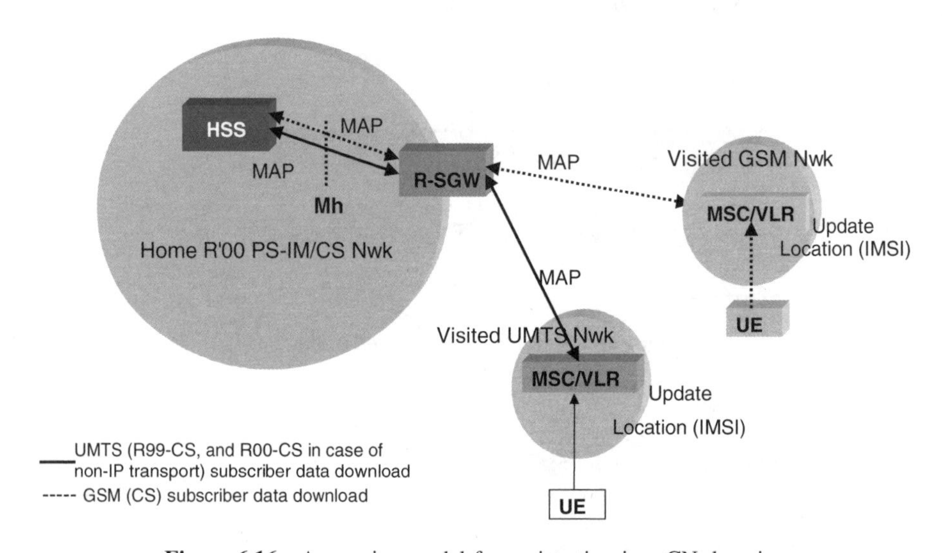

Figure 6.16 A roaming model for registration in a CN domain.

[18]Note that the S-CSCF can be either in the home or a visited network.

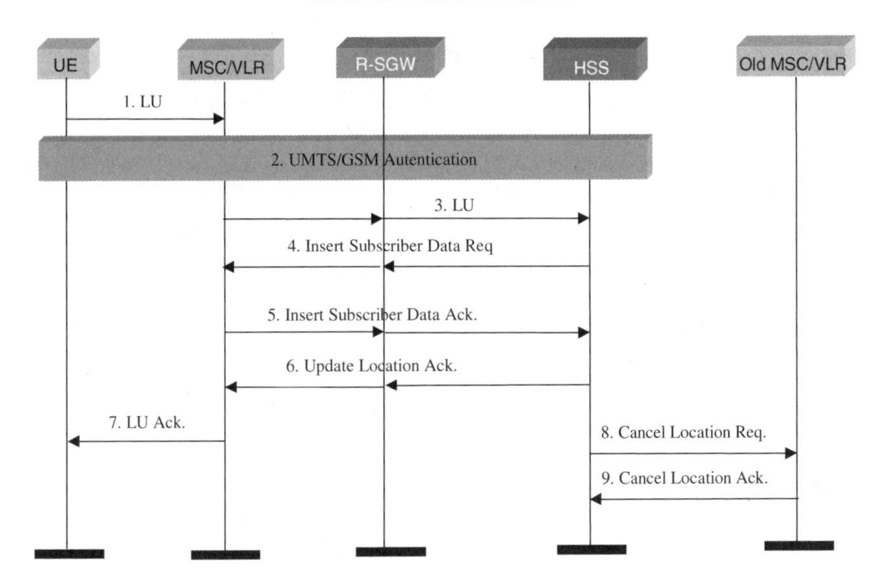

Figure 6.17 Message sequence for roaming into a CN domain [7].

From [7], Figure 6.17 illustrates the detailed message sequence chart for a UMTS R00 subscriber roaming into a CN domain. The sequence can be summarised as follows:

1. The UE initiates the UMTS R99/GSM Location Update (LU) procedure with the MSC/VLR of the visited network, where the LU message contains the IMSI of the subscriber.
2. The UMTS/GSM authentication gets performed as per the existing UMTS R99/GSM specifications.
3. The MSC/VLR initiates the MAP location update procedure towards the HSS of the user via R-SGW. The HSS stores the VLR address, etc. The message contains IMSI and other parameters as defined in UMTS R99/GSM specifications. The message is passed through the R-SGW transparently while the SS7 to/from IP conversion is performed in the R-SGW.
4. The HSS provides the subscriber data for the roaming user to VLR by sending MAP Insert Subscriber Data message via R-SGW. The message contains IMSI and other necessary parameters as defined in the UMTS/GSM specification. The message is passed through the R-SGW transparently while the SS7 to/from IP conversion is performed in R-SGW.
5. The serving VLR then acknowledges the receipt of the subscriber data to the HSS via R-SGW.
6. The HSS acknowledges the completion of location updating procedure to the MSC/VLR via R-SGW.
7. The MSC/VLR acknowledges the completion of location updating procedure to the UE.

8. The HSS sends the MAP cancel location message to the old MSC/VLR (optional procedure).
9. Location cancellation is acknowledged to the HSS by the old MSC/VLR [7].

The steps 8 and 9 above assume that the UE was previously registered to a CN domain. The MAP messages between the MSC/VLR and HSS get passed transparently via the R-SGW. The R-SGW does not interpret the MAP messages in any way, but performs only the lower level conversion between SS7 and IP.

6.4.5 Handover (HO)

For HO of CS services involving the change of CN equipment (only CS-MGW or CS-MGW and MSC-server) the anchor principle applies, i.e.

- The first MSC server involved in a call will become the anchor MSC server for this call during and after HO, and will remain in the call until the call gets released. Every subsequent HO (intra and inter) will then be controlled by this MSC server [7].
- The first CS-MGW involved in a call will become the anchor CS-MGW for this call during and after HO, and will remain in the call until the call is released. The Nc interface gets anchored in the CS-MGW, the correlation between MGW to PSTN and the MGW to UTRAN remain fixed until the call is released [7].

6.5 MULTIMEDIA SIGNALLING

Although multimedia signalling also remains under consolidation within the technical specification bodies of the 3GPP, we can still describe the initial principles based on [7] and extrapolated for R4 and R5 as follows:

- *A single call control between the UE and CSCF* → For multimedia type services delivered via the PS domain within R4 and R4 architectures, we aim to use a single call control protocol between the user equipment UE and the CSCF (over the Gm reference point).
- *Protocols over the Gm Reference Point* → The single protocol applied between the UE and CSCF (over the Gm reference point) within the R4 and R4 architectures will be based on SIP (as defined by RFC 2543, other relevant RFCs and additional enhancements required to support 3GPP's needs).
- *A single call control on the Mw, Mm, Mg* → We aim to use single call control protocol on the call control interfaces between MGCF and CSCF, between CSCFs within one operator's network and between CSCFs in different operators' networks.
- *Protocols for the Mw, Mm, Mg* → We aim to apply SIP[19] for the single call control protocol applied to the interfaces between MGCF and CSCF, between CSCFs within one operator's network and between CSCFs in different operator's networks.

[19]As defined by RFC 2543, other relevant RFCs, and additional enhancements required to support 3GPP's needs.

- UNI versus NNI call control → We may assume that the SIP-based signalling interactions between CN elements may be different than SIP-based signalling between the UE and the CSCF in some cases if not all.

6.5.1 Support of Roaming Subscribers

Here we assume that R4 and R5 architectures will be based on the principle that the service control for a roaming subscriber is designated by the home network. The serving CSCF can then be located either in the home network or in the visited network as illustrated in Figure 6.18. This assignment of the serving CSCF takes place in the home network during the registration of the UE at the visited network.

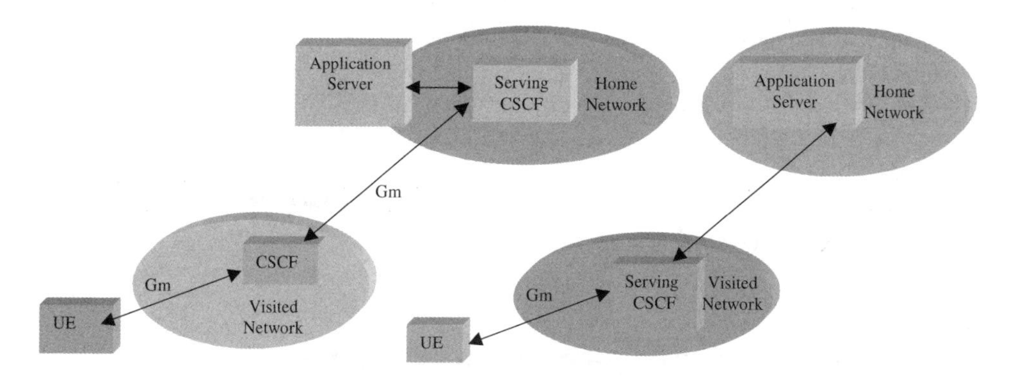

Figure 6.18 Support of the UE via serving CSCF in the home and visited networks.

The R4 and R5 standards will thus support roaming for IP-Multimedia (IM) users between operators. As noted earlier, to achieve roaming the required serving-CSCF can be located in the visited network or in the home network. The decision as to whether the UE is served by a serving CSCF in the home network or a serving CSCF in the visited network occurs at the home network. According to [7] the following takes place:

- When subscribers roam to networks where a serving CSCF does not exist, the roamed to (visited) network will support a proxy CSCF. The proxy CSCF will enable the call control to be passed to the home network-based serving CSCF, which will provide service control.
- When subscribers roam to networks where a serving CSCF exists but the home network decides to use a home-network-based serving CSCF, the roamed to (visited) network will support a proxy CSCF. The proxy CSCF will enable the call control to be passed to the home network-based serving CSCF, which will provide service control.
- When subscribers roam to networks where a serving CSCF exists and the home network decides to use the visited network-based serving CSCF solution, the visited network serving CSCF will be used to provide service control to the roamed subscriber.

While the visited network may support a serving CSCF for inbound roamers, it will usually support proxy CSCF for inbound roamers. Thus, if a visited network decides not to offer serving CSCF capability for inbound roamers, then the home network will provide a serving CSCF to support IM roaming.

The home network may provide a serving CSCF for outbound roamers even when a visited network offers the support of a serving CSCF; if so, the visited network provides the proxy CSCF. On the one hand, when users are within their home network, a home network-based serving CSCF provides service control. On the other hand, if the home operator wishes to use home service control for outbound roamers, then a home network-based serving CSCF will be used for outbound roamers' service control [7].

6.5.2 Assignment of Serving CSCF

The home network designates the serving CSCF in the home network or with the help of the visited network, requests a serving CSCF in the visited network. This selection occurs on a per subscriber basis at registration time based on consideration of at least the following factors [7]:

a. the service capabilities and toolkits supported by the visited network and the home network;

b. the subscription profile of the subscriber.

6.6 IMS PROCEDURES AND SESSION FLOWS

Here we address the main procedures for the service provisioning in IMS. The procedures presentation follows closely the text description as well as information flow diagrams applied in [9], for which additional procedures and details can be found in [21].

Key sessions between mobile users will involve two S-CSCFs (one S-CSCF for each), and a session between a user and a PSTN endpoint involves an S-CSCF for the UE, a BGCF to select the PSTN gateway and an MGCF for the PSTN.

The session flow is decomposed into three parts:

1. *An Origination Part* → covers all network elements between the UE (or PSTN) and the S-CSCF for that UE (or MGCF serving the MGW);
2. *An Inter-Serving-CSCF/MGCF part* and
3. *The termination part* → covers all network elements between the S-CSCF for the UE (or MGCF serving the MGW) and the UE (or PSTN).

6.6.1 CSCF—Establishing PDP Context for IMS Signalling

PDP context activation precedes UE signalling request for IMS services. The UE conveys to the network the intention of using the PDP context for IMS signalling by using the mechanism for 'PDP Context Used for Application Level Signalling Transport' as described in [22]. A signalling flag determines any rules and restrictions that apply at the GGSN for that PDP context, while following the appropriate IMS QoS profile [23] parameters.

6.6.2 Local CSCF Discovery Procedures

The Proxy-CSCF discovery occurs after GPRS attach and/or as part of a successful activation of a PDP context for IMS signalling using one of the following mechanisms:

1. DHCP usage to provide the UE with the Proxy-CSCF domain name and Domain Name Server (DNS) address capable of resolving the Proxy-CSCF name.
2. Transfer a Proxy-CSCF address within the PDP Context Activation signalling to the UE, where the latter requests the P-CSCF address(es) from the GGSN when activating the PDP context. The GGSN sends the P-CSCF address(es)[20] to the UE when accepting the PDP context activation.

6.6.2.1 DHCP/DNS Procedure for P-CSCF Discovery

As illustrated in Figure 6.19, the GGSN acts as a DHCP Relay Agent, relaying DHCP messages between UE and the DHCP server.

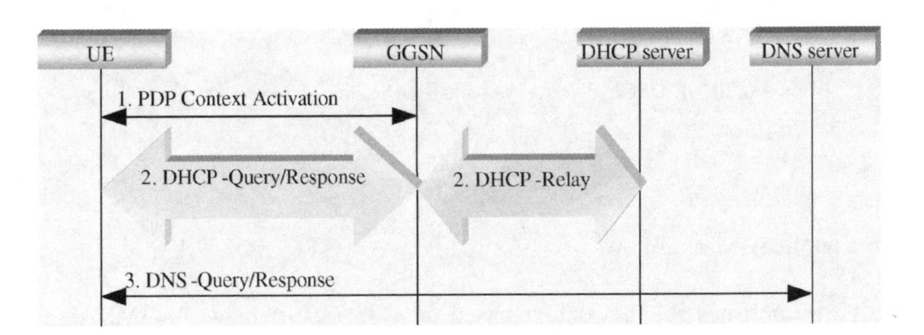

Figure 6.19 P-CSCF discovery using DHCP and DNS.

The interaction sequence goes as follows:

1. Create PDP context bearer through standard the procedure TS 23.060 [23].
2. The UE requests[21] a DHCP server and in addition it requests the domain name of the P-CSCF and IP addresses of DNS servers.

[20]Both, the P-CSCF address(es) request and the P-CSCF address(es) are sent transparently through the SGSN.
[21]It may require a multiple DHCP Query/Response message exchange to retrieve the requested information.

3. The UE performs one or more DNS queries[22] to retrieve a list of P-CSCF(s) IP addresses from which one is selected.

Once the UE gets the domain name and IP address of a P-CSCF, it may initiate communication towards the IMS.

6.6.2.2 *GPRS Procedure for P-CSCF Discovery*

As illustrated in Figure 6.20, this option applies to UE(s) not supporting DHCP, but it may be also used by UE(s) supporting DHCP.

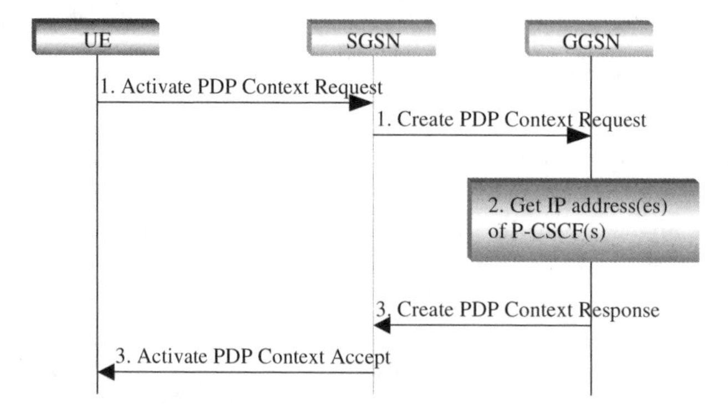

Figure 6.20 P-CSCF discovery using PDP Context Activation signalling.

It may be summarised as follows:

1. The UE establishes a PDP context based on QoS requirements for IMS signalling to request P-CSCF IP address(es) to the GGSN (transparently via the SGSN).
2. The GGSN supplies the P-CSCF(s) IP address(es) based on a mechanism of choice, which is internal configuration implementation dependent.
3. The GGSN may include the P-CSCF(s) IP address(es) in the Create PDP Context Response upon UE request, which are forwarded transparently[23] by the SGSN.

As in the preceding option, once the UE gets P-CSCF IP address, it may initiate communication towards the IMS.

[22] Additional DNS query may be needed to resolve a Fully Qualified Domain Name (FQDN) to an IP address.
[23] Non-transparent for pre-R5 SGSN when using the Secondary PDP Context Activation Procedure [23–43].

6.6.3 Serving-CSCF Assignment Procedures

6.6.3.1 Assigning a Serving-CSCF to a User

Once the UE attaches and makes itself available to IMS through registration, it gets a S-CSCF assigned, which is performed in the I-CSCF with the following information [21]:

1. Required user services capabilities provided by the HSS.
2. Operator preference on a per-user basis provided by the HSS.
3. Home network internal information on individual S-CSCFs capabilities, which may be used in the S-CSCF selection.
4. Home network internal information on user location through topological P-CSCF, which may be used in the S-CSCF selection. The P-CSCF name is received at registration request.
5. S-CSCF location through topological home network internal information, which is obtained by the I-CSCF and may be used in the S-CSCF selection.
6. S-CSCFs availability also as home network internal information is obtained by the I-CSCF and may be used in the S-CSCF selection.

To support the S-CSCF selection just described and to enable the S-CSCF to perform its tasks, the following information must be exchanged between the CSCF and the HSS:

1. The Cx reference point supports the transfer of CSCF-UE security parameters from HSS to CSCF.
   - This affords the CSCF and the UE to communicate in a trusted and secure manner because there is no *à priori* trust relationship between them.
   - The security parameters can be, for example, pre-calculated challenge-response pairs, or keys for an authentication algorithm, etc.
2. Likewise, the Cx reference point supports the transfer of service subscriber parameters from HSS to CSCF.
   - Which may include, e.g. service parameters, application server address, triggers, information on subscribed media etc. in profile identifier[24] form.
3. The Cx reference point supports CSCF transfer capability information from CSCF to HSS.
   - This may include, e.g. supported service set, protocol version numbers, etc.
4. The Cx reference point supports session-signalling transfer of transport parameters from CSCF to HSS. The latter stores the signalling transport parameters utilised for routing mobile-terminated sessions to the Serving-CSCF.
   - The parameters may include, e.g. IP-address and CSCF port number transport protocol, etc.

[24]Allowed media detail parameters associated with the profile identifier are configured in the S-CSCF.

The information in the last 4 items above gets transferred before the CSCF can serve the mobile user, and it may be updated while the CSCF is in service, e.g. new services activation.

6.6.3.2 Serving-CSCF Assignment Cancellation and Reassignment

Assigned Serving CSCF may:

- be initiated at the Serving CSCF itself, e.g. as a result of registration time out;
- occur as result of UE explicit de-activation/de-registration from the IMS;
- take place at HSS request over the Cx interface, e.g. due to subscription changes.

S-CSCF re-assignment may occur:

- due to non-availability of previously assigned S-CSCF during registration;
- at the initial registration, after S-CSCF allocation to an unregistered user.

6.6.4 Interrogating- and Proxy-CSCF Procedures

The IMS architecture supports multiple I-CSCFs, which can be selected by a DNS-based mechanism to allow I-CSCF forward requests, e.g. forwarding node location or identity.

Information flows of SIP routing registration do not take into account previous registrations (i.e. registration state). However, the session routing information flows (e.g. INVITE) do take into account information received during the registration process [21].

6.6.5 Subscription Updating Procedures

After a modification occurs in the subscription data related to the S-CSCF, HSS sends the complete subscription data set to the S-CSCF by using the downloading Push model. Figure 6.21 illustrates this flow, where:

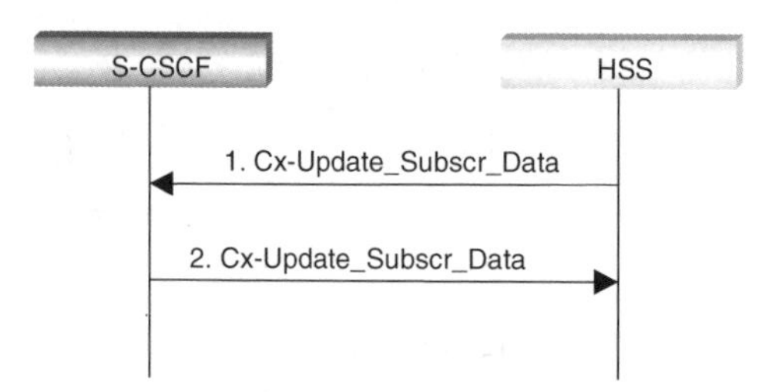

Figure 6.21 S-CSCF data update.

1. The HSS sends the Cx-Update_Subscr_Data with the subscription data to the S-CSCF.
2. The S-CSCF sends Cx-Update_Subscr_Data Resp to the HSS to acknowledge the sending of Cx-Update_Subscr_Data.

6.6.6 Application Level Registration Procedures

Next, we address requirements and information flows related to IMS registration and list the corresponding assumptions according to direct extracts from [21].

6.6.6.1 Registration Requirements

The following points apply to the requirements of the registration procedures:

1. IMS allows the S-CSCFs to have different capabilities or access to different capabilities. For example, a VPN CSCF or CSCFs in different stages of network upgrade.
2. The network operator does not need to inform its network internal structure to another network except for peer agreement basis.
3. A network does not need to expose the explicit IP addresses of the nodes within its infrastructure (excluding firewalls and border gateways).
4. It is preferable that the UE uses the same registration procedure(s) within its home and visited networks.
5. It is desirable that the procedures within the network(s) are transparent to the UE, when it registers with the IMS.
6. The Serving-CSCF understands a service profile and the address of the Proxy-CSCF functionality.
7. The HSS supports the possibility to bar a public user identity from usage for non-registration IMS procedures. The S-CSCF enforces these IMS barring rules.
8. The implicit registration ensures that the UE, P-CSCF and S-CSCF have public user Identity(s) for all IMS procedures after an initial Temporary Public Identity registration.
9. It is possible to register multiple public identities via a single IMS registration procedure from the UE.

6.6.6.2 Implicit Registration

Applies to registering a set of public user identities via a single IMS registration of one of the public user identity's in that set.[25] Figure 6.22 illustrates the implicit registration and public user identities with the following characteristics:

a. HSS contains the set of public user identities that are part of implicit registration.
b. The Cx reference point between S-CSCF and HSS supports download of all public user identities associated with the implicit registration.

[25]No single public identity it is considered as a master to the other public user identities.

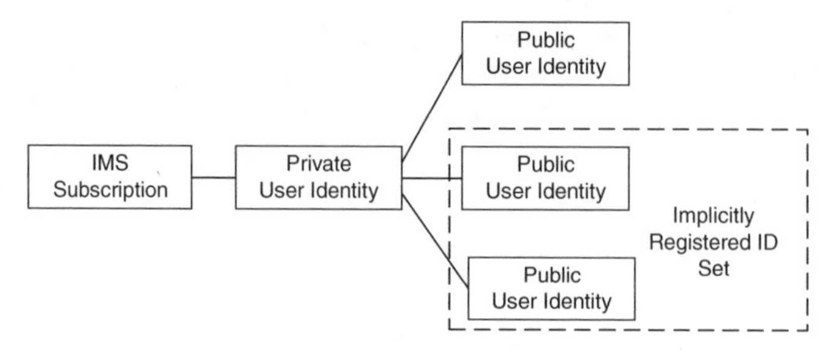

Figure 6.22 Relationship of public user identities when implicitly registered [21].

c. If one of the public user identities within the set gets registered, all public user identities associated with the implicit registration get also registered in parallel.

d. If one of the public user identities within the gets de-registered, all public user identities that have been implicitly registered are de-registered at the same time.

e. Public user identities belonging to an implicit registration set may point to different or the same service profile(s).

f. If a public user identity belongs to an implicit registration set, it cannot be registered/ de-registered individually without removal from the implicit registration list.

g. All IMS related registration timers apply to the set of implicitly registered public user identities.

h. S-CSCF, P-CSCF and UE are notified of the set of public user identities belonging to the implicitly registered function.[26]

i. When a public user identity is barred from IMS communications, only the HSS and S-CSCF shall have access to this public user identity.

6.6.6.2.1 Implicit UE Registration without ISIM
An ISIM-less[27] UE registration receives network assistance to register to at least one public user identity, which is used for session establishment and IMS signalling [21].

6.6.6.3 Registration Flows

A Serving-CSCF assignment at registration does not preclude additional Serving-CSCFs or change of CSCF late on [21]. The key assumptions are:

a. signalling radio bearers are already established with a first REGISTER message mechanism to forward to the proxy;

[26]Session set occurs only when all entities are updated, except for explicitly registered public user identity.
[27]Implicit registration is used as part of a mandatory function for these ISIM-less UEs registration.

b. the I-CSCF uses a Serving-CSCF address determining mechanism based on the required capabilities;

c. the S-CSCF selection decision for the user in the network occurs at the I-CSCF;

d. a role of the I-CSCF is the S-CSCF selection.

Next, we describe name and address mechanism resolution, where the text in the information flows indicates when the name-address resolution mechanism applies. Security aspects for these flows can be found in [24].

6.6.6.3.1 Registration Information Flow—User not Registered

Figure 6.23 illustrates the non-registered user flow with the following characteristics:

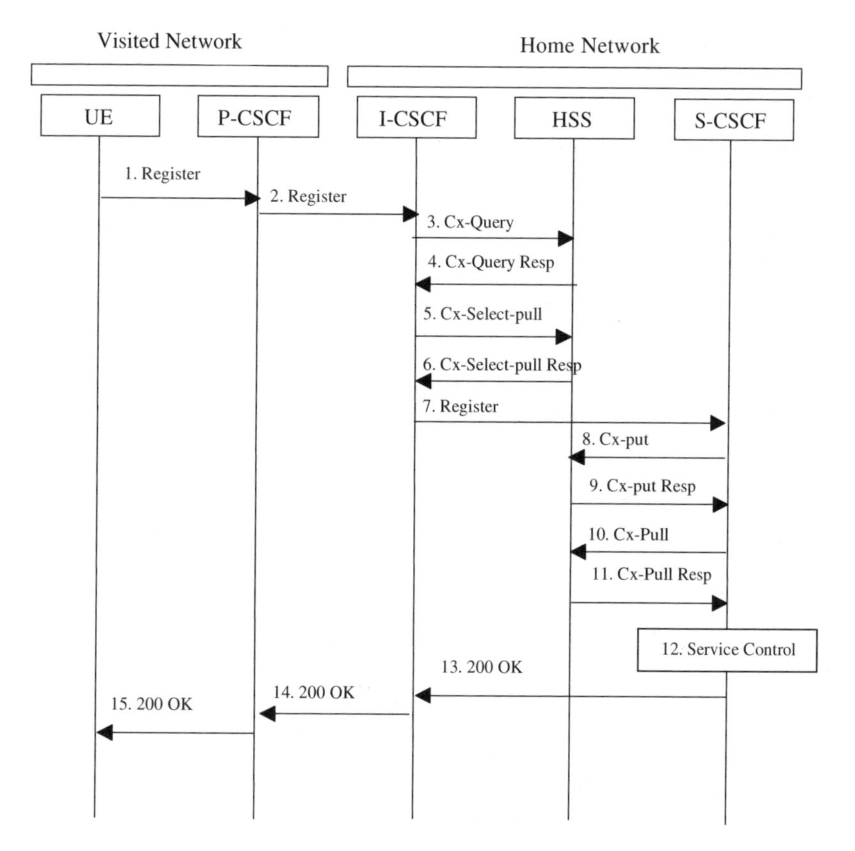

Figure 6.23 Registration—user not registered [21].

1. After the UE obtains a signalling channel through the access network, it performs the IMS registration.[28]

[28]UE sends to proxy: public user identity, private user identity, home network domain name, UE IP address.

2. Upon reception of register information flow, the P-CSCF examines the 'home domain name' to discover the home network (i.e. the I-CSCF) entry point. Then:
   - the proxy sends the *Register information flow* to the I-CSCF (i.e. the P-CSCF address/name, public user identity, private user identity, P-CSCF network identifier, UE IP address);
   - a name-address resolution mechanism determines the home network address from the home domain name identified by the P-CSCF network identifier, where the latter may be the domain name of the P-CSCF network.
3. The I-CSCF sends the Cx-Query information flow to the HSS (public user identity, private user identity, P-CSCF network identifier).

The HSS checks whether the user is registered already and indicates whether the user is allowed to register in that P-CSCF network.

4. HSS sends Cx-Query Resp to the I-CSCF. It contains the S-CSCF name, if it is known by the HSS, and the S-CSCF capabilities, if it is necessary to select a new S-CSCF. Negative HSS response → Cx-Query Resp rejects registration attempt.
5. When the I-CSCF does not get the S-CSCF name, it sends Cx-Select-Pull (public user identity, private user identity) to the HSS requesting S-CSCF capabilities.
6. On Cx-Select-Pull reception, the HSS sends Cx-Select-Pull Resp (required S-CSCF capabilities) to the I-CSCF.
7. The I-CSCF determines[29] the address of the S-CSCF utilising the S-CSCF name through a name-address resolution mechanism.
   - The home network contact point may either be the S-CSCF itself, or a suitable I-CSCF(THIG) when hidden network configuration is desired [21].
8. The S-CSCF sends *Cx-Put* (public user identity, private user identity, S-CSCF name) to the HSS, which stores the S-CSCF name for that user.
9. The HSS sends *Cx-Put Resp* to the S-CSCF to acknowledge the *Cx-Put* sending.
10. After receiving the *Cx-Put Resp* information flow, the S-CSCF sends the Cx-Pull information flow (public user identity, private user identity) to the HSS to retrieve key user profile information.
    - The S-CSCF stores the P-CSCF address/name, as supplied by the visited network, which receives subsequent terminating session signalling for the UE.
11. The HSS returns the user information[30] flow (+ security info) *Cx-Pull Resp* to the S-CSCF. The latter stores the information for the indicated user.
12. Based on the filter criteria, the S-CSCF sends register information to the service control platform and performs the appropriate service control procedures.
13. The S-CSCF returns the 200 OK information flow (home network contact information) to the I-CSCF. If an I-CSCF acts as the home network contact point for implementing

[29]The I-CSCF also determines a suitable home network contact point name from the HESS info.
[30]Which includes one or more names/addresses used to access the platform(s) used for service control.

hiding network configuration, it encrypts [25] the S-CSCF address in the home network contact information.

14. The I-CSCF sends information flow 200 OK (home network contact information) to the P-CSCF. It releases all registration information after sending information flow 200 OK.
15. The P-CSCF stores the home network contact information, and sends information flow 200 OK to the UE.

6.6.6.3.2 Re-Registration Information Flow—User Currently Registered

The UE initiates periodic application level re-registration either to refresh an existing registration or in response to a change EU registration status.[31] It does it to keep active sessions or prevent de-activation. Re-registration follows the same 'Registration Information Flow–User not registered' process, and occurs based on the registration time established during the previous registration.

Figure 6.24 illustrates the re-registration flow with the following characteristics:

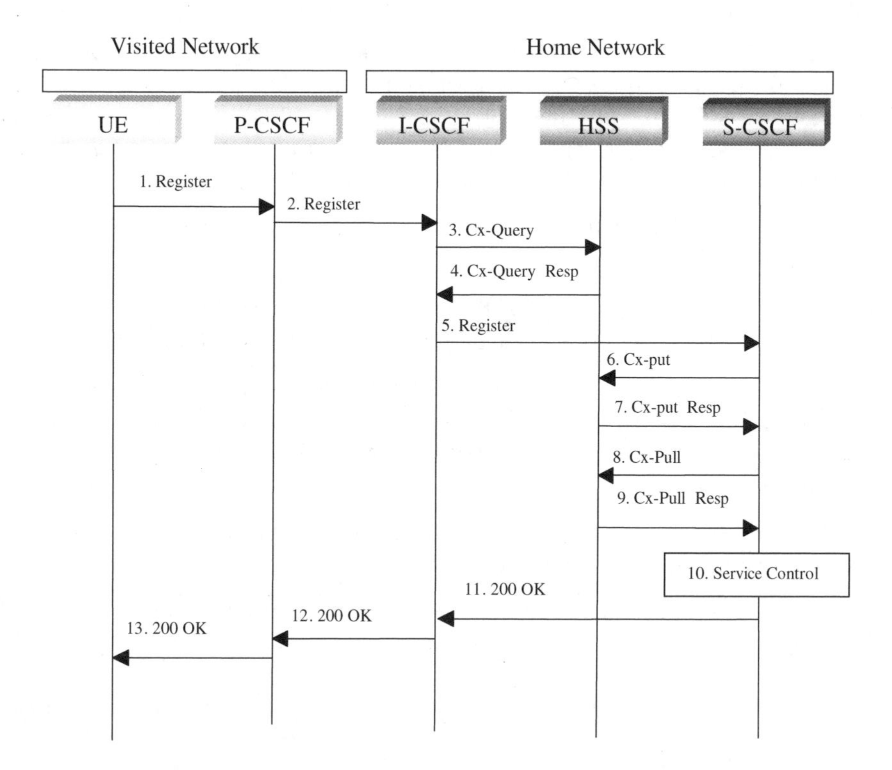

Figure 6.24 Re-registration—user currently registered [21].

[31]The UE keeps a timer shorter than the registration-related timer in the network.

1. UE initiates a re-registration prior to the expiration of the agreed registration timer by sending a new REGISTER request, in which the information flow sent to the proxy includes: public user identity, private user identity, home network domain name, UE IP address.
2. Upon REGISTER information flow reception, the P-CSCF examines the 'Home Domain Name' to discover the new[32] entry point to the home network (i.e. the I-CSCF) and sends it to the I-CSCF. Other features are as in the preceding section.
3. The I-CSCF sends *the Cx-Query information flow* to the HSS.
4. The HSS checks if user registration already exists and returns to the I-CSCF; the S-CSCF assignment indication through the *Cx-Query Resp*.
5. The I-CSCF applies a name-address resolution mechanism and using the S-CSCF name determines the S-CSCF address. It also determines a suitable *home-network contact point* name, which could be either the S-CSCF itself, or a suitable I-CSCF(THIG) in case hiding network configuration is desired. The rest is as in the preceding section.
6. The S-CSCF sends the *Cx-Put* (public user identity, private user identity, S-CSCF name) to the HSS, which stores[33] the S-CSCF name for that user.
7. The HSS sends *Cx-Put Resp* to the S-CSCF to acknowledge the *Cx-Put* sending.
8. Once the S-CSCF receives the *Cx-Put Resp*, it sends the *Cx-Pull*[34] (public user identity, private user identity) to the HSS in order to download key information from the user profile to the S-CSCF. The latter stores the P-CSCF address/name, as supplied by the visited network. This represents the address/name that the home network forwards the subsequent terminating session signalling to for the UE.
9. The HSS returns the *Cx-Pull-Resp* to the S-CSCF, which stores the user information for that indicated user.
10. Applying a filter criteria, the S-CSCF sends re-registration information to the service control platform and performs the appropriate service control procedures.
11. The S-CSCF returns the 200 OK information flow (home network contact information) to the I-CSCF. When the I-CSCF acts as the home network contact point to implement hidden network configuration, it encrypts the S-CSCF address in the home network contact information.
12. The I-CSCF sends information flow 200 OK (home network contact information) to the P-CSCF, and it releases all registration information.
13. The P-CSCF stores the home network contact information, and sends information flow 200 OK to the UE.

Table 6.5 provides an indication of the information stored in the indicated nodes during and after the registration process.

[32]The proxy does not use the entry point cached from prior registrations.
[33]Optionally, the S-CSCF can detect that this is a re-registration and omits the Cx-Put request.
[34]Optionally, the S-CSCF can detect that this a re-registration and omits the Cx-Pull request.

Table 6.5 Information storage before, during and after the registration process [21]

Node	Before registration	During registration	After registration
UE (in local network)	Credentials, home domain proxy name/address	Same as before registration	Credentials, home domain proxy name/address Same as before registration
Proxy-CSCF (in local network)	Routing function	Initial network entry point, UE address Public and private user IDs	Final network entry point UE address Public and private user IDs
Interrogating-CSCF (in home network)	HSS or SLF address	Serving-CSCF address/name P-CSCF network ID Home network contact information	No state information
HSS	User service profile	P-CSCF network ID	Serving-CSCF address/name
Serving-CSCF (home)	No state information	HSS address/name User profile[a] Proxy address/name P-CSCF network ID Public/private user ID UE IP address	May have session state information Same as during registration

[a]Limited, as per network scenario.

6.6.7 Application Level De-registration Procedures

6.6.7.1 Mobile Initiated De-registration

The UE performs application level de-registration to release an IMS connection, which transpires by a registration with an expiration time of 0 sec while following the same path as defined by the 'Registration Information Flow—user not registered' described in Section 6.7.6.3.1. Figure 6.25 illustrates the information flow as follows:

1. The UE initiates de-registration by sending to the proxy[35] a new *REGISTER information flow* request with an expiration value of 0 sec.
2. On reception of the *register information flow*, the proxy examines the 'home domain name' to discover the entry point to the home network (i.e. I-CSCF) as in step 2 of Section 6.7.6.3.1.
3. Then I-CSCF sends the Cx-Query information flow to the HSS (public user identity, private user identity, P-CSCF network identifier).
4. The HSS determines the subscriber registration by looking at the user identity presence and sends the *Cx-Query Resp* (entry point indication, e.g. S-CSCF) to the I-CSCF.

[35] i.e. public user identity, private user identity, home network domain name, UE IP address.

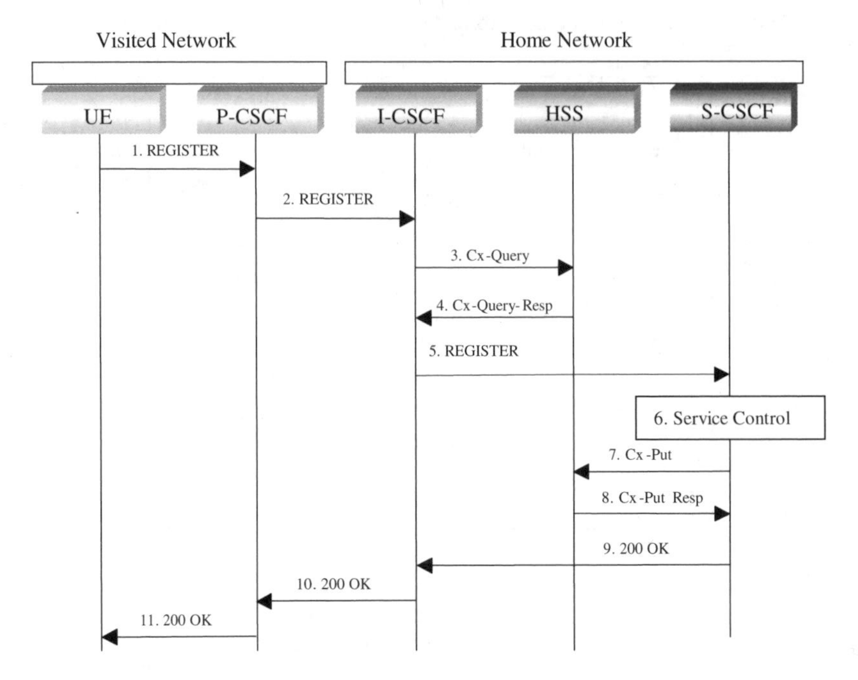

Figure 6.25 De-registration—user currently registered [21].

5. The I-CSCF applies the S-CSCF name and determines the S-CSCF address through a name-address resolution mechanism, then sends the de-register information flow (i.e. P-CSCF address/name, public user identity, private user identity, UE IP address, I-CSCF(THIG) if hidden network configuration is desired) to the S-CSCF.
6. The S-CSCF applies a filter criteria and sends de-registration information to the service control platform to perform appropriate service control procedures, like removing all subscription information related to this specific public user identity.
7. Following service provider preference, the S-CSCF may send either Cx-Put (public user identity, private user identity, clear S-CSCF name) or Cx-Put (public user identity, private user identity, keep S-CSCF name), and the *public user identity* is no longer considered registered in the S-CSCF.
   - The HSS then either clears or keeps the S-CSCF name for that *public user identity* according to request. In both cases the state of the public user identity is stored as unregistered in the HSS. If the S-CSCF name is kept, then the HSS shall be able to clear the serving S-CSCF at any time [21].
8. The HSS sends *Cx-Put Resp* to the S-CSCF to acknowledge Cx-Put sending.
9. The S-CSCF returns the 200 OK information flow to the I-CSCF and may also release all registration information regarding this specific registration of the public user identity after sending information flow 200 OK.
10. The I-CSCF sends information flow 200 OK to the P-CSCF.

11. The P-CSCF sends information flow 200 OK to the UE, and releases all registration information regarding this specific registration of the public user identity after sending information flow 200 OK.

6.6.7.2 *Network Initiated De-registration*

To ensure stable S-CSCF operation and carrier grade service, an SIP protocol[36] level mechanism handles *ungraceful* (i.e. flat battery or mobile leaves coverage) session termination. IMS starts *Network Initiated De-Registration* procedures for the following:

- *Network Maintenance* → force user re-registrations, e.g. in case of data inconsistency at node failure, in case of SIM lost, etc.
- *Network/Traffic Determined* → IMS must support a mechanism to avoid duplicate registrations or inconsistent information storage, e.g. in roaming situations.
- *Application Layer Determined* → offered by IMS as service to the Application Layers, where some IMS parameters may indicate registration removal.
- *Subscription Management* → ability to restrict user access to IMS, e.g. upon contract expiration detection, removal of IM subscription, fraud detection, etc.

Next, we cover de-registrations, where SIP protocols apply, e.g:

- to deal with registrations expirations,
- to allow network forced de-registrations after approved possible causes.

6.6.7.3 *Registration Timeout—(SIP) De-registration*

Figure 6.26 illustrates network-initiated IMS terminal application (SIP) de-registration based on a registration timeout. The flow starts with a timer value provided initially at registration

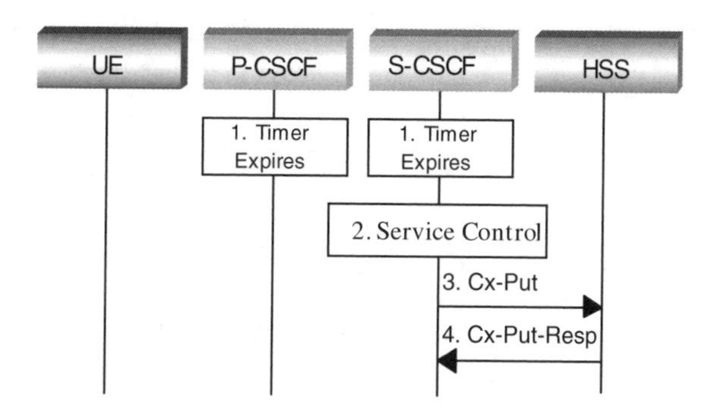

Figure 6.26 Network initiated application de-registration, registration timeout [21].

[36]An SIP protocol level in order to guarantee IMS.

and refreshed by subsequent re-registrations, and assumes that the timer has expired. P-CSCF and S-CSCF locations (i.e. home or visited network) do not show because the scenario remains the same for all cases.

1. The timers of the P-CSCF and S-CSCF, which may be close enough to work without external synchronisation, expire. Then the P-CSCF updates its internal databases to remove the public user identity to prevent registration. The GPRS PDP context cleanup follows independently other process.
2. The S-CSCF follows a filter criteria and sends de-registration information to the *service control platform* and performs the correct service control procedures, and the latter removes all subscription information concerned.
3. The S-CSCF sends to HSS either *Cx-Put* (public user identity, private user identity, clear S-CSCF name) or the same with 'keep S-CSCF name' instead. In both cases, the HSS stores the state of the public user identity as unregistered.
4. The HSS sends *Cx-Put Resp* to S-CSCF to acknowledge *Cx-Put* sending.

6.6.7.4 Administrative Network Initiated Application (SIP) De-registration

Administrative de-registration procedures may reside and initiated in various elements depending on the exact reason (e.g. subscription termination, lost terminal, etc.).

6.6.7.4.1 HSS Network Initiated De-Registration

For example, the HSS (Figure 6.27), which already knows the S-CSCF, uses the Cx-De-register and the S-CSCF uses the Cx-Put to inform the HSS.

Admin de-registrations do not update any subscriber records, EIR records, access authorization, etc; they only address the specific action of clearing the SIP application registration that is currently in effect. The transport layer mechanisms take care of IP transport infrastructure notification (e.g. GGSN, SGSN), or complete packet access is to be denied.

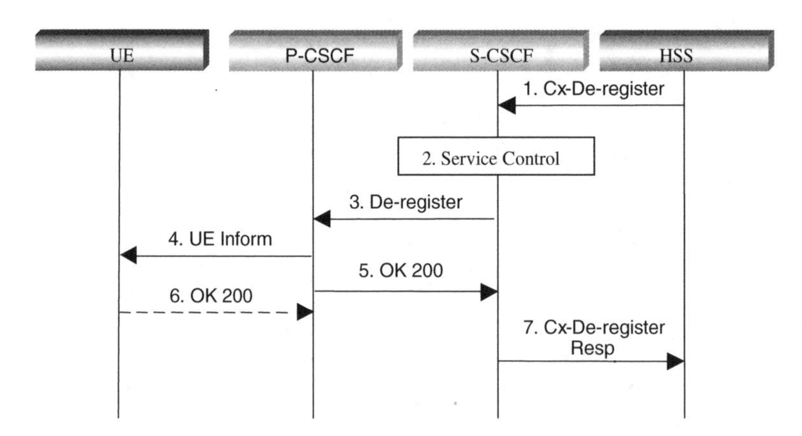

Figure 6.27 Network initiated application de-registration by HSS, administrative [21].

1. HSS initiates de-registration by sending a Cx-De-register (user identity), which may include de-registration reasons.
2. The S-CSCF[37] sends de-registration information to the service control platform and performs appropriate service control procedures.
3. The S-CSCF sends de-registration message, which may include HSS reasons, to the P-CSCF and updates its internal database to prevent the user registration.
4. The P-CSCF informs the UE about de-registration (including reason if any). Nevertheless, because of link lost, the EU may not receive the message.
5. The P-CSCF sends a response to the S-CSCF and updates its internal database to prevent user registration.
6. The UE sends de-registration ACK response to the P-CSCF when feasible.[38] In any case, the P-CSCF will do it and inform the UE by other means.
7. The S-CSCF returns a response to the entity that initiated the process, i.e. HSS.

6.6.7.4.2 S-CSCF Network Initiated De-Registration

Figure 6.28 illustrates a service control flow initiated by the IMS terminal application (SIP) de-registration, where the IP transport infrastructure (e.g. GGSN, SGSN) notification does not occur. All other constrains are as in the HSS case.

1. The S-CSCF receives de-registration information (it may include reasons) from the service platform and executes the appropriate service logic procedures.

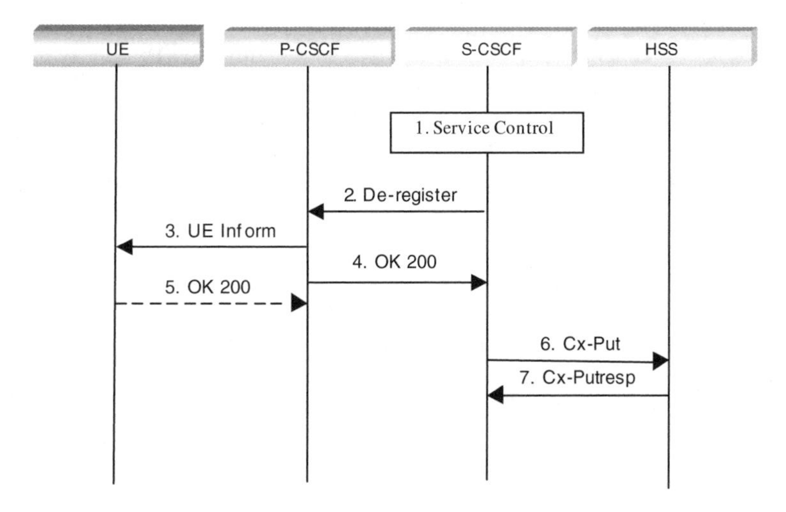

Figure 6.28 Network initiated application de-registration, service platform [21].

[37]Following a filter criteria as mentioned above.

[38]A 'misbehaving UE' or an 'out of P-CSCF coverage UE' will not reply de-registration request messages.

2. The S-CSCF sends de-registration message to the P-CSCF and updates its internal, database, preventing thus user registration (reason passed on if available).
3. P-CSCF informs UE about de-registration and gives reason if available (EU may reply if link still exists).
4. The P-CSCF does also send a response to the S-CSCF and updates its internal database to remove the user from being registered.
5. The UE sends de-registration ACK response to the P-CSCF, at less it is misbehaving UE or out of P-CSCF coverage. The P-CSCF performs de-registration anyway.
6. The S-CSCF sends an update to the HSS to remove itself as the registered S-CSCF for this user.
7. The HSS confirms the update.

6.6.8 Bearer Inter-working Principles

IMS voice bearers need to connect with voice bearers of other networks, e.g. through a Media Gateway Functions (MGW), like transcoding between a codec used by the IMS-UE and the codec used in the other party network.

Use of UE default [25] codecs enables IMS to interwork with other networks on an end-to-end basis or through transcoding.

IMS does also interwork with CS networks (e.g. PSTN, ISDN, CS domain of some PLMN) by supporting AMR to G.711 [26] transcoding in the IMS MGW element. The latter allows as well inter-working between IMS users and IP-multimedia fixed terminals and other implementation dependent codecs.

The UE supports existing network capabilities by sending DTMF tone indications (i.e. inband signalling) through the bearer to terminate *end of a sessions*. DTMF tones also interwork between one network and another, for which they may generate tones on the bearer of one network based on out-of-band signalling on the other network. In the latter case, the MGW provides the tone generation under the control of the MGCF [21].

6.6.9 Inter-working with Internet

According to selected policies, the S-CSCF may forward the SIP request or response to another SIP server located within an ISP domain outside of IMS.

6.6.10 Inter-working with PSTN

The S-CSCF, possibly in conjunction with an application server, determines that the session is forwarding to the PSTN. It forwards the *invite* information flow to the BGCF in the same network.

The BGCF selects the inter-working network pair based on a local policy. When it determines that the inter-working should occur in the same network, the BGCF selects the

MGCF to execute inter-working, otherwise the BGCF forwards the *invite* information flow to the BGCF in the selected network.

The MGCF performs the inter-working to the PSTN and controls the MG for the media conversions, and finally, Figure 6.29 illustrates a high level overview of the network-initiated PSTN inter-working process.

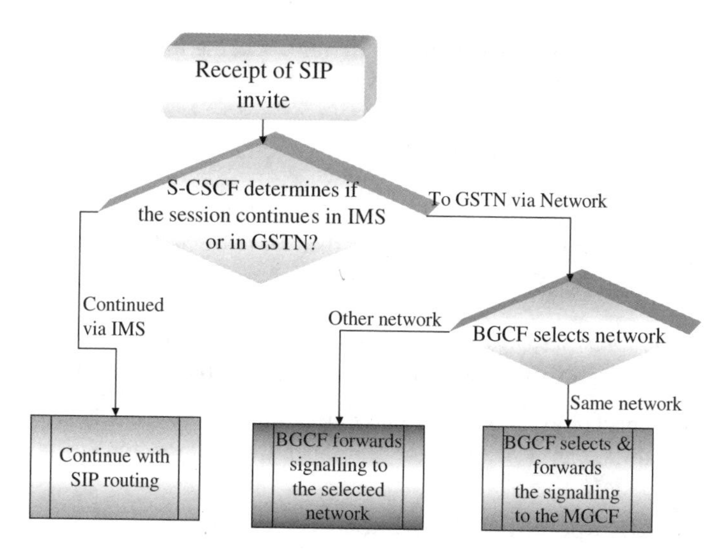

Figure 6.29 Network-based PSTN inter-working breakout process [21].

6.6.11 IMS Session Control Requirements

Carrier-grade IMS service offering requires bearers whose features (e.g. bandwidth) are coherent with the media components negotiated through CSCFs. Thus, the following apply [21]:

1. Both end-session points negotiate (according to service/UE settings) which resources (i.e. which media components) need to be established[39] before destination party alerts.
2. IMS service allows originating party charges for the Access IP-connectivity service of both originating and destination side, or when reverse charging applies to charge the terminating party for the Access IP-connectivity service of both originating and terminating side. That is, Easy CDR correlation held by Access IP-connectivity service (e.g. GPRS) with a session.
3. An operator's CSCF strictly controls session flows (e.g. on source/destination IP address, QoS) established through SIP entering the IMS bearer network from Access IP-connectivity service (e.g. from GPRS).

[39]The session signalling includes IP-connectivity network resources and IP-multimedia backbone resources.

4. Session and bearer control mechanisms allow the session control to decide when user plane traffic between end-points of an SIP session may start or stop. It includes synchronisation start/stop and charging start/stop during the session.
5. The *Access-IP-connectivity* service notifies the IMS session control when it has either modified or suspended or released the bearer(s) of an user associated with a session (e.g. when the user is no longer reachable).
6. IMS complies as well, with the architectural rules relating to separation of bearer level, session control level and service level expressed in [27].

6.6.12 Session Path Information Storing

Session paths, determined during a session initiation, get stored to route subsequent session requests through this determined path, allowing thereby passage through desired nodes, e.g. CSCFs, which perform specific actions, i.e.:

1. *CSCFs (Proxy and Serving)* → store initial session path sections, e.g. 'generate' requests demanding that sessions traverse all elements on a Route path.
2. *P-CSCFs* → check correct header value usage. They overwrite UE header(s) with the appropriate values, e.g. if SIP request need operator policy enforcement.

6.6.13 Terminal Capabilities–End-User Preferences

Because of different capabilities of originating/terminating terminals and non-equal user preferences set-up, it may not be possible to establish all media session requests by the originator. Thus, the following factors related to terminal capabilities and end-user preferences may apply:

- *Terminal capabilities* do affect SDP description in the SIP session flows, because they may support different media types (such as video, audio, application or data) and codecs (audio/video). Capabilities may change when an external device such as a video camera is attached to the terminal. The results are:
  1. terminal configuration (user profile, external attachment, etc.) changes terminal capabilities;
  2. destination user preferences may depend on the user who originates the session and on the situation (e.g. cost, time, day, etc.). Thus, a user may reject/accept a session;
  3. network resource availability plays critical role, as certain media streams, consuming high bandwidth, may be denied. Thus, destination-user alert must be preceded by network resources guarantees after successful link establishment;
  4. consequently, *end-to-end quality of service* provision will include a variety of mechanisms, including *best effort*. Since sessions are affected independently when lack of resources occur, end-to-end QoS renegotiations will be imperative.
- *End-user expectations and preferences* have different interaction options:
  1. the user may select some parameters for outgoing sessions pre-configured as preferences or defined on a per session basis;

2. for incoming sessions, the terminal establishes dialogue with the user, and can manually accept some of the pre-configured proposed parameters (media type, e.g. audio, video, whiteboard, etc.);
3. before establishing or accepting a new session, the user may define or agree on the following parameters, some of which may be pre-configured and others defined on a per session basis:
   a. Type of media (user preferences), i.e. audio, video, whiteboard, etc.
   b. Combination of QoS attributes and codec selection; i.e. media component quality, cost and availability probability of both core and access network resources.
   c. The user may choose the subset of terminal capabilities, e.g. low cost video session with a small window on the screen.
   d. The user may want assured end-to-end QoS for some streams, while may take best effort for others.

6.6.13.1 Bearer Establishment Mechanism

To meet user preferences in the context of terminal capabilities described above, the destination user may get pre-alerts before bearer establishment and negotiation, as well as PDP context activation. The *destination-user* thereby has time to select corresponding media streams and codecs for economic and efficient spectrum usage.

Figure 6.30 illustrates the bearer establishment mechanism, for which the pre-alerting occurs before the initial bearer creation procedures. It also illustrates user interaction after

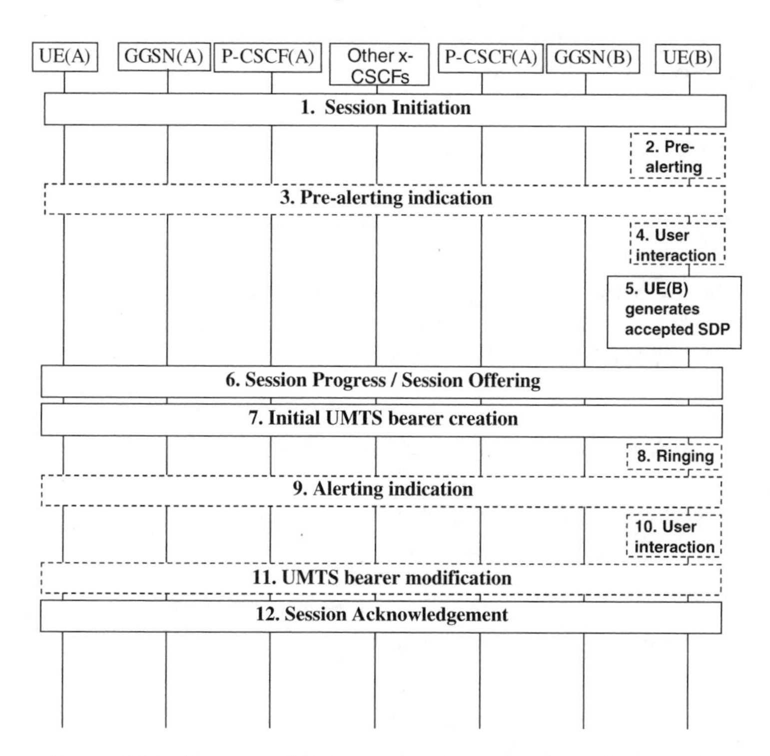

Figure 6.30 Bearer establishment showing optional pre-alerting [21].

the initial bearer creation. In the event of multiple responses, we treat UE resources with 'logical OR' (i.e. pick up the least upper bound) of the resources noted in the multiple responses, preventing thereby un-necessary resource allocation. The UE does not request more resources than was originally proposed in the Original INVITE [21].

We describe the bearer establishment flow sequence in Figure 6.30 as follows [21]:

1. UE(A) starts a *Session Initiation* procedure to UE(B) that includes an SDP proposal.
2. The user at UE(B) is pre-alerted.[40]
3. UE(A) may receive a pre-alerting indication.
4. User at UE(B) will then interact and express his/her wishes about the actual session.
5. UE(B) generates '*accepted SDP*' based on terminal capabilities (i.e. settings/pre-configured profiles) and optionally the user's wishes.
6. The *accepted SDP* passes to the UE(A) in the payload of a reliable SIP response.
7. *Initial Bearer Creation* → UE(A)'s and UE(B)'s access network (and external networks) resources are reserved with PDP context procedures.
8. Terminal at UE(B) begins to ring.[41]
9. The alerting indication goes to UE(A).
10. User at UE(B) may interact and express his/her wishes regarding the actual session.
11. UE(A) and UE(B) may perform bearer modification procedure here with new network resource reservations, if the initial bearers reserved in step 7 and user wishes at UE(B) differ.
12. Session initiation procedure acknowledged.

6.6.13.2 Session Progress Indication to Originating UE

The pre-alerting or alerting indications returned to the originating UE enables the latter to inform the calling user of the session in progress prior to arrival of the incoming media (e.g. originating UE may synthesise local ringing).

6.6.14 Interaction Between QoS and Session Signalling

At PDP context set-up the user shall have access to either GPRS without service-based local policy, or GPRS with service-based local policy. It is operator choice whether to offer both or only one of these alternatives for accessing the IM subsystem.

For the GPRS without service-based local policy case, the bearer is established according to the user's subscription, local operator's IP bearer resource based policy, local operator's

[40] Steps 2–4 are optional and may depend on terminal implementation and/or terminal pre-configured settings.
[41] The steps 8–10 are also optional and may be skipped.

admission control function and GPRS roaming agreements. The establishment of the PDP context bearer shall use the PDP context activation procedure specified in TS 23.060.

For the GPRS with service-based local policy case, service-based local policy decisions (e.g. authorisation and control) are also applied to the bearer.

The description in this subsection is applicable for the case when service-based local policy is employed.

The GGSN contains a Policy Enforcement Function (PEF) that has the capability of policing packet flow into the IP network, and restricting the set of IP destinations that may be reached from/through a PDP context according to a packet classifier. This service-based policy 'gate' function has an external control interface that allows it to be selectively 'opened' or 'closed' on the basis of IP destination address and port. When open, the gate allows packets to pass through (to the destination specified in the classifier) and when closed, no packets are allowed to pass through. The control is performed by a PDF, which is a logical entity of the P-CSCF. (*Note*: If the PDF is implemented in a separate physical node, the interface between the PDF and the P-CSCF is not standardised).

There are eight interactions defined for service-based local policy:

1. authorize QoS resources;
2. resource reservation with service-based local policy;
3. approval of QoS commit for resources authorised in (1), e.g. 'open' the 'gate';
4. removal of QoS commit for resources authorised in (1), e.g. 'close' the 'gate';
5. revoke authorisation for GPRS and IP resources;
6. indication of PDP context release from the GGSN to the PDF;
7. authorisation of PDP context modification;
8. Indication of PDP context modification from the GGSN to the PDF.

These requirements and functional description of these interactions are explained further in the following sections. The complete specification of the interface between the Policy Decision Function and the Policy Enforcement Function is contained in TS 23.207.

6.6.14.1 Authorise QoS Resources

The Authorise QoS Resources procedure is used during an establishment of an SIP session. The P-CSCF(PDF) shall use the SDP contained in the SIP signalling to calculate the proper authorisation. The PDF authorises the required QoS resources.

The authorisation shall include binding information, which shall also be provided by the UE to the GGSN in the allocation request, which enables accurate matching of requests and authorisations. The binding information includes an Authorisation Token sent by the P-CSCF to the UE during SIP signalling, and one or more Flow Identifiers, which are used by the UE, GGSN and PDF to uniquely identify the media component(s). If forking has occurred, the P-CSCF will re-use the same Authorisation Token in all subsequent provisional responses belonging to the same session. If the least upper bound of the requested resources is changed due to a subsequently received response then an update of the authorised resources is performed.

The authorisation shall be expressed in terms of the IP resources to be authorised and shall include limits on IP packet flows, and may include restrictions on IP destination address and port.

6.6.14.2 Resource Reservation with Service-Based Local Policy

The GGSN serves as the Policy Enforcement Point that implements the policy decisions for performing admission control and authorising the GPRS and IP BS QoS Resource request, and policing IP flows entering the external IP network.

Authorisation of GPRS and IP QoS resources shall be required for access to the IP-multimedia subsystem. The GGSN shall determine the need for authorisation, possibly based on provisioning and/or based on the APN of the PDP context.

Resource reservation shall be initiated by the UE, and shall take place only after successful authorisation of QoS resources by the PDF. Resource reservation requests from the UE shall contain the binding information. The use of this binding information enables the GGSN to correctly match the reservation request to the corresponding authorisation. The authorisation shall be 'Pulled' from the PDF by the GGSN when the reservation request is received from the UE. When a UE combines multiple media flows onto a single PDP context, all of the binding information related to those media flows shall be provided in the resource reservation request.

With a request for GPRS QoS resources, the GGSN shall verify the request is less than the sum of the authorised IP resources (within the error tolerance of the conversion mechanism) for all of the combined media flows. With a request for IP QoS resources, the GGSN shall verify that the request is less than the authorised IP resources.

The request for GPRS QoS resources may be signalled independently from the request for IP QoS resources by the UE. At the GPRS BS Level, the PDP Context activation shall be used for QoS signalling. At the IP BS Level, RSVP may be used for QoS signalling.

6.6.14.3 Approval of QoS Commit

The PDF makes policy decisions and provides an indication to the GGSN that the user is now allowed to use the allocated QoS resources for per-session authorisations unless this was done based on service based local policy at the time of the resource reservation procedure. If there is more than one response for the same session, indicating that the session has been forked in the network, the PDF may authorise the 'logical OR' of the resources requested in the responses. When the session established indication has been received, if the PDF earlier have authorised the 'logical OR' of the resources then the PDF will modify the authorisation and commit to resources according to the session established indication.

The GGSN enforces the policy decisions. The GGSN may restrict any use of the GPRS resources prior to this indication from the PDF. The GGSN shall restrict any use of the IP resources prior to this indication from the PDF, e.g. by open the gate and enabling the use of resources for the media flow. Based on local policy, GPRS and/or IP resources may be allowed to be used by the user at the time they are authorised by the PDF.

6.6.14.4 Removal of QoS Commit

The PDF makes policy decisions and provides an indication to the GGSN about revoking the user's capacity to use the allocated QoS resources for per-session authorisations. Removal of QoS Commit for GPRS and IP resources shall be sent as a separate decision to the GGSN corresponding to the previous 'Approval of QoS commit' request.

The GGSN enforces the policy decisions. The GGSN may restrict any use of the GPRS resources after this indication from the PDF. The GGSN shall restrict any use of the IP resources after this indication from the PDF, e.g. by closing the gate and blocking the media flow.

6.6.14.5 Revoke Authorisation for GPRS and IP Resources

At IP-multimedia session release, the UE should de-activate the PDP context(s) used for the IP-multimedia session. In various cases, such as loss of signal from the mobile, the UE will be unable to perform this release itself. The Policy Decision Function provides indication to the GGSN when the resources previously authorised, and possibly allocated by the UE, are to be released. The GGSN shall de-activate the PDP context used for the IP-multimedia session.

6.6.14.6 Indication of PDP Context Release

Any release of a PDP Context that was established based on authorisation from the PDF shall be reported to the PDF by the GGSN.

This indication may be used by the PDF to initiate a session release towards the remote endpoint.

6.6.14.7 Authorisation of PDP Context Modification

When a PDP Context is modified such that the requested QoS falls outside of the limits that were authorised at PDP context activation (or last modification) or such that new binding information is received then the GGSN shall verify the authorisation of this PDP context modification.

If the GGSN does not have sufficient information to authorise the PDP context modification request, the GGSN shall send an authorisation request to the PDF.

6.6.14.8 Indication of PDP Context Modification

When a PDP Context is modified such that the maximum bit rate (downlink and uplink) is downgraded to 0 kbps or changed from 0 kbps to a value that falls within the limits that were authorised at PDP context activation (or last modification) then the GGSN shall report this to the PDF.

This indication may be used by the PDF to initiate a session release towards the remote endpoint.

6.6.15 QoS-Assured Pre-conditions

This section contains concepts for the relation between the resource reservation procedure and the procedure for end-to-end sessions. A pre-condition is a set of constraints about the session, which are introduced during the session initiation. The recipient of the session generates an answer, but does not alert the user or otherwise proceed with session establishment until the pre-conditions are met. This can be known through a local event (such as a confirmation of a resource reservation), or through a new set of constraints sent by the caller.

A 'QoS-Assured' session will not complete until required resources have been allocated to the session. In a QoS-Assured session, the UE must succeed in establishing the QoS bearer for the media stream according to the QoS pre-conditions defined at the session level before it may indicate a successful response to complete the session and alert the other end point. The principles for when a UE shall regard QoS pre-conditions to be met are:

- a minimum requirement to meet the QoS pre-conditions defined for a media stream in a certain direction, is that an appropriate PDP context is established at the local access for that direction;
- segmented resource reservation is performed since the end points are responsible to make access network resource reservations via local mechanisms;
- the end points shall offer the resources it may want to support for the session and negotiate to an agreed set. Multiple negotiation steps may be needed in order to agree on a set of media for the session. The final agreed set is then updated between the end points;
- the action to take in case a UE fails to fulfil the pre-conditions (e.g. failure in establishment of an RSVP session) depends on the reason for failure. If the reason is lack of resources in the network (e.g. an admission control function in the network rejects the request for resources), the UE shall fail to complete the session. For other reasons (e.g. lack of RSVP host or proxy along the path) the action to take is local decision within the UE. It may, for example, (1) choose to fail to complete the session, (2) attempt to complete the session by no longer requiring some of the additional actions (e.g. fall back to establishment of PDP context only).

The flows of Sections 5.5, 5.6 and 5.7 depict the case where both UEs require confirmation from the other of the fulfilment of the pre-conditions. Other cases are possible according to the SIP specifications. For example, the pre-conditions may already be fulfiled (according to the principles above) when the INVITE is sent, or the UE may not require explicit confirmation from the other SIP end point when the pre-conditions are fulfiled. One example of such SIP end point is the MGCF used for PSTN inter-working. In these cases, one or both of the reservation confirmation messages may not be sent.

6.6.16 Event and Information Distribution

The S-CSCF and Application Servers (SIP-AS, IM-SSF, OSA-SCS) shall be able to send service information messages to end points. This shall be done based on an SIP Request/

Response information exchange containing the service information and/or a list of URI(s) pointing to the location of information represented in other media formats. The stimulus for initiating the service event related information message may come from, e.g. a service logic residing in an application server.

In addition, the end points shall also be able to send information to each other. This information shall be delivered using SIP-based messages. The corresponding SIP messages shall be forwarded along the IMS SIP signalling path. This includes the S-CSCF but may also include SIP application servers. The information may be related or unrelated to any ongoing session and/or may be independent of any session. Applicable mechanisms (e.g. routing, security, charging, etc.) defined for IMS SIP sessions shall also be applied for the SIP-based messages delivering the end-point information. The length of the information transferred is restricted by the message size (e.g. the MTU), so fragmentation and re-assembly of the information is not required to be supported in the UE. This information may include, e.g. text message, http URL, etc.

This mechanism considers the following issues:

- The IMS has the capability to handle different kinds of media. That is, it is possible to provide information contained within several different media formats, e.g. text, pictures or video.
- The UE's level of supporting service event related information and its exchange may depend on the UE's capabilities and configuration.
- A UE not participating in the service-related information exchange shall not be affected by a service-related information exchange possibly being performed with another UE of the session (Figure 6.31).

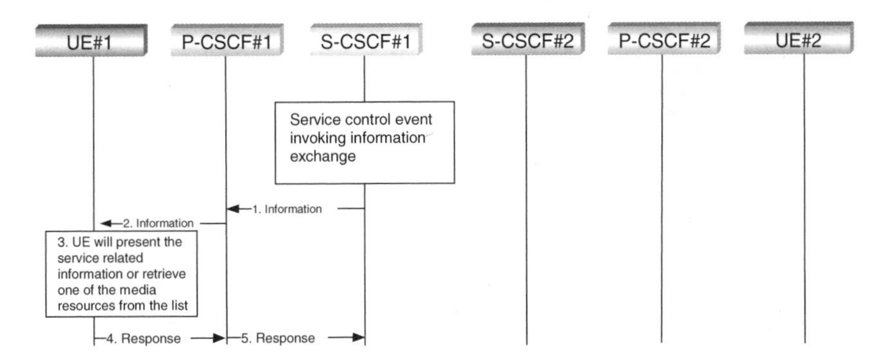

Figure 6.31 Providing service event related information to related end point [21].

Note: The service event related information exchange may either take place in the context of a session, or independently outside the context of any existing session.

1. When a service event occurs, which the S-CSCF or the Application Server wishes to inform an endpoint about, the S-CSCF or the Application Server generates a message request containing information to be presented to the user. The contents may include

text describing the service event, a list of URI(s) or other service modification information.

2. P-CSCF forwards the message request.
3. UE presents the service-related information, to the extent that it conforms to its capabilities and configuration, to the user.
4. Possibly after interaction with the user, the UE will be able to include information in the response to the S-CSCF.
5. P-CSCF forwards the response.

Note 1: The UE may retrieve service event related information using PS Domain or IMS procedures.

Note 2: Transport aspects of the information transfer described above may require further considerations.

6.6.17 Overview of Session Flow Procedures

This section contains the overview description and list of individual procedures for the end-to-end session flows. For an IP-multimedia subsystem session, the session flow procedures are shown in the following Figure 6.32.

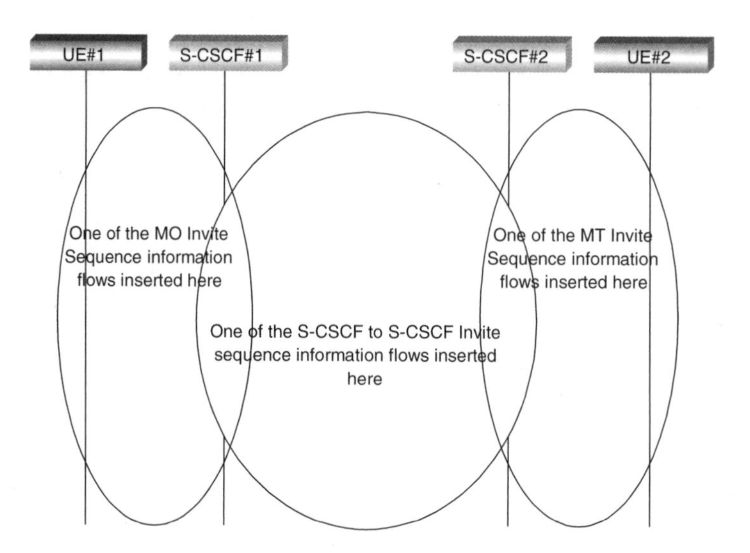

Figure 6.32 Overview of session flow sections.

The following procedures are defined:
For the origination sequence:

- (MO#1) Mobile origination, roaming,
- (MO#2) Mobile origination, home,
- (PSTN-O) PSTN origination.

For the termination sequence:

- (MT#1) Mobile termination, roaming,
- (MT#2) Mobile termination, home,
- (MT#3) Mobile termination, CS domain roaming,
- (PSTN-T) PSTN termination.

For Serving-CSCF/MGCF-to-Serving-CSCF/MGCF sequences:

- (S-S#1) Session origination and termination are served by different network operators,
- (S-S#2) Session origination and termination are served by the same operator,
- (S-S#3) Session origination with PSTN termination in the same network as the S-CSCF,
- (S-S#4) Session origination with PSTN termination in a different network to the S-CSCF.

The media being offered and acknowledged can take multiple negotiation steps or only one negotiation may be used. In these flows, a minimum of two negotiations has been shown. But the subsequent responses may not carry any media information and just confirm the initial media set agreement.

For example, for a non-roaming user initiating a session to another non-roaming user, each a subscriber of the same network operator, it is possible to construct a complete end-to-end session flow from the following procedures:

- (MO#2) Mobile origination, home,
- (S-S#2) Single network operator,
- (MT#2) Mobile termination, home.

There are a large number of end-to-end session flows defined by these procedures. They are built from combinations of origination, serving to serving and termination procedures, as determined from the following Table 6.6. For each row of the table, any one of the listed origination procedures can be combined with any one of the serving–serving procedures, which can be combined with any one of the termination procedures. In addition, several of the procedures give alternatives for network configuration hiding (the number of such alternatives is shown in parentheses).

Service control can occur at any point during a session, based on the filter criteria.

Note that the flows illustrated in Table 6.6 show service control only for the initial INVITE for originating and terminating party as an example.

6.6.18 Signalling Transport Inter-working

A Signalling Gateway Function (SGW) is used to inter-connect different signalling networks, i.e. SCTP/IP-based signalling networks and SS7 signalling networks. The signalling gateway function may be implemented as a stand alone entity or inside another

Table 6.6 Combinations of session procedures

Origination procedure (pick one)	Serving-CSCF–to-serving-CSCF procedure (pick one)	Termination procedure (pick one)
MO# 1 Mobile origination, roaming, home control of services (2). *MO# 2* Mobile origination, located in home service area. *PSTN-O* PSTN origination.	*S-S# 1* Different network operators performing origination and termination, with home control of termination (2). *S-S# 2* Single network operator performing origination and termination, with home control of termination.	*MT# 1* Mobile termination, roaming, home control of services (2). *MT# 2* Mobile termination, located in home service area. *MT# 3* Mobile termination, CS domain roaming.
MO# 1 Mobile origination, roaming, home control of services (2). *MO# 2* Mobile origination, located in home service area.	*S-S# 3* PSTN termination in the same network as the S-CSCF *S-S# 4* PSTN termination in different network than the S-CSCF	*PSTN-T* PSTN termination.

entity [1]. The session flows in this specification do not show the SGW, but when interworking with PSTN/CS domain, it is assumed that there is an SGW for signalling transport conversion.

6.7 IMS TRANSPORT ISSUES

The R00 architecture or more specifically R4 and R5 will support IPv4/IPv6 taking into account the following [7]:

- IP transport between network elements of the IP connectivity services (i.e. RNC, SGSN and GGSN) and IP transport for the CS domain.

At this writing the implementation of R99 does already support both IPv4/IPv6 for IP connectivity; thus, for R00 or R4 and R5 will be mainly a consolidation.

- IP-Multimedia CN Subsystem (IMS) architecture elements (i.e. UE to CSCF and the other elements, e.g. MRF):
  - shall make optimum use of IPv6,
  - shall exclusively support IPv6.
- The R00 (R4 and R5) UE shall exclusively support IPv6 for the connection to R00 (R4 and R5) IMS services.
- Access to existing data services (Intranet, Internet, etc.).
- The UE shall be able to access IPv4 and IPv6 based services.

Clearly, the IP-multimedia sub-network connectivity will emphasise IPv6. However, for access to data services it will need to support both IPv4 and IPv6 to comply with backwards compatibility requirements.

6.7.1 Principles of Mobile IPv4

In this section we highlight some IPv4 characteristics in the context of the mobile IP concept based on [28].

Mobile IP allows a Mobile Station (MS) to maintain connectivity to the Internet or to a corporate network, while using a single and unchanging address (e.g. its home address) despite changes in the link layer point of attachment. Thus, when an MS moves from a home network to a foreign network it registers with its Home Agent (HA) an IP address that the HA can use to tunnel packets to the MS [i.e. a Care of Address (Coa)]. The HA intercepts packets addressed to the MS's home address and tunnels these packets to the CoA. Here we do not require interaction with UMTS location registers.

A CoA may be, for example, a dedicated address each MS gets in the visited network (co-located CoA); if so, the MS is the tunnel end point. Otherwise, the CoA can be an address advertised (or retrieved) by a Foreign Agent (FA); if so, it is a FA-CoA and the FA is the tunnel end point. The FA extracts packets from the tunnel and forwards them to the correct RAN logical link in order to deliver them to the appropriate MS.

6.7.2 Differences Between IPv4 and IPv6

The key differences between protocols MIPv4 [29] and MIPv6 [30] can be summarised as follows [28]:

- Mobile IPv4 allows the use of Foreign Agents (FAs) to forward traffic thus requiring one care of address for multiple mobile stations, or the use of Co-located care-of Addresses (CoA). In contrast MIPv6 supports co-located CoAs only.
- MIPv4 has route optimisation as an add-on, whereas it is an integral part of the MIPv6 specification.
- MIPv4 route optimisation still requires traffic to be tunnelled between the Correspondent Host (CH) and the Mobile Station (MS). In MIPv6 packets can be forwarded without tunnelling, i.e. only with the addition of a routing header.
- In MIPv4 the Home Agent (HA) must get involved in the set-up of optimised routes. In MIPv6 the MS can initiate an optimised route to a CH directly (without involving the HA), and therefore more quickly and efficiently.
- In MIPv4 we obtain a CoA from a FA or via DHCPv4. In MIPv6 we may obtain a CoA via IPv6 stateless or state-full address auto-configuration mechanisms.
- In MIPv4 we require separate mobile IP specific messages to communicate with the FA, HA and CHs (when employing route optimisation). In MIPv6, we can piggyback mobile IP specific information onto data packets.

- MIPv4 has the ability to provide smoother handover as an add-on feature that forms part of the route optimisation protocol. In contrast support for smoother handover is an integral part of the MIPv6 specification.
- In MIPv4 we require reverse tunnelling to avoid ingress filtering problem (where firewalls drop the mobile's outgoing packets) since packets are sent with the home address as the source. In MIPv6 packets may be sent with the COA as the source address, hence there should not be any problem with ingress filtering.
- MIPv4 provides its own security mechanisms whereas MIPv6 employs the IPsec protocol suite.

To adequately assess the evolution and compatibility issues between MIPv4 and MIPv6 when applying to UMTS networks, we have to address each of the above differences. We have to address additional issues when preparing the deployment or migration between IPv4 and IPv6 networks in general [28].

6.7.2.1 Reverse Tunnels

In IPv4 we need reverse tunnels (i.e. tunnels from the FA to the HA), both for remote network secure access and to avoid packet drops due to ingress filtering. Ingress filtering allows tracking of malicious users attempting denial of service attacks based on topologically inconsistent source address spoofing.

In mobile IPv6, we do not need reverse tunnels to avoid problems with ingress filters. However, they may still be beneficial when the ME is concerned about location privacy. The MN may use the care-of-address as sender address but that is not required.

6.7.2.2 Use of Route Optimisation

Route optimisation reduces delays between the CH and ME, and it also reduces the load placed on HAs. Nonetheless, in MIPv4 it adds to the complexity of the HA and requires security associations between the HA and all CHs. Furthermore, it still requires packets to be tunnelled from the CH to the FA-CoA. In contrast, route optimisation in MIPv6 removes the need to tunnel packets, instead we add a routing header to each packet. The ME also has more control to decide when to optimise routes, since it creates the optimised route rather than the HA; thus resulting in simpler MIPv6 HA. When migrating from MIPv4 to MIPv6, we need to make changes to CHs to employ route optimisation. In contrast, all IPv6 CHs will support route optimisation automatically.

6.7.3 IPv4 and IPv6 Inter-operability

Without a doubt IPv4 and IPv6 will co-exist for many years to come. Thus, a range of techniques exist today to facilitate co-existence, the three main categories include [31]:

1. *Dual-Stack* → allows IP-v4/v6 to co-exist in the same devices and networks,
2. *Tunnelling* → allows transport of IPv6 traffic over existing IPv4 infrastructure,
3. *Translation* → allows IPv6-only nodes to communicate with IPv4-only nodes.

6.7.3.1 *Dual Stack Solutions*

Dual Stack (DS) techniques enable complete hybrid IP-v4/v6 nodes, which stand as one or the other or both depending on the infrastructure environment and the desired behaviour. Thus, they have three operating modes, i.e. IPv4-only (v6 disabled), IPv6-only (v4 disabled) and IP-v4/v6 (both protocols enabled). In the latter case there must be at least one address for each version following respective mechanisms and applying DNS functions[42] to resolve IP addresses.

In principle, applications qualified for DS operation use a sensing function to discriminate their IPv4 or IPv6 peers.

To successfully deploy a DS network all routers would have to be enabled to run IP-v4/v6 protocols (i.e. including routing tables) simultaneously. Practically, this would imply the most current Internet networks would need HW/SW or at least SW upgrade if not both; which may be a drawback. Nevertheless, the trade-off would benefit network performance and service flexibility immediately.

6.7.3.2 *Tunnelling Solutions*

Tunnelling facilitates carrying IPv6 traffic by encapsulating it in IPv4 packets and transporting it over an IPv4 infrastructure transparently. As result, packets from a private or corporate IPv6 network may reach other IPv6 networks passing through IPv4 ISPs, e.g. tunnelling can occur either router-to-router, host-to-router, host-to-host or router-to-host.

Figure 6.33 illustrates router-to-router tunnelling, where first the tunnel entry point (R1) decrements the IPv6 hop limit by one, encapsulates the packet in an IPv4 header[43] and sends it through the tunnel. Then the exit point (R2) receives the encapsulated packet and removes the IPv4 headers and routes the IPv6 packet to destination.

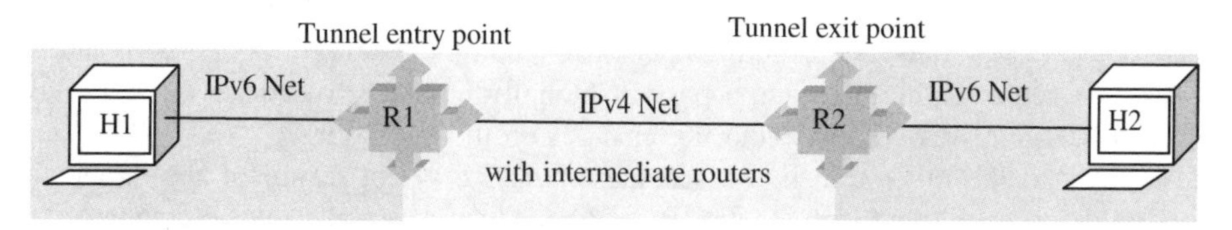

a) Host 1 (H1) sends IPv6 packets to Router (R1)
b) R1 as the Tunnel entry point encapsulates IPv6 packets in IPv4 header and sends to R2
c) R2 as the Tunnel exit point strips off the IPv4 header and forwards packets to Host 2 (H2)

Figure 6.33 Tunnelling and IPv6 encapsulation.

IETF specs, i.e. RFCs 2473, 2893 and 3056 define the following two types of tunnelling:

1. *Manually Configured Tunnelling of IPv6 over IPv4* → point-to-point tunnels manually configure to transport IPv6 packets encapsulated in IPv4 packets.

[42]We use DNS 'A' record to resolve IPv4 addresses and DNS 'A6' record to resolve IPv6 addresses.

[43]If necessary, the composed IPv4 (i.e. IPv6 + IPv4 header) packet may be fragmented and later reassembled.

2. *Automatically Tunnelling of IPv6 over IPv4* → dynamic tunnelling of IPv6 packets using IPv4-compatible IPv6 addresses or 6 to 4 ISATAP[44] addresses.

Configured tunnelling implies assigning the exit point address in the entry point (e.g. R1), which uses it as the destination address in the IPv4 header.

For example, an IPv6/IPv4 host connected to network segments without IPv6 routers can receive a static route to an Internet-IPv6 router at the other side of the IPv4 tunnel and communicate thereby with the remote IPv6 network. This implies that the IPv6 address of the destination IPv6/IPv4 router gets added into the routing table as a default route.

Automatic tunnelling enables IP v6/v4 nodes to communicate over an IPv4 infrastructure dynamically without a need of tunnel destination pre-configuration. This means that the tunnel end point address is determined by the IPv4-compatible destination address, which gets created by taking the IPv4 address for the interface and pre-appending a 96-bit prefix of all zeroes [31].

Combined Automatic/Configured tunnelling applies to hosts that are connected to segments with no IPv6 router. These hosts would have two entries for tunnelling, i.e. one with 96-bit prefix for IPv4-compatible IPv6 destination addresses, and another to an IPv6 router set for automatic tunnelling.

Finally, the RFC 3056 (i.e. *6to4*) enables IPv6 domain connections through IPv4 networks without explicit tunnelling. In this case we treat the wide area IPv4 as a unicast point-to-point link layer, and native IPv6 domains communicate through 6to4 routers also called 6to4 gateways [31].

6.7.3.3 Network Address and Protocol Translation (NAT-PT)

NAT has served (and serves still) to overcome address space limitations in corporate networks, which use IPv4 addresses from a private range and an NAT router at their borders translates the private addresses to a single or limited number of public addresses.

NAT-PT provides transparent routing nodes while communicating between IPv6 and IPv4 networks. It achieves this by using a pool of globally unique IPv4 addresses and binding them to IPv6 addresses without requiring changes on the end nodes.

Compared to the other translation schemes, NAT-PT is not aware of the applications traversing because it looks only at the IP headers. Thus, it permits multiple end points to appear as a single IP address.

To allow communication of IPv6 nodes with IPv4 nodes, NAT-PT assigns dynamically IPv4 addresses to IPv6 nodes. It also maps IPv6 addresses to IPv4 nodes to allow communication of IPv4 nodes with IPv6 nodes. To achieve these assignments NAT-PT manages a pool of IPv4 addresses.

The two main NAT-PT types include:

1. an IPv6 host initiates communications, i.e. an IPv6 client wants to communicate with server on an IPv4 network;
2. an IPv4 host initiates communications, i.e. and IPv4 client attempts to communicate with server in an IPv6 network.

[44]Intra-site automatic tunnel addressing protocol.

In case one, the host uses its IPv6 address as the source and IPv4-mapped IPv6 address as the destination address. NAP-PT translates the IPv6 packet into IPv4 by replacing the original IPv6 source address with the assigned IPv4 address and extracts the IPv4 address from the IPv4-mapped IPv6 destination address. It also tabulates the address relationships to serve IPv4 packets returning from the server.

The translation procedure in the 2nd case is similar to that of the 1st one. So when NAT-PT works in conjunction with a DNS, it uses the IPv6 host name to look up the IPv4 address. NAP-PT registers the assigned IPv4 address for the IPv6 machine in the DNS server.

6.7.3.4 Key NAT-PT Issues

The type of address translation mentioned above implies the assignment of an IPv4 to each IPv6 host in liaisons with an IPv4 host. But due to the scarcity of IPv4 addresses, NAT-PT extends the translation notion also to the transport identifier (e.g. TCP or UDP port numbers, and ICMP query identifiers). This allows the transport identifiers of number of IPv6 hosts to be multiplexed into the transport identifiers of a single assigned IPv4 address. NAT-PT enables thus a set of IPv6 hosts to share a single IPv4 address.

For outbound IPv6 network packets NAT-PT translates the source IP address and the source transport identifier, which can be either a TCP/UDP port or ICMP query ID. For inbound packets, the destination IP address, destination transport identifier, the IP and transport header checksum need translation.

Although TCP and UDP should be independent as protocols above the IP layer, because they use IP sources and destinations in the pseudo-header checksums, a change of the IP address implies the recalculation of their checksum. In addition, because the ICMPv6 includes a pseudo-header checksum not available in ICMPv4, checksum in ICMP need modification by the translation.

IPv6 demands Maximum Transmission Unit (MTU) discovery and it allows only end-to-end fragmentation, in contrast, MTU discovery in IPv4 is optional. Thus, the translator must warrant that the packet does not exceed the MTU on the IPv6 path by fragmenting the IPv4 packet into a 1280 byte IPv6 packet, which is the minimum IPv6 required packet.

As indicated earlier, NAT-PT is transparent to applications, i.e. it is unaware of what applications are transmitting packets. Yet some applications (e.g. FTP) carry IP addresses in their payloads, and logically they would need translation. To achieve this, we use an Application Level Gateway (ALG) interposed between the peers.

6.7.3.5 NAT-PT Limitations and Drawbacks

While ALGs may resolve some limitations, they create others, e.g. they basically stop or delay the transport protocol resolution (thereby penalising performance), and may also modify the data stream altogether before forwarding it.

NAT-PT acts as state sensitive device, i.e. it performs efficiently only when all factors follow expected patterns in each *session.*[45] In a combined IPv4/IPv6 environment such a mechanism may not always have the one-to-one mapping. Therefore, total accuracy may not necessarily be guaranteed for the following reasons:

[45]The latter includes a queue of packets containing the same source host and port, destination host and port, as well as transport protocol type, all with the corresponding sequence number.

- Some IPv4 fields have changed meaning in IPv6 and seamless translation is not always evident, e.g. NAT-PT:
  - does not translate some IPv6 relevant option headers;
  - complicates the use of multi-homing by a site for the sake of reliability increase of its Internet connectivity;
  - places constraints on the deployment of applications with IP addresses in data stream and;
  - operates with the assumption that each session is independent. However, apps or protocols, which on the other hand assume end-to-end addressing integrity will simply fail when crossing NAT-PTs points;
  - may adjust TCP sequence numbers prior forwarding packets, which may have negative impact on the application characteristics;
  - solves problems to NAT-PT (e.g. TCP, UDP and other port multiplexing), but it would fail when the assigned NAT-PT pool of IPv4 addresses runs out; thus limiting new IPv6 sessions;
  - prevents end-to-end security mechanisms because it intervenes at the header level. It becomes a single point of failure;
  - inhibits IPSec implementation, i.e. breaks the flexible end-to-end model of the Internet;
  - does not guaranteed *Authentication Header* even if IPSec functionality is maintained;
  - does not scale very well in large networks;
  - is a single point of failure for all ongoing connections;
  - causes additional forwarding delays due to further processing, when compared to normal IP forwarding;
  - difficulties with source address selection due to the inclusion of a DNS ALG on the same node [32];
  - requires application level GWs for applications that embed IP addresses in their payload.

6.7.4 Conclusions on IPv4 and IPv6 Inter-operability

From the preceding Inter-operability techniques, i.e. Dual Stack (DS), Tunnelling and NAT-PT can be compared as follows:

- DS stands as easy to use and flexible technique. Hosts can communicate with IPv4 hosts using IPv4 or communicate with IPv6 hosts using IPv6. Once full upgrade takes place, IPv4 can be removed or turn off. DS demands or has the following limitations:
  - additional hosts CPU power and memory (some new investment),
  - double command control or protocol interface (a little extra ops),
  - dual stack DNS and FW resolver (evolved IPnet architecture).
- Tunnelling facilitates gradual IPv6 migration without restriction of network component upgrade order (i.e. routers 1st, or hosts, or subnets). It does not demand IPv6 ISPs and can

use IPv4 infrastructure to reach IPv6 network interaction. Limitations demands amount to:

- ○ need of more power and time on routers to handle extra load due to encapsulation and de-capsulation on entry and exit points;
- ○ complex trouble shooting due to large count of hops and fragmentation problems.

- As stated earlier in the preceding section, NAT-PT should be used only if the other techniques do not apply.
  - ○ It should be used only as a temporary solution.
  - ○ It does not allow advanced IPv6 techniques like end-to-end security.
  - ○ It limits network design topology because it demands replies through the same NAT router (i.e. creates a single point of failure).
  - ○ As mentioned earlier, it limits applications because it does not discriminate packet type.

In conclusion, apply DS where and as much as possible. Apply tunnelling with moderation, avoid or never use NAT-PT if you can do it otherwise for IPv4 and IPv6 inter-operability.

Recommendations: IPv6 is mature enough for corporate and commercial use, do not longer make large investments on setting up complex IPv4 configurations anymore, move into IPv6 gradually without even touching the running IPv4 systems if not required.

6.7.5 Advanced IPv6 Mobility Features

IPv6 and SIP will become the corner stones to warrant mobility in IMS. SIP blends easily with IPv6 to secure personal mobility by enabling ubiquitous user communications (i.e. person-to-person and person-to-machine interaction) at all times.

As a simple fact, today's Interned addressing model limits mobility. If we consider the two key IP address functions, i.e. routing and host identification, they work well for fixed nodes but become complex and un-necessarily difficult for moving nodes. Current addressing, where routing prefix forms the initial bits of IP address and remaining bits indicate specific host identity among all nodes identified by the routing prefix, prefers fixed node location for transparent routing.

6.7.5.1 Mobile IP

The main stream Internet traffic uses TCP connections, which are defined by the combination of IP address and port number of both end points of the communication. Thus, if one of these four entities changes communication flow disruption occurs. This implies that if a Mobile Node connects to different network access point, it will require a new IP address.

Mobile IP aims to solve aforementioned limitation by introducing the 'care-of-address' (CoA) concept for the moving node. The moving node receives a CoA association as it moves from network to network while keeping its 'Home Address' or originally assigned IP address.

A dedicated router called 'Home Agent' (HA) forwards packets to the moving node based on the CoA updates it receives. *Binding*, i.e. the association between the moving node's home address and its CoA includes a lifetime value indicating CoA expiration for deletion. A 'Binding Update' (BU) occurs when the CoA must be kept or refreshed.

The BU stands as the key tool in the implementation of Mobile IP. This mechanism facilitates Internet addressing by keeping packet-forwarding scheme towards the moving node's home address despite its location. Thus, to applications the moving node does not reflect change since the TCP does not detect the CoA. Figure 6.34 illustrates mobile IP agents flow.

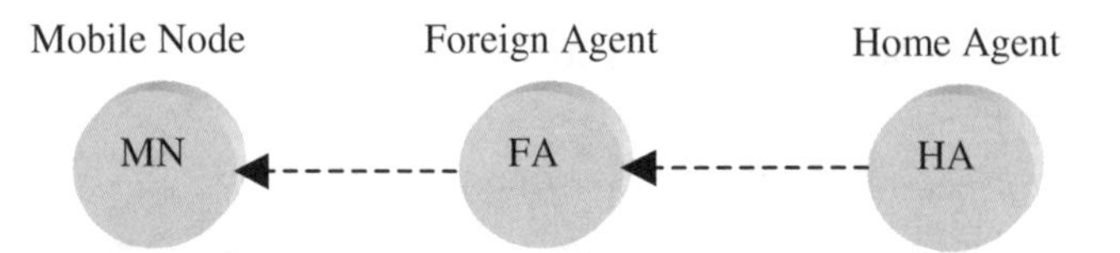

- The *MN* needs to register its current *CoA* with the *HA*
- The *HA* collects al packets addressed to the home address and forwards them to the *MN*'s *CoA*

Figure 6.34 The mobile IP agents [31].

- The HA, when redirecting a packet to the MN's home address, it changes the destination address to the CoA of the MN.
- When the MN receives the datagram, the IP address gets changed again so that it appears to have the MN's home address as destination IP address.

6.7.5.2 Key Difference between Mobile-IPv4 and -IPv6

Basically, the principles above apply to the both IPv4 and IPv6. However [31],

- IPv6, due to its advanced features, does not need FA because stateless auto-configuration and neighbour discovery offer the needed functionality and are built into IPv6;
- with Mobile IPv4, all packets sent to the MN always go to the HA resulting in a triangular routing effect;
- IPv6 uses extension headers to allow the sending of packets to the MN's CoA directly by caching the binding of a MN's home address with its CoA.
  - As a result, the MN attached to a foreign network now uses its CoA as a source address when sending packets.
  - The MN's home address gets carried in a Destination Option header, which implies that the session control information is *piggybacked* onto the same packet.
  - In Mobile IPv4, those control messages had to be sent in separate UDP packets.
- CoA usage as the source address also simplifies multi-cast packet routing, which with Mobile IPv4, these packets would have to be tunnelled to the HA in order to use the home address as multicast packet source address.
  - Mobile IPv6 includes this info in the Destination Option header to be processed by the receiving node.

- In Mobile IPv4, all packets sent to the home address while an MN is attached to a foreign network, must be encapsulated and cannot use Routing header.
  - In Mobile IPv6 packets sent to the MN's home address, although still encapsulated, they can be sent using a Routing header, which minimises overhead.
- Thus, Mobile IPv6 includes optimised routing for direct peer-to-peer[46] mobility signalling. It enables MNs to send and receive packets directly through the optimal route.[47]
  - In Mobile IPv4, MNs communicate via the HA while maintaining bi-directional tunnelling through the HA during the whole session.
- Route-optimised traffic towards MN containing the HA in a special routing header is not misuse because it does not get forwarded out of the receiving node.
- In Mobile IPv4, an MN detects HA by sending a broadcast request, which leads to the reception of separate answers from all the HAs on its segment.
  - Mobile IPv6 sends requests using an *anycast* address, which results in one reply only.

6.7.5.3 Real-Time Transport Optimisation

In mobile networks, transport optimisation is a necessity because:

- in traffic with stringent timing characteristics like during a rich call, an MN movement adapting to a network may cause short pauses generating a series of latencies;
- unexpected latencies[48] acting as communications discontinuities may trigger excess signalling validation and saturate allocated bandwidths or prevent timer responses;
- these inefficiencies may cause combined delay effects with non-negligible impacts on the user perception;
- today Mobile IPv4 does not possess the key enablers as IPv6 does. Therefore, this drawback will prevent optimum introduction of real-time services;
- IMS is all about a new generation of IP-based real-time services, which may very well not be as effective with IPv4 as it would be with IPv6.

6.7.6 IPv6–IPv4 Inter-operability Scenarios in IMS

In the preceding sections, we have outlined the key features of IPv4 and IPv6 inter-operability at the backbone level independent from any interactions with dedicated sub-networks such as GPRS, which serves IMS.[49] The latter will house or will enable a new generation of services exploiting the aforementioned features.

[46] Thus, the routing infrastructure can live without optimisation, nevertheless for security IPSec would apply.
[47] Thereby fully using the routing fabric.
[48] Due to unplanned network congestion or late router response to Mobile IP agents.
[49] IMS, it is not confined to GPRS-GSM access, it may also use other solutions, e.g. WLANs.

In this section, we will briefly outline the interaction options while IMS services go into operation, i.e. when the UE access services through IMS, based on IETF internal working drafts, e.g. RFC2026.

6.7.6.1 *Interactions through GPRS and IMS*

The GGSN (Figure 6.35) serves as an anchor point for the GPRS mobility management and as default router for a UE integrated with GPRS/IP stack. The peer node represents the element with which the EU communicates. The liaison logic can be expressed as follows:

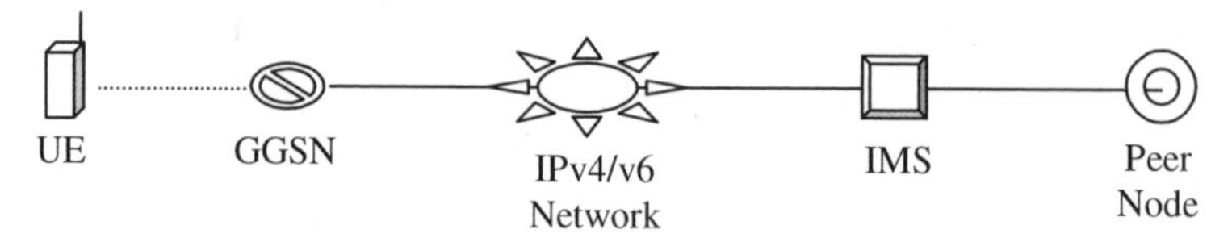

Figure 6.35 A simplified EU-GPRS—IP network—peer-node link.

- We call Packet Data Protocol (PDP) context to the dedicated link between the UE and the GGSN.
- This link results from the PDP activation process, through which the UE gets an IP address configuration and information to maintain the IP connection.
- From the inter-operability interaction of this analysis we then have three different types of PDP Context: IPv4, IPv6 and Point-to-Point Protocol (PPP).
  - The EU may have one or more simultaneous PDP contexts open with the same or different GGSNs.
  - The PDP Context can be either of the same or different types.
- An IMS capable EU, with an activated PDP Context, uses therefore the GPRS network as an access to reach IMS.

Since the IMS is at present exclusively IPv6 based, the PDP context is IPv6 type. In addition, its SIP server and proxy servers are also IPv6 based.

Therefore, if the traffic type would be only IPv6, i.e. signalling and content exchanged by IPv6 EUs through IPv6 networks to reach IMS, there would be seamless communications without a need of inter-operability.

However, that is not the case at present, and as noted already earlier there is a need of IPv4 and IPv6 co-existence because today's Internet is still predominantly IPv4. Hence, we need Transition scenarios to reach IMS solutions and services.

6.7.6.2 *Transition Scenarios for GPRS and IMS*

Plain GPRS EUs or devices communicating with their Internet applications servers such as E-mail, web, etc., pass through the GGSN with the following IETF RFC2026 [33] scenarios:[50]

[50]These scenarios apply also to PDP contexts with Point-to-Point Protocols (PPP) terminating in the GGSN.

1. Dual Stack UE connecting to IPv4 and IPv6 nodes
2. IPv6 UE connecting to an IPv6 node through an IPv4 network
3. IPv4 UE connecting to an IPv4 node through an IPv6 network
4. IPv6 UE connecting to an IPv4 node
5. IPv4 UE connecting to an IPv6 node

The IMS paths include primarily:

1. UE connecting to a node in an IPv4 network through IMS;
2. two IPv6 IMS islands connected via IPv4 network.

A description following the type of PDP configuration will be summarised next for completeness, more details can be found in [33] since the work is still ongoing.

6.7.6.3 DNS Inter-Working in IMS

Even though current IMS specs are IPv6 based only, it is recommended that every recursive DNS server should be either IPv4-only or dual stack and every single DNS zone should be served by at least an IPv4 reachable DNS server assure transition requirements. In particular, to perform IMS DNS resolution, we should configure the UE as a stub resolver pointing to a recursive dual stack DNS resolver.

6.7.6.4 Dual Stack (DS) UE Connecting to IPv4 and IPv6 Nodes

In this case, the UE can communicate with both IPv4 and IPv6 nodes by activating an IPv4 or IPv6 PDP context as illustrated in Figure 6.36. It requires IPv4/v6 capable GGSN and the DS UE may have only one stack or both stacks *active* simultaneously. When we need 'IPv6 in IPv4' tunnelling we aim to activate an IPv6 PDP context and make encapsulation/de-capsulation in the network.

If the GGSN does not support IPv6, and UE application needs to communicate with an IPv6 node, the UE may activate an IPv4 PDP context and tunnel IPv6 packets in IPv4

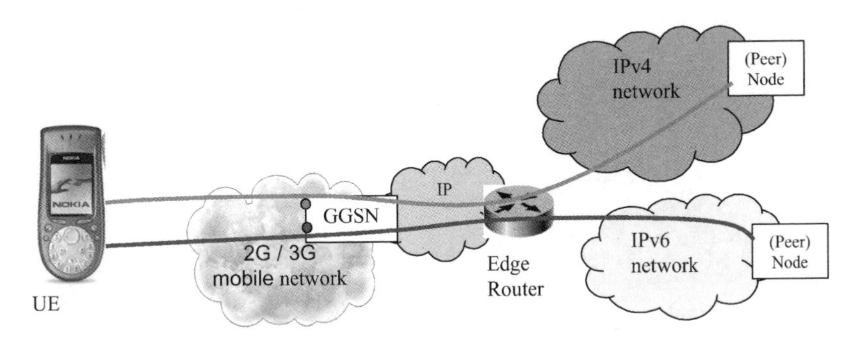

Figure 6.36 Dual stack UE connecting to IPv4/6 nodes [33].

packets through a tunnelling mechanism.[51] Tunnelling assumes DS UE capabilities and sufficient availability of public IPv4 addresses for optimum operation.

On the other hand, the use of EU private IPv4 addresses depends on the support of these addresses by the tunnelling flow and the deployment scenario. For example, if the tunnel end points are in the same private domain or the tunnelling flow works through IPv4 NAT, private IPv4 addresses can apply [33].

A DS UE can discriminate whether the end point required is an IPv4 or IPv6 capable node by examining the address to discover as to which address family category it falls into. The DNS may also contain records sufficient to identify which protocol should be used to initiate connection with the end point. Since the UE can natively communicate with both protocols, a key concern will be then to obtain the correct address space and routing management, i.e. maintain address spaces for both IPv4/v6 protocols.

Because insufficient[52] supply of public IPv4 addresses, allocating a globally unique IPv4 address to each UE continually will be impossible. Now utilising private IPv4 addresses means use of NATs (Network Address Translators) when communicating with a peer node outside the operator's network. Furthermore, in large networks, NAT systems can become very complex, costly and difficult to maintain. Therefore, it is recommended that IPv6 communication (native or tunnelled from the UE) be preferred instead IPv4 communication going through IPv4 NATs to the same dual stack peer node. In this scenario, the UE communicates with the DNS resolver using the IP version available via the activated PDP context [33].

To hold the Internet name space un-fragmented, i.e. to keep any record in the public Internet unmodified to any (IPv4 or IPv6) nodes, every recursive DNS server should be either IPv4-only or *dual stack* and every single DNS zone should be served by at least an IPv4 reachable DNS server[53] [34].

6.7.6.5 *IPv6 UE Link to an IPv6 Node through an IPv4 Network*

In this case, 'IPv6 in IPv4' tunnelling provides the ideal solution as illustrated in Figure 6.37, where an IPv6 PDP context gets activated between the UE and the GGSN.

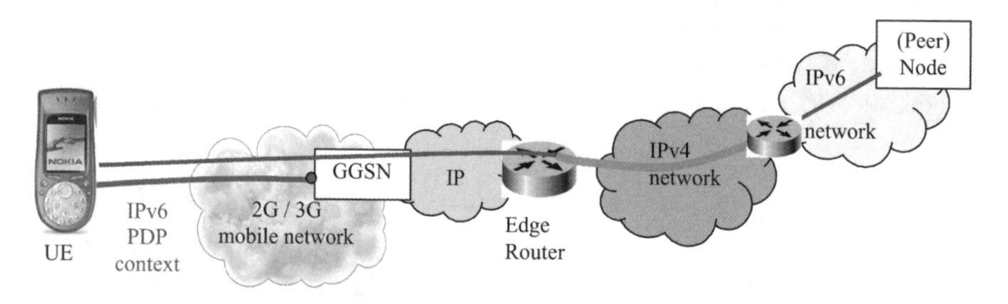

Figure 6.37 IPv6 UE connecting to IPv6 node through an IPv4 network [33].

[51]For example, 3G UE modem and IPv6 packets encapsulated in IPv4 packets → IPv4 PDP context gets activated.

[52]The insufficient IPv4 address makes then also very expensive and inaccessible to the grand public.

[53]This rules out IPv6-only recursive DNS servers and DNS zones served by IPv6-only DNS servers.

Since the non-DS IPv6 UE cannot cope with tunnelling, it is handled in the network and encapsulation can take place in the GGSN, the edge router between the border of the operator's IPv6 network and the public Internet, or any other DS node within the operator's IP network. The same node can then handle the encapsulation (uplink) and de-capsulation (downlink). In general, the tunnelling, which enables end-to-end IPv6 connectivity for applications, handled by the network elements is transparent to the UEs and the IP traffic looks like native IPv6 traffic to them [33].

The operator can choose either static or dynamic 'IPv6 in IPv4' tunnelling between the IPv6 islands. In the context of 3GPP operators and [33], two types of tunnelling may take place, i.e.:

6.7.6.5.1 Tunnelling Internal to the 3GPP Operator's Network

Assuming IPv4 backbone for GPRS operators, we would expect gradual migration or soft introduction IPv6 islands. With this small scale of IPv6 deployment, a reduced number of 'IPv6 in IPv4 tunnels' would be configured manually[54] to integrate IPv6 islands over predominantly IPv4 networks.

Dynamic tunnelling mechanisms such as '6to4' [35] and IGP/EGP routing protocol [36,37] based tunnelling mechanisms may also apply.[55] BGP routing, e.g. runs BGP protocol between the neighbouring router tunnel end points and using multi-protocol BGP extensions to exchange reachability information of IPv6 prefixes. Then routers create IPv6 in IPv4 tunnel interfaces and route IPv6 packets over the IPv4 network.

6.7.6.5.2 Tunnelling External to the 3GPP Operators

When the peer node is outside the operator's network, the 'IPv6 in IPv4' tunnel starting point may occur at the operator's encapsulating node, e.g. the GGSN or the edge router. It is easier if the upstream ISP provides native IPv6 connectivity to the Internet; otherwise, an 'IPv6 in IPv4' tunnel can be configured, e.g. from the GGSN to the dual stack border gateway in order to access the upstream ISP.

In the case of only IPv4 connectivity, we transport 3GPP-IPv6 initiated traffic to the ISP tunnelled in IPv4, where the tunnel end point depends on the deployment scenario. Manually configured 'IPv6 in IPv4' tunnelling is sensible to the number of the tunnels. For example, 10–15 may be enough for 3GPP network towards the ISP/Internet.

6.7.6.6 IPv4 UE Link to an IPv4 Node Through an IPv6 Network

Because of the low spread of IPv6 networks, it will be easier to let IPv4-only UEs to use an IPv4 link (PDP context) to connect to the GGSN without the need to pass over an IPv6 network. Therefore, as illustrated in Figure 6.38, tunnelling of IPv4 in IPv6 will not be required from the GGSN to external IPv4 networks. Besides, 3GPP operators will gradually phase out IPv4 UEs and the IPv4 transport network leaving only IPv6 UEs and IP environment. Hence, transitions involving an IPv4 UE communicating with an IPv4 peer through an IPv6 network will very unlikely in 3GPP networks.

[54]Manually configured tunnels can be an administrative burden with a large number of IPv6 islands.

[55]For example, evolving 3GPP scenarios may use manually configured tunnels or EGP/IGP based tunnelling.

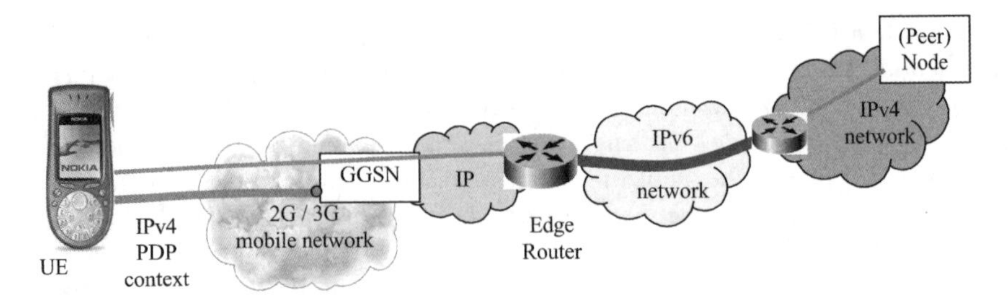

Figure 6.38 IPv4 EU connecting to IPv4 node through an IPv6 network [33].

6.7.6.7 IPv6 UE Link to an IPv4 Node

As indicated in the preceding sections, IPv6 nodes can communicate with IPv4 hosts by making use of translators, e.g. SIIT [38], NAT-PT [39]. However, since common set-ups take care primarily of the infrastructure and do not act effectively at the application level NAT-PTs need to be placed on the GGSN external (Gi) interface (e.g. on the edge of the operator's network and the public Internet) to minimise drawbacks. The Application Level Gateway does the translation as seen in Figure 6.39.

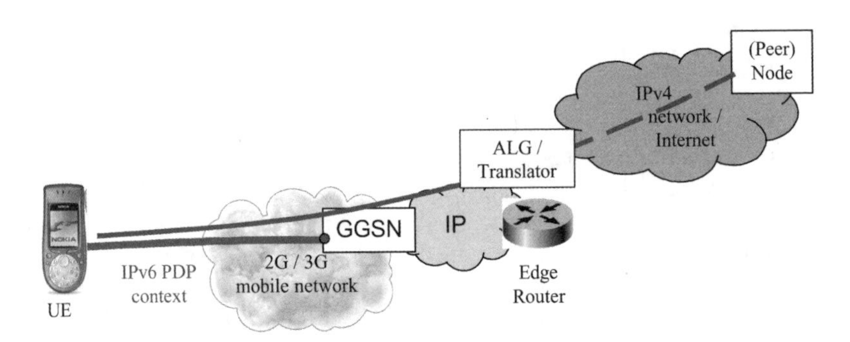

Figure 6.39 IPv6 UE connecting to an IPv4 node [33].

Despite the special arrangements, NAT-PT limitations to the 3GPP architecture are non-negligible as listed in Section 6.7.3 and [33,40].

Since 3GPP networks operating with IMS will need to handle a very large number of subscribers on a single GGSN (i.e. default router), e.g. each handling hundreds of thousands of connections, they become a single point of failure and affect the overall network performance.

Furthermore, IMS-IPv6 delay-sensitive applications will need minimum forwarding latencies within the IP backbone. Therefore, to support the unprecedented number of connections handled by the default routers (GGSN), optimisation of the translation mechanisms is required. Thus, to minimise the problems associated with NAT-PT, IETF [33] recommends the following actions:

1. separate DNS ALG from the NAT-PT node (in the 'IPv6 to IPv4' case);
2. ensure (if possible) that NAT-PT does not become a single point of failure;
3. allow for load sharing[56] between different translators, i.e. it should be possible for different connections to go through different translators.

6.7.6.8 IPv4 UE Link to an IPv6 Node

Current or legacy IPv4 nodes support today's popular Internet applications (i.e. E-mail, web-browsing, instant messaging, etc.), which will be also supported by IPv6 networks, which drive the Internet evolution. On the other hand, IPv4 EUs or legacy handsets will not necessarily support IPv6 dedicated applications; because these applications aim or will aim to exploit IPv6 enablers in all environments or where possible.

In addition, IPv4 UEs will rely primarily on proxies or local IPv4 and IPv6 enabled servers to communicate between private address space networks and the Internet; and thereby they will not attempt alone to link up with IPv6 nodes. Therefore, this scenario as seen in Figure 6.40, will have low occurrence or none. Besides, according to IETF [33], the DNS zones containing AAAA records for the IPv6 nodes will need to be served by at least one IPv4 accessible DNS server [34] and perform the necessary translations for the services that apply or the IPv4 EU can support from the IPv6 node or application server.

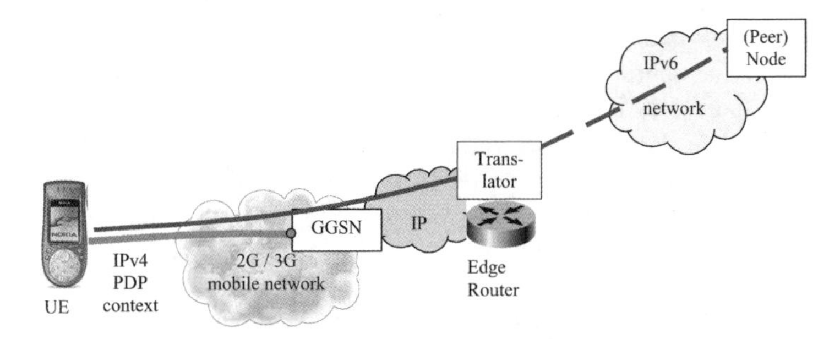

Figure 6.40 An IPv4 UE connecting to an IPv6 node [33].

6.7.6.9 Transition Characteristics in IMS

Until now we have indirectly referred to the interactions of IPv4/IPv6 with IMS, here we will summarise the direct transition scenarios outlined by EITF [33], i.e. *UE connecting to a node in an IPv4 network through IMS* and *Two IMS islands connected over IPv4 network.*

6.7.6.9.1 UE Link to a Node in an IPv4 Network through IMS

Here we assume an IMS UE_{IPv6} connects to a node in the IPv4 Internet[57] through the IMS, or vice versa. Because the IMS is exclusively IPv6 based [9,42], translators are required for

[56]Note that load sharing alone does not prevent NA(P)T-PT from becoming a single point of failure, see [41].

[57]A non-3GPP node, e.g. a fixed PC, with only IPv4 capabilities.

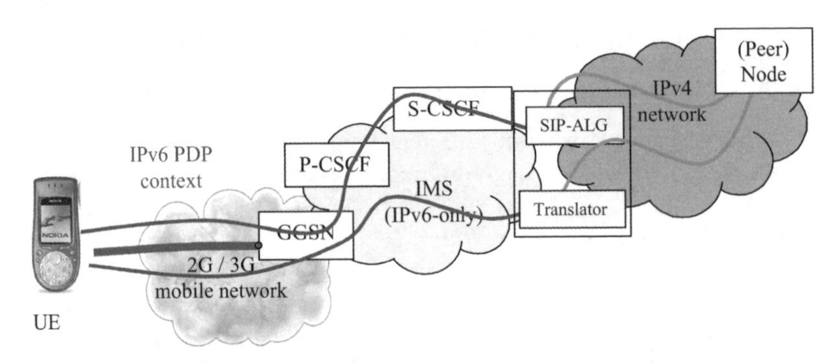

Figure 6.41 UE connecting to a node in an IPv4 network through IMS [33].

communication between the IPv6 IMS and legacy IPv4 hosts, i.e. a dual stack inter-operability does not apply to the IPv4 node.

In this case, which is illustrated by Figure 6.41, since signalling and user traffic separate, the IMS traffic translation follows two levels, i.e. Session Initiation Protocol (SIP) [43] for signalling in the Mm interface, and Session Description Protocol (SDP) [44,45] for the actual user data traffic in the Mb interface.

SIP and SDP transition takes place in the SIP/SDP Application Level Gateway (ALG), which changes the IP addresses transported in the SIP messages and the SDP payload of those messages to the appropriate version assuming inter-operability for DNS queries.

User data transport goes IPv4 to IPv6 protocol translation, where the transported traffic translates from IPv6 to IPv4, and vice versa. The user traffic translator through the SIP/SDP ALG (applying, e.g. Megaco protocol [46]) allocates an IPv4 address to the IMS UE, enabling it thus to route the user traffic within the legacy IPv4 network to the correct translator.

6.7.6.9.2 Two IMS Islands Communicating over IPv4 Network

Gradual introduction of IPv6 IMS solutions may imply that they will need to communicate through legacy IPv4 networks such as the Internet. In this case (Figure 6.42), end-to-end SIP connections for UE_{IPv6} and IMS_{IPv6} would be based on IPv6 passing through the IPv4 network. The steps of the *IPv6 UE Connecting to an IPv6 Node through an IPv4 Network* in the GPRS mode would apply.

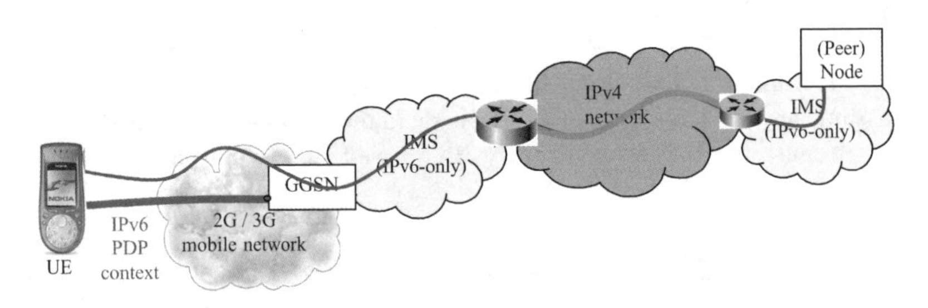

Figure 6.42 Two IMS islands connected via an IPv4 network [33].

6.7.7 Analysis of Inter-operability Scenarios

In the preceding sections we have summarised the key analyses made by the EITF team [33] regarding transitions between IMS sub-networks operating in mixed IPv4/IPv6 infrastructures. We have included all the options as to document all possible cases under study to find the most optimum solutions will allow seamless IMS *inter-operability* between existing and future or IPv4 and IPv6 Internet environments.

It is clear from the presentation of the transition scenarios and the principles of IPv4/IPv6 inter-operability outlined earlier, that there is not magic to resolve all cases. At present the limitations are related to physics, i.e. they will not change at less we operate them in totally new conditions. It is like the law of gravity, as long as it operates on Earth the phenomenon will be the same; however, if would operate it on the Moon the phenomenon will have different impacts on moving objects.

IMS services co-existing in IPv4 and IPv6 backbones will depend on the inter-operability techniques for optimum performance. Finding the ideal optimisation will be a serious matter and the work needs to start early to find the most suitable deployment modes. So far there is not an ideal model of inter-working, it will vary from operator to operator and it may even vary from application to application. Nevertheless, the following conclusions, which some have been already stated, may always apply:

1. The Dual Stack inter-operability mechanism stands as the best solutions to support IMS services in co-existing IPv4/IPv6 environments.
2. Tunnelling will be the next best mechanism which may go through different optimisation phases as the need arise or applications demand it.
3. Avoid using NAT-PT at all times if all possible.
4. Plan smooth migration to IPv6 early enough and start exploiting its features on advance without having to wait for the widespread penetration of IPv6 native networks. Today most IP equipments are coming with DS options.
5. Do not regress to an IMS with IPv4, the lack of public IPv4 address will eventually slow down the widespread usage of real-time IMS services.

6.8 DEPLOYING IMS SOLUTIONS

As described in the preceding sections, the SIP [43] based IMS will enable mobile network service providers with the capability to support and control sessions for voice, video and multimedia communication over IP as well as related services and applications like Instant Messaging, Push-to-Talk, Video Conferencing and many others. Figure 6.43 illustrates an IMS overview from commercial perspective where the main building blocks take part in the system integration and service offering process [47].

In Figure 6.43, the CSCF (call session control function) stands as the main IMS control server responsible for service control and session control, and acting as SIP registrar. It controls the bearer packet flow at the GGSN, e.g. a flow of IP-based voice packets. Other main IMS servers include the HSS (Home Subscriber Server), which deals with the subscriber mobility management, and the AS (Application Server) to support SIP-based applications and services.

6.8.1 IMS Commercial Implementation Issues

As covered earlier in the preceding sections, IMS technical specifications implies recommendations of; IPv6-based network infrastructure, new terminal equipment or SIP-enabled devices, new servers and new applications all at once, yet without suggesting transition steps in operating mobile networks. Therefore, it remains to describe a practical way forward to incorporate IMS to live networks, weather they are combined 2.5G and 3G or simply 2.5G or 3G networks. After all, IMS would work with all of them as illustrated in Figure 6.43.

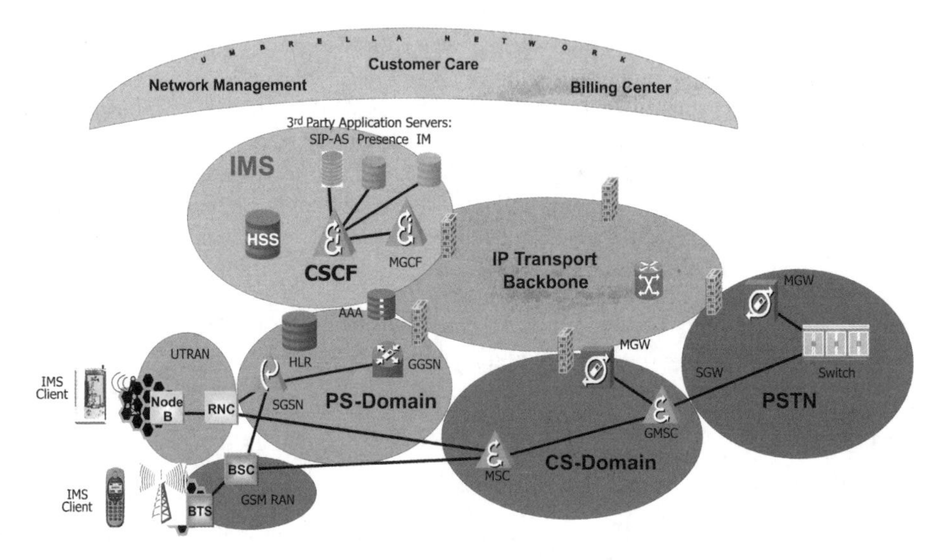

Figure 6.43 IMS solution overview, after [47].

Henceforth, the forthcoming sections illustrate one pragmatic IMS deployment option based on the contributions from Siemens's *'Early-Start'-IMS-System* proposal [47]. This approach could enable service-providers to offer rapidly the 1st phase of IP-based multimedia services within existing mobile network environments with modest investments, yet without preventing full-scale migration to wholly capable IMS system once IMS users grow, adopt and demand innovative real- and non real-time IP-multimedia-based applications.

6.8.1.1 Criteria for IMS Early Introduction

Early IMS deployment must satisfy at least the following criteria:

1. Subscribers adopting IMS shall not be obliged to buy a new device or terminal, but should be able to use IMS services with their existing equipment as much as possible.
2. Service provider would not need to install a complete IPv6-based network infrastructure just for IMS services.

3. Non-UMTS subscribers, e.g. GPRS, or fixed network, or WLAN subscribers should also be able to use IMS services.
4. Seamless migration from an early IMS solution to a full IMS with all capabilities specified by 3GPP should be possible.

6.8.1.2 Overcoming IMS-Enabled Device Limitations

While IMS may exploit existing UMTS or GPRS infrastructure to support IP multimedia, a subscriber of a full IMS System would typically have to purchase new Terminal Equipment, mainly because of the 3GPP requirement to run an IMS-specific user authentication. A re-use of existing equipment is therefore not possible, since IMS does not allow usage of existing SIM cards for GPRS or UMTS (USIM) to perform user authentication for IMS. As a result, the need to use new terminal equipment becomes non-negligible limitation and cost factor to promote early IMS adoption.

To practically enable IMS Services through existing terminals, it should be possible to download the IMS Client SW capable of handling SIP protocol into these devices. Current tests indicate the possibility to specify vendor-independent IMS Client SW stacks for existing GPRS/UMTS terminal equipment like PDAs, Pocket PCs, Symbian-based equipment and smart phones, since most of these terminals already allow application (e.g. games, user interface apps, etc. through the air interface or fixed ISP channels, e.g. PC and mobile phone synchronisation features.

6.8.1.3 Overcoming New IP Infrastructure Requirements

Technical specifications, i.e. 3GPP recommendations imply IMS solely based on IPv6 with implicitly IPSec features, which would mean that to deploy full IMS capabilities, and network service provider would have to implement a new IPv6-based control infrastructure able to inter-work with existing IPv4-based Internet. Likewise, it would have to ensure that IMS applications can work on top of IPv6 system.

To facilitate the transition to a new IMS control infrastructure based on IPv6, existing IPv4-based GPRS/UMTS core network may be used while applying double IPv4/IPv6 stacks where possible. Therefore, IPv4 and new IPv6-routers can co-exist. Likewise, switches or other existing IPv4-based equipment allowing access to application servers can remain. The latter also implies that there would not be rush to IPv6 SW upgrade in the device side.

6.8.1.4 Going Beyond and Before UMTS

Technical specifications also imply primarily a UMTS framework even though many IMS-provided services can be channelled, e.g. through GPRS networks, WLANs or even fixed networks for that. The implications for service provider, which has not advanced UMTS deployment plans or access to such, are delays on IMS service offers. The latter would be un-necessary restriction limiting early return-of-investment.

Figure 6.44 illustrates an early configuration IMS, where one central IMS sub-network connects to an existing PS-Domain network, and where the main IMS control servers

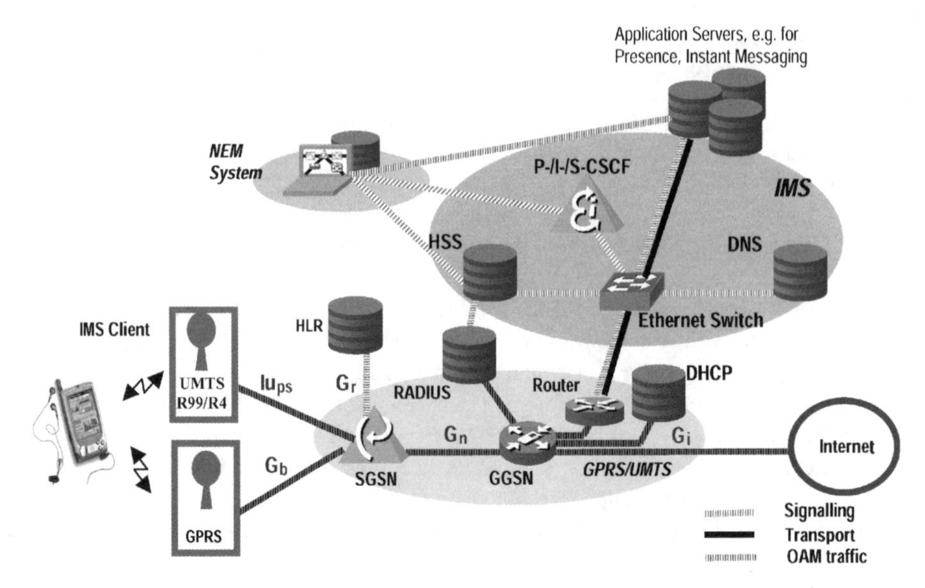

Figure 6.44 An early IMS Deployment Topology option, after [47].

communicate through a local Ethernet Switch. In an early IMS deployment, the different CSCF 'roles' specified by 3GPP (i.e. proxy CSCF, interrogating CSCF and serving CSCF) might physically implemented on the same CSCF server.

Figure 6.44 also shows IMS users access the IMS via GPRS or UMTS with their signalling messages (e.g. SIP session establishment) and related transport information (e.g. voice-over-IP packets) getting transferred through the network. In addition, we also see the basic Network Element Management (NEM) and a number of application servers, providing SIP-based applications, e.g. related to presence, or instant messaging [47].

6.8.1.5 Assuring Network Security

To maximise synergies with existing network security solutions, an Early-Start IMS security option may use existing GPRS/UMTS network security approach.[58] In such scenario, we protect the network at its boundaries against unauthorised access, but assume secure internal exchange of information between local network elements. This implies IMS traffic within the IMS, and between the IMS servers and the PS domain of the UMTS/GPRS does not have to be encrypted by mechanisms like IPSec, but are secured from external intrusions through firewalls at the network borders without direct access by 3rd party application servers.

6.8.1.6 Subscriber Authentication in Early IMS

If we assume a user has subscribed to IMS by downloading the Early-Start IMS SW into the GPRS/UMTS Terminal Equipment, the following would be expected [47] (see Figure 6.45):

[58]Often referred to as the 'Walled Garden Model'.

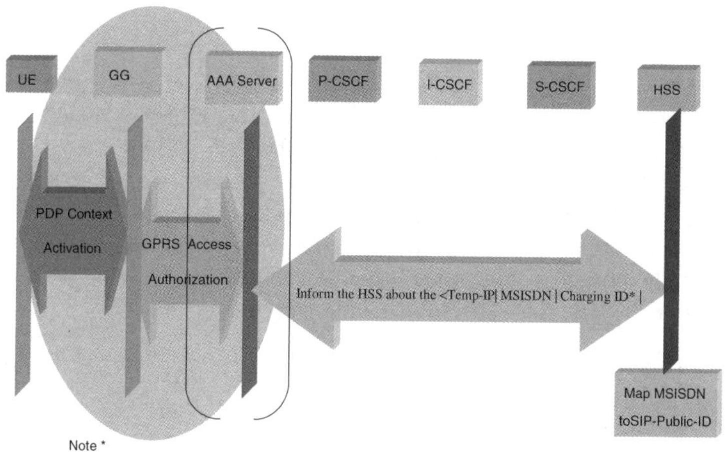

Figure 6.45 GPRS/UMTS and IMS correlation [47].

- Before the Early-Start-IMS subscriber gets authenticated at the IMS level, he will be authenticated at the GPRS/UMTS bearer level.[59] During this authentication, the GGSN will assign a temporary IP (Temp-IP) address to the IMS device.
- After a GPRS/UMTS successful authentication, the access network sends a RADIUS[60] message to the HSS containing an information vector [Temp-IP | MSISDN | Charging-ID | GGSN address | SGSN address], which will be stored there, together with the related SIP Public User ID of that user.
- Registration at the IMS level takes place following the 3GPP-standardised IMS specs, where the S-CSCF receives the SIP Public User ID and the Temp-IP address from the user, and will compare these values with the related information stored in the HSS for authentication. If they match, the S-CSCF will retrieve a related 'token' from the HSS,[61] forward the token to the IMS user and store it for further checks.
- The token is then included by the IMS user in the header of each subsequent SIP message (e.g. SIP Invite) for authentication.
- During re-registration [9] we recommend a re-authentication of the IMS user (and the issuing of a new token).

The above security level and authentication will match the current WAP/HTTP access, which can serve until higher security level of the IMS-AKA mechanism for the full 3GPP-compatible IMS gets implemented.

On the other hand, as per [47] the token-based mechanism may also serve as a single-sign-on mechanism for further authentication functions at the application level, e.g. if a subscriber wants to configure his supplementary services like session forwarding and session barring, or if he wants to modify his Buddy List on a Presence Server.

[59] Following GPRS/UMTS specs.
[60] Must have the necessary extra IMS authentication parameters besides GPRS/UMTS.
[61] i.e. random number corresponding to that SIP Public User ID.

6.8.2 Accessing Early-Start IMS Solutions

While GPRS/UMTS-based networks stand as ideal access for Early-Start IMS, it may be extended to be accessible from other networks in the context of the so-called fixed-mobile convergence, Bluetooth for personal area coverage and Wifi WLAN.

In the case of Bluetooth, a local gateway from the personal coverage area to the fixed public network would provide access to mobile network services. In the second case, IPSec-based VPN tunnel from the WLAN terminal via the WLAN access point through a packet data gateway at the border of the mobile network, can transfer the authentication messages to the Early-Start IMS via that VPN tunnel and can serve as an access.

6.8.3 Seamless Migration to Fully Capable IMS

It may appear that an early IMS introduction will create a series of inconveniences, yet if it paves the way for user pioneer experience and final acceptance of innovative IP-multimedia services, it sets the way of seamless migration to fully capable 3GPP specified IMS.

An IMS solution deployed entirely according to standard specifications will exploit in full UMTS capabilities with all its releases in view and benefits from IPv6 (i.e large address space, avoid NATs, auto-configuration, etc.) and provide higher security level while using the standardised IMS AKA (Authentication and Key Agreement) mechanism as briefly illustrated in Figure 6.46.

A smooth and graceful transition from an Early-Start IMS would take place with maximum re-use of already installed IMS-dependent elements (i.e. CSCF, HSS, AS, etc.) already implemented in the Early-Start IMS, as well as the related interfaces and protocol software, with the exception of authentication as noted early.

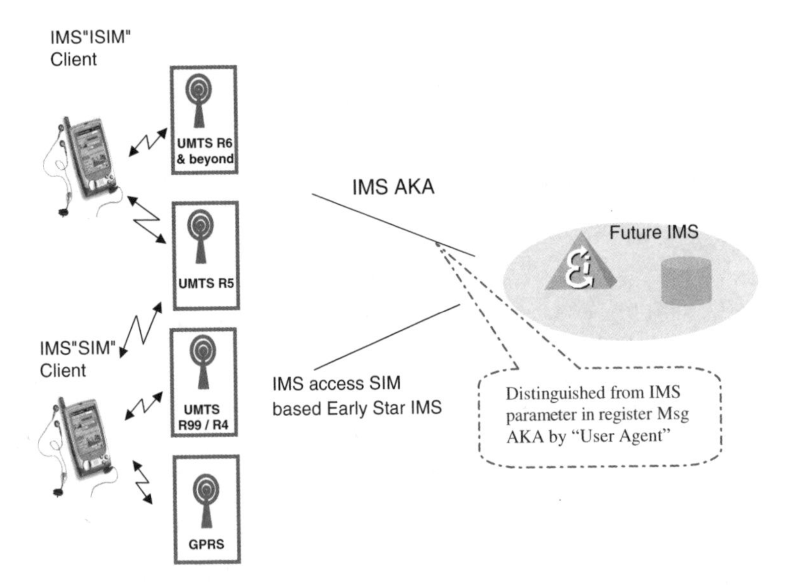

Figure 6.46 Evolution of IMS terminal capabilities, after [47].

To prevent rapid expansion of HW changes in the existing IMS system, the transition to the full 3GPP-IMS would start in the IMS core network through SW upgrades or serve upgrades, whilst using the existing terminal equipment from the Early-Start IMS. Thereby minimising user experience uncertainties. A 'User Agent' parameter in the terminal SIP Register Message can indicate the attachment options, i.e.:

- terminal equipment type (pocket PC, Smartphone, etc.);
- access network (GPRS, UMTS R99, R4, etc.);
- IMS Software package (i.e. early version, R5, R6, etc.).

6.8.3.1 Migration to IPv6 Backbone

In Section 6.7.6 we have outlined the different IPv4/IPv6 inter-working scenarios. Thus, there exist several transition paths, which would take place as the number of element become IPv6 compatible. Find also more IPv4/IPv6 transition details in [48]. On the other hand, commercial deployment of full 3GPP-IMS will most likely support both IPv4-based terminals and IPv6-capable terminals simultaneously through dual stacks [49].

Thus, as along as the migration path starts with the IMS signalling, transport and services first, other parts of the network such as OAM, data provisioning, terminals can follow without much service interruption. Especially if IPv4-only terminals continued to be supported while new terminals might use IPv6 from the start [47].

6.9 CONCLUSIONS

In this chapter we have provided an overview of the IMS architecture. We have also described its main functional elements and its key interfaces. In addition, we covered the essential items of the mobility management, application level registration, QoS, transport issues, etc.

Likewise, we have illustrated commercial possibilities of pioneer IP-based multimedia services introduction in mobile networks using a stepwise approach, starting with only limited investment for an *Early-Start IMS* based on Siemens [47] proposal, which can stand as basis for similar approaches.

Key advantages of an Early IMS introduction over existing GPRS and UMTS terminals and networks will have non-negligible impact in the immediate and medium term process to build IP-multimedia-based services end to end. In particular, if seamless migration to full IMS is possible. The fact that we can enable secure user authentication for current mobile terminals and enable IMS subscriber access in the context of fixed-mobile convergence through WLAN and Bluetooth makes early IMS introduction an attractive opportunity for innovation in 'all IP' scenarios.

Notwithstanding, the ultimate goal is to arrive to a full IMS solution, which exploits IPv6 and maximises the use of the IMS features at all levels, in particular at the terminal point, where the large flexibility and vast availability of IP addresses will facilitate all types of innovative applications.

REFERENCES

1. Drafting Team, Vision and Road Map for UMTS Evolution, TSGS#8(00)0337, 2000.
2. 3GPP, Study on Release 2000 SERVICES and Capabilities (3G TS 22.976), V2.0.0, 2000–06.
3. TS 22.003, 3rd Generation Partnership Project, Technical Specification Group Services and System Aspects, Circuit Teleservices Supported by a Public Land Mobile Network (PLMN).
4. TS 22.002, 3rd Generation Partnership Project, Technical Specification Group Services and System Aspects, Circuit Bearer Services (BS) Supported by a Public Land Mobile Network (PLMN).
5. TS 22.004, 3rd Generation Partnership Project, Technical Specification Group Services and System Aspects, General on Supplementary Services.
6. TS 22.060, General Packet Radio Service (GPRS) Stage 1.
7. 3GPP, Architecture Principles for Release 2000 (3G TR 23.821), V1.0.1, 2000–07.
8. 3GPP TS 23.002: Network Architecture.
9. 3GPP TS 23.228 V5.7.0(2000–12), IP Multimedia Subsystem (IMS) Stage 2,
10. 3GPP TS 23.002 V5.7.0 (2002–6), Network Architecture R5.
11. IETF RFC3261: SIP: Session Initiation Protocol.
12. ITU-T Recommendation H.323: Packet-Based Multimedia Communications Systems.
13. ITU-T Recommendation H.248: Gateway Control Protocol.
14. 3GPP TS 32.200: Telecommunication Management; Charging Management; Charging Principles.
15. 3GPP TS 32.225: Telecommunication Management; Charging Management; Charging Data Description for IP Multimedia Subsystem.
16. 3GPP TS 29.078: Customised Applications for Mobile network Enhanced Logic (CAMEL) Phase 3; CAMEL Application Part (CAP) specification.
17. 3GPP TS23.218 V5.3.0 2002–12, IP Multimedia (IM) Session Handling & IP Multimedia (IM) Call Model, Stage 2.
18. 3GPP TS 24.229: IP Multimedia Call Control Protocol Based on SIP and SDP; Stage 3, 2002–12.
19. IETF RFC 3265: Session Initiation Protocol (SIP) Event Notification.
20. 3GPP TS 32.225 V5.10, Charging Data Description for the IP Multimedia Subsystem (IMS), R5, Dec 2002.
21. 3GPP TS 24.228: *Signalling Flows for the IP Multimedia Call Control Based on SIP and SDP*
22. 3GPP TS 23.207: *End-to-end QoS concept and architecture.*
23. 3GPP TS 23.060: *General Packet Radio Service (GPRS)*; Service description; Stage 2.
24. 3GPP TS 33.203: *Access Security for IP-based services.*
25. 3GPP TS 26.235: *Packet Switched Multimedia Applications*; Default Codecs.
26. ITU Recommendation G.711: *Pulse code modulation (PCM) of voice frequencies.*
27. 3GPP TS 23.221: *Architectural Requirements.*
28. 3GPP, Combined GSM and Mobile IP Mobility Handling in UMTS IP CN 3G TR 23.923 Version 3.0.0, 2000–05.
29. IETF RFC 2002 (1996): IP Mobility Support, C. Perkins.
30. Internet draft, Johson and Perkins, Mobility Support in IPv6, October 1999. http://www.ietf.org/internet-drafts/draft-ietf-mobileip-ipv6-09.txt.
31. Silvia Hagen, IPv6 Essentials, O'Reilly, July 2002.
32. Durand, A.: *Issues with NAT-PT DNS ALG in RFC2766*, January 2003, draft-durand-v6ops-natpt-dns-alg-issues-00.txt, work in progress.
33. J. Wiljakka (ed. Nokia), Analysis on IPv6 Transition in 3GPP Networks, draft-ietf-v6ops-3gpp-analysis-03.txt, http://www.ietf.org/internet-drafts/draft-ietf-v6ops-3gpp-analysis-03.txt.
34. Durand, A.: IPv6 DNS Transition Issues, October 2002, draft-ietf-dnsop-ipv6-dns-issues-00.txt, work in progress.
35. Carpenter, B., Moore, K.: *Connection of IPv6 Domains via IPv4 Clouds*, RFC 3056, February 2001.

36. De Clercq, J., Gastaud, G., Ooms, D., Prevost, S., Le Faucheur, F.: *Connecting IPv6 Islands across IPv4 Clouds with BGP*, October 2002, draft-ooms-v6ops-bgp-tunnel-00.txt, work in progress.
37. Cristallo, G., Gastaud, G., Ooms, D., Galand, D., Preguica, C., Baudot, A., Diribarne, G.: *Connecting IPv6 Islands within an IPv4 AS*, February 2002, draft-many-ngtrans-connect-ipv6-igp-02.txt, work in progress, the draft has expired.
38. Nordmark, E.: Stateless IP/ICMP Translation Algorithm.
39. Tsirtsis, G., Srisuresh, P.: Network Address Translation-Protocol Translation (NAT-PT), RFC 2766, February 2000.
40. Huitema, C., Austein, R., Dilettante, B., Satapati, S., van der Pol, R.: *Evaluation of Transition Mechanisms for Unmanaged Networks*, November 2002, draft-huitema-ngtrans-unmaneval-01.txt, work in progress.
41. Durand, A.: *NAT64 - NAT46*, June 2002, draft-durand-ngtrans-nat64-nat46-00.txt, work in progress, the draft has expired.
42. 3GPP TS 23.221 V5.7.0, *Architectural requirements* (Release 5), Dec 2002.
43. RFC 3261 - Rosenberg, J., *et al.*: SIP: Session Initiation Protocol, June 2002.
44. RFC 2327 - Handley, M., Jacobson, V.: SDP: Session Description.
45. RFC 3266- Olson, S., Camarillo, G., Roach, A. B.: Support for IPv6 in Session Description Protocol (SDP), June 2002.
46. RFC3015 - Cuervo, F., *et al*: Megaco Protocol Version 1.0, RFC 3015, November 2000.
47. Bernhard Petri, *Early Introduction of IP Based Multimedia Services in Mobile Networks*, Siemens, AG, 2003.
48. http://www.ietf.org/html.charters/ngtrans-charter.html
49. 3GPP TS 33.203: Access Security for IP Based Services, 2002–12.

7

Deploying 3G Networks

7.1 BACKGROUND

Logically deploying 3G networks implies dimensioning and implementing corresponding elements within a geographical area, where an operator would desire to offer advanced mobile communications services, e.g. voice, mobile Internet, video telephony, etc.

In the preceding six chapters we have outlined the service requirements and technical specifications of the UMTS solution. In this chapter we aim to describe the application of the proposed solutions and go through the process of designing a network to provide UMTS services.

Before describing the results of field study with reference parameters based on real scenarios, we provide the necessary principles for dimensioning and implementing a 3G network using the UMTS technology. We then present results of dimensioning and introduce the functional capabilities of the selected elements.

7.2 NETWORK DIMENSIONING PRINCIPLES

Figure 7.1 identifies non-exhaustively the major areas to dimension a 3G network. It summarises as well the essential tasks to obtain the necessary count of elements for implementation and network deployment.

To simplify the whole process we will group the dimensioning tasks to four key iterative actions, i.e.

- radio coverage and traffic flow identification,
- system dimensioning,
- network configuration and verification (i.e. radio, core and transmission),
- implementation and deployment.

In the first action, *radio coverage* depends on both propagation environment, (i.e. service population areas) and the traffic flow expected. Through a computerised process and

All IP in 3G CDMA Networks J. Castro
 ISBN: 0-470-85322-0

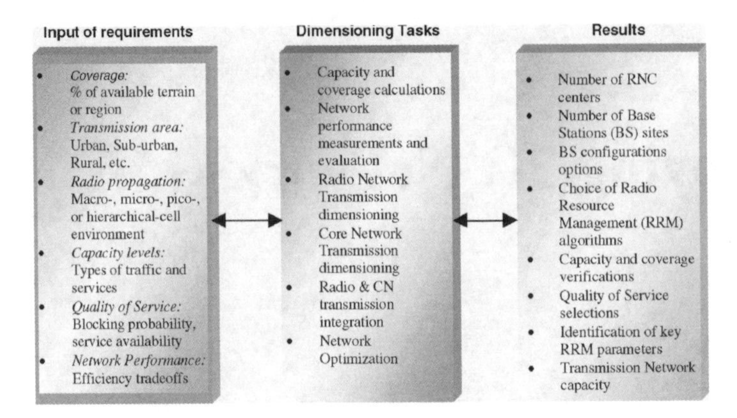

Figure 7.1 Essential network dimensioning tasks.

classical optimisation, the main output consists of the identification of sites for BS (or Node B) location. The latter will depend on the projected service strategy and the BS range and capacity. The service strategy will take into account the traffic flow generated based on the subscriber profiles of service utilisation levels and population densities. The radio coverage task will include or use the multi-path channel models, and reference service rates illustrated in Chapter 2.

System dimensioning involves the optimisation of coverage and capacity based on macro-cells and micro-cell in densely populated areas. It aims to take into account the asymmetry of traffic in the UL and DL and includes in the optimisation the TDD mode to maximize capacity and flexibility in micro- and pico-cells.

Network configuration and verification consolidates the coverage and site location exercise by starting a process for the integrated solution of radio and core elements. Based on the capacity and service targets requirements sets the 3G system architecture for the Node Bs and CS and PS elements in the core network side. It also looks at the impact on the transmission subsystem.

Implementation and deployment completes the 3G-network design process by realising the projected site locations, service target requirements and time to service. It takes into account the solution adopted for the network deployment, e.g. sharing sites with existing 2G BSs and evolution of CN elements, or a complete new overlay network on the top of existing 2G system. It may also apply to totally green field network, i.e. a new deployment. It will also take into account the hierarchy of the network, i.e. the macro- and micro-layers where applicable.

When deploying in the macro-cell environment primarily with the FDD mode or the WCDMA technology, the implementation will take into account the coverage dependency on the transmission rates and technology availability in terms of antenna configuration and interference minimizing features. Thus, the four actions or steps outlined above do have an iterative process.

7.2.1 Coverage and Capacity Trade-off in the FDD Mode

From the practical side, as mentioned in earlier chapters and Section 7.3 of this chapter, in the FDD mode, which uses WCDMA techniques, the interference increases with the number of active users, thereby limiting capacity. Within this soft limitation, the system quality decreases continuously until service performance degrades to an intolerable state. This state leads to the *breathing cells* phenomenon, i.e. when user numbers get too high, the quality of users at the *cell-edge* degrades rapidly to the point to drop the link or the call. Such event implies that cell radio coverage shrinks. On the other hand, when call drops occur, interference decreases for the remaining users' cell area coverage grows again. This is what we call the trade-off between capacity and coverage in the FDD mode.

Cell coverage and capacity depend thus on the received bit-energy-to-total-noise-plus-interference ratio $E_b/(N_0 + I_0)$ on each cell part for the DL and in the BS for the UL. This means that any parameter, which affects the signal level and/or the interference,[1] or reduces the $E_b/(N_0 + I_0)$ requirements,[2] has an impact on cell coverage and capacity, as well as on the overall system.

7.2.1.1 Soft Handover and Orthogonality

We described soft handover in Chapter 4 from the design side; here we look at it from the performance and dimensioning side. In this context, an MS performs handover when the signal strength of a neighbouring cell exceeds the signal strength of the current cell with a given threshold. In *soft handover* position, an MS connects to more than one BS simultaneously. Thus, the FDD mode uses *soft handover*[3] to *minimize interference* into neighbouring cells and thereby *improve performance* through macro-diversity, i.e. we combine all the paths together to get a better signal quality. We also reduce power originating from two or more BSs to reach the same mobile's E_b/N_o requirement while we combine the paths.

We separate the information signal of different users by assigning to each one a different broadband and time limited, user specific carrier signal derived from orthogonal code sequences (e.g. OVSF codes). When completely orthogonal,[4] we can perfectly separate synchronously transmitted and received signals. However, this does not occur in the UL, for example, due to different propagation paths, i.e. different distances with different time delays. In the DL even if all signals originate from a single point and the parallel code channels can be synchronised, there is still no perfect signal separation. As a result, we cannot maintain complete orthogonality due to multi-path propagation, and we have to use orthogonality compensation factors as noted in, e.g. Chapter 2.

7.3 PARAMETERS FOR MULTI-SERVICE TRAFFIC

While some earlier[5] 2G mobile systems measure network quality mainly for one service, e.g. speech, UMTS has many different bearer services with varying quality requirements. We

[1]Interference = intracell interference and intercell interference.

[2]Interference here implies intracell interference and intercell interference.

[3]Softer handover is a soft handover between two sectors of a site.

[4]Two function orthogonality, e.g. $g(t)$ and $s(t)$ occurs when their cross-correlation function equals to zero.

[5]Today GSM has evolved to a more than just speech network, it does also GPRS and HSCSD.

characterise these differing services by parameters such as the bit-rate, the maximal delay, connection symmetry and tolerable maximum BER. As a result, to accurately dimension or design a network for multiple services, we need to use different traffic models and settings. We have to plan the BS numbers to handle the expected service mix. The multiple set of services will have *different impact on capacity and coverage*. For example, user bit rate will have large impact on coverage as illustrated in Figure 7.2. On the other hand, we can often adjust all services to the same cell range by individually adjusting the emitted power of each service.

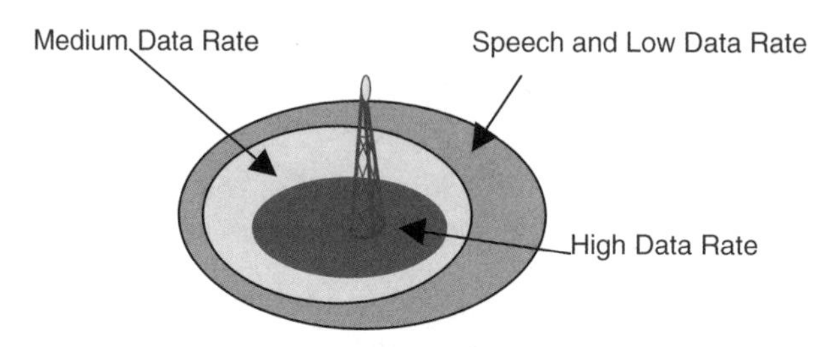

Figure 7.2 Transmission rates and coverage.

7.3.1 Circuit- and Packet-Switched Services

When dimensioning a 3G network in the FDD mode, e.g. the number of concurrent channels derived to cope with the different service requirements becomes the main input of the link budget analysis. Thus, if we have to manage traffic beyond a cell loading of 30%, any small load variation will have direct impact on the cell radius. We then have to achieve a dimensioning to *meet the peak traffic during the busy hour* in order to obtain a stable network. This stability will depend on how we treat the different type of service, i.e. Real-Time (RT) or circuit-switched types and Non-real Time (NRT) or packet-switched types.

7.3.1.1 Circuit Switched (CS)

To dimension capacity for CS services we can follow the classical approach, i.e. given the offered load (Erlangs) and the blocking rate, we derive from the traffic assumptions the offered traffic at the busy hour per cell (Erlang). Here we would assume that the cell radius gets optimised iteratively with the link budget. Then, from Erlang B table we would determine the number of concurrent channels required during the busy hour for a given blocking rate. Although the traditional solution may allow us to estimate CS capacity easily, it may also over-dimension the required number of channels. Thus, it seems imperative that we use the multi-service Erlang B formulation and pool the resources for better availability on demand. This implies that we offer the CS channels depending on the required number, e.g. if one service requires 2 channels and the other 10, both can benefit from the pool, which may contain 20 channels. The latter would also imply that we could use different blocking rates for each service. For example, voice calls can tolerate more degradation than video calls.

7.3.1.2 *Packet-Switched Services*

As in the CS, although with more sophistication, we also need to estimate the number of concurrent channels required for PS traffic. This number of channels will correspond to the peak traffic during a Busy Hour (BH), which as in the CS, we determine also from the traffic assumptions of the offered load during the busy hour per cell expressed in kbits. In general, we treat each service independently to meet the different grade of service or asymmetry required.

We calculate the number of PS service channels by accounting a duration window corresponding to an acceptable delay, (e.g. $d \approx 0.5$–0.7 s) for a given service. From the principles outlined in Chapter 2, we can illustrate the calculation for WWW application[6] as follows.

We take 384 kbps service with packet length $\zeta = 480$ bytes. From the total BH traffic for a given reference area we calculate the mean offered data rate m in kbps. Translating this into a mean packet arrival p rate, i.e. $p = (m \times d)/\zeta$. Then assuming a Poisson packet arrival distribution for all users, with a mean p, we obtain the Probability Density Function (PDF), as well as the Cumulative Density Function (CDF). Figure 7.3 illustrates the peak packet arrival rate h at 95% time probability.

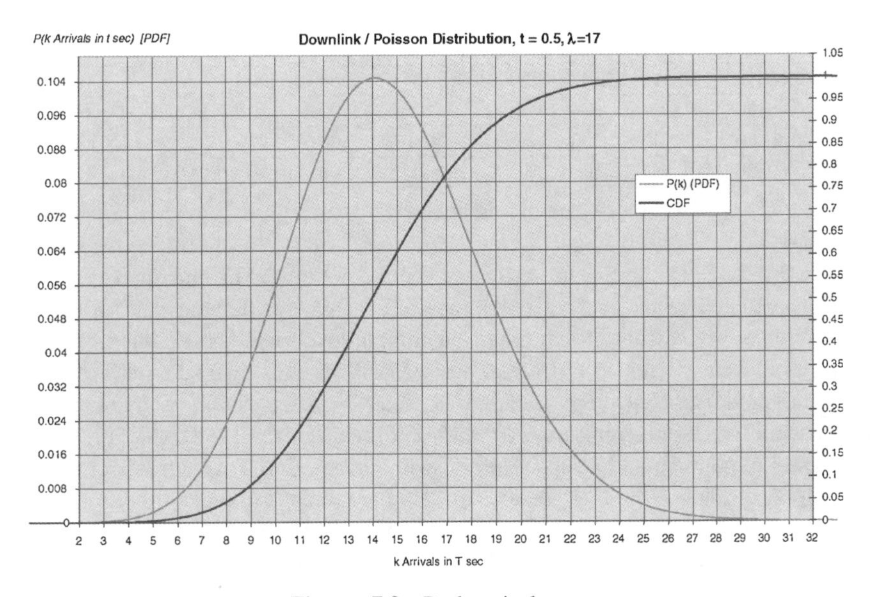

Figure 7.3 Peak arrival rate.

Utilising the upper 95% time probability of the packet arrival rate (Figure 7.3) and applying the typical packet length we translated back into kbps. We then calculate the number of channels (ch) dividing by the service bearer rate r, i.e. ch = h (kbps)/r. We can summarize the process as Chs = (1/Serv rate) × (1/Serv delay) × CDF $P\{$(m/Serv delay $\times \zeta$), 95%$\}$, where CDF $p(x, y)$ corresponds to the point of probability on the CDF associated to the Poisson law of mean x, and where m represents the mean offered data rate

[6]For example, e-commerce, on-line banking, file transfer, information DB access, etc.

in kbps. We should note here that this process can be inefficient with low traffic in the cell, resulting in over-dimensioning for PS services. Thus, other distributions, with fluid-flow type models shall be applied.

7.4 ESTABLISHING SERVICE MODELS

Before deploying new elements in a mobile telecommunications network, weather it is an existing system based on Second Generation (2G) technology like GSM, or it is a new one like UMTS, we will need a projection for the potential number of subscribers. In this chapter, we consider a field study example to extrapolate some subscriber numbers from two growth forecasts in[7] central Europe. Although these projections will not necessarily apply to particular deployment scenario, it will serve to illustrate network dimensioning based on the split of voice only and combined (voice + data) services.

In Table 7.1 we illustrate estimations for a 10 year period where 2G values correspond primarily to GSM voice services and 3G values to data starting with GPRS in the first 2 years. Thereafter, full multimedia services expand rapidly at the introduction of UMTS in existing GSM networks. A major breaking point occurs around 2005 with high predominance of 3G type services.

Table 7.1 Subscriber growth within a 10 year period (subs in 1000s)

Year subs	2000	2001	2002	2003	2004	2005	2006	2007	2008	2009
2G	750	1000	900	600	400	300	200	150	100	50
3G	0	0	300	700	1000	1200	1400	1500	1600	1700
Total	750	1000	1200	1300	1400	1500	1600	1650	1700	1750

In Table 7.2 we illustrate the subscriber growth beginning 2002 when penetration of data has already reached about 30% of the total traffic. Here we assume that GPRS carrying wireless IP type services has grown to non-negligible levels right before the introduction of UMTS. Despite the stretch to a 15 year period, 2005 stands again as the breaking point towards full predominance of multimedia services. Nonetheless, as in the projections of 10 year period, voice-only services will remain a good fourth of all traffic.

After 2005 in both cases the subscriber growth appears low. This can reflect the fact that the overall penetration of mobile services in the region begins to reach the limits or that the market share between operators starts to stabilise. Thus, for all practical purposes, in

Table 7.2 Subscriber growth with a 15 year period (subs in 1000s)

Year subs	2002	2003	2004	2005	2007	2009	2011	2013	2015	2017
Voice	900	600	400	300	150	100	50	0	0	0
Voice + Data	300	700	1000	1200	1450	1550	1650	1750	1800	1850
Total	1200	1300	1400	1500	1600	1650	1700	1750	1800	1850

[7]The forecast has harmonised numbers, which does not apply to any operator or service provider in particular.

particular for the network dimensioning exercise in this field study we will consider primarily the data from 2002 to 2005 from Table 7.2.

7.5 PROJECTING CAPACITY NEEDS

Based on the preceding section dimensioning in this field study would then begin for about 1.5 million subscribers all using either voice-only or multimedia services. The proportion will depend on the business strategy and the type of service products offered. Business strategy will have strong relationship with the market segment addressed and the penetration of type of services proposed. If we take Switzerland, for example, penetration of mobile services will reach 60% in all segments by the time we complete this writing. Clearly, the voice appears as the predominant service, although data through SMS and HSCSD and early GPRS may grow. This means that the market for multimedia services remains quite open even up to 100%. Thus, following a pragmatic approach, network dimensioning and capacity projections will imperatively be done for multimedia services addressing all segments.

Now for all practical purposes we identify three main segments, i.e. business, residential and mass market (see Chapter 6). The traffic distribution among these segments will depend on the subscriber demand, operator's service[8] offer and qualitative thinking. Nevertheless, looking at the data in Tables 7.1 and 7.2, about 70% of the market stands open for multimedia type services. If we distribute the latter as 40% mass market and 15% business and residential, respectively; then dimensioning should follow conventional wisdom.

Conventional wisdom will tell us that residential and business segments will tend to use larger transmission rates (e.g. 384 kbps) in sub-urban and urban areas, while mass-market subscribers will use medium rate services (144 kbps) from everywhere.

7.6 CELLULAR COVERAGE PLANNING ISSUES

Before discussing the fundamental parameters, assumptions and planning methodology, we select Switzerland as region in central Europe representing a typical subscriber population and complex geographical area for cellular planning. The mountainous landscape with large canyons and valleys, as well as hilly cities can make Switzerland an interesting point of reference.

7.6.1 The Coverage Concept

As illustrated in Figure 7.4 the ideal UMTS coverage concerns all types of environments, i.e. in-building (pico-cells), urban (micro-cells), sub-urban (macro-cells), and global (global cells). However, at this time we cover mainly pico-cells to macro-cells. While FDD coverage

[8]The operator's initiative and creativity on new services offers, product packages and business approach altogether will make a large difference. It will not depend only on the Internet traffic.

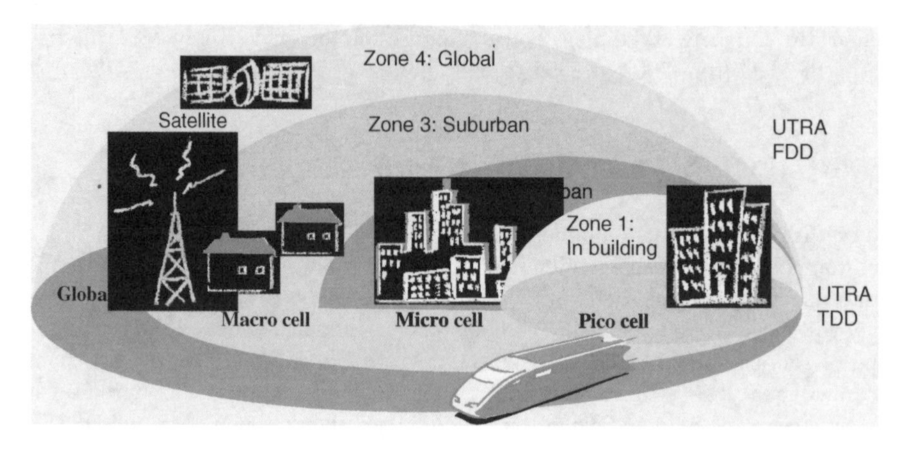

Figure 7.4 Service coverage summary.

here may apply primarily[9] to macro-cells, the TDD solution applies more to pico- and micro-cells. Figure 7.5 shows an option for combining the UTRA technologies for maximum coverage.

Hotspots stand as the main target to complement FDD with TDD. This seems justified if we consider that intensive wide-band demanding asymmetric services will take place to a greater extent in dense urban areas. Placing TDD over FDD will handle additional traffic, thereby adding capacity. The absence of TDD in dense areas would imply denser FDD site deployment or simply a micro-cell FDD implementation and/or additional carriers on the Node B hardware.

As indicated in Chapter 6, the motivation behind the introduction of 3G networks lays innovative applications; a 3G subscriber will not sign-up just because of the exiting UMTS network technology. At the end, service technology gets transparent, if a user's request for service gets satisfied; few will notice the underlying technology. Now, most innovative applications tend towards transmission asymmetry (e.g. multimedia Internet). Thus, we can meet areas with small cell sizes and a high data asymmetry more appropriately[10] with TDD than FDD. The lower efficiency of FDD results from larger number soft handovers in small cells, and the dynamic DL/UL resource allocation of TDD.

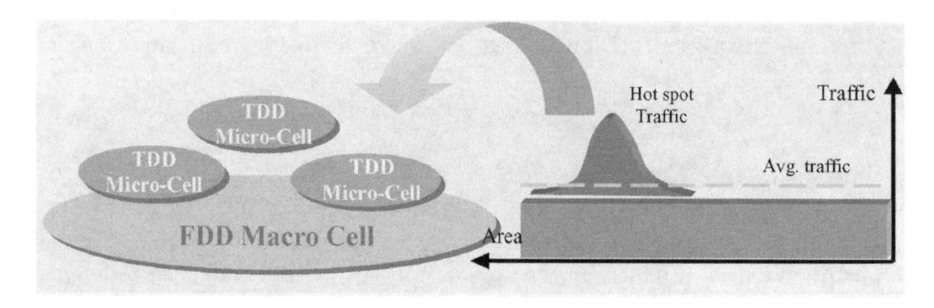

Figure 7.5 Combining FDD and TDD coverage.

[9]The FDD does also apply to micro-cells, and it is not precluded for usage in pico-cells.
[10]FDD needs 2 × 5 Mhz instead of 1 × 5 Mhz as for TDD in small cells with asymmetric traffic.

Thus, it seems reasonable to think that we do not need to wait for TDD until we need overall network capacity; we can exploit it at the introduction of UMTS in hot spot areas. This means that 3G coverage planning can benefit from TDD from the start. In this field study, we will assume that TDD can apply to dense urban areas and concentrate on macro-cell dimensioning for FDD or WCDMA.

7.6.2 Radio Network Parameter Assumptions

Figure 7.6 illustrates the population (shadowed parts) and geographical areas. Logically, an operator or service provider will aim to have 99% coverage for the populated area while maximising the geographical coverage. On the other hand, the penetration of UMTS at the introduction will not necessarily include all populated[11] environments. Thus, starting in the main cities and sub-urban areas, 3G network coverage can progress in three phases, i.e. 50%, 75 (80)% and 99%. For business strategic reasons in, e.g. Switzerland it would be expedient to cover also major ski resorts even if these areas do not have permanent population, but transitory during one-fourth of the year, i.e. the winter season. Which means a sound business case for the introduction of UMTS would start with more than just 50% coverage of the populated area.

With the assumptions above, in the sequel we will outline key issues when designing a macro-cellular network based on the FDD mode or WCDMA.

Figure 7.7 illustrates the conversion of population density to area coverage, where 50% of the population corresponds to about 10% of the coverage area. Thus, we can tailor coverage depending on strategy or demand once basic coverage has been achieved.

Table 7.3 illustrates the morphology distribution of the 50% and 75% population coverage. It indicates area coverage proportion in km^2 of the different service environments, i.e. Dense Urban (DU), Urban (U), Commercial/Industrial (CI), Sub-Urban (SU), Forest (FO) and Open (OP). It also indicates the service area proportions in of the total area corresponding to the 50% or 75% population density. These proportions serve as the points of reference to establish the number of subscriber per service area and plan accordingly the

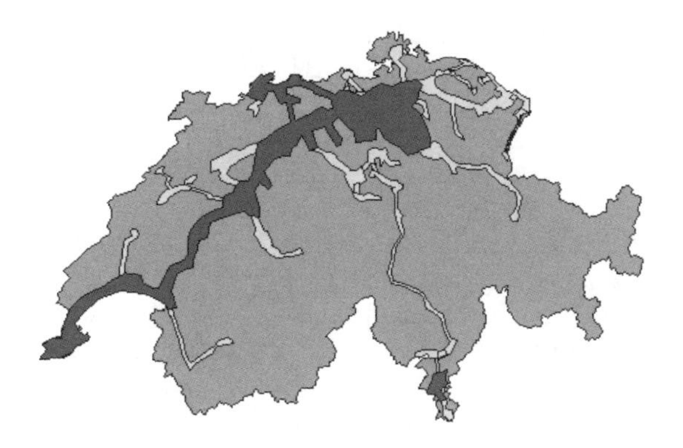

Figure 7.6 Population coverage example.

[11]Regulators in some countries are demanding only 50% coverage for initial coverage.

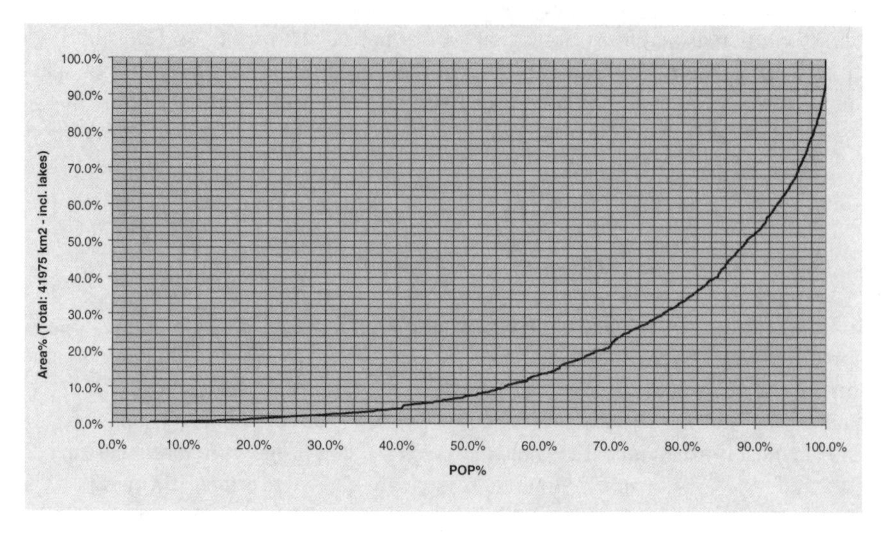

Figure 7.7 Population density conversion to area coverage.

number of sites or cells required for each service environment. It will also allow estimation of RF unit number according with the number of sectors per site.

Table 7.4 illustrates the service quality assumptions for projected radio bearer services in UMTS. The transmission rates or bearers corresponding to the service environments represent the most common services. On the other hand, we do not necessarily exclude speech, LCD 384, LCD 2048 and UDD 2048. For example, voice service may take the following assumptions following suggestions: Adaptive Multi-Rate (AMR) codec with a bit-rate of 12.2 kbps and with 50% voice activity factor. We can also assume 20 mE/subs with the following average holding times per subscriber:

Table 7.3 Morphology distribution of the population density

Coverage Area	50% POP	75% POP
Total size (km^2)	4067.00	6741.00
Morphology distribution (sqkm)		
Dense urban	2.33	2.37
Urban	9.90	10.60
Commercial/industrial	101.00	138.00
Sub-urban	387.00	617.00
Forest	1270.00	1961.00
Open	2297.00	4012.00
Morphology distribution (%)		
Dense urban	0.06%	0.04%
Urban	0.24%	0.16%
Commercial/industrial	2.48%	2.05%
Sub-urban	9.52%	9.15%
Forest	31.23%	29.09%
Open	56.48%	59.52%

Table 7.4 Service quality requirements

Area/Bearer Service	LCD 64	LCD 144	UDD 64	UDD 144	UDD 384
Dense Urban	INDOOR LCP95%	INDOOR LCP95%	INDOOR LCP95%	INDOOR LCP95%	INDOOR LCP95%
Urban	INDOOR LCP95%	INDOOR LCP95%	INDOOR LCP95%	INDOOR LCP95%	INDOOR LCP95%
Commercial/ Industrial	INDOOR LCP95%	INDOOR LCP95%	INDOOR LCP95%	INDOOR LCP95%	INDOOR LCP90%
Sub-urban	INDOOR LCP90%	INDOOR LCP90%	INDOOR LCP90%	INDOOR LCP90%	
Forest	IN-CAR LCP 90%	IN-CAR LCP 90%	IN-CAR LCP 90%	IN-CAR LCP 90%	
Open	IN-CAR LCP 90%	IN-CAR LCP 90%	IN-CAR LCP 90%	IN-CAR LCP 90%	

- Holding time of a mobile-originated call is 75s.
- Holding time of a mobile-terminated call is 90s.

The traffic distribution is estimated:

- Proportion of call attempts that is mobile-originated is 0.60 and mobile-terminated is 0.40.

LCD 384 and LCD 2048 can be considered for indoor transmission with LCP95%. The number of subscriber with these rates in each cell will not exceed couple of users.

The traffic data example illustrated in Table 7.5 portrays a possible distribution of the different types of bearer services. Note that it does not include voice services.

Table 7.5 Traffic data example for 50% and 75% population coverage

Area	DU	U	IND	SU	FO	OP
Active subs. @ 50% POP Cov	6000	21 000	80 000	265 000	70 000	30 800
Active subs. @ 75% POP Cov.	7000	22 000	110 000	350 000	110 000	401 000
Busy hour traffic/sub. UL						
Bearer UDD64 (kbps)	0.079	0.079	0.079	0.08	0.08	0.08
Bearer UDD144 (kbps)	0.060	0.060	0.060	0.07	0.07	0.07
Bearer UDD384 (kbps)	0.015	0.015	0.015			
Bearer LCD64 (mErl)	0.50	0.50	0.50	0.50	0.50	0.50
Bearer LCD144 (mErl)	0.25	0.25	0.25	0.25	0.25	0.25
Busy Hour Traffic/sub. DL						
Bearer UDD64 (kbps)	0.120	0.120	0.120	0.15	0.15	0.15
Bearer UDD144 (kbps)	0.18	0.18	0.18	0.24	0.24	0.24
Bearer UDD384 (kbps)	0.08	0.08	0.08			
Bearer LCD64 (mErl)	0.50	0.50	0.50	0.50	0.50	0.50
Bearer LCD144 (mErl)	0.25	0.25	0.25	0.25	0.25	0.25

The traffic data, i.e. Un-restricted Delay Data (UDD) and Low delay Circuit switch Data (LCD) for the different environments (DU, U, Industrial (IND), SU, FO and OP), represent the possible traffic flow in the 3G network. We provide them here, only as reference to make realistic projections.

Note that the traffic in the DL is higher than in the UL due to the fact that the users do download more information than they upload. We can also see that a good part of the subscriber base remains in the open areas in this particular density distribution.

Consolidating 3G BS areas will vary from region to region. Some regions have already strict regulations for the implementation of sites besides the high costs in dense areas. This means that site acquisition will exceed the minimum requirements. Thus, Table 7.5 shows the necessary margins projected for subscriber growth assuming that sites can be available within a short term. The turn around to prepare sites to increase coverage and capacity may not necessarily match a rapid subscriber growth. If we apply the 50% of the population coverage to the first case and the 75% to the second case, we then have about 750 K UMTS subscriber for the initial phase and about 1000 K for the latter. This means we dimension the 3G network initially with enough margin for growth towards the latter phase where the subscriber base approaches the predicted numbers for 2005 in Table 7.1 when adding the 2G subscribers, i.e. $\approx$1500 K subscribers.

7.6.3 Circuit-Switched Data Calls Assumptions

From [1] for 64 kbps UDI we assumed that 25% of the UMTS subscribers will also be CS data subscribers. We also assume that 50% of the calls will be UL + DL, 25% of the calls will be UL only and 25% of the calls will be DL only. This means that one call will occupy two channels (one for DL and one for UL) but with a 75% usage each.

CS data users may use Multimedia with the following traffic mix:

- 1 data call per 24 h, with duration of 30 min. We assume that half of these calls occur during busy hour (BHCA = 0.5). 3% of the CS data users use this service.
- 1 data call per 3 h, with duration of 5 min. It is assumed that two-third of these calls are made during busy hour. (BHCA = 0.67). 6% of the CS data users use this service.

CS Data users may use other UDI services with the following traffic:

- 1 data call per 3 h, with duration of 5 min. It is assumed that two-third of these calls are made during busy hour. (BHCA = 0.67). 3% of the CS data users use this service.

7.6.4 Packet-Switched Applications

Packet data traffic will have different requirements on delays, packet loss, etc. The recommended classes include streaming, conversational, interactive and background. On this basis Table 7.6 illustrates the traffic mix of users and total traffic that may be applied.

Table 7.6 Packet traffic mix

	SCENARIO				
Traffic classes	% Users	% Traffic	Traffic BH (Kbytes)		
			DL	UL	Total
Background	59	21	49	16	65
Interactive	156[a]	39	110	20	130
Streaming	4	18	50	10	60
Conversational	5	22	38	38	76
Total		100	247	84	331

[a]Note that each subscriber may use several applications.

7.6.5 Characteristic of CDMA Cells

The factors affecting CDMA cell size, capacity and co-channel parameters in the forward and reverse links include 'same cell interference' and 'other cell interference'. These events also have an impact on the link power budgets.

7.6.5.1 Theoretical Capacity

Here we look capacity from the user interference side. To illustrate a basic case, we will use the link reference parameter, i.e. E_b/N_o, or energy per pit per noise power density, which later will apply to the link budget framework.

Picking it up from eqn (2.6) in Chapter 2, we consider the generic reverse-link capacity in CDMA[12] as the limiting factor. Thus, assuming perfect power control for this instance, the received powers from all mobile users are the same. Then

$$\frac{S}{N} = \frac{1}{M-1} \tag{7.1}$$

where M is the total number of active users in given band, and where the total interference power in the band equals the sum of powers of single users. Now equating the energy per bit to the average modulating signal power we defined

$$E_b = ST = \frac{S}{R} \tag{7.2}$$

where S is the average modulating signal power, T is bit time duration and R is the bit rate, i.e. $1/T$. Then, incorporating the noise power density N_o, which is the total noise power N divided by the bandwidth B (i.e. $N_o = N/B$), we get

$$\frac{E_b}{N_o} = \frac{S}{N}\frac{B}{R} = \frac{1}{(M-1)}\frac{B}{R} \tag{7.3}$$

[12]Mainly in rural areas; in urban area the downlink may/will become the limiting factor.

Solving for M it yields

$$M - 1 = \frac{(B/R)}{(E_b/N_o)}$$

and for large M we get

$$M = \frac{(B/R)}{(E/N_0)} = \frac{G_p}{(E/N_0)} \tag{7.4}$$

where G_p corresponds to system processing gain defined in eqn (2.3) in Chapter 2, and M defines the number of projected users in a single CDMA cell with omni-directional antenna without interference from neighbouring cell users transmitting continuously.

7.6.5.2 *The Cell Loading Effect*

Since in real 3G mobile network there always exists more than one cell and more than one sector, we need to introduce a loading effect due to interference from neighbouring cells as follows:

$$\frac{E_b}{N_o} = \frac{1}{(M-1)} \frac{B}{R} \left(\frac{1}{1+\beta} \right) \tag{7.5}$$

where β is the loading factor (ranging from 0% to 100%) as introduced in eqn (2.29) in Chapter 2. Typical β values will range from 45% to 50%. The inverse of $(1+\beta)$ has often been defined as the frequency reuse factor, i.e. $F = 1/(1+\beta)$. The ideal single cell CDMA value of $F = 1$ (i.e. $\beta = 0$) decreases as the loading of multi-cell environments increases.

Sectorisation can decrease interference from other users in other cells. Thus, instead of deploying only omni-directional antennas with 360° a majority (if not) of all sites can bear at least three sectors (e.g. 120°), and allow thereby the sectorised antenna rejected interference from users outside its antenna pattern. Such event will decrease the loading effect and introduce a sectorisation gain λ, which is expressed as

$$\lambda = \frac{\int_0^{2\pi} I(\theta) d\theta}{\int_0^{2\pi} \left(\frac{A_G(\theta)}{A_G(0)} \right) I(\theta) d\theta} \tag{7.6}$$

where $A_G(0)$ is the peak antenna gain occurring generally at the boresight (i.e. $\theta = 0$), $A_G(\theta)$ is the horizontal antenna pattern of the sector antenna and $I(\theta)$ represents the received interference power from users of other cells as a function of θ. In practice $\lambda = 2.5$ for a three sector configuration and about 5 for a six sector one. Then, incorporating the sectorisation gain in the loading effect we get

$$\frac{E_b}{N_o} = \frac{1}{(M-1)} \frac{B}{R} \left(\frac{1}{1+\beta} \right) \lambda \tag{7.7}$$

Initially for the single cell case, we have assumed continuous transmission. However, this does not occur for voice and some multimedia services; although it does for data. Thus, we will now introduce an activity factor $1/\nu$ to reflect this events in the UL loading effect. Then, we get

$$\frac{E_b}{N_o} = \frac{1}{(M-1)}\frac{B}{R}\left(\frac{1}{1+\beta}\right)\lambda\left(\frac{1}{v}\right) \tag{7.8}$$

where v may range between 40 and 50% for voice and 1% for data. Therefore, the value of v reduces the overall interference of UL loading effect equation. For the Downlink (DL) we need an additional parameter ε to reflect the orthogonality of the transmission. Thus, empirically, we can express it as

$$\frac{E_b}{N_o} = \frac{1}{(M-1)}\frac{B}{R}\left(\frac{1}{(1-\varepsilon)+\beta}\right)\lambda\left(\frac{1}{v}\right)$$

7.6.6 Link Budgets

Link budget aims to provide the steps to calculate the ratio of the received bit energy to thermal noise (i.e. E_b/N_o) and the interference density I_o. It considers transmitted power, transmitted and received antenna gains, channel capacity factors, propagation environment and receiver noise figure.

Based on the channel models introduced in Chapter 2 here we present the background for link budgets. Following the guidelines from [2] the formulation assumes that path loss formulas help to determine the maximum range and the coverage area. We also assume that in case of hexagonal deployment of sectored cells, the area covered by one sector is

$$S = 3\sqrt{3}x\left(\frac{R}{2}\right)^2 \Big/ 2 \tag{7.9}$$

where R is the range obtained in the link budget. This implies that we use hexagonal sectors with base stations placed in the corners of the hexagons. Coverage analysis can thus apply to tri-sectored antennas for macro-cells and with omni-directional antennas for micro-cells and pico-cells coverages.

Before describing the actual reference parameters for the link budgets on Table 7.7, in the following we provide a generic background of the analysis steps for the forward and reverse links.

7.6.6.1 The Forward Link

Applying the logic for the traffic channels analysis in [3], to the Dedicated Physical Control Channel (DPCCH) and Dedicated Physical Data Channel (DPDCH) we can formulate a generic E_b/N_o for the forward link in multi-cellular environment.

Table 7.7 Link budget reference template

Elements	Forward link	Reverse link
Reference: environment, services, multi-path channels	See Section 2.2 (Chapter 2)	See Section 2.2 (Chapter 2)
(a_0) Average transmitter power per traffic channel	dBm	dBm
(a_1) Maximum transmitter power per traffic channel	dBm	dBm
(a_2) Maximum total transmitter power	dBm	dBm
(b) Cable, connector and combiner losses, etc.	2 dB	0 dB
(c) Transmitter antenna gain (e.g. 18 dBi vehicular, 10 dBi pedestrian, 2 dBi indoor)	Will vary	0 dBi
(d_1) Transmitter e.i.r.p. per traffic channel $= (a_1 - b + c)$	dBm	dBm
(d_2) Total transmitter e.i.r.p. $= (a_2 - b + c)$	dBm	dBm
(e) Receiver antenna gain (e.g. 18 dBi vehicular, 10 dBi pedestrian, 2 dBi indoor)	0 dBi	Will vary
(f) Cable and connector losses	0 dB	2 dB
(g) Receiver noise figure	5 dB	5 dB
(h) Thermal noise density (H) (linear units)	-174 dBm/Hz 3.98×10^{-18} mW/Hz	-174 dBm/Hz 3.98×10^{-18} mW/Hz
(i) Receiver interference density (I) (linear units)	dBm/Hz mW/Hz	dBm/Hz mW/Hz
(j) Total effective noise plus interference density $= 10$ Log $(10^{((g+h)/10)} + I)$	dBm/Hz	dBm/Hz
(k) Information rate (10 log (R_b))	dBHz	dBHz
(l) Required $E_b/(N_o + I_o)$	dB	dB
(m) Receiver sensitivity $= (j + k + l)$		
(n) Hand-off gain	dB	dB
(o) Explicit diversity gain	dB	dB
(o') Other gain	dB	dB
(p) Log-normal fade margin	dB	dB
(q) Maximum path loss $= \{d_1 - m + (e - f) + o + n + o' - p\}$	dB	dB
(r) Maximum range	m	m

Starting from single cell with a single Mobile Station (MS)

$$\frac{E_b}{N_o} = \frac{P_o L_o A_G}{I_b + I_n + N} G \tag{7.10}$$

where P_o is the BS sector traffic channel ERP in the direction of the MS within a give antenna patter with its angle θ_o, L_o equals the path loss from the home BS in the direction of θ_o within a given distance, A_G is the received antenna gain of the MS, I_n equals the interference power received at the MS from non-CDMA origins, N is the thermal noise power, G is the processing gain and I_b can be defined as

$$I_b = (1 - \varepsilon) P' L_o A_G \tag{7.11}$$

where ε is the orthogonality factor, P' is the home BS excess ERP (e.g. paging, sync powers, etc.) in the direction of the MS under consideration. In the presence of many cells and single MS, interference originates from the powers of the surrounding BSs, in addition to the excess powers of its own cell. Thus, we introduce the interference from the surrounding as I_o

$$\frac{E_b}{N_o} = \frac{P_o L_o A_G}{I_b + I_n + I_o + N} G$$

When looking at single cell with many MSs, the BS serves all MSs plus the MS under consideration. Therefore, the latter gets the interference from the DL powers aimed at the other MSs. We denote this additional interference as

$$I_m = (1 - \varepsilon) A_G L_o \sum_{i=1}^{I} P_i \qquad (7.12)$$

where again ε is the orthogonality factor and P_i is the forward traffic channel ERP aimed for MS i, but radiated to the desired MS measuring E_b/N_o. P_i may also denote the traffic channel ERP aimed for MS i but captured by the desired MS. Then

$$\frac{E_b}{N_o} = \frac{P_o L_o A_G}{I_b + I_n + I_m + N} G \qquad (7.13)$$

When an MS measuring E_b/N_o finds itself among many other MSs and many other cells, there is an additional interference term I_t, i.e. the total traffic channel power received from all other BSs. It can be define as

$$I_t = A_G \sum_{k=1}^{K} P_k L_k \qquad (7.14)$$

where P_k is the total traffic channel ERP from BS k. Thus, I_t represents the sum of all traffic channel powers receive by the desired MS from all other BS, but excluding its own. K is the total number of cells or sectors in the system under consideration. We can define P_k as

$$P_k = \sum_{j=1}^{J_K} P'_j \qquad (7.15)$$

The P_k expression indicates that, for each BS k, we sum the forward traffic channel ERPs for all MSs corresponding to that of BS k.

The expression also implies that P'_j is the traffic channel power aimed at MS j but captured by the MS calculating E_b/N_o. J_k is the total number of MS served by BS k.

Then E_b/N_o for the MS among many MS within many cells can be defined as

$$\frac{E_b}{N_o} = \frac{P_o L_o A_G}{I_b + I_n + I_o + I_m + I_t + N} G \qquad (7.16)$$

This latter expression will be the most likely environment when calculating the forward link budget.

7.6.6.2 The Reverse Link

In the reverse link or uplink, i.e. MS to BS connection, a single cell serving a single MS has the following E_b/N_o expression

$$\frac{E_b}{N_o} = \frac{P_R L_R A_{GR}}{I_{nR} + N} G \tag{7.17}$$

where P_R equals the reverse traffic channel ERP of the desired MS assuming an omni-directional transmit pattern, LR is the reverse path loss from the desired MS in the direction of θ_o to the home BS at given distance, A_{GR} equals the receive antenna gain of the home BS in th direction of θ_o to the desired MS and I_{nR} is the power received at the home BS from other interference of non-CDMA sources.

When considering a single cell with many mobiles one BS serves many MSs, and the MS measuring E_b/N_o gets extra interference (I_{mR}), which can be expressed as

$$I_m = \sum_{j=1}^{J} P_{R_j} L_{R_j} A_{GR} \tag{7.18}$$

where P_{R_j} corresponds to the reverse traffic channel ERP of MS j, L_{R_j} is the reverse path loss from MS j in the direction of θ_j back to the home BS at given distance and A_{GR} is the received antenna gain of the home BS in the direction of θ_j to MS j. Thus, I_{mR} represents the total reverse link interference generated by MS served by home BS. P_{R_j} dynamically changes based on the power control algorithm. Then, the reverse link E_b/N_o for single cell with many MS is

$$\frac{E_b}{N_o} = \frac{P_R L_R A_{GR}}{I_{nR} + I_{mR} + N} G \tag{7.19}$$

In scenarios involving many MSs and multiple cells, the MS measuring E_b/N_o gets additional interference from MSs served by BSs from neighbouring cells. We can express this interference as

$$I_{tR} = \sum_{k=1}^{K} P_{R_k} \quad \text{with} \quad P_{R_k} = \sum_{j=1}^{J_k} P_{R_{k,j}} L_{R_{k,j}} A_{GR} \tag{7.20}$$

where I_{tR} is the total interference from the reverse link generated by MSs served by BSs other than the home BS of the MS measuring E_b/N_o, P_{R_k} is the total reverse link traffic power received from MSs served by BS k and K is the total number of BSs excluding the home BS of the concerned MS. We get P_{R_k} by adding the powers of the traffic channels from MSs served by BS k, where for this BS $P_{R_{k,j}}$ is the reverse traffic channel ERP of MS j. Similarly for BS k, $L_{R_{k,j}}$ is the reverse path loss from MS j in the direction of $\theta_{R_{k,j}}$ at a given distance. A_{GR} is the receive antenna gain of the home BS in the direction of $\theta_{R_{k,j}}$ to MS j served by BS k. Then

$$\frac{E_b}{N_o} = \frac{P_R L_R A_{GR}}{I_{nR} + I_{mR} + I_{tR} + N} G \tag{7.21}$$

The sum of the interfering elements divided by the thermal noise power N gives rise to the reverse link factor. This factor η_r represents the rise of the interference level above the thermal noise level. We can define it as

$$\eta_r = \frac{I_{nR} + I_{mR} + I_{tR} + N}{N} \tag{7.22}$$

Through the value of η_r we can determine the BS loading level. Thus, higher η_r values indicate that the BS can no longer support additional users or MSs.

With generic analytical background of the preceding sections, i.e. the forward and reverse link estimation for E_b/N_o, in the sequel we outline the main elements for link budgets.

7.6.6.3 Link-Budget Elements

Table 7.7 illustrates reference elements typically utilized in the calculations of link budgets. The template of [4] applies to both forward and reverse links unless specifically stated otherwise. In the forward link the BS acts as the transmitter and the MS as the receiver. In the reverse link the MS acts as the transmitter and the BS as the receiver. For completeness the elements are re-defined as follows:

- (a_0) *Average transmitter power per traffic channel* (dBm) is the mean of the total transmitted power over an entire transmission cycle with maximum transmitted power when transmitting.
- (a_1) *Maximum transmitter power per traffic channel* (dBm) is the total power at the transmitter output for a single traffic[13] channel.
- (a_2) *Maximum total transmitter power* (dBm) is the aggregate maximum transmit power of all channels.
- (b) *Cable, connector and combiner losses (transmitter)* (dB) is the combined losses of all transmission system components between the transmitter output and the antenna input (all losses in + dB values).
- (c) *Transmitter antenna gain* (dBi) is the maximum gain of the transmitter antenna in the horizontal plane (specified as dB relative to an isotropic radiator).
- (d_1) *Transmitter e.i.r.p. per traffic channel* (dBm) is the sum of the transmitter power output per traffic channel (dBm), transmission system losses (-dB) and the transmitter antenna gain (dBi) in the direction of maximum radiation.
- (d_2) *Transmitter e.i.r.p.* (dBm) is the sum of the total transmitter power (dBm), transmission system losses (-dB) and the transmitter antenna gain (dBi).

[13]We define a traffic channel as a communication path between an MS and a BS used for information transfer and signalling traffic. Thus, traffic channel implies a forward traffic channel and reverse traffic channel pair.

- (e) *Receiver antenna gain* (dBi) is the maximum gain of the receiver antenna in the horizontal plane, it is specified in dB relative to an isotropic radiator.
- (f) *Cable, connector and splitter losses (receiver)* (dB) is includes the combined losses of all transmission system components between the receiving antenna output and the receiver input (all losses in + dB values).
- (g) *Receiver noise figure* (dB) is the noise figure of the receiving system referenced to the receiver input.
- (h), *(H) Thermal noise density* N_o (dBm/Hz) is the noise power per Hertz at the receiver input. Note that h is logarithmic units and H is linear units.
- (i), *(I) Receiver interference density* I_o (dBm/Hz) is the interference power per Hertz at the receiver front end. This corresponds to the in-band interference power divided by the system bandwidth. Note that i is logarithmic units and I is linear units. Receiver interference density I_o for forward link is the interference power per Hertz at the MS receiver located at the edge of coverage, in an interior cell.
- (j) *Total effective noise plus interference density* (dBm/Hz) is the logarithmic sum of the receiver noise density and the receiver noise figure and the arithmetic sum with the receiver interference density.
- (k) *Information rate* (10Log(R_b)) (dBHz) is the channel bit rate in dBHz; the choice of R_b must be consistent with the E_b assumptions.
- (l) *Required* $E_b/(N_o + I_o)$ (dB) is the ratio between the received energy per information bit to the total effective noise and interference power density needed to satisfy quality objectives.
- (m) *Receiver sensitivity* $(j + k + l)$ (dBm) is the signal level needed at the receiver input that just satisfies the required $E_b/(N_o + I_o)$.
- (n) *Hand-off gain/loss* (dB) is the gain/loss factor (+ or −) brought by hand-off to maintain specified reliability at the boundary.
- (o) *Explicit diversity gain* (dB) is the effective gain achieved using diversity techniques. If the diversity gain has been included in the $E_b/(N_o + I_o)$ specification, it should not be included here.
- (o') *Other gain* (dB) is additional gains, e.g. Space Diversity Multiple Access (SDMA) may provide an excess antenna gain.
- (p) *Log-normal fade margin* (dB) is defined at the cell boundary for isolated cells corresponds to the margin required to provide a specified coverage availability over the individual cells.
- (q) *Maximum path loss* (dB) is the maximum loss that permits minimum SRTT performance at the cell boundary. Maximum path loss $= d1 - m + (e - f) + o + o' + n - p$.
- (r) *Maximum range–*R_{max} (km) – is computed for each deployment scenario is given by the range associated with the maximum path loss (see Chapter 2 for details).

7.6.6.4 Link Budget for Multi-Services

Here we consider how the environment of WCDMA in the FDD mode will influence multi-service provision. In multi-service link budget the analysis process to calculate the *interference degradation* or the *loading factor* takes into account the interference contribution of all the users with their different services. This results in a common link budget, which aims to provide the same cell radius for all the services by trying to match all the acting UE TX powers. It also aims to balance the two links (i.e. UL and DL) without any *a priori* knowledge on the limiting link in terms of coverage. This process permits us to estimate the actual system interference degradation without dependency on margins, which may lead to over-dimensioning.

7.6.7 Coverage Analysis

After providing the background to calculate the E_b/N_o values and the link budget in the last two sections, we now look at the practical design factors having impact on coverage.

Coverage may not be an issue at the introduction of UMTS in some regions, because requirements about it will be gradual. However, from the service side, to back a pragmatic business case, a network will most likely start with about 50% coverage of populated areas as mentioned at the beginning of this chapter. Thus, such coverage will depend to a good degree on service strategy. From the network design side, this implies that good indoor coverage for high rate services will require dense sites in the urban areas with downlink limitation and less dense in rural areas with uplink limitation. The latter implies that coverage and capacity trade-off will go hand to hand even at the beginning of UMTS service. Here we are mainly concerned with coverage.

7.6.7.1 UpLink (UL) and DownLink (DL) Coverage

DL coverage depends primarily on the load because the transmission power remains the same despite the number of MSs active in a given BS, where all share the same power. This means that DL coverage will decrease as function of the number of MSs and their transmission rates. The latter implies that additional power will afford better coverage for higher rates in the DL.

In WCDMA higher transmission rates imply more spreading, which results in lower processing gain, thereby smaller coverage. On the other hand, higher bit rates (demanding more transmission power) require lower E_b/N_o because the extra power allows better channel estimation, thereby compensating for larger[14] coverage. In relation to the physical channels, i.e. DPCCH/DPDCH, the dependency of the bit rate for E_b/N_o has to do with the mode of channel operation. Figure 7.8 shows that there is a difference in the power utilization for each channel, as it is also overhead difference depending on the transmission rates. When assuming the same E_b/N_o for all rates, e.g. the over head for 384 kbps does not

[14]In particular this applies to higher rates in data packet transmission.

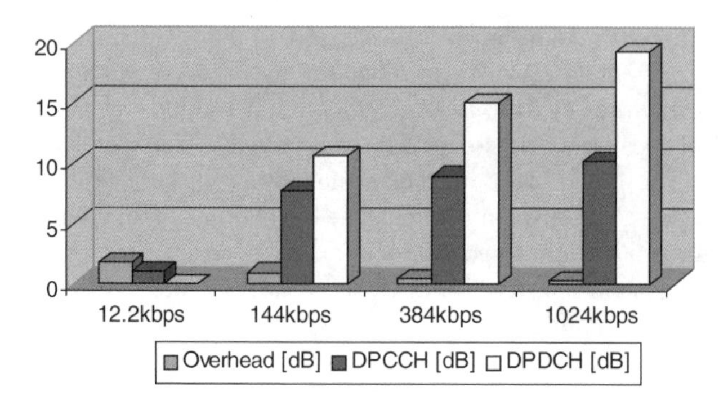

Figure 7.8 DPCCH/DPDCH and overhead power distribution.

exceed 6% of the total power in the DPCCH if we define DPCCH overhead as $10\log_{10}(1 + 10^{(\mathrm{DPDCH}\,-\,\mathrm{DPCCH})/10})$.

Thus, when looking at the power differences for the reference service rates, logically we can conclude that to support 384 kbps we will need a denser site deployment than we would for 144 kbps.

Other factors having impact on the uplink E_b/N_o values are multi-path diversity, macro-diversity gain, advanced BS signal processing techniques and receiver antenna diversity.

In the first case, when looking at characteristics of the reference multi-path channels in Chapter 2, we will see that the vehicular channels have more taps that those for the pedestrian ones. More taps implies higher multi-path diversity gain and thereby larger coverage.

In the second case, in the absence of high multi-path diversity gain during *soft handover*, i.e. when the MS receives signal from at least two BSs, the probability of accurate signal detection increases resulting in higher micro-diversity gain.

Better baseband processing, e.g. adaptive filters for fading environments, will improve error rates and thereby lower E_b/N_o values, which in turn will increase coverage.

Finally, through antenna diversity techniques we can also get a coverage gain of 2–3 dB. For example, transmit diversity can use two independent transmit paths from the base station to the mobile, in order to mitigate the effect of fading. The two paths may come from using two spatially separated antennas, or by using the two orthogonal polarisations of one cross-polarised antenna [5,6]. On the uplink, two-branch diversity combining or Maximal Ratio Combining (MRC) is optimal when the traffic consists of voice users only. However, when individual high-data-rate users are also present a fully adaptive two-branch Minimum Mean Squared Estimate (MMSE) algorithm will provide improved performance by cancelling the interference due to these users. This cancellation results in a gain of the order of 1.5 dB.

As mentioned earlier, in the DL we can add power gradually when necessary, thereby we can increase coverage for higher rates. However, this may not be the case in the UL because the MS has limited power. For example, a handset with an average power capacity of 21 dBm will have a maximum of 26 or 27 dBm power. The latter if we assume the MS gains 5–6 dBm at the BS due to the high reception sensitivity, antenna diversity and lower noise figures. High rate data terminals[15] or data terminals in general will have 3 dB lower E_b/N_o. Thus, DL coverage for high rates will depend on the DL power amplifier rating, the UL cell dimensioning and most likely the adjacent cell loading as noted in the preceding section.

7.6.8 Capacity Analysis

In WCDMA capacity impacts apply to the DL and UL. In the first case it has to do with dense areas for high rates as well as subscriber number. In the second case it has to do with rural areas in the context of coverage for high rates. On the other hand, due to the asymmetry of traffic flow, we expect more download information than upload. Hence, DL capacity appears more critical at least at the beginning of UMTS.

Orthogonal codes make the DL more robust against intra-cell interference. However, inter-cell interference does still affect DL capacity, which depends on the load of the neighbouring cells and the propagation environment. For example, short orthogonal codes are more vulnerable to multi-path channel than are to single path channels; hence, in the micro-cell environment orthogonality gets preserved better that it does in macro-cell environment. Consequently, loading despite the E_b/N_o values on adjacent macro-cells should not exceed 75% in the DL and about 55% in the UL. On the other hand, micro-cells can probably take 65% UL and 85% loading, respectively. This means we need to apply the appropriate orthogonality factors when utilizing the load equations described generically in Section 7.6.5.

The number of orthogonal codes does also have impact on DL capacity despite a good propagation environment and good load sharing. The maximum number of orthogonal codes depends on the Spreading Factor (SF). For example, in general only one scrambling code and thus only one code three gets used per sector in the BS, where common and dedicated channels share the same three. On the other hand, the number of orthogonal codes does not imply complete[16] limitation when enabling DL capacity, because we can apply a second scrambling code. However, the first and second codes will not remain orthogonal to one another, and channels with the second code interfere more with the channels with the first code.

7.7 DIMENSIONING RNC INTERFACES

When dimensioning RNC Iub interface, i.e. the connection between the Node B and RNC, we also consider the traffic mix in order to determine the number of RNCs required. Thus, RNC interface dimensioning will take into account the number of Node Bs and the projected type of services with the forecasted subscribers and their traffic profiles. Figure 7.9 illustrates the UTRAN interface configuration.

7.7.1 Dimensioning the Iub

The average traffic per Node B provides the *total traffic* based on the service mix statistics, the soft handover traffic and overheads, signalling and O&M traffic.

[15]Speech terminals have about 3 dB body loss.
[16]Often referred as 'hard-blocking'.

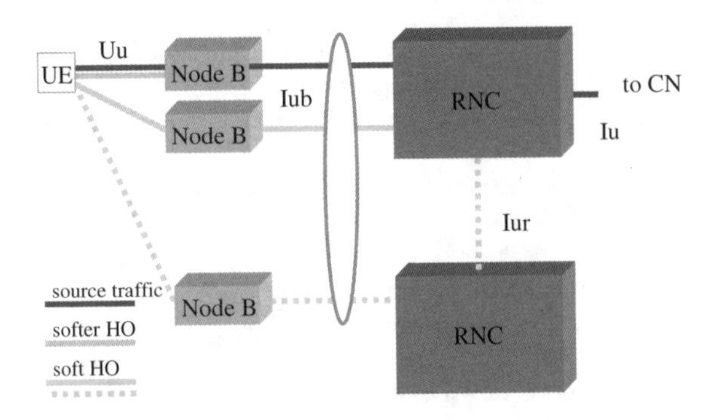

Figure 7.9 The UTRAN interface configuration.

Thus, to determine the total traffic passing through the Iub we consider first the *peak* aggregate traffic mix calculated analytically taking into account the service parameters, e.g. the number of subscribers (S_i), subscriber bit rate (R_i), session time length (t_i), session inter-arrival time length ($1/\lambda_i$), activity factor α_i, plus signalling overheads and O&M margins. Here we assume that the ratio *peak traffic* over *average traffic* corresponds to the burstiness factor β.

We then calculate the overall PDF (R_a) and CDF(R_a), where R_a corresponds to the aggregate bit rate to determine the outage probability for each value of the user bit rate. Afterwards we obtain a set of outage probabilities, which corresponds to each user bit rate R_b. At the end we dimension the channel capacity by fixing a common outage, probably value P_0 for each service i.

7.7.1.1 *Iub Total Traffic*

As indicated in the preceding section, after we calculate the peak traffic per Node B, we take into account additional overheads and signalling loads. Thus, we obtain the *total traffic* at the Iub interface from the user information traffic, soft handover traffic, burstiness factor and overheads as well as signalling margins. Typical assumptions for the margins include O&M = 10%, signalling = 20% and ATM overhead = 40% of the Iub peak user traffic, respectively. In summary, we can define the total Iub traffic as

$$\text{total Iub traffic} = \text{peak traffic} + \text{O\&M} + \text{signalling} + \text{overhead} \qquad \text{or}$$

$$\text{total Iub traffic} = \text{average traffic} \times \beta \times (1 + 0.4 + 0.2 + 0.1).$$

7.7.2 RNC Capacity

To practically determine the number of RNCs required for a network, we generally consider the maximum number of Node Bs to be managed; the maximum Iub, Iu and Iur connections supported; and the maximum throughput of both CS and PS services.

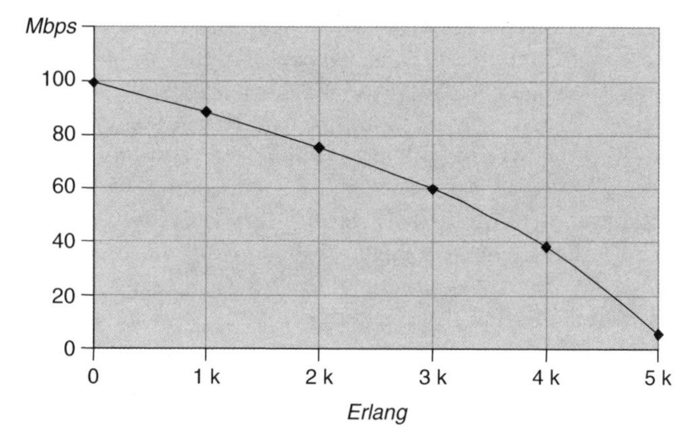

Figure 7.10 Nominal RNC traffic loads in Mbps vs. Erlang (000's).

We determine the total number of Node Bs based on the type of services offered, the number of subscribers projected and area of coverage desired among other parameters. We express the average traffic forecasted in Erlangs for CS and Mbps for PS.

Plotting nominal values of PS versus CS traffic, e.g. 64 kbps for both services we can see in Figure 7.10 that proportion of PS and CS traffic depends on the desired load for either service. In any case, it seems that we cannot have half and half of each service type.

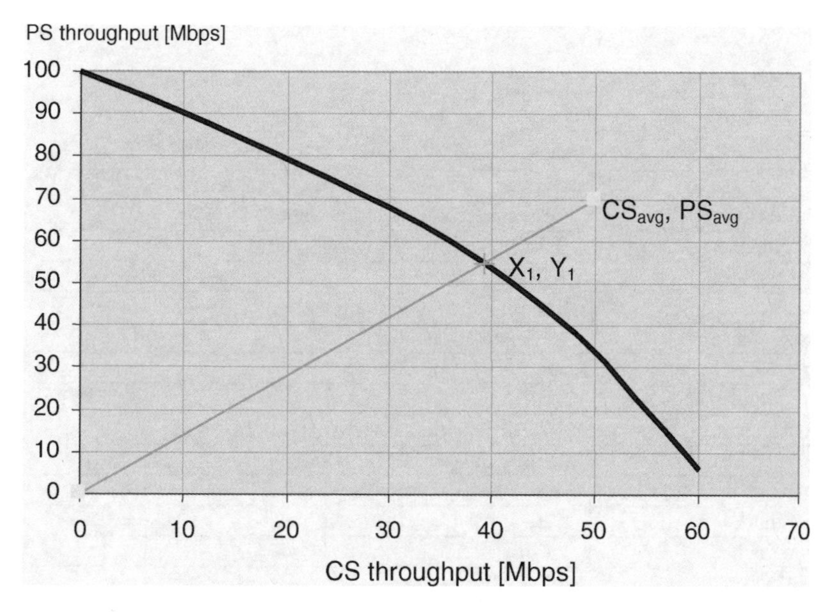

Figure 7.11 Estimation of the RNC throughput.

The Iub, Iu and Iur interfaces will in general support sufficient capacity margins, and the overheads will not exceed peak rates. Thus, the key RNC dimensioning parameters include the number of Node Bs in the coverage area, and the average traffic in this given area. The first parameter gives $\text{RNC}_{No\_Node\_B} = \textit{total no Node Bs/no node Bs supported per RNC}$, and the second one allows to calculate throughput capacity, i.e. $\text{RNC}_{\text{throughput}} = \max\{\lceil CS_{\text{avg}}/X_1 \rceil, \lceil PS_{\text{avg}}/Y_1 \rceil\}$. We can obtain the initial value of $\text{RNC}_{\text{throughput}}$ from the CS and PS average traffic uniformly distributed in the target area. We can modify the PS (Mbps) versus CS (Erlang) output of Figure 7.10 to a PS versus CS (Mbps) output by translating the CS traffic from Erlang to Mbps (i.e. Erlang $\times$ 12.2 kbps AMR voice codec). Then we can illustrate the $\text{RNC}_{\text{throughput}}$ in terms of average value of the CS and PS traffic in Mbps as shown in Figure 7.11. However, the average traffic will not take into account the traffic burstiness. Thus, peak values should be projected iteratively by using the Gaussian law. In general, the peak values will mean that the proportion of PS traffic will increase.

7.8 RADIO NETWORK DIMENSIONING FIELD STUDY

The analysis assumptions for the projected subscriber growth and traffic flow illustrated in Table 7.5 take the values shown in Table 7.8. These assumptions provide the generic set of information to calculate the number radio network elements required for the coverage aforementioned in Section 7.5.

Note, for example, that we set the blocking characteristics to 1%, however, it is often set to 2%. The design assumes UL limitation by setting the load to 50%. With optimised interference techniques, UL load can reach up to 65%. Antenna heights for the MS can vary from 1.5 to 1.7 m and for the BS from 27 to 30 m.

Table 7.8 Analysis assumptions for lower-bound traffic flow

	DU	U	IND	SU	FO	OP
Multi-path channels	Reference clutter distributions					
Capacity growth	Based on Table 7.5					
Loading	50% UL load					
Features	Simplified, excludes, e.g. MHA					
Blocking on air interface	1% (network quality)					
Average antenna heights	MS 1.55 m, BS 27.5 m					
MS noise figure	8 dB					
MS output power	+24 dBm for dataspeeds ≥64 kbps					
MS antenna gain	2 dB data speeds ≥64 kbps					
Body loss	0 dB					
BS noise figure	5 dB					
BS antenna gain	18 dBi					
BS output power	43 dBm × 20 Watts					
Soft handover	40%					

7.8.1 Lower Bound Results

As noted in Section 7.6.2, the reference values illustrated in Table 7.5 represent the parameters to obtain projection for lower-bound dimensioning, i.e. the minimum number of elements to meet coverage and traffic requirements. Thus, Table 7.9 shows the results taking into account the morphology distribution of the population density in Table 7.3, the service quality requirements in Table 7.4 and the minimum traffic flow in Table 7.5.

We can observe that the Node B configuration in Tables 7.9 and 7.10 have only one carrier per sector. In practice, in DU areas we would most likely need two carriers per sector in some cities. However, because the results project UL limited design, it does not take into account the DL limitation in dense areas. In Tables 7.9 and 7.10 we can also see that the predominant services correspond to those with high rates, e.g. 384 kbps in the PS mode within the DU and U areas. As mentioned in Section 7.6.8, the cell range decreases in areas where high transmission rates become predominant, e.g. DU and U.

Table 7.9 Lower-bound results for 50% coverage

Planned population coverage	50%					
Area type	DU	U	SU	IND	FO	OP
Subscribers	6000	21 000	265 000	80 000	70 000	308 000
Area size (km^2)	2.33	9.9	387	101	1270	2297
Node B 1 + 1 + 1	66	70	191	60	103	111
Sites total	601					
Cells total	1803					
RF units total	1803					
Dominating service class	PS384	PS384	PS144	PS144	PS144	PS144
Cell range (km)	0.155	0.285	1.065	0.870	2.355	3.250

Table 7.10 Lower-bound results for 75% coverage

Planned population	75%					
coverage Area type	DU	U	SU	IND	FO	OP
Subscribers	7000	22 000	350 000	110 000	110 000	401 000
Area size (km^2)	2.37	10.6	617	138	1961	4012
Node B 1 + 1 + 1	78	79	306	102	201	173
Sites total	939					
Cells total	2817					
RF units total	2817					
Pre-dominant service class	PS384	PS384	PS144	PS144	PS144	PS144
Cell range (km)	0.155	0.285	1.065	0.870	2.355	3.250

7.8.2 Upper-Bound Results

Realising a radio network for the minimum traffic and for the minimum subscriber growth in the immediate term will minimise investment costs. However, it may limit the flexibility in service strategy if rapid increase in subscriber and traffic occurs. Hence, in the following we assume upper limit bounds and project network based in higher traffic profiles primarily to illustrate the network element configuration limits. *This does not imply that deployed networks will be operating with the illustrated upper limits, specially at the introduction of UMTS.*

To simplify the propagation environments we group the six areas into three as follows: urban (DU + U); sub-urban (SU + IND) and rural (OP + FO). We increase traffic profiles at the Busy Hour (BH) for every service and harmonise the subscriber base for the three areas as well. Clearly, we keep the assumptions of traffic patterns for the DL and UL, i.e. we assume higher volumes in the DL than in the UL. Such asymmetric traffic will allow us to exploit the benefits of the TDD technology in the dense areas by incorporating micro- and pico-cells already at the introduction of UMTS. The latter assuming that the technology has sufficient commercial maturity. Table 7.11 illustrates the consolidated traffic for the three aforementioned propagation environments. In addition, it does also show the total traffic generated per area and the overall traffic proposed for the total number of subscribers proposed and reference. The activity factor for LCD 12.2 (e.g. voice with AMR codec) is 0.5, for the other LCD and UDD services it is 1.0.

The upper-bound results illustrated in Table 7.12 take into account the reference inputs from Table 7.11 and dimension the radio network using this time, the two UTRA technologies, i.e. FDD and TDD. We consider micro- and pico-cells of the urban and

Table 7.11 Upper-bound traffic flow with consolidated areas

75% Population coverage reference projection 2005									
Areas	Urban (DU + U)			Sub-urban (SU + IND)			Rural (OP + FO)		
Subs no	127,000			663,000			210,000		
Bearer (kbps)	kbps		mErl	kbps		mErl	kbps		mErl
	UL	DL	UL/DL	UL	DL	UL/DL	UL	DL	UL/DL
	UDD64	0.388	0.578		0.388	0.578		0.388	0.578
UDD144	0.306	0.901		0.306	0.901		0.306	0.901	
UDD384	0.066	0.415		0.066	0.415		0.066	0.415	
LCD12.2	0.122	0.122	20.0	0.122	0.122	20.0	0.122	0.122	20.0
LCD64	0.154	0.154	2.41	0.154	0.154	2.41	0.154	0.154	2.41
LCD144	0.174	0.174	1.21	0.174	0.174	1.21	0.174	0.174	1.21
Total traffic	153,731	297,749	47.24	802,548	1,554,390	47,24	254,201	492,341	47,24
Total traffic per area	451,480			2,356,938			746,542		
Total traffic from all areas (kbps) per Busy Hour (BH)									3,554,960

Table 7.12 Upper-bound results for 75% population coverage in 2005

	Representative propagation environments		
Node B type	Urban	Sub-urban	Rural
Macro-FDD (1 + 1 + 1)	480	160	280
Macro-cells	1440	480	840
RF units in macro-cells	1140	480	280
Micro-FDD	310	100	0
Micro-TDD	40	90	0
Pico-TDD	160	170	0
Total Node Bs	990	520	280
Area coverage (km^2)	800	1200	19 500
Total existing area (km^2)	800	1600	38 800

sub-urban environments. Thus, enabling the network with 3G services with transmission rates beyond 384 kbps in urban areas.

The TDD nodes aim to cover hot spots from the introduction of UMTS, e.g. shopping and educational centres, railway stations, office conglomerates, airports, etc. This approach will reduce traffic in FDD Macro-cells and allow coverage extension with overall increase capacity throughout all the network layers, i.e. pico-, micro- and macro-layers. The capacity and coverage optimisation results outlined in Table 7.12 shows also the utilisation of MUD and intelligent antenna techniques. The first has been implemented already in the TDD technology as part of its interference solution, and the latter as part of the UL/DL capacity optimisation to increase end-to-end performance in terms of coverage and capacity by 3.

The TDD micro-cell will range from approximately 40 m to about 250 m assuming medium mobility. The TDD pico-cell will cover only indoors within a range of about 10–15 m with low mobility (e.g. below 8 km/h), but will have capabilities to offer transmission rates up to 2 Mbps. Thus, the pico-cell coverage will reach mainly office and home environment.

7.8.2.1 Upper-Bound Examples FDD

To illustrate the radio network dimensioning based on the assumptions and principles of the preceding sections, here we take the upper-bound values shown in Table 7.11. Thus, using the figures from the aforementioned table we can deduce average data volume, per subscriber, at busy hour, on the DL as shown in Table 7.13.

Table 7.13 Data volumes for PS and CS traffic

	kbps
Average PS[a] traffic per subscriber at BH (DL)	7445
Average CS traffic per subscriber at BH (DL)	555
Average total voice CS traffic per subscriber at BH (DL)	527

[a]We assume LCD 144 as packet for the CN.

Table 7.14 Reference scenario for 75% coverage

	Sites		
GSM sites usable for UMTS	1400		
New UMTS sites (D-urban and urban for capacity):	200		
*New sites replacing GSM sites not capable to support UMTS	200		
Total	*1800*		
		Cells	RF carriers
Sites in urban	250	750	750
Sites in sub-urban	850	2550	2550
Sites in rural	700	2100	2100
Total	*1800*	*5400*	*5400*

To have a concrete point of reference in terms of coverage area and number of sites we illustrate values for an existing[17] 2G 1800 MHz GSM network. In such network, if we can assume that about 70% sites can be re-used, then we will have co-siting of GSM and UMTS sites. The remaining sites would then be UMTS sites only. Table 7.14 illustrates hypothetical values as examples. Note the 3 sector sites with $1 + 1 + 1$ carrier configuration; although, some dense areas may need $2 + 2 + 2$ or $3 + 3 + 3$. On the other hand, it is believed that capacity will not be an issue at the introduction of the UMTS.

To calculate the multi-service link budget we use the WCDMA parameters assumed in Table 7.15.

The assumptions for the transmission parameters include TMA (MHA), 18 dBi antenna gain, 27 m antenna height and tri-sectored antenna in all sites. Other parameters concern the

Table 7.15 Selected link-budget parameters for WCDMA

FDD dedicated parameters	Downlink	Uplink
Chip rate	3.8 Mch/s	3.8 Mch/s
Path loss imbalance	2 dB	2 dB
Avged. PSCH Ec/It synchronisation channels	−15 dB	2 dB
Activity factor in synchronisation channel	10%	
P-CCPCH E_b/N_o	6 dB	
P-CCPCH bit rate	30 kbps	
Orthogonality factor	0.4	
DL ratio of MSs at cell edge	30%	
DL inner radius/cell-edge ratio	50%	
I_o other-cell/I_o same-cell DL at inner cell radius	0.06	
I_o other-cell/I_o same-cell at cell edge	2.3	0.7
Shadowing margin	2 dB	6.35 dB
Loading factor	75%	75%

[17] A reference network for illustration only.

Table 7.16 Reference parameters (LCD) in the urban environment (*I*)

Service type parameters	Speech service		LCD 64		LCD 144	
	UL	DL	UL	DL	UL	DL
E_b/N_o	3.7 dB	5.8 dB	2.2 dB	5.1 dB	0.6 dB	4.1 dB
Bit rate	12.2 kbps	12.2 kbps	64 kbps	64 kbps	144 kbps	144 kbps
Penetration margin	15 dB	15 dB	15 dB	15 dB	15 dB	15 dB
Body loss	3 dB	3 dB	0 dB	0 dB	0 dB	0 dB
Number of active channels/carrier	3.8	3.2	0.8	0.7	0.4	0.3
Activity factor	60%	60%	100%	100%	100%	100%
Maximum MS Tx power	21 dBm		24 dBm		24 dBm	
Soft HO gain	0 dB	2.5 dB	0 dB	2.5 dB	0 dB	2.5 dB

UE with 0 dB cable and connector losses, 0 dBi antenna gain and 8 dB receiver noise. The characteristics for the Node B are 18 dBi antenna gain, max Tx power 43 dB, cable and connector loss 3 dB and noise figure of 3.5 dB. The TRUE has MHA noise figure of 2.5 dB, MHA gain of 15 dB and a global receiver noise figure of 2.8 dB.

While the subscriber response or traffic mix does not depend on the environment, the number of subscriber per cell does depend on the service area or environment. Based on the inputs shown in Table 7.11, the number of subscribers per cell in the urban environment is about 180 in 2005. Tables 7.16 and 7.17 illustrate reference parameter for the urban environment.

Based on the preceding tables we can now calculate the maximum allowable path loss for the traffic mix in the urban environment and determine the cell range. Tables 7.18 and 7.19 illustrate uplink and downlink values to serve as comparative reference. Some of these values are round off from earlier dimensioning exercises to provide generic illustrations of multi-service budget-link results.

Table 7.17 Reference parameters (UDD) in the urban environment (II)

Service type parameters	UDD 64		UDD 144		UDD 384	
	UL	DL	UL	DL	UL	DL
E_b/N_o	1.7 dB	4.5 dB	1.1 dB	3.8 dB	1.0 dB	3.0
Bit rate	64 kbps	64 kbps	144 kbps	144 kbps	384 kbps	384 kbps
Penetration margin	15 dB	15 dB	15 dB	15 dB	15 dB	15 dB
Body loss	0 dB	0 dB	0 dB	0 dB	0 dB	0 dB
Number of active channels/carrier	1.9	2.4	0.7	1.7	0.8	2.0
Activity factor	100%	100%	100%	100%	100%	100%
Maximum MS Tx power	24 dBm		24 dBm		24 dBm	
Soft HO gain	0 dB	2.5 dB	0 dB	2.5 dB	0 dB	2.5 dB

Table 7.18 Link-budget example, uplink

	Units	Uplink					
		Speech	LCD 64	LCD 144	UDD 64	UDD 144	UDD 384
Radio channels/total chs		49%	10%	5%	25%	9%	2%
Activity factor		60%	100%	100%	100%	100%	100%
DL orthogonality factor							
Chip rate	Mchip	3.84					
P-CCPCH bit rate	kbps						
Service bit rate	kbps	12.2	64	144	64	144	384
Processing gain		314.6	60	26.5	60	26.5	10
Target E_b/N_o	dB	3.7	2.2	0.6	1.7	1.1	0
Required P-CCPCH E_b/N_o							
Required SCH E_c/I_t	dB						
Tx information							
Cable and combiner losses	dB	0	0	0	0	0	0
Tx Antenna gain	dB	0	0	0	0	0	0
DL Tx power	dBm						
SCH Tx power	dBm						
P-CCPCH Tx power	dBm						
Total Tx power	dBm	16	18	20	18	21	24
Total Tx EIRP	dBm	16	18	20	18	21	24
Rx information							
Rx antenna gain	dB	18	18	18	18	18	18
Cable and connector losses	dB	0	0	0	0	0	0
Receiver noise figure	dB	2.8	2.8	2.8	2.8	2.8	2.8
Thermal noise	dBm/Hz	−174	−174	−174	−174	−174	−174
Service Rx sensitivity	dBm	−126	−120	−118	−121	−118	−115
Synchronisation Rx sensitivity	dBm						
Pilot Rx sensitivity	dBm						
Gains							
Log normal fade margin	dB	6.5	6.5	6.5	6.5	6.5	6.5
Penetration margin	dB	15	15	15	15	15	15
Body loss	dB	3	0	0	0	0	0
Soft HO gain	dB	0	0	0	0	0	0
Max path loss for services	dB	134	134	134	134	134	134
Max pilot and synchronisation ch DL							
Cell range	km	0.75	0.75	0.75	0.75	0.75	0.75
Loading factor		25%	25%	25%	25%	25%	25%
Number of active channels per service		3.8	0.7	0.38	1.9	0.7	0.1
Total number of active ch/cell		8	8	8	8	8	8

Table 7.19 Link-budget example, downlink

	Units	Uplink					
		Speech	LCD 64	LCD 144	UDD 64	UDD 144	UDD 384
Radio channels/total chs		37%	7%	4%	28%	19%	4%
Activity factor		60%	100%	100%	100%	100%	100%
DL orthogonality factor		0.4	0.4	0.4	0.4	0.4	0.4
Chip rate	Mchip	3.84					
P-CCPCH bit rate	kbps	30	30	30	30	30	30
Service bit rate	kbps	12.2	64	144	64	144	384
Processing gain		315	60	27	60	27	10
Target E_b/N_o	dB	3.4	2.7	1.7	2.0	1.4	0.9
Required P-CCPCH E_b/N_o	dB	6	6	6	6	6	6
Required SCH E_c/I_t	dB	−15	−15	−15	−15	−15	−15
Tx information							
Cable and combiner losses	dB	3	3	3	3	3	3
Tx antenna gain	dB	18	18	18	18	18	18
DL Tx power per service	dBm	24	20	19	25	26	23
SCH Tx power	dBm	28.7	28.7	28.7	28.7	28.7	28.7
P-CCPCH Tx power	dBm	28.6	28.6	28.6	28.6	28.6	28.6
Total Tx power	dBm	34.5	34.5	34.5	34.5	34.5	34.5
Total Tx EIRP	dBm	49.5	49.5	49.5	49.5	49.5	49.5
RX information							
Rx antenna gain	dB	0	0	0	0	0	0
Cable and connector losses	dB	0	0	0	0	0	0
Receiver noise figure	dB	8	8	8	8	8	8
Thermal noise	dBm/Hz	−174	−174	−174	−174	−174	−174
Service Rx sensitivity	dBm	−122	−115.5	−113	−116	−113	−109.5
Synchronisation Rx sensitivity	dBm	−115.1	−115.1	−115.1	−115.1	−115.1	−115.1
Pilot Rx sensitivity	dBm	−115.2	−115.2	−115.2	−115.2	−115.2	−115.2
Gains							
Log normal fade margin	dB	2	2	2	2	2	2
Penetration margin	dB	15	15	15	15	15	15
Body loss	dB	3	0	0	0	0	0
Soft HO gain	dB	2.5	2.5	2.5	2.5	2.5	2.5
Max path loss for services	dB	134	134	134	134	134	134
Max pilot and synchronisation ch DL	dB	136	136	136	136	136	136
Cell range	km	0.75	0.75	0.75	0.75	0.75	0.75
Loading factor		45%	45%	45%	45%	45%	45%
Number of active channels per service		3.2	0.6	0.3	2.5	1.7	0.4
Total number of active ch/cell		9	9	9	9	9	9

7.8.2.2 *RNC Dimensioning for 75% Coverage*

Following the principles to dimension the RNC, here we illustrate some results based on the traffic figures illustrated in Table 7.11. We calculated the average throughput per city, at busy hour, both for PS and CS traffic. Table 7.20 illustrates the traffic split in the reference cities.

Table 7.21 illustrates results of the reference examples based on the site distribution illustrated in Table 7.20 and the aforementioned steps for RNC dimensioning.

Table 7.20 Subscriber traffic and site split reference 75% (2005)

Location	User traffic split	Site split	Average total PS throughput (Mbps)	Average total CS throughput (Mbps)
City A	44%	800	910	132
City B	29%	520	600	87
City C	27%	480	560	81
Total	*100%*	*1800*	*2070*	*300*

Table 7.21 RNC Dimensioning results of reference examples

	RNC-A count	Traffic per RNC-A (Iu + Iur) Mbps	STM 1s per RNC-A	RNC-B count	Traffic per RNC-B (Iu + Iur) Mbps	STM 1s per RNC-B
City A	3	230	2	4	330	3
City B	3	226	2	2	290	2
City C	1	210	2	3	270	2

7.9 CORE NETWORK (CN) DESIGN

The basic input for the CN dimensioning resides also in Tables 7.5 and 7.11. In addition, it includes the results illustrated in Tables 7.10 and 7.12. For all practical purposes, we illustrate the dimensioning for both the evolving and the layered architecture, which we will describe in the forthcoming sections.

7.9.1 CN Analysis Assumptions

Figure 7.12 illustrates key CN elements which we will refer to throughout the dimensioning process. Note that here we imply a co-existence with second generation elements, e.g. GSM. The next chapter will describe the essential functions of this network.

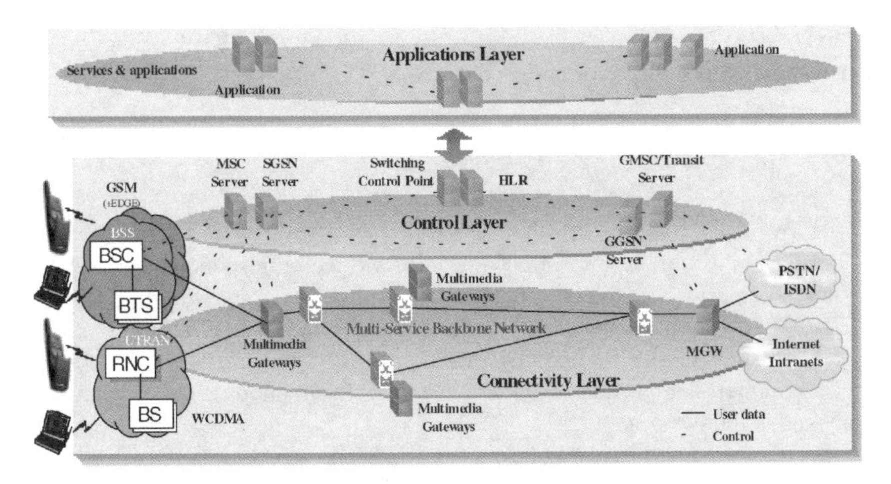

Figure 7.12 The core network layer architecture reference.

7.9.1.1 *Locating CN Nodes*

Logically, the starting point of the CN design and dimensioning will start by locating the main switching centres. In this field study we consider three sites, let us say they are within cities A–C of our country example in this chapter. Thus, we locate CN nodes in centres A, B and C.

7.9.1.2 *Locating RNC Nodes*

The number of RNC nodes will depend on the traffic load and number of BS sites connected. Their location in general will vary according to chosen topology; thus, we can have a centralised or distributed network layout. The centralised implies placing the RNCs at the switching centres together with the CN nodes. The distributed option will in general expand beyond the CN sites. Each option has its trade-offs and will depend on the availability of transmission infrastructure as well. For example, some of the advantages of a centralised RNC network, i.e. co-located with the CN nodes include the following:

- Rapid implementation and integration process, besides the lower investment on the RNCs themselves. Today technology allows high capacity systems as better choices than small ones, because a higher grade of statistical multiplexing is gained in larger nodes than in smaller ones.
- Lower overhead costs on O&M and spare parts besides less restriction of site space.
- With the trend towards more packet data traffic in the UTRAN, a transport network with centralised RNCs will require less bandwidth due to lower overhead over the Iub interface.
- About half of the traffic can be dropped locally at the RNC/MGW site thereby reducing the traffic in the core backbone network to a minimum.

- We do not need Iu transmission resources between RNC and MGW, neither do we need Iur transmission resources for co-located RNCs.

7.9.1.3 Network Overheads

While looking at the traffic flow, we also need to see the corresponding overheads in the various interfaces.

In the RNC different overheads apply to the Iu and Iub interfaces. Table 7.22 illustrates a non-exhaustive summary of the RNC overheads.

We can summarise more details on the overhead assumptions considered for the CN dimensioning as follows:

- Packet data over Iu (GTP-U/UDP/IP/AAL5/ATM) has 72% OH, assuming average packet size 125 bytes, and about 10% signalling OH on top of user data on the Iu and CN. In addition, there is 21% OH on circuit data over Iu and core network, AAL2/ATM.
- Overhead in PS interfaces: about 35% OH on packet data for Gb over ATM network, including BSSGP/FR/AAL5/ATM, about 35% OH on packet data on Gn interface, including GTP-U/UDP/IP. Very low OH on Gi interface, on top of the IP address application.
- We assume 0.2% grade of service for UTRAN links with 90% load factor on the physical payload on ATM links, and 0.1% grade of service for voice and circuit data over the CN with also 90% load factor on physical payload on ATM links.
- We add 10% re-transmission OH on packet data over the Iub on PS user data, assuming 70–75% load factor on IP links in the CN.
- We estimate 66% OH, i.e. 10.35 kbps per voice channel over Iub, for AMR 12.2 voice coding and 50% voice activity factor, including frame overhead UP/AAL2/ATM.
- There exists also 43% OH, i.e. 8.91 kbps per voice channel over Iu and CN for AMR 12.2 voice coding with 50% activity factor, including frame overhead UP/AAL2/ATM.

Table 7.22 Summary of RNC overhead assumptions

	User data	Iu	Iub
Voice activity factor	50%	50%	50%
Voice bit rate	6.22	8.91	10.35
Voice codec (AMR)	12.2	12.2	12.2
Re-transmission	10%	10%	10%
Signalling		10%	10%
Frame OH, CS		21%	23%
Frame OH, PS		70%	24%
O&M			2%

- The CS and PS may have 20–25% OH on CS over the Iub. This includes TCP/IP header compression and UP/AAL2/ATM. For PS, the average packet length is 125 bytes.
- The signalling OH amounts to 10% on top of framed user data, and 2% O&M OH on top of the framed user data on Iub, both with a minimum of 64 kbps per link. In addition, there is 25% frame overhead on signalling O&M traffic.

7.9.1.4 Analysis Assumptions

Grade-of-Service (GoS) applies to the calculation for delay sensitive traffic, and delay varying services, i.e. voice and LCD services, where calculations estimate the required bandwidth to obtain a specific GoS. For ATM links carrying multiple services, the narrow-band Erlang formula applies while considering different amounts of multiple rate traffic. For regular TDM links carrying a single service, i.e. voice between, e.g. BSCs and MSCs, the narrow-band Erlang formula applies.

We assume a load factor for ATM and IP links, i.e. 90% for ATM and 75% for IP links, which describe the maximum allowed theoretical loading on the links. However, we do not assume load factor for TDM links. Thus, for links carrying both real-time and background type of traffic, the real-time bandwidth results from GoS calculations, and the average packet data originates from the load factor. On Iub and Iur interfaces both voice, LCD and LDD data require the same quality. Thus, we include GoS calculations for LDD data.

For all practical purposes, we assume that packet data terminates in, e.g. an ISP or third party network at the local switch site in nodes A, B or C, which means that no packet data transfer occurs between switch sites. We also assume that one-third or 30% of voice and CS data generated in the BSS and UTRAN connected to one switch site terminates in another switch site, thereby, generating inter-MSC and UMTS inter-MGW traffic. We assume that the remaining traffic terminates locally to PSTN, PLMN or stays within the UMTS network.

7.9.2 Reference Outputs in CN Dimensioning

The final number of elements will depend very much on the type of configuration and capacities a supplier will offer for each CN node. Hence, in the following we will provide simple reference results for completeness.

7.9.2.1 RNC Configuration

Tables 7.23 and 7.24 contain reference values for the UTRAN RNC design in terms of total number of RNCs and total traffic volumes per area. The reference traffic flows reside on values provided in Table 7.5.

The reference values will vary based on the type and configuration of the RNC proposed. Here we provide generic numbers in order to allow a starting point to illustrate a RNC and CN dimensioning process. The configuration includes growth surge margins and takes into account the overheads described in the preceding sections.

Table 7.23 RNC configurations and locations for years 2002 and 2005 (750 k and 1000 k subscribers)

Location	Capacity configuration[a]		Node B's supporting	
	2002	2005	2002	2005
City A	2 × L	3 × L	450	380
City B	1 × L, 1 × M	2 × L	290	405
City C	1 × L, 1 × S	1 × L, 1 × M	280	620
Total (Accum.)	*4 × L, 1 × M, 1 × S*	*6 × L, 1 × M*	*1020*	*1405*

[a]Large (L) = 375 k subscriber, Medium (M) = 240 k subscriber, Small (S) = 66 k subscribers.

Table 7.24 Total traffic volumes through RNC for years 2002 and 2005 (750 k and 1000 k subscribers)

Location	User data (Mbps)		Iu traffic (Mbps)		Iub traffic (Mbps)	
	2002	2005	2002	2005	2002	2005
City A	206	275	365	487	397	530
City B	134	178	237	316	258	344
City C	126	168	224	298	243	324
Total	466	621	826	1101	898	1198

7.9.2.2 Core-Network Elements

As illustrated in Figure 7.12, our reference CN architecture has a layered structure with total separation of the pay load transport and traffic control into the user plane and control plane, respectively. The user plane builds on the MGWs to the CN through the transmission backbone. The control plane contains the common data bases (e.g. HLR/AuC) besides the

Table 7.25 MGW PS and CS, GGSN and SGSN configurations for years 2002 and 2005

Location	Configuration GGSN (minimum 300 K subscriber)		Configuration SGSN (minimum 300 K subscriber)		Simultaneous data sessions (e.g. 5 k SDS package)		MGW configuration 8400(E) 540 (kbps)	
	2002	2005	2002	2005	2002	2005	2002	2005
City A	1	1	1	2	67.5	90	1	1
City B	1	1	1	1	45	60	1	1
City C	1	1	1	1	37.5	50	1	1
Total	3	3	3	4	150	200	3	3

different servers (e.g. MSC, SGSN, GGSN; GMSC servers) which handle traffic. If we assume decentralised Media Gateway (MGW) configurations, one MGW in each switch centre (i.e. A, B and C) will handle capacities $\geq$ 8000 Erlang CS traffic and $\geq$ 500 kpps packet-switched traffic in each city configuration as needed. Table 7.25 illustrates the value for the assumptions.

Initially, to consolidate the functionality and operation of MGW (new on CN) we can centralise it in one city where there is high-density traffic, e.g. city A. Then we can expand it to other cities as need arises. Or we can start with one MGW in each city to facilitate network layout. In this study we shall choose the latter for the first year starting 2002. We also estimate the total Simultaneous Data Sessions (SDS) and distribute them throughout the three cities. The SDS allows us to estimate the traffic interaction, e.g. in the PS network, which occurs based on the number of subscribers attached. The latter depends on the GGSN and SGSN capacity for handling PDP contexts and throughput. For example, a GGSN capable of handling 500 k PDP context and a throughput of 500 kpps with a SGSN also handling 500 k PDP contexts and supporting 400 k simultaneous attached users can serve as reference here. We can also assume that the proportion of traffic passing through the cities are: City A 45%, City B 30% and City C 25%.

The SGSN and GGSN servers control the PS traffic. Thus, the number of GSN servers depends on the number of simultaneous packet data sessions. Because the data packets are switched in the media gateways, the data volumes in the network do not influence the dimensioning of the GSN servers.

Likewise, the MSC server controls the circuit-switched calls, and only the total circuit-switched traffic determines the MSC server dimensioning. Packet-switched traffic is controlled by the GSN servers and does not impact the MSC Server. The MSC server capacity is influenced by the number of call attempts for circuit-switched services and by the total circuit-switched Erlangs in the network.

Assuming co-located MSC/VLR/HLR/AuC and signalling functionality in to one physical node, Table 7.26 illustrates 3G MSC/HLR locations. The software on the MSC servers is dimensioned according to the Simultaneous Call Capacity (SCC). The SCC equals the CS total traffic (in Erlang).

Table 7.26 Number of 3G MSC/HLR for years 2002 and 2005

Location	MSC/VLR HLR/AuC		SW for SCC for CS (Erl)		Signalling HW/SW, e.g. 2 Mb boards		Projected HLR capacity (K subscribers)	
	2002	2005	2002	2005	2002	2005	2002	2005
City A	1	1	7911	9375.7	2	4	330	450
City B	1	1	4674	6250.5	2	2	200	300
Total	3	3	15 580	20 835.0	6	8	750	1050

7.10 TRANSMISSION NETWORK ASSESSMENT

To adapt or implement a new transmission system for 3G networks implies taking into account existing new factors such as wider-band applications and larger base of subscribers, besides existing 2G traffic if any. In the following, before illustrating the dimensioning of transmission network for our field study we will outline some of the major issues that concern building a 3G transmission network.

7.10.1 Building 3G Transport Systems

Some of the factors which will have non-negligible impact on the design of a 3G transmission network include new services with higher average traffic per subscriber, which will increase the traffic density in urban areas. The latter resulting in denser BS network, new dedicated sites and new transport links. In addition, service evolution will also play important role because a large number of 3G networks will also continue to serve 2G subscribers. These subscribers will continue utilising evolving applications such as SMS, E-mails, ftp, Internet, interactive games, video calls and WAP, e-commerce, video telephony, etc.

Due to the increase on the multimedia applications, traffic can easily increase from today's 0.10 Mbyte/user/month to about 35 Mbyte/user/month by 2005. Real-Time (RT) services with tighter delay requirements will augment gradually, as will non-real-time services with low packet loss demands.

Thus, a 3G-network transmission will require higher flexibility to increase capacity, incorporate new sites dynamically and support broader set of protocols (e.g. ATM, IP, etc.). In addition, QoS will be an inherent requirement, while demanding higher reliability and security. On the other hand, overall transmission operational cost would still need to diminish, despite the increase of functionality. As a result, the challenging tasks to build a 3G transmission network include access capacity increase to cope with GMS (+GPRS), EDGE[18] and WCDMA sites, increase CN capacity and efficiency from the start of 3G. This will keep the transmission network as a strategic asset, so it can evolve gradually and get optimised for packet traffic all the way from the core.

7.10.1.1 Present State and Future Evolution

Today micro-wave has become the preferred choice for BS network access. It is cost efficient and we can deploy it rapidly with significant savings through integrated solutions. In the CN transmission side, i.e. connections from BSCs to MSCS sites, 2 Mbps leased lines, or protected micro-wave at 16×2 Mbps or owned or leased SDH transport is common.

In the future, sites will evolve to multi-band multi-standard sites, e.g. plain GSM to GSM-EDGE and/or WCDMA. Macro-cell transmission capacity will thus increase from typical 2 to 10 Mbps as illustrated in Figure 7.13. The level of increase clearly justifies new ways to implement transmission network to maximise efficiency and to add flexibility. High capacity

[18]Although it may not apply to all networks, specially if some choose direct path to UMTS (e.g. WCDMA).

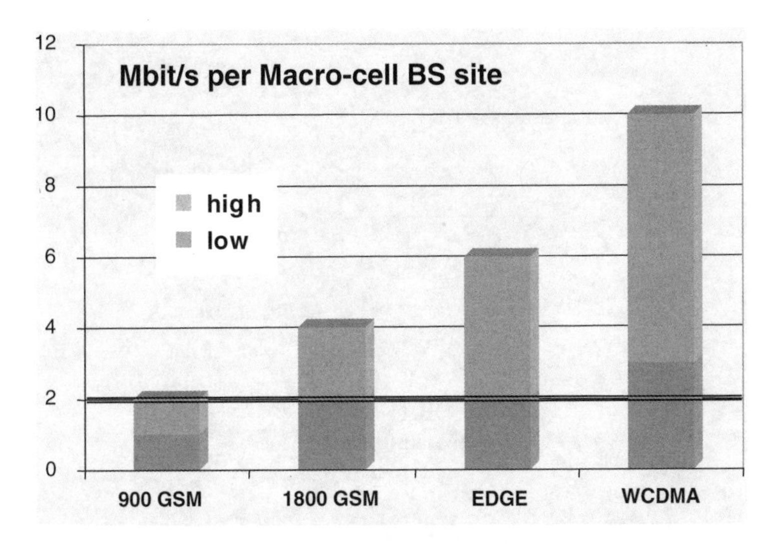

Figure 7.13 Macro-cell traffic growth example.

SDH (STM-4–STM-16), e.g. can become the core transport network while complementing the access network with STM-1 to enter the backbone and using $n \times 2$M for the access connections. The latter would be mainly micro-wave radios.

Capacity increases will accumulate primarily around the controller sites, e.g. an area served by a 16×2 M radio link will saturate as it takes 3G traffic. Thus, density will be a major challenge with radio links, consequently we will need traffic grooming in the field near the BSs.

In mixed environment, e.g. 2G and 3G, each BSC/RNC will concentrate capacity to the NSS and GPRS elements. Thus, even if the total volume will be lower than that in the access network, an integrated transport network will add efficiency and performance overall as illustrated in Figure 7.14. Consequently, an SDH based transport CN will provide flexible capacity and base for multi-protocol support besides basic connectivity.

The access network has more stringent delay requirements, e.g. WCDMA traffic needs ATM transport carried over TDM circuits. Thus, 3GPP technical specs recommend ATM over semi-permanent point-to-point connection for the RAN network. On the other hand, in the CN transmission efficiency gets more important. Thus, packet traffic can travel either on the top of ATM, or IP packets get carried directly over TDM infrastructure based on, e.g. fibre.

7.10.2 Transmission Reference Network

Coming back to our field study we will make the following assumptions:

- The transmission network will serve an existing 2G network (e.g. 1800 MHz GSM network).
- About 70% of existing BS sites can house a WCDMA BS. New UTRA sites will replace non-usable GSM sites and/or stand for coverage where required, e.g. dense areas.

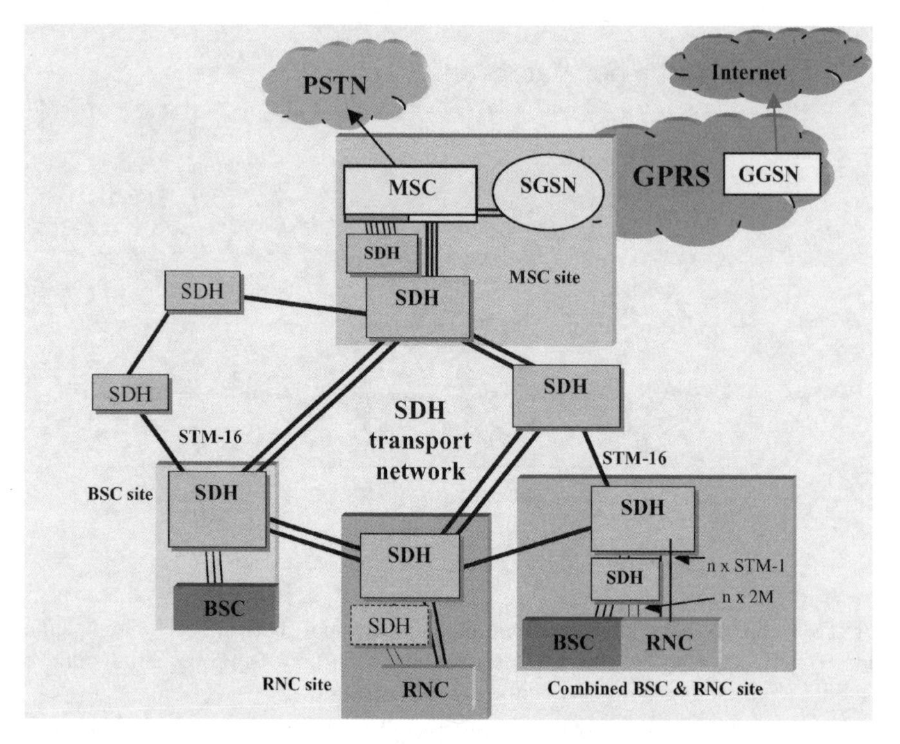

Figure 7.14 An integrated core transport network.

- Figure 7.15 illustrates a typical 2G-access network model, which will serve to project the links required in years 2002 and 2005.
- In the model, 30% GSM BTSs connect through leased lines for the last mile access, 35% use MW radio links in cascaded structures and 35% use micro-wave radio links in hub structures.

In Figure 7.15 leased lines terminate point-to-point between BTS and BSC, while cascaded micro-wave structure uses three sites connected in a cascade, and the structure hub serves 5 BTSs including itself. In the example links, about half of hub-to-BSC links are 4×2 Mb and other half is 8×2 Mb ($4 \times (2 \times 2$ M)).

7.10.2.1 Co-siting Transmission Sites

In a hybrid 2G/3G environment, we would aim to integrate the UMTS network with the existing GSM network infrastructure to reuse the latter to the maximum. Thus, in the macro-cell coverage, e.g. we would to co-locate WCDMA BSs with existing GSM BSs as much as possible. To achieve this Figure 7.16 illustrates two key options, excluding an initial approach of simply carrying 3G traffic over the existing GMS transmission infrastructure.

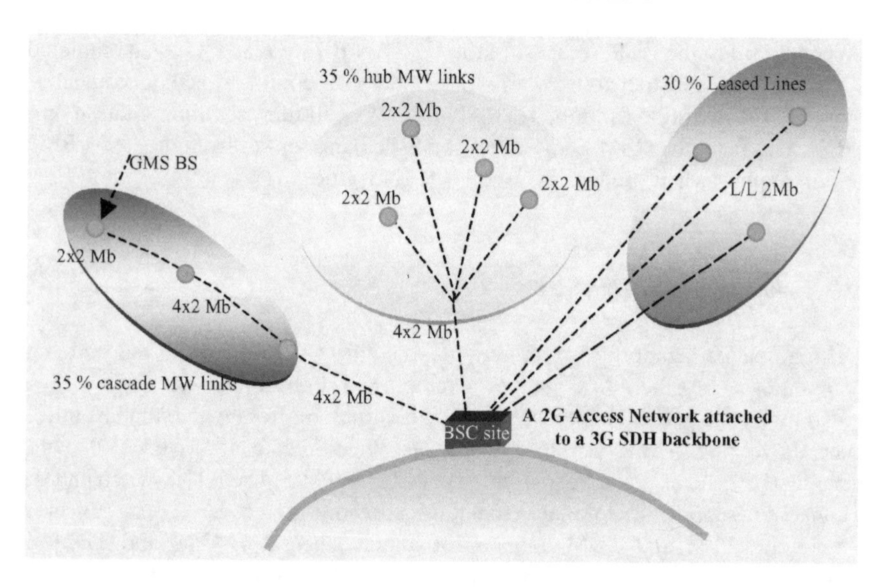

Figure 7.15 Typical 2G-access network attached to an SDH Backbone.

In the first option, carrying 3G traffic over the BSS, the ATM transmission to/from the 3G BS gets mapped from/to an E1 2 Mbps through 64 kbps time slots. We call this 'ATM on fractional E1'. Here we assume the time slot handling DXC function is included in the GSM BS. At the GSM BSC, we connect the UTRAN time slots to the RNC semi-permanently. This solution may apply while 3G traffic remains low.

In the second option, all radio-access network transmission gets carried by a common ATM backbone network composed by UTRAN nodes (e.g. RNCs and BSs). The GSM BSs

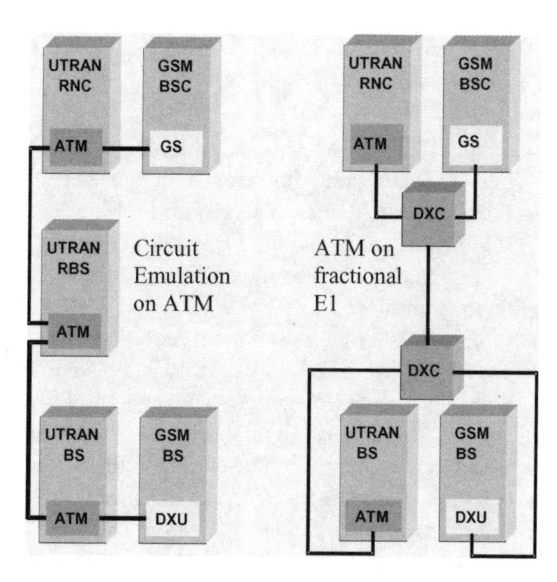

Figure 7.16 2G and 3G transmission site sharing.

get also connected to the UTRAN BSs. In the UTRAN BS we map E1s to an emulated ATM based 2 Mbps line, and transmit towards the RNC on a common backbone together with ATM based UTRAN traffic from the UTRAN BSs. We call this 'circuit emulation on ATM'. In the RNC E1s from the GSM BS gets re-established and sent to/from the GSM BSC node. This option applies when projecting large UTRAN traffic.

7.10.2.2 The Backhaul Network

An SDH regional backhaul infrastructure aims to connect the BSCs with the switch sites in the main cities (i.e. cities A–C in our example). The backhaul network will provide distributed high capacity and high availability infrastructure, from where the last mile access can proceed over micro-wave links or leased lines. We co-locate an ATM/AAL2 switch with Frame Relay (FR) encapsulation capabilities at each BSC site. The ATM switch and the BSC inter-connection supports fractional ATM flow, allowing thereby AAL1 circuit emulation traffic onto the ATM and TDM domain for a pure separated ATM and TDM domain upstream in the network. GSM voice traffic uses TDM transport form the BSC whereas UTRAN traffic and GPRS packet data share common ATM transport resources. We may assume that on an average each regional ring terminates 4–6 BSCs to the switch sites. We can also assume that for the size of our example network, six regional rings can terminate in switch site of city A, and four rings can terminate in the switch sites of city B and city C, respectively.

Capacity projections in our example shows an average of 15 E1s required per ring in 2000 based on GSM traffic, and about 80 E1s in 2005 including 3G traffic. Thus, any rings with STM-1 capacity will need an upgrade before 2005. Consequently, a mix of STM-4 rings for larger traffic and STM-1 rings for lower traffic areas will result in the ideal-backhaul network.

7.10.2.3 The National Backbone Network

To take the traffic flow from STM-1–STM-4 rings, a high capacity STM-16 ring will inter-connect the main BSC sites of each region with the switching sites in cities A–C. In addition, we would expect a high-capacity backbone ATM switch in each switch site.

At the switch sites BSC-to-MSC traffic would terminate to the MSC from the SDH DXC. The UTRAN traffic would terminate over the regional rings to the ATM switch located in the switch site, while aggregating traffic to the RNCs. As noted earlier, RNCs and CN nodes reside co-located at the switch sites hard wired without requiring any transmission resources. Inter-site MSC traffic flow would use STM transport over the SDH backbone, and UMTS inter-switch site traffic would get routed over the ATM switching layer.

Capacity projections show a higher bandwidth need in 2000 than in 2005 despite the higher number of subscribers in the latter. The reason lies in the better bandwidth utilisation for UMTS traffic, e.g. codec at the edge for voice traffic means lower capacity need. For example, in 2000 we would need 160 E1s, about 210 E1s in 2002 and 118 E1s in 2005. This means that national ring capacity requirements can drop to STM-4 capacity rather than STM-16.

Table 7.27 Reference number of sites for years 2002 and 2005

	DU		U/IND		SU		FO/OP	
Year	2002	2005	2002	2005	2002	2005	2002	2005
Co-located 2G/3G sites	5	8	230	230	100	100	490	490
Stand-alone 3G sites	4	6	60	160	30	75	100	350
Stand-alone 2G sites	2	4	80	80	35	35	165	165

7.10.3 Transmission Dimensioning Results

In the following, for completeness, we outline some dimensioning exercise results assuming 750 k UMTS subscribers and 900 k GSM subscribers around 2002. We also assume about 1000 k UMTS subscribers and 300 k GMS subscribers by 2005. These assumption are mainly to create enough margin and do not necessarily reflect the actual evolution of subscriber population. We do still use traffic patterns shown in Table 7.5 and subscriber population in Table 7.2. The dimensioning exercise does include the margins for unexpected subscriber growth.

7.10.3.1 Traffic Per Site

Table 7.27 illustrates the reference number of sites for this transmission analysis, and the values in Table 7.28 indicate the average amount of bandwidth required per site on the Abis and Iub interfaces for the different service areas. The values include frame OH and load factor, ready for mapping onto the payload on the physical PDH or SDH links.

In Table 7.27 the number of sites are cumulative. The totals can be summarised as follows: 2002, co-located = 825; stand-alone 3G = 194 and stand-alone 2G = 282. Total for 2002 = 1301. 2005, co-located = 828; stand-alone 3G = 591 and stand-alone 2G = 284. Total for 2005 = 1703. Therefore, the 2002 sites grow to 1703 by 2005 in this example.

7.10.3.2 Micro-Wave Network

Based on the assumptions of the site traffic in Tables 7.28 and 7.29 illustrates the number of micro-wave links projected for 2002 and 2005. Values are absolute figures.

Table 7.28 Projected traffic per site for years 2002 and 2005 (Mbps)

	DU		U/IND		SU		FO/OP	
Year	2002	2005	2002	2005	2002	2005	2002	2005
Co-located 2G/3G sites	3	2.0	1.9	1.3	4.7	4.0	2.5	2.0
Stand-alone 3G sites	1.5	1.4	1.0	.9	3.8	3.4	1.6	1.6
Stand-alone 2G sites	1.0	.5	1.0	.5	1.0	.5	1.0	.5

Table 7.29 Micro-wave links for years 2002 and 2005

	DU		U/IND		SU		FO/OP	
Year	2002	2005	2002	2005	2002	2005	2002	2005
2 × 2 Mbps	5	5	165	235	35	100	330	500
4 × 2 Mbps	3	2	70	95	40	0	75	155
8 × 2 Mbps	0	0	20	0	15	35	120	45
16 × 2 Mbps	0	0	0	0	25	10	0	0
Total	8	7	255	330	115	145	525	700

The MW frequencies will vary from region and/or country to country. For example, in Switzerland, the last mile access links in dense urban, urban and industrial areas will most likely use 38 GHz. In sub-urban, forest and open areas links will use 23 GHz. However, these frequencies are subject to change depending on the radio frequency availability.

7.10.3.3 Leased Line Network

Table 7.30 illustrates the number of projected leased lines for 2002 and 2005. Values for each site category shown are absolute figures. Taking into account the information of Tables 7.29 and 7.30, we illustrate the projected capacities for the regional rings in Table 7.31.

7.10.3.4 SDH Backbone Network

Table 7.32 projects the number of E1s between the switch sites city A, city B and city C. MSC-MSC values indicate GSM voice traffic over the TDM domain. MGW-MGW values indicate UMTS voice and CS data figures over the ATM domain. All packet data get terminated locally; thus, packet data transfer does not exist between switch sites.

7.10.3.5 Local Switch Site Traffic

The traffic flow in the local switch sites corresponds to the interface load between UMTS nodes at the switch sites. Values in Table 7.33 include all overheads as well as load factor,

Table 7.30 Leased line projections (E1's) for years 2002 and 2005

	DU		U/IND		SU		FO/OP	
Year	2002	2005	2002	2005	2002	2005	2002	2005
Co-located 2G/3G sites	3	2	70	70	90	60	290	295
Stand-alone 3G sites	1	1	20	45	20	45	35	100
Stand-alone 2G sites	1	1	20	25	10	10	50	50

Table 7.31 Projected capacity regional rings (E1s) for years 2002 and 2005

	City A		City B		City C	
	2002	2005	2002	2005	2002	2005
A1	70	75				
A2	60	70				
A3	80	85				
A4	60	70				
A5	80	85				
A6	80	90				
B1			85	90		
B2			70	80		
B3			60	55		
B4			60	70		
C1					80	90
C2					50	60
C3					70	75
C4					65	65
Total	430	475	272	295	265	290

Table 7.32 Inter-switch site traffic (E1s) for years 2002 and 2005

	MSC-MSC		MGW-MGW	
	2002	2005	2002	2005
City A – City B	75	40	20	1 STM-1
City B – City C	50	20	10	1 STM-1
City C – City A	70	35	20	1 STM-1
Total	195	95	50	3 STM-1

Table 7.33 Local switch traffic for years 2002 and 2005

Local switch traffic (Mbps)	Interface	City A		City B		City B	
		2002	2005	2002	2005	2002	2005
UTRAN – RNC	Iub	500	660	300	430	320	405
RNC – MGW	Iu	425	570	260	370	275	350
MGW – GGSN	Gn	235	310	145	200	155	191
MGW – PSTN (external net)		312	420	190	270	200	255
GGSN – ISP (external net)	Gi	174	230	105	150	115	141
MGW – Other switch sites	G	95	125	60	80	60	80
MGW – MSC – S (sign)	GCP	8	10	5	7	5	7
MGW – SGSN – S (sign)	GCP	16	20	10	15	9	14
SGSN – S – GGSN (sign)	Gn	16	20	10	15	9	14

ready to get mapped onto the payload of the physical PDH/SDH/Ethernet links. Values in Table 7.33 reflect total figures.

7.11 CO-LOCATING AND SHARING SITES

With the wide spread of 2G mobile network like GSM, by the time we start deploying UMTS most countries or regions will already have at least one 2G mobile system in operation. While in some place restrictions for new sites will not affect 3G-network deployment, in others it will become a key factor to maximise re-use of sites and push site sharing to the limits. In Switzerland, for example, regulations on power levels, cost of sites and environmental protection regulations do make the acquisition of new sites difficult. Thus, co-locating 2G and 3G equipments becomes imperative. To address these issues in the following we illustrate co-location primarily with UMTS and GSM 1800 MHz.

Site sharing depends on the relative coverage of the existing network when compared to UTRA-FDD.[19] Here we compare at the relative UL coverage of GSM 1800 with full-rate speech service and WCDMA speech and 144 kbps data service. Table 7.34 illustrates the basic assumptions and reference results.

In Table 7.34, we assume 2.0 dBi gain for UTRA-FDD data terminals, and 3 dB body loss for the speech terminals. The interference margin in FDD may apply to 50% loading. The FDD sensitivity assumes 5.0 dB BS noise figure with $E_b/N_0 = 5.0$ dB for 12.2 kbps speech (AMR) and 1.5 dB for 144 kbps data. GSM sensitivity $= -108$ dBm. The FDD fast fade margin includes the macro-diversity gain against fast fading. Co-siting assumes three-sector antenna configuration for both GSM and FDD.

From the reference values in Table 7.34, we can extrapolate that the coverage of 144 kbps data service with 21 dBm mobiles may correspond to the GSM 1800 speech coverage. Thus, we can provide 144 kbps data with WCDMA when re-using or co-siting GSM 1800 sites with the same coverage as GSM1800 speech. In the DL we have less power restrictions; hence, higher 3G data rates may be possible with the GSM 1800 coverage. On the other hand, we should be aware that coverage comparison and site GSM 1800 re-utilisation for WCDMA will depend on the receiver sensitivity values and other system parameters, e.g. HO, FH, power control, etc. In the sequel, we will expand the interference levels and criteria for co-siting, as well as de-coupling issues.

Table 7.34 Common path losses for GSM 1800 MHz and UTRA-FDD

	GSM 1800/speech	WCDMA/speech	WCDMA/144 kbps
Mobile transmission power	30 dBm	21 dBm	24 dBm (21 dBm)
Receiver sensitivity	−108 dBm	−123 dBm	−116 dBm
Interference margin	0.0 dB	2.5 dB	2.5 dB
Fast fade margin	2.0 dB	2.0 dB	2.0 dB
Antenna gain	18.0 dBi	18.0 dBi	18.0 dBi
Body loss	3.0 dB	3.0 dB	–
Mobile antenna gain	0.0 dBi	0.0 dBi	12.0 dBi
Maximum path loss	153.0 dB	155.0 dB	154.0 dB

[19]The UTRA FDD uses the WCDMA technology as indicated in preceding chapters.

7.11.1 Interference Levels and De-coupling

Radio site sharing solutions imply either utilisation of multi-band antennas or co-locating 2G/3G HW/SW. To assess these needs we first list the interfering levels given by the standard specifications.

7.11.1.1 GSM/UMTS Interference

Considering the GSM recommendations (05.05 and 11.21) and the UMTS recommendation (25.104) we can calculate the necessary de-coupling between:

- two UMTS systems,
- one GSM 900 system and one UMTS system,
- one GSM 1800 system and one UMTS system.

The analysis of co-location includes wide-band noise, spurious emissions, blocking and intermodulation. The solutions may come from antenna isolation, use of filters (e.g. diplexer or triplexer). We assume the following reference output powers: 43 dBm output power at the antenna port of GSM 900/1800 BTS and 45 dBm output power at the antenna port of UMTS BS.

Wide-band noise. The frequency separation between the GSM 900 and the UMTS, as well as between the GSM 1800 and the UMTS bands, is sufficiently large. Thus, it is frequently assumed that the transmit part of GSM 900/1800 does not generate wide-band noise in the receive part of UMTS.

Spurious emissions. GSM recommendation 05.05 specifies a maximum level of spurious emissions in the frequency band 1–12.5 GHz, including the UMTS receiving band. This maximum power should not be greater than −30 dBm at the base station RF output port, power measured in 3 MHz bandwidth, what is equivalent to −95 dBm/Hz.

Looking at the maximum tolerable interference level in the Rx channel of −118 dBm, (i.e. −185 dBm/Hz, over 5 MHz worst case), we can calculate the required isolation to avoid interference for the Tx GSM/DCS Rx UMTS. Thus, assuming specified spurious emissions over 3 MHz–30 dBm, specified spurious emissions per Hz–95 dBm and maximum level of unwanted signal per Hz–185 dBm, the necessary isolation between antenna connectors is 90 dB.

GSM and UMTS band analysis indicates that a second and third order intermodulation product of GSM 900 and 1800 transmission frequencies, respectively, fall into the UMTS uplink band. Thus, GSM 05.05 recommendation specifies that when intermodulation power level of interfering signal gets injected into the antenna connector at a level of 30 dB lower than that of the wanted signal, it shall not exceed the spurious emission requirements, i.e. −30 dBm in the UMTS receiving band at the BS RF output port measured in 3 MHz bandwidth, which is equivalent to −95 dBm/Hz. Since this is at the same level as the one specified for the spurious emissions, we need the de-coupling between antenna connectors.

Table 7.35 UMTS GMS isolation requirements

System spec requirements	UMTS Tx to GSM 900 Rx	UMTS Tx to GSM 1800 Rx	GSM 900/1800 Tx to UMTS Rx
Spurious emissions and intermodulation products	34 dB	34 dB	90 dB (39 for BTS R99 or later)
Blocking	37 dB	45 dB	58 dB

Blocking. UMTS 3GPP TS 25.104 recommendation specifies that out-of-band, the maximum level of interfering signals (CW carrier) for blocking is equal to −15 dBm. The GSM power is assumed to be 43 dBm at antenna connector. This power is over 200 kHz, rather than a CW interfering source as defined in TS 25.104. Comparing this value with the UMTS blocking point, we can calculate the necessary isolation between the two antennas. Then, assuming that the jamming system is Tx GSM/DCS and the victim the Rx UMTS, with a GMS/DCS transmit output power 43 dBm, the blocking point out of receiving band is −15 dBm. Thus, the necessary isolation between antenna connectors is 58 dB.

In conclusion, Through the preceding analysis, we see that we need an isolation of 37 and 45 dB to protect GSM 900 and 1800 MHz from UMTS interference, respectively. We can obtain these de-coupling values by physical separation between the GSM 900 and/or 1800 antenna and the UMTS antenna. Table 7.35 summarises the required isolation, in case of co-location.

Through the preceding analysis, we see that we need an isolation of 37 and 45 dB to protect GSM 900 and 1800 from UMTS interference, respectively. We can obtain these de-coupling values by physical separation between the GSM 900 and/or 1800 antenna and the UMTS antenna.

Obtaining the isolation of 90 dB to protect UMTS from GMS900 or/and 1800 interference gets more difficult through only space separation. We would need specific filters at GSM side to reduce the spurious emission levels, and in UMTS side to protect the UMTS BTS from GMS emissions. The latter will be imperative when using:

- dual-band antenna or tri band for which the isolation is only about 30 dB between GSM 900 or/and 1800 and UMTS bands,
- wide-band antenna for which there is no isolation between GSM 900 or/and 1800 and UMTS bands.

Other de-coupling needs. For the GSM 1800 and UTRA-FDD frequencies, i.e. 1710–1785 and 1920–1980 MHz UL, respectively, and 1805–1880 and 2110–2170 MHz DL, respectively, there is a need of 68 dB air de-coupling.

7.11.1.2 Power-Level Limitations

Maximum power-level requirements contribute also to the limitations for site co-location. Although the limits may vary from region to region, there are limits everywhere. Thus, unnecessary power rise in co-located sites will minimise power emission concerns. On the other hand, with the advent of WCDMA power calculations should not follow the same strictness of GSM, because a WCDMA carrier will not be transmitting at full power all the time.

7.12 CO-LOCATING ANTENNA SYSTEMS

Another key issue with co-siting or site sharing is the integration of 2G/3G antenna solutions. Thus, to complete the analysis of co-siting in the following we will summarise the main aspects implementing antennas for UMTS and 2G systems like GSM.

While single-band antennas can meet co-siting needs, by far multi-band antennas are the ideal solutions for integrated antenna systems in co-siting spaces for 2G/3G. Multi-band antennas (i.e. dual-band or triple-band), paired with additional diplexer or triplexer minimise the amount of antennas and feeder cables per site. On the other hand, adding diplexers to an existing system will increase losses thereby decreasing coverage. Thus, the trade has to be taken into account when completing radio planning.

7.12.1 Co-siting GSM 1800 and UMTS

7.12.1.1 Single Band

Due to demands for, e.g. separate tilting, single-band antennas may be ideal. Figure 7.17 illustrates the air-decoupling configuration for single-band antennas, where separate feeder cables connect the Node B and BTS, respectively.

The single-band antennas get separated either by a vertical distance or by a horizontal distance. Side by side antennas, e.g. GSM 1800 and UMTS may de-couple already at 40 dB. Thus, to apply to most relevant frequencies (with better than 42 dB) recommended coupling distance (including security/error margin) will lie between 0.9 and 1.0 m horizontal separation. Likewise in the vertical case, to reach a decoupling of more than 42 dB in the UMTS frequency range, a coupling distance of 1.8–2.0 m (including security/error margin) of vertical separation is recommended.

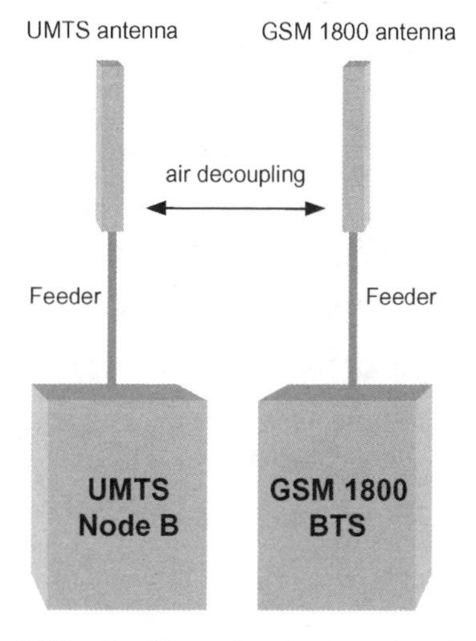

Figure 7.17 Air decoupling in single-band antennas.

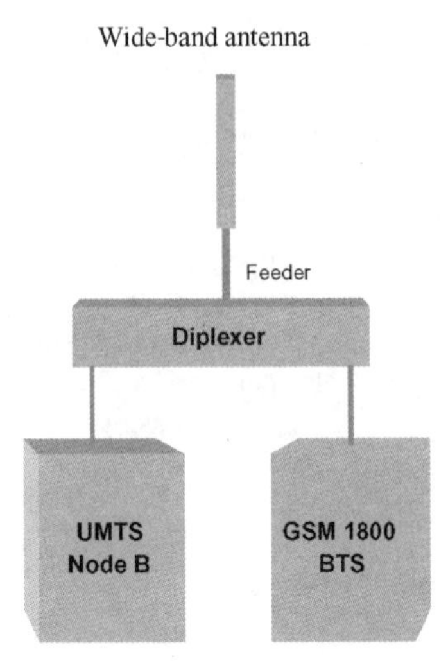

Figure 7.18 Wide-band antenna configuration with one diplexer.

7.12.1.2 Wide-band Antenna

Figure 7.18 illustrates a wide-band solution[20] for GSM 1800 and UMTS using a diplexer, one feeder cable. This solution appears advantageous because it needs only one feeder cable and one antenna panel. However, it has limitations because it imposes the same antenna characteristics to the GSM 1800 and the UMTS bands. Such limitation, e.g. prevents different electrical downtilt for the two systems.

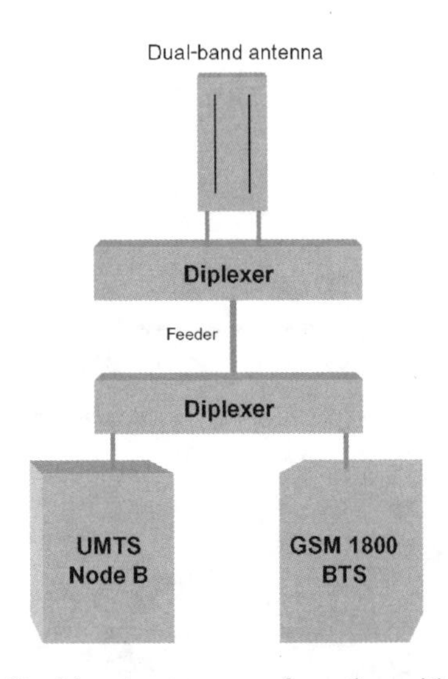

Figure 7.19 Dual-band antenna configuration with two diplexers.

[20]This combination gets doubled for the second antenna branch.

7.12.1.3 GSM 1800/UMTS Dual-band Antenna With Double Diplexers

Double diplexer dual-band antennas as illustrated in Figure 7.19 include one BTS-side diplexer, one feeder cable, one antenna side and a GSM 1800/UMTS dual-band antenna[21] consisting of two antennas within one panel. The BTS-side diplexer will generally afford 45–48 dB of decoupling from GSM 1800 transmit port to UMTS receive port, while the antenna side diplexer provides a decoupling value of 30–32 dB, which is generally sufficient.

This dual-band antenna solution with two diplexers does allow different gain and electrical tilt selection for GSM 1800 and UMTS. On the other hand, it demands the implementation of two diplexers.

7.12.1.4 Feeder Cables

As noted in the preceding sections we can realise dual-band systems, either with separated single-band antennas or dual-band antennas. On the other hand, when an antenna system supports diversity, we require at least two additional antenna branches per BTS sector and mobile system. The latter implies four antenna branches for a dual-band BTS sector.[22] Hence, to prevent the usage of four feeder cables, e.g. we can apply feeder sharing with additional diplexers and use two feeders only. Figure 7.20 illustrates this logic with a cross-polarised dual-band antenna.

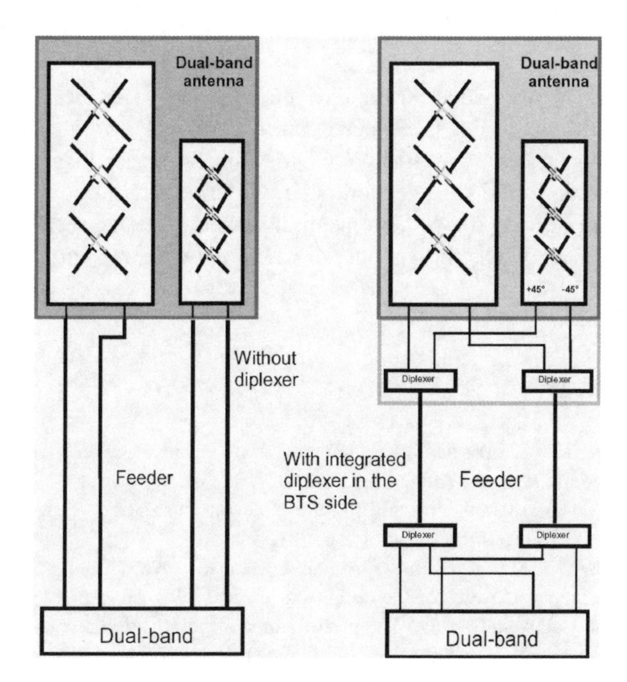

Figure 7.20 Dual-band antenna with and without diplexer.

[21]As in the wide antenna, this combination has to be doubled for the second antenna branch.
[22]This does not apply to the solution with broadband antennas for GSM 1800 and UMTS.

Dual-band antennas are characterised by being suitable for both frequency ranges with separate input connectors. Although dual-band antennas suit ideally both GSM 1800 and UMTS frequencies ranges, they lead to double number of antenna connectors when compared to corresponding single-band antennas if sufficient diplexers are not used. Thus, it seems more efficient to upgrade dual-band antennas with additional diplexers in order to decrease the number of antenna connectors by a factor of two. In this case the required feeder system will remain the same as for a single-band antenna system.

The trade-off of extra investment for additional diplexers and expense on the feeder system will depend at the end on the losses due to feeder attenuation. In any case, when doing a transition from single-band to a dual-band system, the existing feeder system can be used ensuring a fast installation during retrofit if the necessary diplexers are in place.

7.13 CONCLUSIONS

In this chapter we have covered essential aspects for 3G-network deployment. We have brought together the main issues of network dimensioning on the radio, core network, as well as transmission. We have used a field study to illustrate key parameters and introduce some relevant assumptions when trying to size a network for a given population within a particular region.

The data examples correspond to harmonised values making a case for the method used. However, they do not necessarily reflect the actual growth of network or exact projection of network size. Nevertheless, do make realistic reference to the given concepts. Thus, they can serve as examples.

The analytical process provided for dimensioning the radio side aimed to set the basis to focus on key WCDMA issues that will need optimisation as networks get implemented for mixed services. The factors involved in WCDMA appear larger than the ones in current CDMA systems today. In the core network side, we have also introduced new factors not typically seen in 2G networks. On the transmission side, we have seen the traffic increase will demand larger bandwidths throughout the whole network and not only in the core, but also the radio side.

REFERENCES

1. Ericsson - Orange, UMTS Traffic Model Dimensioning, Guidelines and Assumptions, Input on UMTS dimensioning examples, August 3, 2000.
2. TS 101 111 (UMTS 21.01), Universal Mobile Telecommunication System (UMTS); Overall Requirements on the Radio Interface(s) of the UMTS.
3. Yang, S. C., *CDMA RF System Engineering*, Artech House, 1998.
4. High level requirements relevant for the definition of the UTRA concept, V3.0.1, 1998–2010.
5. Lempiäinen, J. and Laiho-Steffens, J., The performance of polarisation diversity schemes at a base-station in small/micro-cells at 1800 MHz, *IEEE Trans. Vehic. Tech.*, **47**(3), 1087–1092, 1998.
6. Laiho-Steffens, J. and Lempiäinen, J., Impact of the Mobile Antenna Inclinations on the Polarisation Diversity Gain in DCS1800 Network, Proceedings of PIMRC'97, Helsinki, Finland, September 1997, pp. 580–583.

8

High Speed Downlink Packet Access

8.1 BACKGROUND

UMTS releases prior to Release 5 (R5) downlink packet data, uses three channels, i.e. Dedicated Channel (DCH), Downlink Shared Channel (DSCH) and Forward Access Channel (FACH).

- DCH applies to all types of services with a fixed downlink Spreading Factor (SF), while reserving code space capacity depending on the connection peak data rate.[1] It has power control and can operate in soft handover.
- The DSCH operates always in conjunction with a DCH to use channel properties defined to meet packet data needs with tight delay budget (e.g. speech or video). It has a dynamically varying spreading factor passed on a 10 ms frame-by-frame basis with the TFCI signalling carried on the associated DCH, shares code resources between several users, may employ either single-code or multi-code transmission, can use fast power control associated DCH and does not support soft handover.
- The FACH[2] applies also to downlink packet data usage. It operates on its own with a fixed spreading factor and high power level to reach all cell users because of physical layer feedback lack in the uplink. It does not use fast power control or soft handover.

8.1.1 HSDPA Principles

The High Speed Downlink Packet Access (HSDPA) increases packet data throughput with link adaptation and fast physical layer re-transmission combination. An evolving architecture takes into account new memory management requirements and air-interface link control adaptation. The High Speed Downlink Shared Channel (HS-DSCH) transports HSDPA user data. It has the following characteristics:

- does not support variable Spreading Factor (SF),
- does not use fast power control,

[1]For example, up to 2 Mbps reserving the code tree for a very high peak rate with low actual duty cycle (inefficient).
[2]Carried on the Secondary-Common Control Physical Channel (S-CCPCH).

All IP in 3G CDMA Networks J. Castro
 ISBN: 0-470-85322-0

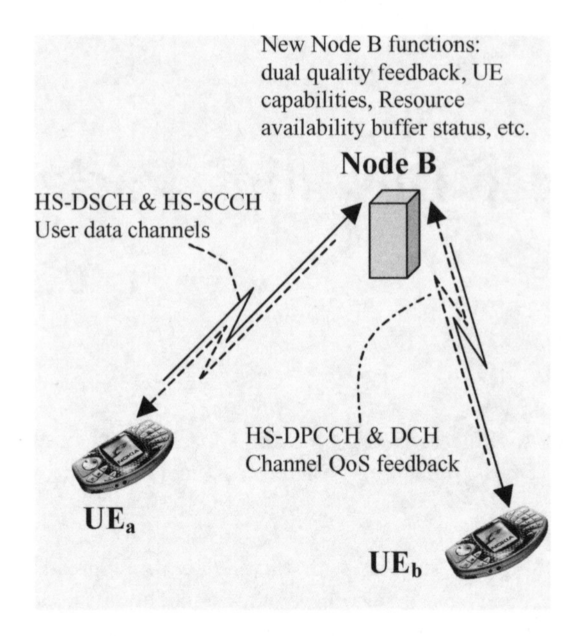

Figure 8.1 HSDPA operation principle.

- it does use Adaptive Modulation and Coding (AMC),
- does apply extended multi-code operations and
- utilises fast Layer 1 (L1) Hybrid Automatic Repeat Request (HARQ).

Figure 8.1 illustrates the HSDPA operation principle where the Node B estimates user channel quality applying, e.g. power control, ACK/NACK ratio, QoS and HSDPA dedicated user feedback. Link adaptation and scheduling takes place at a fast pace following active scheduling algorithm and user prioritisation scheme.

In HSDPA Adaptive Modulation and Coding (AMC), extended multi-code operation and fast-spectrally efficient retransmission functions replace variable spreading factor and fast power control. WCDMA works with downlink power control of 20 dB dynamics range, while the uplink power control has 70 dB dynamics range. Thus, intra-cell interference and Node B implementation limit the downlink dynamics, which in turn may impact power control management for users close to the Node B. HSDPA exploits this limitation by applying link adaptation function and AMC to select coding and modulation to meet higher E_c/I_{or} requirements, which is available for users near the Node B, who can also benefit[3] from the combination of up to 15 multi-codes in parallel.

Layer 1 packet combination implies storing received data packets in the terminal's soft memory when decoding has failed. Then subsequent transmissions may combine old packets before channel decoding, and re-transmission may contain the first transmission or different bits compared to the channel encoder output that was received during the last transmission;

[3]Robust coding, fast HARQ and multi-code operation make up for the need of variable Spreading Factor (SF).

thereby achieving through this incremental redundancy strategy diversity gain and improved decoding efficiency.

8.1.2 HSDPA Supporting Technology Overview

HSDPA as an evolution of the UMTS brings together several radio-access techniques, which includes:

8.1.2.1 Adaptive Modulation and Coding – AMC

AMC changes the modulation and coding format following channel condition variations while conforming to system restrictions. Channel conditions can be determined by the receiver feedback. In solutions with AMC, users in close to the cell site[4] will get higher order modulation with higher code rates (e.g. 16QAM[5] with $R = 3/4$ Turbo codes), while users close to the cell boundary will get lower order modulation with lower code rates (e.g. QPSK with $R = 1/2$ Turbo codes). The AMC benefits are thus: first, availability of higher data rates to users in favourable positions, which in turn increases cell average throughput; and second, reduced interference variation due to link adaptation based on variations in the modulation/coding scheme instead of variations in transmit power [1].

8.1.2.2 Hybrid ARQ (HARQ) Techniques

HARQ stands as an implicit link adaptation technique, where we use link layer acknowledgements for re-transmission decisions. It autonomously adapts to the instantaneous channel conditions and is insensitive to the measurement error and delay. Thus, combining AMC with HARQ leads to the best of both worlds – AMC provides the coarse data rate selection, while HARQ provides for fine data rate adjustment based on the channel conditions [1].

Recommendations in [1] proposed an HARQ scheme based on N-channel stop-and-wait protocol in order to reduce receiver-buffering requirements. For example, in an ideal solution signalling for the sub-channel number may be fully asynchronous as illustrated in Figure 8.2, where packets get transmitted using four parallel ARQ processes for UE1 and one ARQ process for UE2, each using stop-and-wait principle. Each packet gets acknowledged during the transmission of other packets, thereby keeping downlink channel occupied all the time when there are packets to transmit.

When N-channel HARQ supports asynchronous transmission:

- we may schedule different users freely without waiting completion of a given transmission, assuming correct HARQ process for packet ownership identification exists;

[4]Close to Node B = favourable position, closer to the cell border = unfavourable position.
[5]Quadrature Amplitude Modulation.

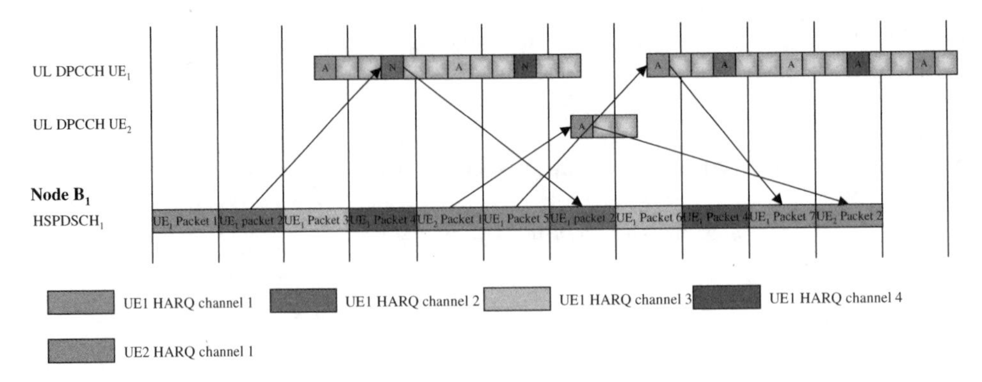

Figure 8.2 Principle of N-channel stop-and-wait HARQ (N = 4) [1].

- the transmission for a given user is assumed to continue when the channel is again allocated.

8.1.2.3 Fast Cell Selection

FCS facilitates the HSDPA UE in the choosing of its best downlink-serving cell through the uplink signalling. As a result, even though multiple cells may be members of the active set, only one of them transmits at any time, while decreasing interference and increasing system capacity. Identifying the best cell depends also on available resources like power and code space for the cells in the active set, besides radio propagation conditions.

FCS physical layer requirements are conceptually similar to physical layer aspect of Site Selection Diversity Transmission (SSDT) included in R99 [1]; therefore, we will not repeat it here.

8.1.2.4 Multiple Input Multiple Output (MIMO) Antenna Processing

As in [1] here the open-loop MIMO implementation stands a representative technology to outline the potential HSDPA performance enhancement.

On one hand, in a traditional single antenna HSDPA, a set of N downlink physical channels (codes) would share transmission among many users. On the other hand, when using an open-loop MIMO architecture with M transmit antennas, while using the same set of codes; each code is re-used M times and each modulates distinct data sub-streams. More precisely, a high rate data source gets coded, rate-matched and interleaved. Figure 8.3 illustrates the coded data stream, which is then de-multiplexed into MN substreams, and the nth group ($n = 1, \ldots, N$) of M substreams spread by the nth spreading code. The mth substream ($m = 1, \ldots, M$) of each group gets summed and transmitted over the mth antenna so that the sub-streams sharing the same code are transmitted over different antennas. We also add mutually orthogonal dedicated pilot symbols to each antenna's Common Pilot Channel (CPICH) to enable coherent detection. For $M = 2$ or 4 antennas we may use the pilot symbol sequences 2 or 4 antenna antenna STTD close-loop transmit diversity.

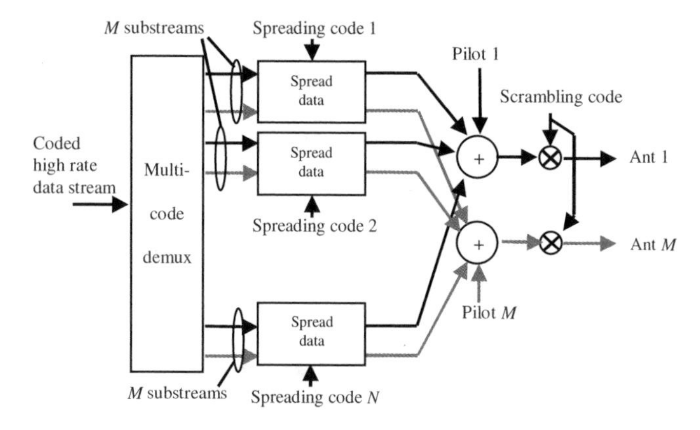

Figure 8.3 Block diagram of a MIMO transmitter [1].

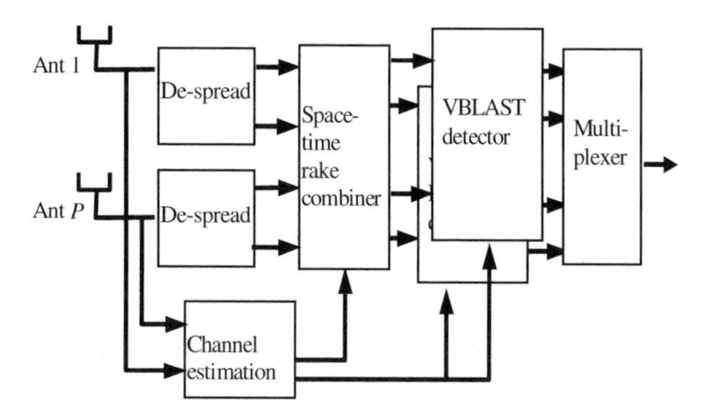

Figure 8.4 Block diagram of a MIMO receiver [1].

The User Equipment (UE) uses multiple antennas and spatial signal processing to discriminate the M sub-streams sharing the same code. Figure 8.4 illustrates a representative MIMO receiver with P antennas, where for coherent detection at the UE, complex amplitude channel estimates, are required for each transmit/receive antenna pair. MP complex channel coefficients characterise a flat fading channel, while LMP coefficients (where L is the number of rake receiver fingers) characterise frequency selective channels. We obtain the channel estimates by correlating the received signals with the M orthogonal pilot sequences, which compared to a conventional single antenna receiver, they have higher complexity by a factor of M. To detect data, a bank of filters matched to the N spreading codes follow each antenna, e.g. an LN de-spreaders per antenna. A space–time rake operation, i.e. the LP corresponding de-spreader outputs get weighed by the complex conjugate of its corresponding channel estimate and summed together to form a sufficient statistic, applies to each of the MN distinct data sub-streams.

Sub-streams transmitted on the other codes, i.e. Multi-Access Interference (MAI), does affect the sufficient statistics of the M sub-streams sharing the same code in flat fading channels, because the code orthogonality gets maintained by the channel, and for each group of M co-code sub-streams, a multi-user detector is used to remove the effects of the MAI [1].

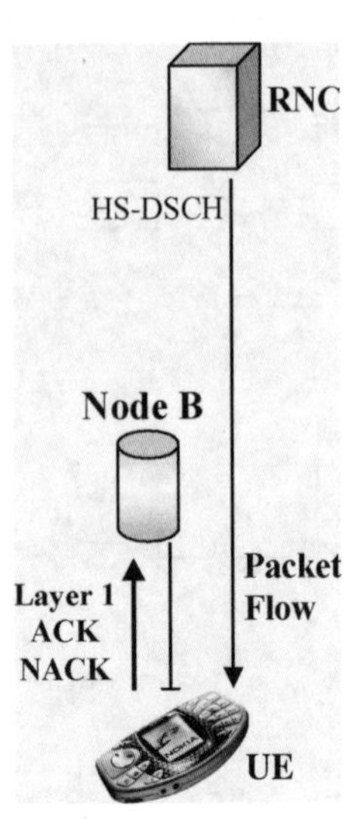

Figure 8.5 HSDPA transmission channel control.

8.2 HSDPA AND RADIO-ACCESS NETWORK ARCHITECTURE

The HSDPA Medium Access Control (MAC) layer in the Node B takes advantage of the HS-DSCH to directly control re-transmissions, which leads to minimum packet delays. The flow mechanism illustrated in Figure 8.5 assumes that Node B buffers packets appropriately to prevent over-flows during the ACK/NACK events.

Figure 8.6 illustrates the HSDPA MAC protocol layer architecture, where the RNC still retains the RLC and other classical R99 functionalities, e.g. re-transmission when Node B HS-DSCH transmission fails due to maximum number of L1 re-transmissions. The new Node B MAC functionality (MAC-hs) handles the ARQ functionality and scheduling as well as priority handling. The RLC layer fulfils Ciphering functions with a mask that stays identical for each re-transmission, enabling thus combining of physical layer re-transmission.

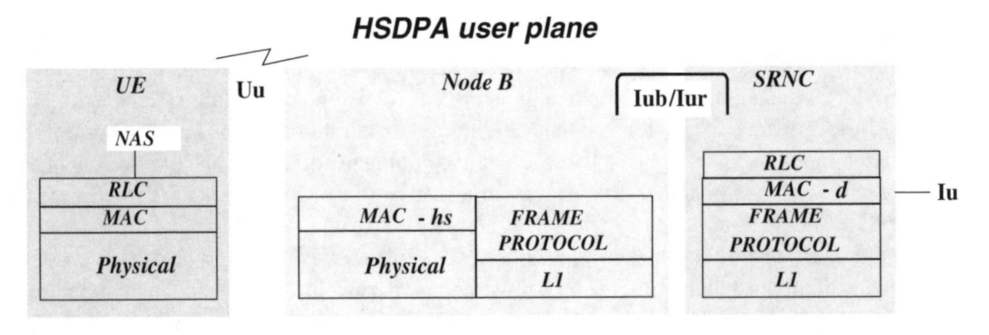

Figure 8.6 Protocol architecture in HSDPA.

8.3 STRUCTURE OF THE HSDPA PHYSICAL LAYER

HSDPA exploits three channels in the physical layer, these include:

1. *High Speed Downlink Shared Channel* – HS-DSCH – carrying downlink user data with peak rates up to 10 Mbps while applying 16QAM.
2. *High Speed Shared Control Channel* – HS-SCCH – carrying L1 control info enabling decoding on HS-DSCH data and performing re-transmission combining.
3. *Uplink High Speed Dedicated Physical Control Channel* – HS-DPCCH – carrying uplink control info (e.g. ARQ ACK/NACK) and downlink quality feedback info.

8.3.1 High Speed Downlink Shared Channel – HS-DSCH

The HS-DSCH Transmission Time Interval (TTI) or inter-leaving period has 2 ms (3 slots) achieving thereby shorter[6] operation roundtrip retransmission-delay between terminal and Node B. It applies higher order modulation scheme, 16 QAM with lower encoding redundancy increasing thus instantaneous peak data rate. It has fixed (always 16), spreading factor and multi-code transmission multiplexing for different users. Depending on the UE capabilities, 5, 10 or 15 codes may be allocated. Figure 8.7 illustrates two subscribers using the same HS-DSCH while checking HS-SCCH information for HS-DSCH code de-spreading and detection parameters.

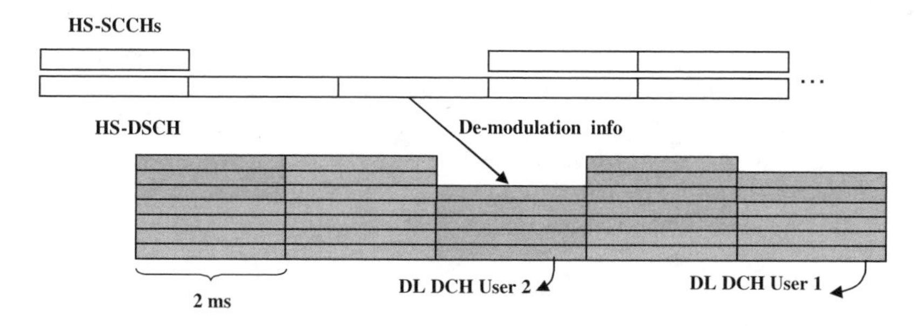

Figure 8.7 Code multiplexing example with two users active.

8.3.2 HS-DSCH Modulation

Figure 8.8 illustrates 16QAM, which doubles transmission rate compared to QPSK and allows up to 10 Mbps peak data rate with 15 codes of SF 16. This modulation demands[7] accurate phase and amplitude estimation to separate the constellation points. A HS-DSCH

[6]Shorter TTI compared to the 10, 20, 40 or 80 ms TTI sizes supported in Release 1999.
[7]QPSK modulation in R99 required only phase estimation.

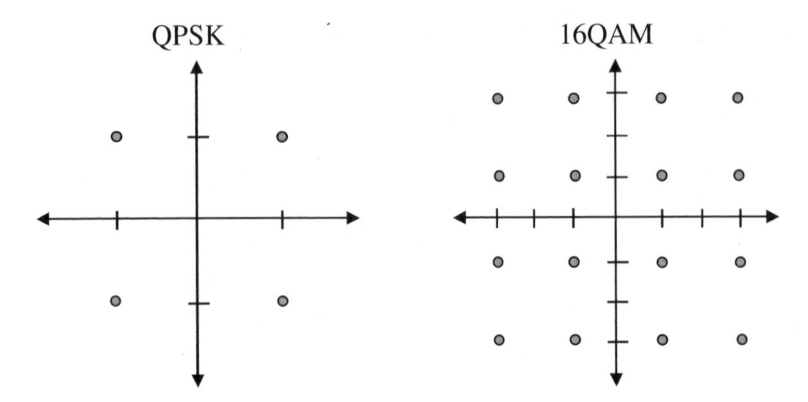

Figure 8.8 QPSK and 16QAM constellations examples.

capable terminal obtains relative amplitude ratio estimates of the DSCH power level compared to the pilot power level while the Node B does not adjust the HS-DSCH power between slots.

8.4 CODING IN THE HS-DSCH CHANNEL

The HS-DSCH channel has only one transport channel active, since channel multiplexing blocks for the same users may be left out and its inter-leaving spans only over a single 2 ms period without separate intra-frame or inter-frame inter-leaving. It uses only Turbo channel coding with varying transport block size, modulation scheme and number of multi-codes. Thus, its effective variable code includes rates from 1/4 to 3/4, where the number of bits per code may increase at the expense of reduced coding gain. As illustrated in Figure 8.9, HS-DSCH coding incorporates the Hybrid ARQ (HARQ) functionality, applies two parallel (identical) channel inter-leavers for 16QAM and the Node B selects the corresponding (modulation/number of codes) transport format available at its scheduler.

8.4.1 Hybrid Automatic Repeat Request – HARQ

Figure 8.10 illustrates the 2-stage rate matching HARQ functionality. The implementation includes a redundancy-tuning virtual[8] buffer between the rate matching stages. The HARQ can send identical retransmissions, i.e. soft combining, and different parameters with incremental redundancy, i.e. first transmission with systematic bits, then transmission with only parity bits, which may perform better while using more memory. The EU default memory settings take place based on soft combining and at maximum terminal data rate. Thus, at highest data rates only soft combining applies, while at lower data rates incremental redundancy may also apply.

[8]Practical rate matching implementation would consist of a single rate matching, no buffering at first stage.

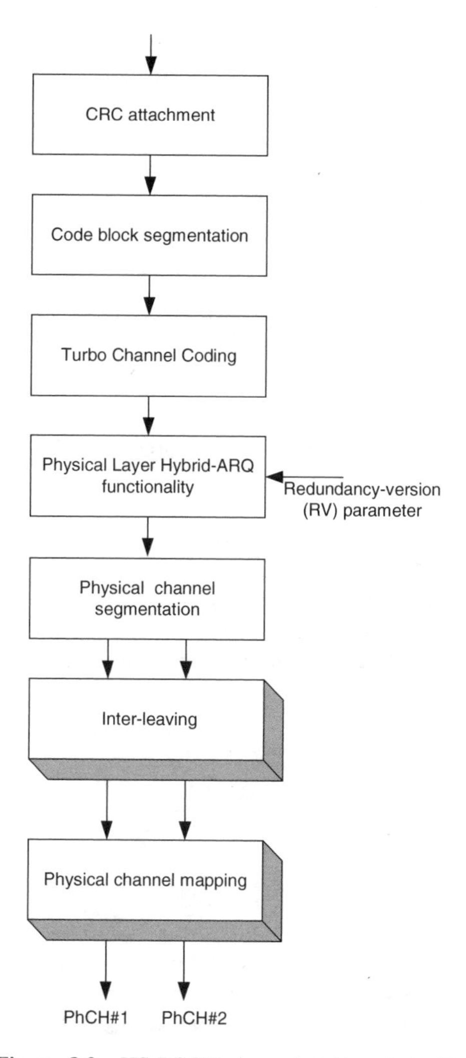

Figure 8.9 HS-DSCH channel coding chain [2].

Since the different bits mapped to the 16QAM symbols have different reliability, they are compensated in connection with the ARQ process through a constellation re-arrangement method. Through this method different re-transmissions use slightly different mapping of the bits to 16QAM symbols to maximise performance [3].

8.4.2 First Rate Matching Stage Parameters

As in [4] following Figure 8.10 and with directs from [3], the parameters of the first rate matching stage include the number of soft bits available at the UE for soft combining of HARQ channel n as $N_{UE,n}$ and the number of coded bits in a TTI before rate matching on TrCh i with transport format l as $N_{i,l}^{TTI}$.

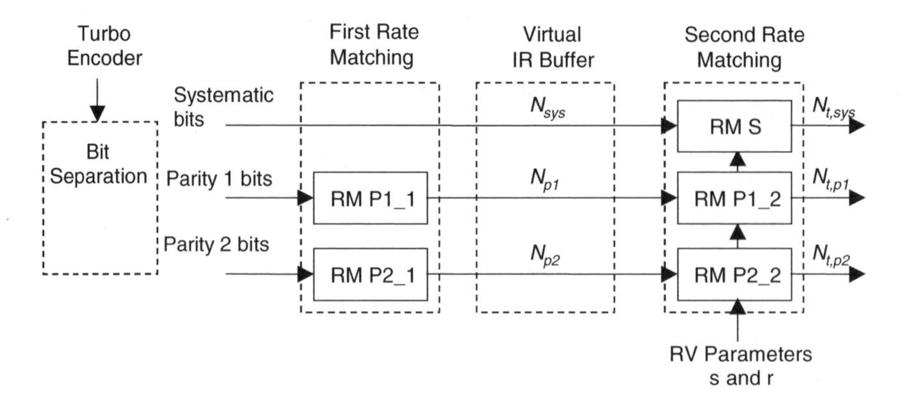

Figure 8.10 HARQ function principle, after [3].

When $N_{UE,n} \geq N_{i,l}^{TTI}$, i.e. if all Turbo-coded bits of the corresponding TTI can be stored, the first rate matching stage is transparent, which gets achieved by setting $e_{\text{minus}} = 0$ without repetition.

If $N_{UE,n} < N_{i,l}^{TTI}$, the parity bit streams are punctured by setting the total number of punctured bits per TTI to:

$$\Delta N_{il}^{TTI} = N_{UE,n} - N_{i,l}^{TTI} \tag{8.1}$$

8.4.3 Second Rate Matching Stage Parameter

Also following Figure 8.10, the second rate matching stage parameters depend on the value of the Redundancy Version (RV) parameter s, which can take values 0 or 1 to distinguish non self-decodable (0) and self-decodable (1) transmissions, as well as the RV parameter r (range 0 to $r_{\max}$), which changes the initial error variable e_{ini} [3].

N_{sys} denotes the number of bits for the systematic bits before second rate matching, N_{p1} for the parity 1 bits and N_{p2} for the parity 2 bits, respectively. N_{data} represents the number of available physical channel bits per TTI. We use bit separation and the rate matching parameters are determined as follows:

For $N_{\text{data}} \leq N_{sys} + N_{p1} + N_{p2}$, puncturing is performed in the second rate matching stage. The number of transmitted systematic bits in a re-transmission is

$$N_{t,\text{sys}} = \min\{N_{\text{sys}}, N_{\text{data}}\} \tag{8.2}$$

for a transmission of self-decodable type $(s = 1)$ and

$$N_{t,\text{sys}} = \max\{N_{\text{data}} - (N_{p1} + N_{p2}), 0\} \tag{8.3}$$

in the non self-decodable case, i.e. $s = 0$.

Table 8.1 Parameters for second rate matching stage in case of puncturing

	X_i	e_{plus}	e_{minus}
Systematic RM S	N_{sys}	N_{sys}	$\lvert N_{sys} - N_{t,sys} \rvert$
Parity 1 RM P1_2	N_{p1}	$a \cdot N_{p1}$	$a \cdot \lvert N_{p1} - N_{t,p1} \rvert$
Parity 2 RM P2_2	N_{p2}	$a \cdot N_{p2}$	$a \cdot \lvert N_{p2} - N_{t,p2} \rvert$

For $N_{data} > N_{sys} + N_{p1} + N_{p2}$ repetition is performed in the second rate matching stage. Likewise, repetition rate in all bit streams gets achieved by setting the number of transmitted systematic bits to

$$N_{t,sys} = N_{sys} + \left\lceil \frac{N_{data} - (N_{sys} + N_{p1} + N_{p2})}{3} \right\rceil \tag{8.4}$$

The available room for parity bits in a transmission is then:

$$N_{t,p1} = \left\lfloor \frac{N_{data} - N_{t,sys}}{2} \right\rfloor \quad \text{and} \quad N_{t,p2} = \left\lceil \frac{N_{data} - N_{t,sys}}{2} \right\rceil \tag{8.5}$$

for the parity 1 and parity 2 bits, respectively.

Table 8.1 summarises the resulting parameter choice for the second rate matching stage, where parameter a gets chosen as in [4], i.e. $a = 2$ for parity 1 and $a = 1$ for parity 2.

The rate matching parameter e_{ini} [4] is calculated for each bit stream according to the e_{ini} variation parameter r using

$$e_{ini}(r) = \{[X_i - r \cdot e_{minus} - 1] \bmod e_{plus}\} + 1 \tag{8.6}$$

in the case of puncturing, i.e. $N_{data} \leq N_{sys} + N_{p1} + N_{p2}$, and

$$e_{ini}(r) = \{[X_i - (2 \cdot s + r) \cdot e_{minus} - 1] \bmod e_{plus}\} + 1 \tag{8.7}$$

for repetition, i.e. $N_{data} > N_{sys} + N_{p1} + N_{p2}$.

Note: The equations above as direct extracts from [3] are for concept illustration only and they need verification.

8.5 ASSOCIATED SIGNALLING FOR HSDPA

8.5.1 Key HS-DSCH Downlink Characteristics

From [4], Figure 8.11 shows the basic HS-DSCH physical-channel structure for the associated downlink signalling perceived from the UE side, which includes downlink DPCH and a number of HS-SCCHs.

For each HS-DSCH TTI, a Shared Control Channel (HS-SCCH) carries HS-DSCH downlink signalling information for one UE, and when there is HS-DSCH transmission to

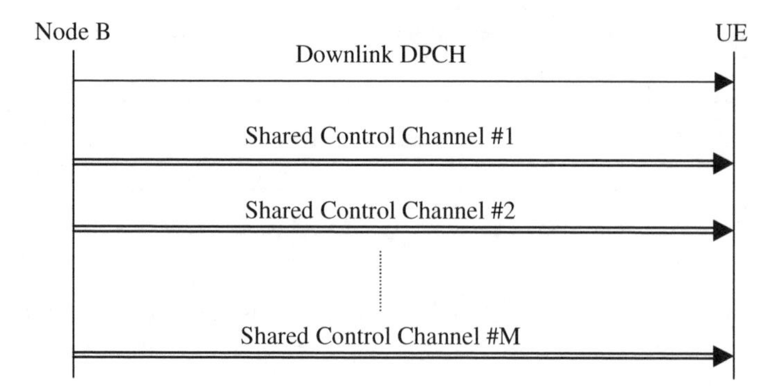

Figure 8.11 HS-DSCH physical-channel structure for associated downlink signalling [4].

the same UE in consecutive HS-DSCH TTIs, we use the same HS-SCCH for the corresponding associated downlink signalling.

Table 8.2 summarises the different HS-DSCH and DSCH physical characteristics, where HS-DSCH HARQ operation may also apply at the RLC level when the physical layer ARQ timers or the maximum number of re-transmissions exceed.

8.5.1.1 High Speed Shared Control Channel – HS-SCCH

Whenever there is HS-DSCH data, the HS-SCCH transports key HS-DSCH de-modulation information corresponding to the maximum number of users, which will be code multiplexed. However, even though there may be a high number of HS-SCCHs allocated per cell, the EU will consider only four HS-SCCHs[9] at a given time while following the network signalling sequence.

The HS-SCCH block has a three-slot duration divided into two independently ½-rate convolution encoded functional parts.[10] The first slot—1st part, carries time critical

Table 8.2 Characteristic summary of DSCH channels

Channel	HS-DSCH	DSCH
HARQ	Packet combining at L1	RLC level
Modulation	QPSK/16QAM	QPSK
Power control	Fixed/slow power setting	Fast (associated .DCH)
Spreading factor	Fixed, 16	Variable (256-4)[a]
Inter-leaving	2 ms	10–80 ms
Channel coding	Turbo coding	Turbo & convolutional
Soft handover	Associated (DCH)	Associated (DCH)
Channel multiplexing	No	Yes

[a]Frame by frame.

[9]While four may be too much, more than one HS-SCCH will be ideal to better match available codes to EUs.
[10]Both parts use terminal-specific masking to discriminate control channel ownership.

Table 8.3 HS-SCCH summary of carried parameters

HS-SCCH Part 1	HS-SCCH Part 2
Codes to de-spread → indicating EU de-spreading capability (maximum 5, 10, 15 codes)	Redundancy version info → allowing decoding and combining possibly with earlier transmissions
Modulation indication → allowing QPSK or 16QAM selection	ARQ process number → indicating the ARQ process data ownership
	First transmission or retransmission indicator → indicating transmission combination[a] of flushing

[a]With existing buffer data (if not successfully decoded earlier) or flush buffer and filled with new data.

information required to start de-modulation while avoiding chip level buffering; the next two slots – 2nd part, transports less time critical parameters, which includes HS-SCCH validity checking CRC and HARQ process information. Table 8.3 summarises key HS-SCCH transported parameters; other parameters, e.g. channel coding rate gets determined from the transport format and block size information.

Considered HS-SCCH contents at the time of [4] include:

- Transport-Format and Resource related Information (TFRI)
  - Channelisation-code set: 7 bits
  - Modulation scheme: 1 bit
  - Transport-block size: 6 bits
- Hybrid-ARQ-related Information (HARQ information)
  - HARQ process number: 3 bits
  - RV: 2 bits
  - New-data indicator: 1 bit
  - UE ID: 10 bits implicitly encoded (see also section on CRC attachment)

The HS-SCCH information is split into two parts:

Part-1. Channelisation code set and modulation scheme (8 bits) and
Part-2. Transport-block size and HARQ-related information (12 bits).

Figure 8.12 illustrates the single slot duration terminal *that determines*,[9] which codes to de-spread from the HS-DSCH. The downlink DCH timing does not depend on the HS-SCCH, neither does it from the HS-DSCH timing.

As seen in Figure 8.11, when multiple HS-SCCHs apply to a UE, the downlink DPCH carries an HS-DSCH Indicator (HI), besides the non-HS-DSCH-related physical-layer signalling and DCH transport channels. The HI consists of two information bits that indicate the HS-SCCH that carries the HS-DSCH-related signalling for the corresponding UE. The HI is transmitted in every third slot. If no HS-SCCH carries HS-DSCH-related signalling to the UE in a TTI, the HI is not transmitted (DTX) during the corresponding TTI.

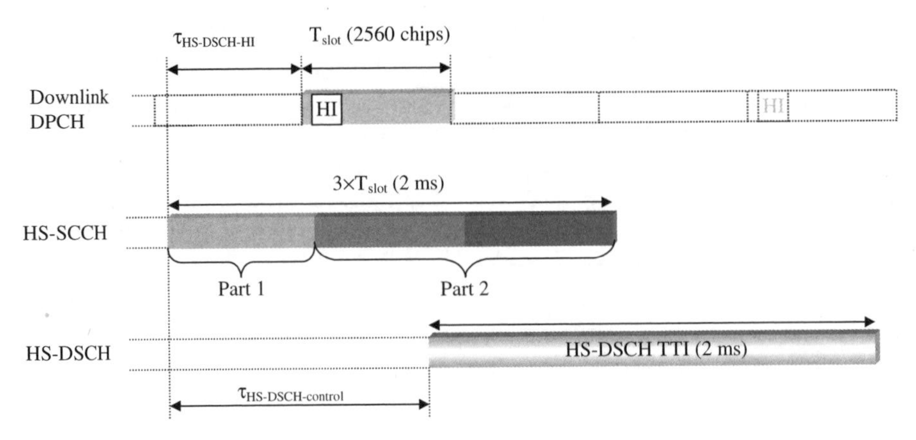

Figure 8.12 Timing in HS-SCCH and HS-DSCH slots, after [4].

In the HS-DSCH timing structure seen in Figure 8.12, the fixed time offset between the HS-SCCH information and the beginning of the corresponding HS-DSCH TTI, equals $\tau_{\text{HS-DSCH-control}}$ $(2 \times T_{\text{slot}} = 5120\,\text{chips})$. Likewise, the time offset between the DL DPCH slot carrying the HI and the beginning of the HS-SCCH information ($\tau_{\text{HS-DSCH-HI}}$) is such that the start of the HI overlaps with the first slot of the HS-SCCH transmission [4].

Finally, as indicated earlier, when consecutive HS-DSCH TTIs transmissions to one UE occur, we use the same HS-SCCH for the corresponding associated downlink signalling. As result, the UE considers only the HI value if there was no HS-SCCH/HS-DSCH transmission in the previous[11] TTI. Otherwise, the UE may use the same HS-SCCH as in the previous TTI.

8.5.2 The Uplink HS-DPCCH Characteristics

The HS-DSCH uplink signalling uses HS-DPCCH with spreading factor of 256, i.e. code multiplexed with the existing dedicated uplink physical channels. The signalling between the UE and Node B consists of HARQ ACK/NAK and channel quality indicator.

From [4], the HS-DSCH associated uplink Dedicated Physical Control Channel (DPCCH-HS) carries the following information:

1. *HARQ acknowledgement.* HARQ acknowledgement uses 1-bit ACK/NACK indication, which is repetition coded to 10 bits and transmitted in one slot. The HARQ acknowledgement field gets DTX'ed in the absence of ACK/NACK transmission information.

2. *Measurement feedback information.* The measurement feedback information contains channel quality indicator aimed to transport format and resource selection by the HS-DSCH serving Node B; it is transmitted in two slots as described above. Upon upper

[11]Or if a CRC-1 of an earlier HS-SCCH transmission checked incorrect (previous TTI HI was incorrect).

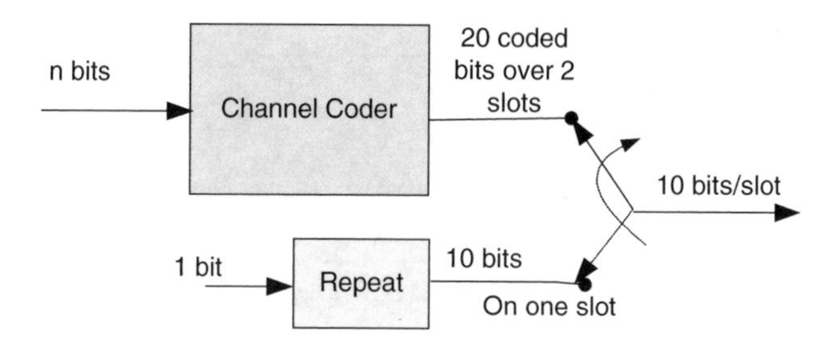

Figure 8.13 Format for additional DPCCH for HS-DSCH-related uplink signalling [4].

layer request, the UTRAN determines and signals the transmission cycle and timing for channel quality indicator, which consists of a TFRC selection provided by the UE to Node B from a reference list.

Figure 8.13 [4] illustrates the additional DPCCH, where the ACK/NACK message gets transmitted with an upper layer power parameter offsets ΔP_{AN} for ACK and ΔP_{NAN} for NACK relative to the Release 1999 uplink DPCCH.

8.5.2.1 Feedback Measurements Procedures

Here for completeness, the following includes UE internal procedures used in adaptive modulation and coding operation at the time of [4].

The EU Procedures. The UE measures a DL channel quality indicator from the P-CPICH, otherwise from the S-CPICH whenever beam-forming with S-CPICH applies. The EU also takes into account the default power offset, P_{hs}, of the HS-DSCH code channel relative to P-CPICH (or S-CPICH), and checks in its internal mapping table, which TFRC from the TFRC reference list would be able to receive a given channel condition.

The $BLER_{threshold}$ serves the UE as criteria to determine the ideal TFRC within specific channel conditions. If the channel quality, measured from P-CPICH (or S-CPICH in the case beam-forming with S-CPICH), indicates that:

- the TFRC x would be received with BLER $< BLER_{threshold}$ and HS-DSCH code channel power offset $< y$ dB in addition to the default power level, but the
- TFRC $x + 1$ would be received with BLER $> BLER_{threshold}$, then the
- UE will report TFRC x and the required power offset y back to the network.

The UE reports a reference-list TFRC together with a power offset at the transmission timing defined by higher layer signalling [4].

The higher layer specific parameters passed to the EU as outlined in [4] include:

1. P_{hs}, default power offset between HS-DSCH code channel and P-CPICH;[12]

[12]Or S-CPICH in case beam-forming with S-CPICH applies.

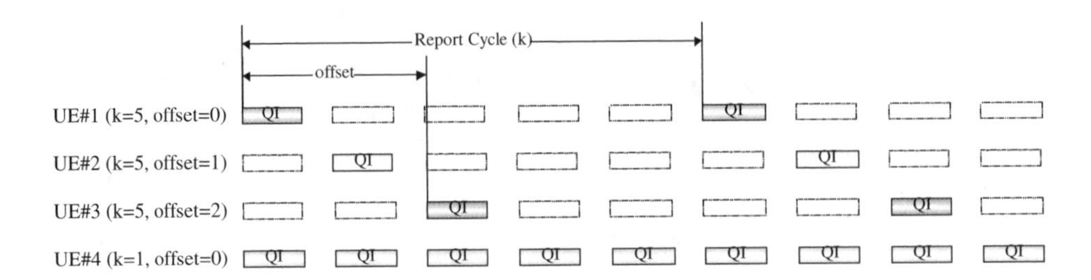

Figure 8.14 Feedback measurement transmission timing example [4].

2. $BLER_{threshold}$, BLER value used by the UE for selecting the TFRC;
3. measurement feedback cycle k, with possible value of (1, 5, 10, 20, 40, 80) corresponding to the feedback cycle of (2, 10, 20, 40, 80, 160) ms. The indication value $k = 0$ shuts off completely;
4. measurement feedback offset l. The exact definition of l at the time of [4] was pending.

Figure 8.14 illustrates the use for the measurement feedback cycle k and feedback offset l, the timing relations in the figure do not reflect the differences in DPCH frame offset.

The TFRC Reference List. The EU uses only one specified TFRC reference list to create the measurement report, where each TFRC contains modulation, transport block set size and number of code channels. Table 8.4 illustrates a reference list with six TFRCs.

Definition of Measurement Feedback. The measurement report can take up to (32) values each containing a power offset value and one reference-list TFRC at a time.

8.5.2.2 Uplink DPCCH/HS-DPCCH Timing Relation

Figure 8.15 illustrates the timing offset between the downlink-associated DPCH and the uplink DPCCH. Notice that the code-multiplexed uplink HS-DPCCH begins with $m \times 256$

Table 8.4 TFRC reference list example

TFRCs	Modulation	Transport block set size	No. of code channels
TFRC1	QPSK	1200	5
TFRC2	QPSK	2400	5
TFRC3	QPSK	3600	5
TFRC4	16QAM	4800	5
TFRC5	16QAM	6000	5
TFRC6	16QAM	7200	5

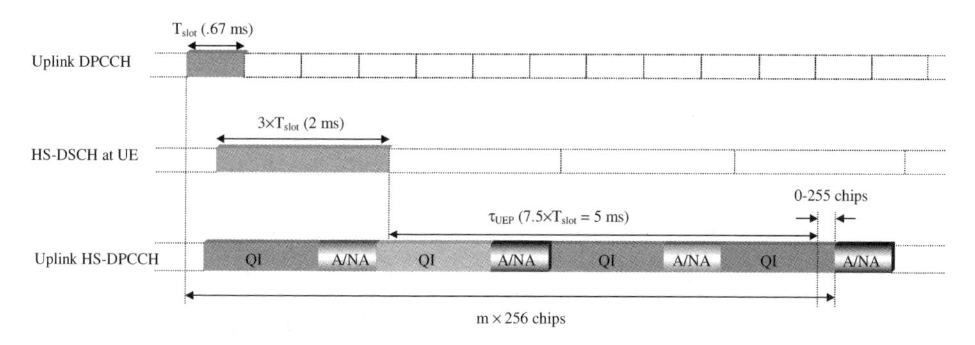

Figure 8.15 Timing structure at UE for UL HS-DPCCH control signalling [4].

chips after the uplink DPCCH starts with an m selected by the UE, which has an ACK/NACK transmission (of 1 time-slot duration) beginning within the first 0–255 chips after 7.5 slots at the end of the received HS-DSCH [4].

We maintain thus the UE processing time at 7.5 slots (5.0 ms) as the offset between DPCCH and HS-DPCCH varies. The first slot carries the ACK bit of the code multiplexed uplink HS-DPCCH, which leaves about 4.5 slots − 512 chips (prop delay) − 256 chips (HS-DPCCH offset) = 2.8 ms ($T_{\text{Node B}}$) for Node B to perform scheduler and signal processing functions [4].

We reserve every first slot on the HS-DPCCH (based on the parameters above) for ACK/NACK signalling. We then use the remaining HS-DPCCH slots for quality indicator transmission following the measurement feedback cycle and offset parameter provided by UTRAN and DTX slots not utilised for ACK/NACK or Quality indicator signalling.

For backwards compatibility, the existing uplink channel structure remains unchanged and HSDPA gets new information elements on a parallel code channel, named the Uplink High Speed-Dedicated Physical Control Channel (HS-DPCCH). This channel carries the following information and gets divided into two parts as seen in Figure 8.16:

- *ACK/NACK transmission* → reflecting the CRC-check after the packet decoding and combining.

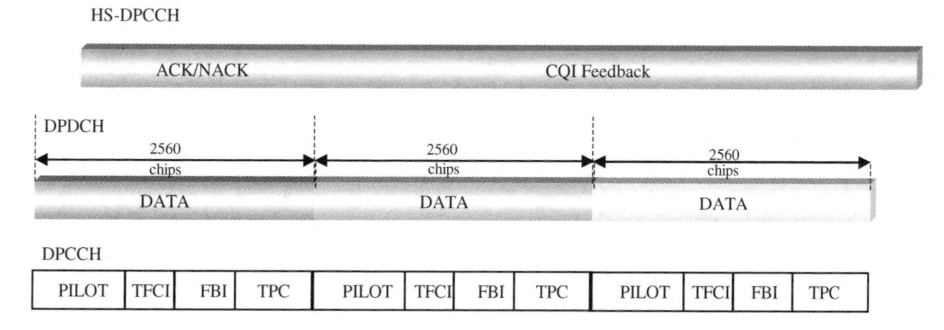

Figure 8.16 HS-DPCCH structure [4].

- *Downlink Channel Quality Indicator (CQI)* → indicating the ideal[13]estimated transport block-size, modulation type and number of parallel codes.

The feedback information consists of 5 bits carrying quality related information, where one signalling state indicates: 'do not bother to transmit' and other states represent EU transmission capabilities at the given time. These states range in quality from single code QPSK transmission up to 15 codes 16QAM transmission (+ various coding rates).

8.6 HSDPA TECHNOLOGY EVALUATION AND EU FEATURES

Here for completeness we cover the HSDPA driving technologies and comparative performance outlined first in [1] as design reference and to support the 3GPP technical specifications of HSDPA. While R99 exploits fast power control and variable spreading factor, HSDPA as part of R5 does not depend on these WCDMA characteristics, but aims to exploit features such as Adaptive Modulation and Coding (AMC), Hybrid ARQ, Fast Cell Selection (FCS) and Multiple Input Multiple Output (MIMO) antenna processing. Thus, radio-access performance metrics for R5 using HSDPA will differ from that of R99 brought in at the introduction of UMTS.

8.6.1 EU Key Characteristics

HSDPA EUs may have up to 10 different terminal categories with transmission rates ranging from 1.2 to 10 Mbps. Table 8.5 illustrates the different category classes that vary according to the maximum number of parallel codes they support and required 2 ms TTI. The minimum inter-TTI interval equals 3, 2 and 1 for categories 1–2, 3–4 and 5–10, respectively.

The maximum theoretical data rate of 14.4 Mbps corresponds to the 10th category while applying 1/3 Turbo coding with high puncturing approaching code rate close to 1. Category 9

Table 8.5 HSDPA terminal capability categories

Category	Max Num Parallel HS-DSCH codes	Transport channel bits per TTI	Achievable maximum data rates (Mbps)
1	5	7300	1.2
2	5	7300	1.2
3	5	7300	1.8
4	5	7300	1.8
5	5	7300	3.6
6	5	7300	3.6
7	10	14 600	7.2
8	10	14 600	7.2
9	15	20 432	10.2
10	15	28 776	14.4

[13]That is, which combination can be received correctly (with reasonable BLER) in the downlink direction.

takes into account the maximum turbo R99 encoding block size, which results 10.2 Mbps peak user data rate value with four turbo encoding blocks.

8.6.2 HSPDA Technology Assessment

As this writing pre-commercial evaluations for HSPDA is still in process; therefore, we may envisage to illustrate its results in later editions. However, here for completeness, we will incorporate selected performance evaluation extracts from [1] illustrating key design issues of the proposed technology. Furthermore, to provide a reference framework for an integral view of the different performance aspects, we will first outline the link and system simulation assumptions.

8.6.2.1 Reference Assumptions for Link Level AMC and HARQ Assessment

Single cell (link level) simulations provide the input data for the initial multi-cell (system level) analysis and to evaluate the wireless link performance of different AMC schemes and fast HARQ solutions.

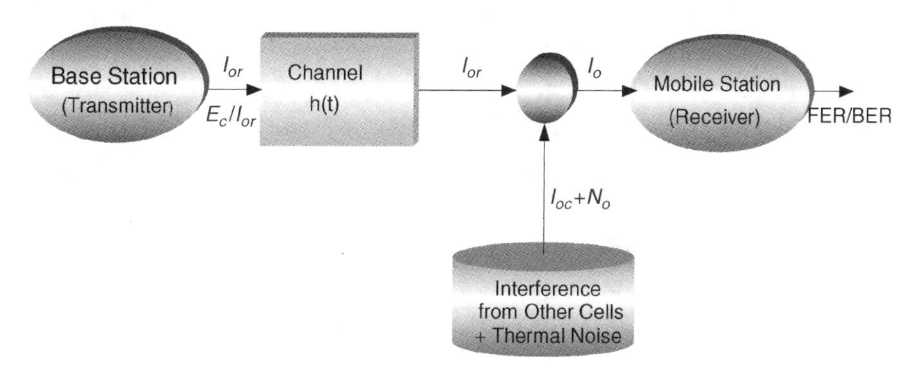

Figure 8.17 Simulation block diagram [1].

The block diagram illustrated in Figure 8.17, as symbol level downlink simulator enables evaluation of higher order modulation schemes and HARQ with the following terminology: I_{or} is the total transmitted power density by a BTS, $\hat{I}_{or}$ is the post-channel transmitted power density, $I_{oc} + N_o$ is the other cell interference plus noise power density and I_o is the total received power density at the Mobile Station (MS) antenna. Note, that the ratio $\hat{I}_{or}/(I_{oc} + N_o)$ is fixed in this simulation model, assuming that the Base Station (BS) has a fixed amount of power (set by the BTS power amplifier size), i.e. the average BTS transmitted (often called allocated) power to the MS that determines the user capacity of the forward link. We call this fraction of allocated power, the average traffic channel E_c/I_{or}, which is inversely proportional to the forward link capacity [1].

- As seen in Figure 8.18, in the case of 8-PSK modulation, every three binary symbols from the channel inter-leaver output gets mapped to an 8-PSK modulation symbol

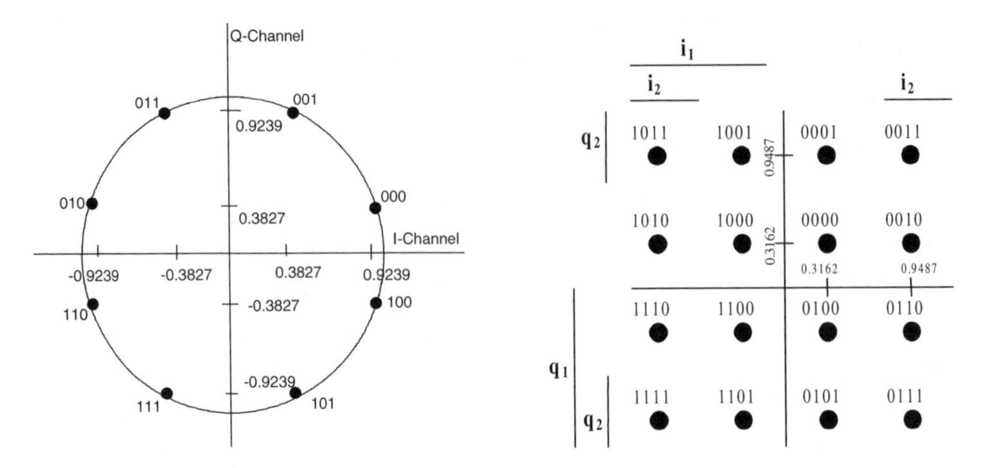

Figure 8.18 Signal constellation for 8-PSK and 16QAM modulation [1].

according; and in the case of 16QAM modulation, every four binary symbols of the block interleaver output gets mapped to a 16QAM modulation symbol.

- Also, as direct extract from [5], we note the reference-decoding sequence for the simulation model, i.e. an M-ary QAM de-modulator generating soft decisions as inputs to the Turbo decoder. Then define a baseline method with soft inputs to the decoder generated by an approximation to the log-likelihood ratio function as follows:

$$\Lambda^{(i)}(z) = K_f \left[\underset{f \in S_i}{\mathrm{Min}}\{d_j^2\} - \underset{j \in \bar{S}_i}{\mathrm{Min}}\{d_j^2\} \right], \quad i = 0, 1, 2, \ldots, \log_2 M - 1 \tag{8.8}$$

where M is the modulation alphabet size, i.e. 8, 16, 32 or 64 and

$$z = A_d A_p \alpha \hat{\alpha} e^{-j(\theta - \hat{\theta})} x + n \tag{8.9}$$

x is the transmitted QAM symbol, A_d is the traffic channel gain, A_p is the pilot channel gain, $\alpha e^{j\theta}$ is the complex fading channel gain and $A_p \hat{\alpha} e^{j\theta}$ is the fading channel estimate obtained from the pilot channel,

$$S_i = \{\forall j : ith \text{ component of } y_j \text{ is '0'}\} \tag{8.10}$$

$$\bar{S}_i = \{\forall j : ith \text{ component of } y_j \text{ is '1'}\} \tag{8.11}$$

and K_f is a scale factor proportional to the received signal-to-noise ratio. The parameter d_j is the Euclidean distance of the received symbol z from the points on the QAM constellation in S or its complement. We assume that the pilot/data gain is known at the receiver so the distance metric is computed as follows:

$$d_j^2 = |A_p z - Q_j \beta \gamma^2|^2 \quad Q_j \in S_i \text{ or } \bar{S}_i \tag{8.12}$$

where $\beta = A_d$ and $\gamma = A_p \hat{\alpha}$ is an estimate formed from the pilot channel after processing through the channel estimation filter.

- The performance metrics uses the following notation:
  1. FER versus E_c/I_{or} (for a fixed $\hat{I}_{or}/(I_{oc}+N_o)$) or FER versus $\hat{I}_{or}/(I_{oc}+N_o)$ (for a fixed E_c/I_{or})
  2. Throughput versus E_c/I_{oc}, where throughput measured in term of bits per second is

$$T = R\left(\frac{1 - FER_r}{\overline{N}}\right) \tag{8.13}$$

where T is the throughput, R is the transmitted information bit rate and FER_r is the residual Frame Error Rate (FER) beyond the maximum number of transmissions and $\overline{N}$ is the average number of transmission attempts.

- Finally, Table 8.6 illustrates the reference simulation parameters for performance assessment of the link level.

More detailed assumptions and information for the HSDPA link-level analysis can be found in [5]. Here, we will next illustrate the results based on the simulations assumptions outlined above.

Table 8.6 Link level simulation parameters

Parameter	Value	Comments
Carrier frequency	2 GHz	
Propagation conditions	AWGN, Flat, Pedestrian	
Vehicle speed for flat fading	A (3 km/h) 3/30/120 km/h	
CPICH relative power	10% (−10 dB)	
Closed loop power control	OFF	Power control possible
HSDPA frame Length	10, 3.33, 0.67 ms	
I_{or}/I_{oc}	Variable	
Channel estimation	Ideal/non-ideal (using CPICH)	
Fast fading model	Jakes spectrum	by Jakes or filtering approach
Channel coding (PCCC)	Turbo code rate 1/4, 1/2, 3/4, etc.	
Tail bits	6	
Max. no. of iterations Turbo coder	8	
Metric for Turbo coder	Max.	sensitive to SNR scaling
Input to Turbo decoder	Soft	
Turbo interleaver	Random	
Number of rake fingers	No. of taps in the channel model	
Hybrid ARQ	Chase combining	Other may apply
Max. No. of frame HARQ transmissions		Specify the value used
Information bit rates (kbps)	As defined	
No. of multi-codes simulated	As defined	
TFCI model	Receiver random symbols ignored	Assume error-free reception
STTD	On/off	
Other L1 parameters	as specified in R99 specification	

8.6.2.2 Link Level Evaluation Results

The notation to understand the outputs can be summarised as follows [1]:

CE	Channel estimation: 0 = ideal, 1 = non-ideal	IBM	Ray imbalance
STTD	Transmit diversity: 0 = off, 1 = on	ECIOR	Power allocation (set 80%) E_c/I_{or}
nc	Number of codes	SPEED	Vehicle speed
ARQ	HARQ – chase combining: 0 = off, 1 = on	*Q or q*	QAM modulation level (4/8/16/64)
np	Number of paths	cr	Turbo code rate
FC	Carrier frequency (2 GHz)	MR	Maximum number of repeats

Figure 8.19 illustrates a FER versus I_{or}/I_{oc} curve for different MCS levels obtained with a fixed power allocation of −1 dB. In this output, we can see that as the order of the Modulation and Coding Scheme (MCS) increases (e.g. for a 16QAM modulation, as we augment the number of codes (nc) and the Turbo code rate (cr)), the I_{or}/I_{oc} requirements also increase in order to achieve the same FER.

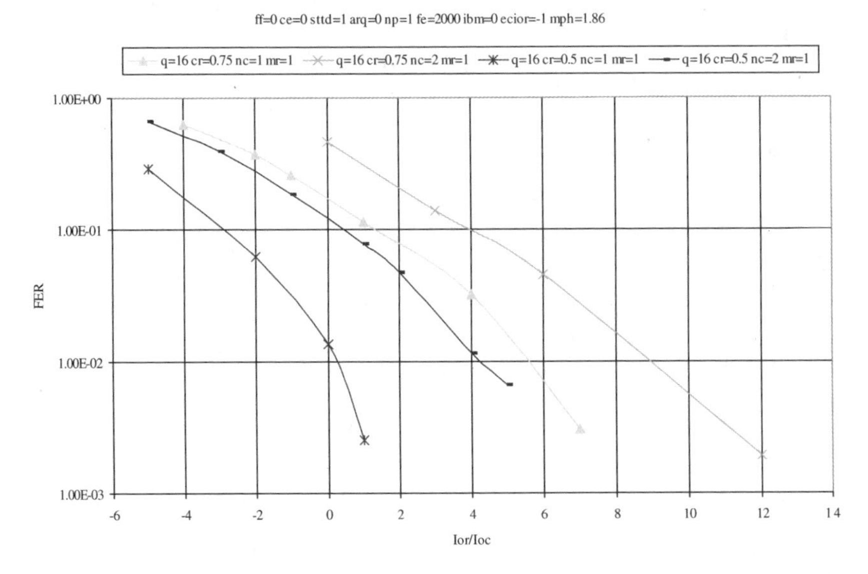

Figure 8.19 FER vs. I_{or}/I_{oc} for 16 QAM-flat fading 3 km/h [1].

Figure 8.20 illustrates the average throughputs with and without HARQ. Clearly, AMCs with HARQ provide higher throughput than AMCs without it, particularly seen at lower speeds. An HARQ system with fast feedback ensures that extra redundancy get sent only when necessary and still meet delay constraints.

Figure 8.21 illustrates the effect of multi-path in the channel estimation accuracy. With Ideal Channel Estimation (ICE), the FER performance at higher vehicle speed appears better than that at lower vehicle speed for a 16QAM[14]due to more un-correlated errors at higher

[14]The same applies for QPSK.

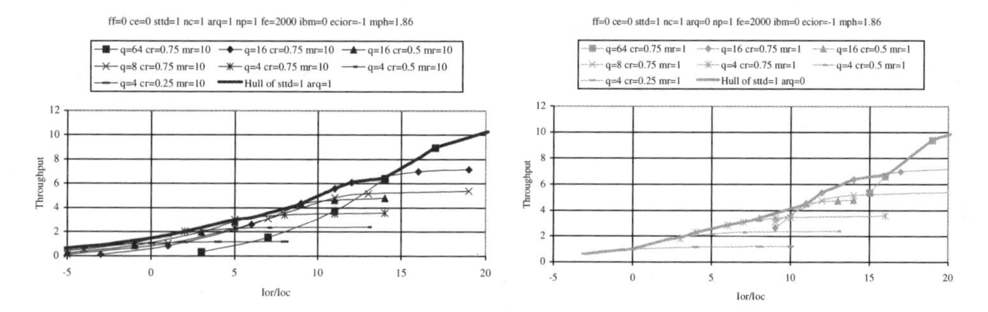

Figure 8.20 Throughput curves at 3 km/h with and without HAQR [1].

speed. Notwithstanding, at higher speeds channel estimation errors are also higher due to the time-variant behaviour of the channel, thereby degrading these ideal results. Thus, channel estimation accuracy for higher modulations such as 16QAM or 64QAM, becomes more critical because of the close signal constellation and both phase- and magnitude-related soft input computation [1].

In conclusion, the above reference simulation examples concerning the impact of channel estimation error on HSDPA link level performance, indicates that higher order modulation is very sensitive to the channel estimation applied for channel compensation. Hence, accurate channel estimation will be essential for EUs operating at high vehicle speeds.

8.6.2.3 System Level Simulation Assumptions

The multi-cell analysis will also enable us to illustrate the potential HSPDA performance gains while applying Adaptive Modulation and Coding scheme (AMCs), fast HARQ and Fast Cell Selection (FCS). Table 8.7 illustrates the basic system level assumptions.

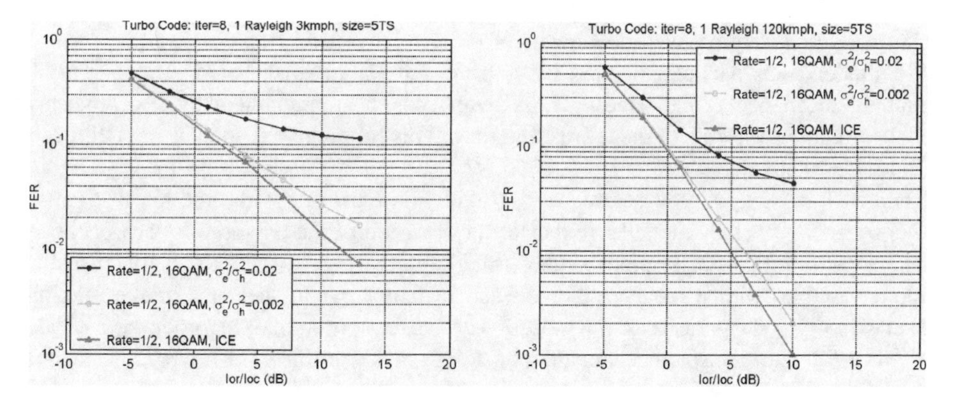

Figure 8.21 FER vs. I_{or}/I_{oc}, rate ½ 16QAM for 3 and 120 km/h in Rayleigh fading environment.

Table 8.7 System level reference simulation assumptions [1]

Parameter	Value	Comments
Cellular layout	Hexagonal grid, 3-sector sites	Provide your cell layout picture
Site to site distance	2800 m	
Antenna pattern	As proposed in TR 29.456. V3.1.0	Only horizontal pattern specified
Propagation model	$L = 128.1 + 37.6\ \mathrm{Log}_{10}(R)$	R in kilometres
CPICH power	−10 dB	
Other common channels	−10 dB	
HSDPA allocated power	Max. 80% of total cell power	
Slow fading	As in UMTs 30.03, B 1.4.1.4	
Standard Deviation of slow fading (SD)	8 dB	
Sector correlation	1.0	
Correlation between sites	0.5	
Slow fading correlation	50 m	Correlation distance
Carrier frequency	2000 MHz	
BS antenna gain	14 dB	
UE antenna gain	0 dBi	
UE noise figure	9 dB	
Max. no. of retransmiss...	Specify the value used	Retransmissions by fast HARQ
Fast HARQ scheme	Chase combining	For initial evaluation of fast HARQ
BS total Tx power	Up to 44 dBm	
Active set size	3	Maximum size
Fast fading model	Jakes spectrum	Generated Jakes or filter approach

8.6.2.4 Traffic Model Assumptions for Data

The reference data-traffic model simulates bursty web traffic with TCP/IP rate adaptation mechanisms to pace the packet arrival process of packets within a packet call. The model assumes also that all UEs drops occur in an active packet sessions, which consist of multiple packet calls representing Web downloads or other related activities.

We model each packet call size by a truncated Pareto distributed random variable, producing mean packet call sizes of 25 Kbytes, which are separated by a geometrically modelled random variable (5 mean) reading time. This time starts as soon as the UE receives the entire packet call.

Then each packet call gets segmented into individual packets with time interval between two consecutive packets modelled either as an open-loop or as a closed-loop process.

The open-loop process models the timer interval as a geometrically distributed random variable, with a mean packet inter-arrival time set to the ratio of the maximum packet size divided by the peak link speed. The closed-loop model incorporates the 'slow-start' TCP/IP rate control mechanism for pacing packet traffic. The TCP ACK feedback assumes a total round-trip network delay of 100 ms. Table 8.8 summarises the data-traffic parameters.

Table 8.8 Data-traffic model parameters [1]

Process	Random variable	Parameters
Packet calls size	Pareto with cut-off	$A = 1.1$, $k = 4.5$ Kbytes, $m = 2$ Mbytes, $\mu = 25$ Kbytes
Time between packet calls	Geometric	$\mu = 5$ s
Packet size	Segmented: MTU size	(e.g. 1500 octets)
Packets per packet call	Deterministic	Based on packet call size and packet MTU
Packet inter-arrival time[a]	Geometric	μ = MTU size /peak link speed
Packet inter-arrival time[b]	Deterministic	TCP/IP slow start

[a]Open loop → Parameter example: [1500 octets * 8]/2 Mbps = 6 ms.
[b]Closed loop (fixed network delay of 100 ms).

8.6.2.5 *User Equipment (UE) Model*

A static or dynamic UE mobility model with fixed or distributed speed applies. Figure 8.22 illustrates the latter case with values shown in Table 8.9, from which we assign a speed to each user at the beginning of the simulation and maintain it throughout. For stationary UEs [1], we assume signal paths with Rician distribution using a k factor of 12 dB and 2 Hz Doppler spread.

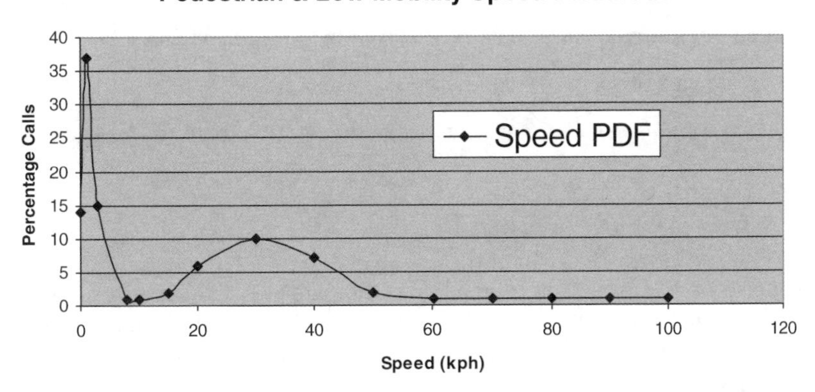

Figure 8.22 Pedestrian and low mobility speed distribution [1].

Table 8.9 Speed distribution low [1]

Speed (km/h)	0	1	3	8	10	15	20	30	40	50	60	70	80	90	100
Percentage	14	37	15	1	1	2	6	10	7	2	1	1	1	1	1

8.6.2.6 Scheduling Packets

In this reference example, [1] uses a (*C*/*I* based) scheduler, which provides maximum system capacity at the expense of fairness; and a Round Robin (RR) scheduler, which provides a more fair sharing of resources (frames) at the expense of a lower system capacity.

Both scheduling options obey the following principles:

- Scheduling gets performed on a frame-by-frame basis.
- The HSPDA frame = 0.67 ms (1 slot), 3.33 ms (5 slots) or 10 ms (15 slots).
- 'Non-empty' queue containing at least 1 octet of information.
- Packets received in error are explicitly rescheduled after the ARQ feedback delay.
- High priority queues expedite re-transmission of failed packet transmission attempts.
- Low-priority packet transmission occurs only after high-priority queues are empty.
- Transmission during a frame cannot be aborted or pre-empted for any reason.

The *C*/*I* scheduler follows these additional rules:

- All non-empty source queues get rank ordered by *C*/*I* for transmission during a frame.
- The scheduler will keep data transfer to the UE with highest *C*/*I* until queue completion.[15]
- Both high- and low-priority queues are ranked by *C*/*I*.

The RR scheduler obeys the following rules:

- Non-empty source queues are serviced in an RR fashion.
- All non-empty source queues must be serviced before re-servicing a user.
- A next frame cannot service a same user as current frame.[16]
- The scheduler is allowed to group packets from the selected source queue within the frame.

8.6.2.7 System Level Performance Metrics

The percentage of users stands as a function of throughput for different loading levels. We measure this throughput on a per packet basis, which equals to the number of information bits divided by the total transmission time, thereby accounting re-transmissions to reduce peak data rate statistics.

We define the total transmission time to include the time to transmit the initial attempt and each subsequent retry. For example, the packet rate m takes the following logic:

- Packet m contains I_m information bits.

[15]Even if data arrives for another UE with higher *C*/*I*, or a re-transmission is schedules taking higher priority.
[16]Unless there is only one-empty source queue.

- Packet requires three attempts to transmit.
- Packet m takes $T_{m,j}$ seconds to transmit for attempt j.

$$R(m) = \frac{I_m}{\sum_{j=1}^{3} T_{m,j}} \tag{8.14}$$

The following statistics as a function of offered load may also be provided [1]:

1. *Throughput per sector* → total number of bits successfully transferred divided by the total number of sectors and simulation duration.
2. *Average and variance of packet call completion time* → measured from first packet arrival at the BS's queue to last packet reception by the UE station within a packet call.
3. *Average and variance packet call transfer rate* → payload size of a packet call divided by the transfer time (measured as above).
4. *Service rate* → the number of completed packet calls per second.

8.6.2.8 Reference Simulations

Case 1 permits to evaluate Adaptive Modulation/Coding (AMC) and fast HARQ. Case 2 enable us to access not only fast HARQ and AMC but also FCSS with the same parameters of Case 1 plus additional parameters as illustrated in Table 8.10.

Table 8.10 Reference simulation cases

Case 1	Case 2
MCS selected based on CPICH measurement[a]	Cell selection rate: once per 3.33 ms
MCS update rate: once per 3.33 ms (5 slots)	Cell selection error rate: 1%
CPICH measurement transmission delay: 1 frame	FCSS transm request & cell selection delay: 2 frames
Selected MCS = 1 frame delay after measurement rpt	Power allocated to overhead channels (CPICH, PICH, SCH, BCCH, dedicated): 30%
SD of CPICH measurement error: 0, 3 dB	Maximum power allocated to DSCH: 70%
CPICH measurement rate: once per 3.33 ms	Maximum number of retries: 15
CPICH measurement report error rate: 1%	Cell maximum power: 17 WattsTx Que/Priority Que: 5 frame intervals/30 frame intervals
Frame length for fast HARQ: 3.33 ms	E_b/N_t implementation loss: 0 dB
Fast HARQ feedback error rate: 0, 1 or 4%.	SD of CPICH measurement error: 0

[a]For example, RSCP/ISCP, or power control feedback information.

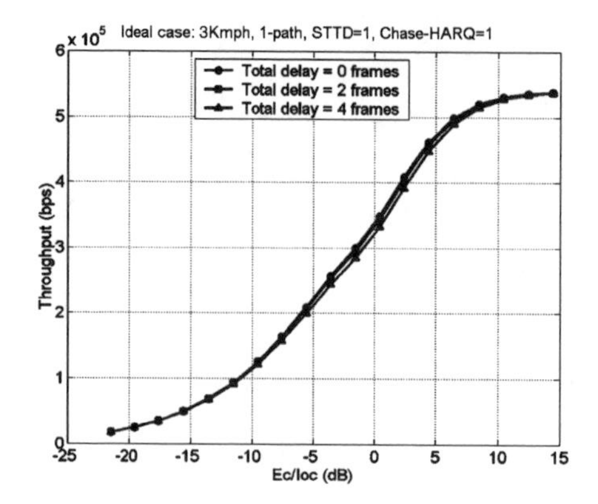

Figure 8.23 Ideal measurement and feedback (speed = 3 km/h) [1].

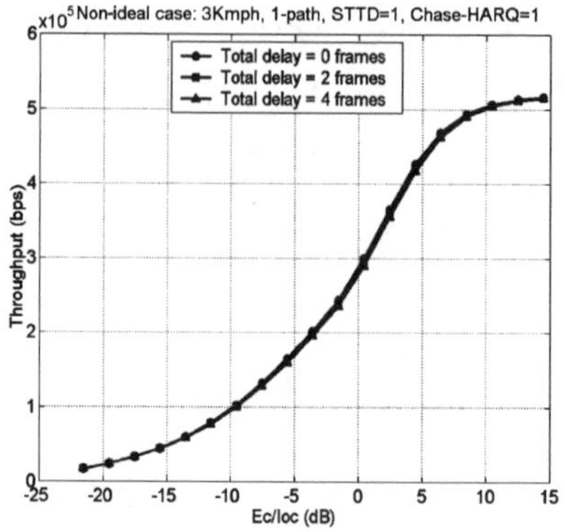

Figure 8.24 Non-ideal measurement and feedback (speed = 3 km/h) [1].

8.6.2.9 MCS Selection Delay Effects on HSDPA Performance

Figures 8.23 and 8.24 illustrate the throughput versus E_c/I_{oc} with a single code for HSDPA at 3 km/h for different MCS selection delays in ideal and non-ideal measurement and feedback case, respectively. Here, total MCS selection delay implies the time difference between the CPICH measurement at UE and MCS selection applied at Node B. Delays may occur due to EU time processing, Node B time processing, transmission delay and multiplexing and scheduling delay. In the ideal case, the CPICH standard deviation measurement error = 0 dB, while we assume that the CPICH measurement report and HARQ feedback are both error free. In the non-ideal case, the CPICH standard deviation measurement error = 3 dB, the CPICH error rate measurement report and HARQ feedback = 1 and 4%, respectively [1].

Clearly, at very low vehicle speeds, limited MCS selection delays do not have relevant performance loss as result of the slow changing channel.

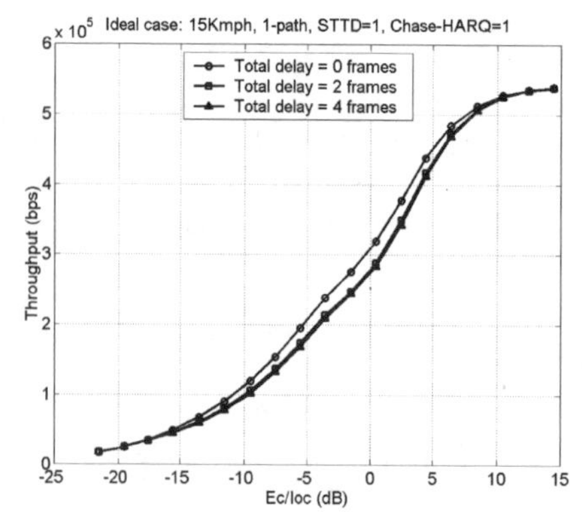

Figure 8.25 Ideal measurement and feedback (speed = 15 km/h) [1].

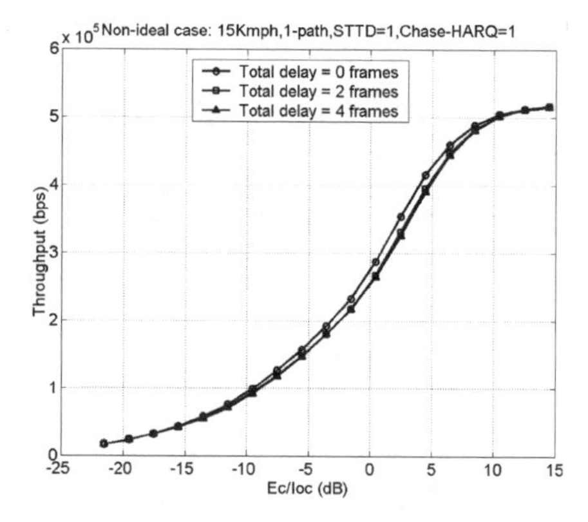

Figure 8.26 Non-ideal measurement and feedback (speed = 15 km/h).

Figures 8.25 and 8.26 illustrate simulation results at higher vehicle speed (i.e. 15 km/h). Contrarily, to the slow moving vehicle here throughput loss due to MCS selection delay reach about 1 dB or 22% throughput loss. Extrapolation, to higher delays compared with the channel correlation time, indicates that the performance loss would be larger. Thus, optimum channel prediction techniques will minimise performance loss due to MCS selection delay and will improve the channel estimation.

In conclusion, performance loss due to the MCS delay has low impact at very slow vehicle speed, but increases at higher vehicle speeds and for larger MCS selection delays. Thus, highly efficient channel-prediction techniques, in conjunction with the MCS selection rule, become imperative to maximise AMCS and HARQ performance in HSDPA at high speeds.

8.6.2.10 Integrated Voice Data Performance Analysis

Next we summarise the analysis of a sector data throughput in an integrated voice and HSDPA environment based on link level inputs.

The integrated elements include a data throughput characteristic (Thruput(x)) as illustrated in eqn (8.15) available from link level simulations for a given channel condition with the achievable Carrier-to-Interference Ratio ($C/I\ (x)$), which is also available from system simulations for the coverage area of several representative sectors of the voice and data system [1].

$$\text{Ave Sector Thruput} = \int_{-20}^{20} \text{Thruput}(x) \times P(x) \times \mathrm{d}x \tag{8.15}$$

The predicted sector throughput is therefore calculated from a combination of link level and system simulations. Figure 8.27 shows the throughput in b/s versus the ratio of energy per chip over other cell interference (E_c/I_{oc}) at the mobile receiver. We now use E_c/I_{oc}

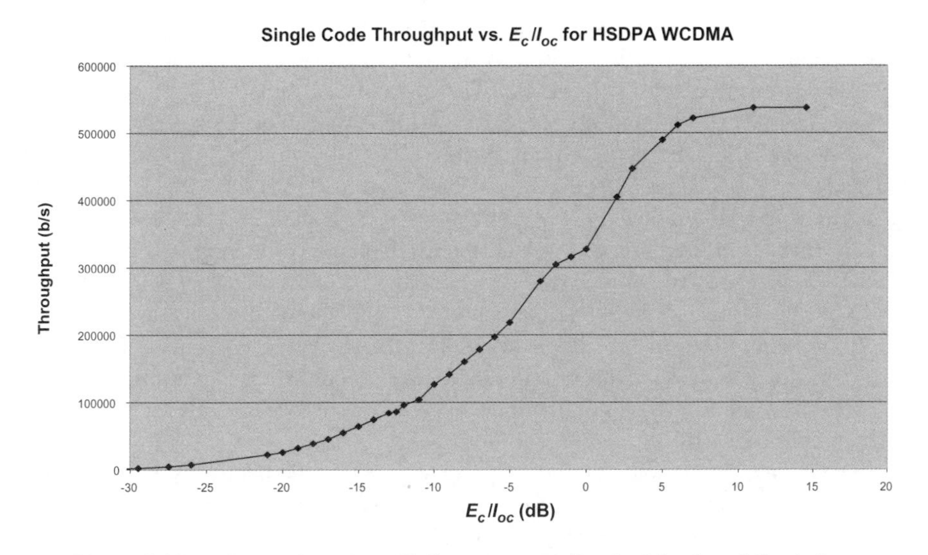

Figure 8.27 HSDPA throughput Hull curve vs. E_c/I_{oc} for 3 km/h and flat fading.

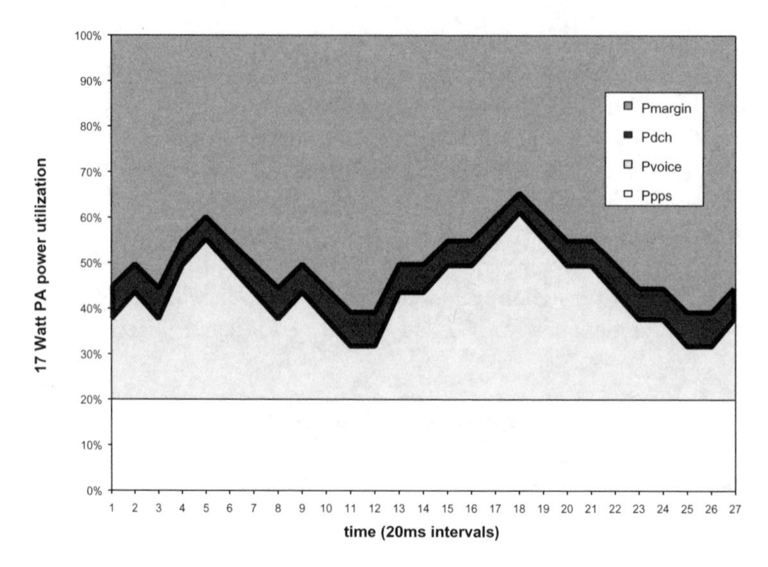

Figure 8.28 Percentage utilisation of 17 W PA for a given sector in the system. Note at all times 17 W is transmitted for each sector.

instead of C/I. Both hybrid ARQ and transmit diversity (STTD) are enabled. Each curve plotted is in fact a composite of several link simulations for 64QAM $R = 3/4$, 16QAM $R = 1/2$ and $R = 3/4$, and QPSK using Turbo coding (see [5]). Hybrid ARQ in these simulations uses max-ratio combining of successive attempts (Chase combining). All curves have been simulated at 3 km/h one-ray Rayleigh fading channel model at a carrier frequency of 2 GHz. Link adaptation switches between the modulation and coding levels to maximise the throughput for given E_c/I_{oc} value.

The area probability for a given E_c/I_{oc} has been calculated from a system simulation of two hexagonal rings comprising 19 3-sector cells with log-normal standard deviation of 8.0 dB and a 50% site-to-site correlation. A full set of radio (E_b/N_t vs. FER) curves is used for modelling the 12.2 kbps voice users. These curves account for 1, 2 or 3 rays with imbalances from 0 to 12 dB and speeds from 0 to 120 km/h and geometries ($\hat{I}_{or}/I_{oc}$) ranging from −6 to +12 dB. The system is assumed to be 100% loaded resulting in the base transceivers having a constant 100% Linear Power Amp (LPA) load of 17 W. By always transmitting with constant power (17 W in this case, see Figure 8.28, the voice users will not see abrupt changes in interference levels as the available power margin is allocated to data. Of the LPA load, up to 70% of the power can be allocated to the data channel constructed up to 20 (or 28) multi-codes with spreading factor 32 depending on the voice users (12.2 kbps) loading. The other 30% of the LPA load is allocated to overhead channels (such as pilot (CPICH), paging (PICH), Synchronisation (SCH), etc.) and dedicated control channels. The E_c/I_{oc} area distribution is based on the inner ring sectors and centre cell sectors in order to exclude system edge effects.

The sector data throughput for 'equal average power' scheduler may be calculated by integrating the throughput from the link simulations against the area pdf for E_c/I_{oc} derived from the system simulation [see eqn (2) below]. The E_c/I_{oc} is determined from the available power margin left-over after power is allocated to overhead channels (such as pilot (CPICH), paging (PICH), synchronisation (SCH)), dedicated control channels and voice user channels.

The number of size 32 OVSF codes, and hence the peak rate that can be allocated, depends on the code tree left-over after the overhead and voice channels have been allocated their codes. The equal average power scheduler assigns equal BS power to all users throughout the coverage area achieving the maximum possible throughput for each location. For HSDPA, equal average power scheduling would be achieved by cycling through all users in the coverage area, assigning one 3.33 ms frame with up to 20 (or 28) size 32 OVSF codes and up to 70% of the LPA power while using the optimum modulation and coding level. Over time, each user would receive an equal number of frames and therefore an equal average power allocation from the serving BTS. However, the average data received per user would be biased by the user's location. Users closer to the base site would receive more data than those towards the cell edge.

Equation 3—note the power margin could be up to 80% in the system simulation. Therefore, in the equations the computed E_c/I_{oc} is reduced by 0.6 dB to limit the maximum available power for HSDPA data to 70%—as follows:

$$\frac{E_c}{I_{or}} = \frac{P_{\text{margin}}}{P_{\text{margin}} + P_{\text{voice}} + P_{\text{ovhd}}} = \frac{P_{\text{margin}}}{P_{\text{cell}}(j)} \quad \text{for cell } j \text{ at time } t$$

$$\frac{E_c}{I_{oc}} = 10\ \log_{10}\left(\frac{E_c}{I_{or}} \cdot \lambda/(I_{on} - \lambda)\right)$$

$$I_{on} = \sum_{i=1}^{N\ \text{cells}} P_{\text{cell}}(i)T(i,k)/P_{\text{cell}}(j)T(j,k)$$

where,

$P_{cell}(j)$—total power in Watts for cell j (always 17 W);
$T(i,k)$—transmission gain from cell i to probe mobile at location k;
λ—fraction of total available power recovered (FRP);
I_{on}—best serving cell to total power ratio for location k.

$$\text{Thrupt} = N_{multi\text{-}codes} \cdot \text{Thruput\_Hull\_Curve}\left(\frac{E_c}{I_{oc}} - N_{multi\text{-}codes\_dB} - 0.6\right)$$

8.6.2.11 Conclusion

Figure 8.29 below presents sector throughput of the equal average power scheduler for increasing voice loading and for different (FRP) fractions of total recovered power due to delay spread. For FRP = 0.98 and 20 codes, the achieved data only throughput is approximately 2.5 Mbps which then drops almost linearly (see also Figure 8.30) as voice erlangs per sector increases. For a voice user (12.2 kbps) load of approximately 35 erlangs/sector the data 'equal power' sector throughput is still approximately 1 Mbps.

An FRP of 0.98 results in about a 10% loss in throughput relative to an FRP = 1.0 while an FRP of 0.92 results in about a 35% loss.

High data sector throughput is maintained by simply allocating the available power margin to data users. This approach is effective as long as the delay from measuring C/I and scheduling for a given user is small. Also the less slots (power control updates), an HSDPA frame encompasses the less margin needs to be set aside for voice users to guarantee them a

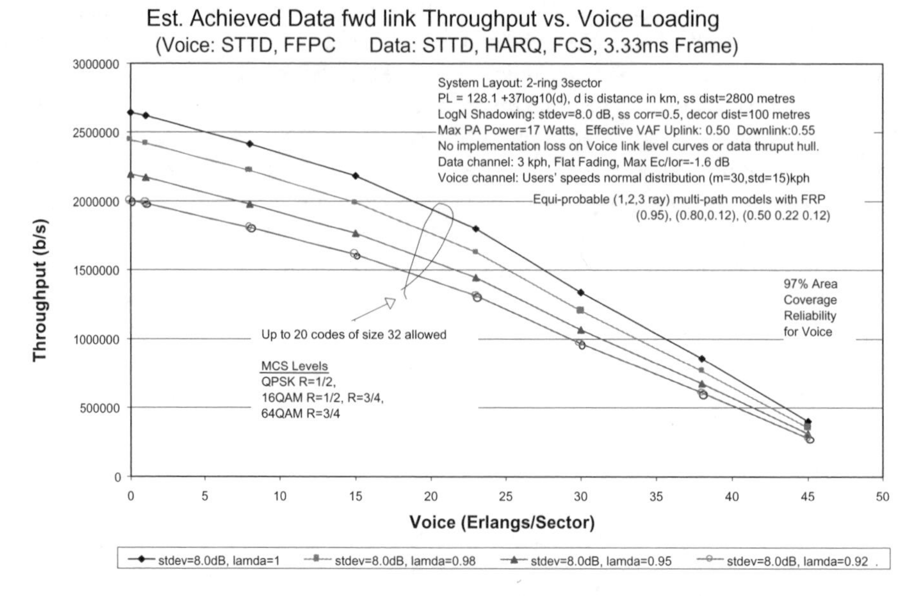

Figure 8.29 HSDPA throughput vs. voice loading for different FRP (lamda) where an equal average power scheduler is assumed.

minimum performance level during a scheduled burst. Alternatively, the power control rate for voice users could be reduced to 500 Hz to minimise the power margin needed for voice users over a data frame interval such as 3.33 ms.

The drawbacks of the kind of simulation-analysis presented are that the effects of voice activity and fast FPC are not adequately modelled and such effects may degrade C/I

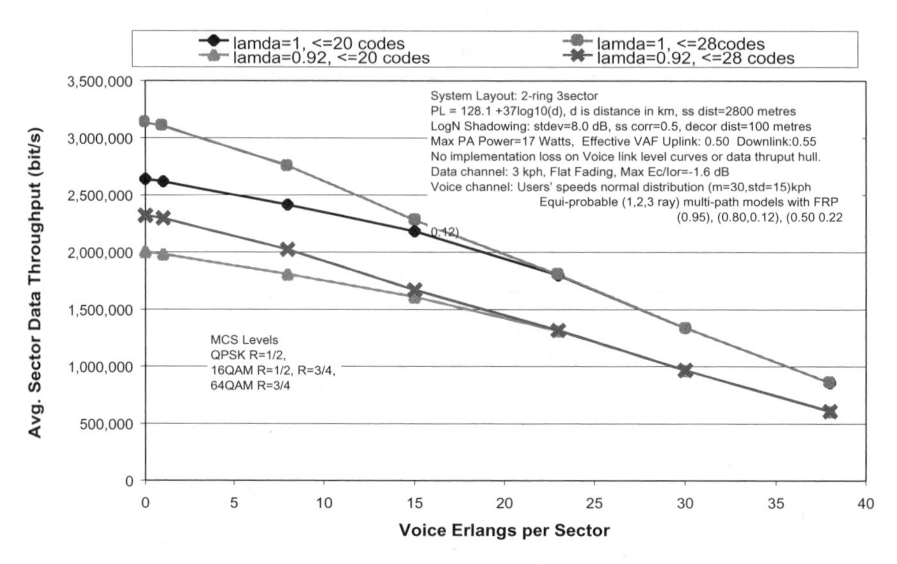

Figure 8.30 HSDPA throughput vs. voice loading for different FRP (lamda) and for 20 and 28 OVSF codes where and equal average power scheduler is assumed.

estimation and hence degrade data throughput. Effects of voice activity and fast FPC will be addressed in later analysis and future quasi-static system simulations. The reverse link will also be modelled.

8.7 TERMINAL RECEIVER ASPECTS

The terminal receiver aspects were discussed earlier in the chapter since one of the new challenges is the need for amplitude estimates for the 16QAM detection. However, there are other challenges coming from the use of 16QAM as well. A good quality voice call in WCDMA typically requires a C/I of −20 dB compared to 10 dB for GSM. Since the interference, including the inter-symbol interference, can be 20 dB above the signal level, the WCDMA voice signal is very robust against interference and does not benefit significantly from equalisers. However, for the high peak data rates provided with HSDPA service, higher C/I (E_b/N_o) values above 0 dB are required and, consequently, the signal becomes less robust against inter-symbol interference.

Hence, the HSDPA concept with 16QAM transmission potentially benefits from equaliser concepts that reduce the interference from multi-path components. The multi-path interference cancellation receiver shown in Figure 8.31 was discussed and analysed in [2]. The same receiver front-end as employed in the rake receiver is used as a pre-stage to provide draft symbol estimates. Those estimates are then used to remove the multi-path interference from the received signal and new symbol estimates can be obtained with the same matched filter. After a few iterations, the final symbol estimates are calculated. Another type of advanced receiver is linear equaliser. The advanced receiver algorithms are discussed in more detail in Section 12.6.

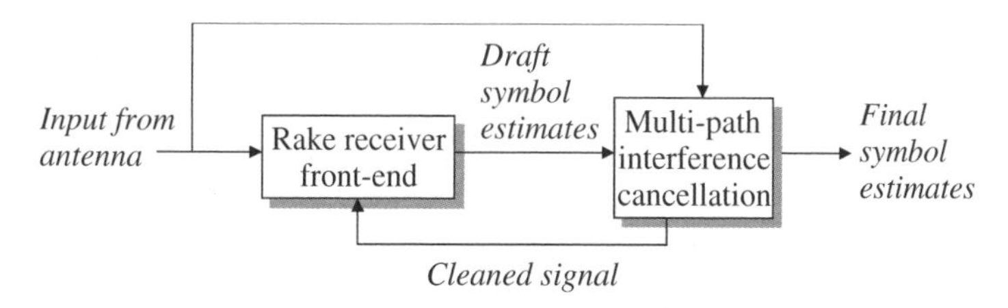

Figure 8.31 Example multi-path interference cancellation.

Advanced receivers make it possible to provide higher bit rates in multi-path channels compared to what is achievable with normal rake receivers. On the other hand, the complexity of such receivers is significantly higher than for the standard rake receiver.

In 3GPP standardisation there is no intention to specify any receiver solutions but only performance requirements in particular cases [6–16].

Evolution Beyond Release 5. As described previously, the HSDPA concept of Release 5 is able to provide a clear increase in the WCDMA downlink packet data throughput. It is obvious that further enhancements on top of the HSDPA feature can be considered for increased user bit rates and cell throughput. Possible techniques raised previously include antenna techniques and fast cell selection, which are briefly discussed in this section.

8.7.1 Spectral Efficiency, Code Efficiency and Dynamic Range

In WCDMA, both spectral efficiency and code efficiency are important optimisation criteria to accommodate code-limited and power-limited system states. In this respect, HSDPA provides some important improvements over Release 1999 DCH and DSCH:

Spectral efficiency is improved at lower SIR ranges (medium to long distance from Node B) by introducing more efficient coding and fast HARQ with redundancy combining. HARQ combines each packet retransmission with earlier transmissions, such that no transmissions are wasted. Further, extensive multi-code operation offers high spectral efficiency; similar to variable spreading factor but with higher resolution. At very good SIR conditions (vicinity of Node B), HSDPA offers higher peak data rates and thus better channel utilisation and spectral efficiency.

Code efficiency is obtained by offering more user bits per symbol and thus more data per channelisation code. This is obtained through higher order modulation and reduced coding. Further, the use of time multiplexing and shared channels generally leads to better code utilisation for bursty traffic as described in Chapter 10.

8.7.2 Cell Throughput and Coverage

The HSDPA cell throughput depends significantly on the interference distribution across the cell, the time dispersion, as well as the multi-code and power resources allocated to HSDPA.

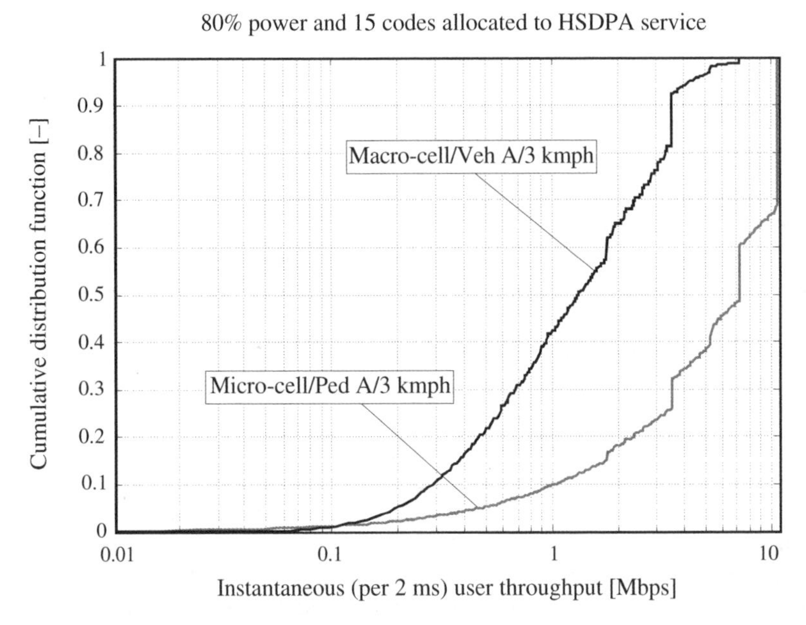

Figure 8.32 Instantaneous user throughput Cumulative Distribution Function (CDF) for micro-cell and macro-cell scenarios.

In Figure 8.32, the Cumulative Distribution Function (CDF) of instantaneous user throughput for both macro-cell outdoor and micro-cell outdoor–indoor scenarios is considered. As mentioned in Chapter 10, the chosen packet scheduling method has a significant impact on the overall cell throughput and the end-user perceived Quality of Service (QoS). The shown CDFs correspond to the case where the fair time scheduling is employed. The fair time scheduling means that the same power is allocated to all users so that users with better channel conditions experience a higher throughput. Figure 8.32 assumes that the available capacity of the cell is allocated to the studied user and that other cells are fully loaded. Note that for the micro-cell case, 30% of the users has sufficient channel quality to support peak data rates exceeding 10 Mbps. The reason is the assumption of a limited time dispersion as well as a favourable I_{or}/I_{oc}-distribution due to high intercell isolation Figure 8.32. The mean bit rate that can be obtained is more than 5 Mbps. For the macro-cell case, the presence of time dispersion and high levels of othercell interference widely limits the available peak data rates. Nevertheless, peak data rates of more than 512 kbps are supported 70% of the time and the mean bit rate is over 1 Mbps. For users located in the vicinity of the Node B, time dispersion limits the maximum peak data rate to around 6–7 Mbps. Depending on the HSDPA code and power allocation, the 16QAM selection probability is on the order of 10–20% for macro-cell and 50–70% for micro-cell scenarios.

So far, the discussion has assumed fair time scheduling. As described in Chapter 10, the C/I scheduling method can be used to obtain more cell throughput at the expense of user fairness. Since the HSDPA concept supports very high peak data rates, the C/I scheduling method provides a 50–100% higher average cell throughput performance than the fair time scheduling. The average cell throughput may be on the order of 4 Mbps for the macro-cell case and exceeding 8 Mbps for the micro-cell case. However, since this throughput would only be enjoyed by few users located very close to the Node B, this type of scheduling is believed to be impractical for most scenarios. In Figure 8.33, the average cell throughput performance of fair throughput and fair time scheduling have been compared. As described in Chapter 10, the cell throughput of best effort techniques is higher than for techniques that facilitate a high user fairness. As shown, the cell throughput gain of going from fair throughput to fair time scheduling is on the order of 70–80%. For the fair throughput scheduler, a target cell coverage of 90% has been assumed (remaining users must be served by other means) but within this region all users will be given the target bit rate (0% out-age). For the fair time packet scheduling, the full cell is served but it is accepted that up to 10% of the users do not get the target bit rate. By reducing the cell coverage level for the fair throughput packet scheduler to 70–80%, it performs equally to the fair time scheduling method. In the macro-cell case, the fair throughput scheduling leads to an average cell throughput around 930 Mbps while the fair time scheduling yields 1.5 Mbps. For the micro-cell scenario, these numbers are increased to 3.2 and 5.8 Mbps, respectively. From the results depicted in Figure 8.33, it is seen that the HSDPA performance is robust up to medium terminal speeds on the order of 50 km/h. This is mainly a benefit of the short packet size and fast HARQ.

The cell throughput performance difference between the Release 1999 DCH/DSCH and the Release 5 HS-DSCH is listed in Table 8.11. The DCH/DSCH performance is described in Chapter 10 and the HSDPA performance is now listed including the effect of a significant AMC error and delay. As the interference distribution is worse for HSDPA due to a lack of soft handover, the gain for the macro-cell environments is mainly achieved due to the spectral efficiency improvement inherent to HARQ, a more robust coding efficiency, a high

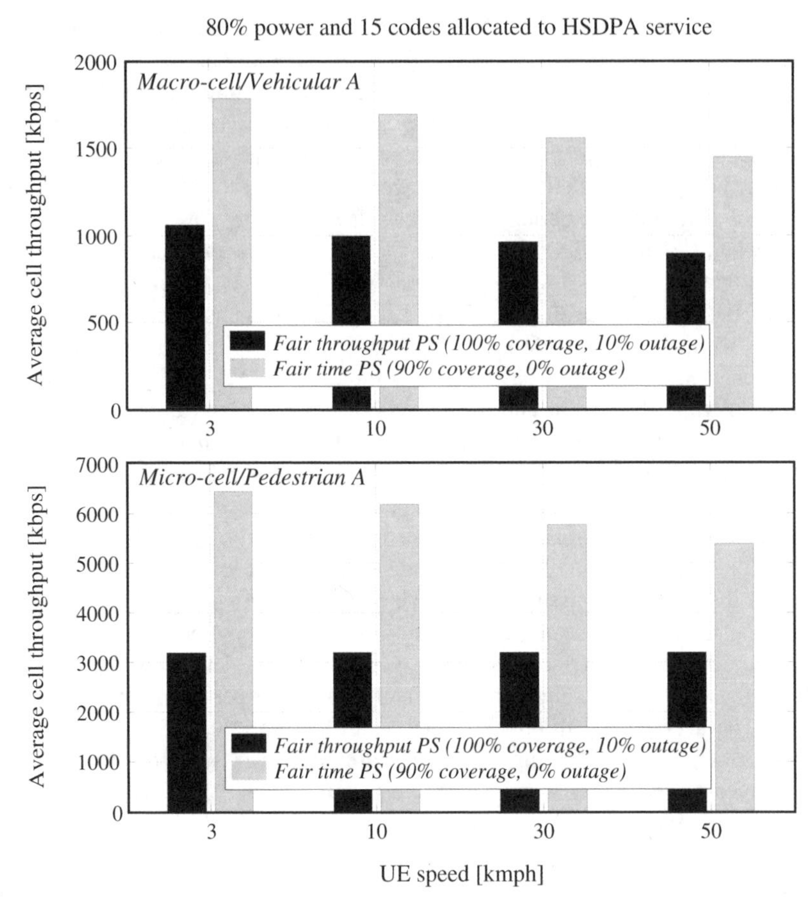

Figure 8.33 Average cell throughput results for different scenarios and packet schedulers.

resolution multi-code operation, and a reduced power control power rise. For more details on the power rise, see Section 9.2.1. As seen from the numbers, HSDPA increases the cell throughput up to 70% compared to the Release 1999 DSCH. For the micro-cell case, the gain of HSDPA exceeds 200% due to the availability of very high user peak data rates. However, for this favourable case, practical imperfections associated with the terminal and Node B hardware, link adaptation, and packet scheduling may limit the achievable cell throughput in practice. Further, it is assumed that in favourable conditions, a user will always utilise the available throughput. Nevertheless, with the higher peak data rates of HSDPA the favourable channel conditions of the micro-cell environment are better utilised compared to the DCH/DSCH which is hard limited in this case (at best, it can give 2 Mbps). The throughput values in Table 8.11 are lower than in Figure 8.33 because Table 8.11 includes AMC delay and error.

Ultimately, the used scheduling method is likely to be a hybrid of the different variants to accommodate changing QoS requirements among users' and operators' strategic interests. In this sense, the coverage becomes important. As the HS-DSCH does not employ a fast power control, the coverage is defined as the area over which the average user throughput is of

Table 8.11 Cell throughput with Release 1999 DSCH and Release 5 HSDPA (including 2 dB and 6 ms AMC error/delay). HSDPA fair throughput scheduling operates with a 90% coverage area

Environment	Time dispersion	Packet scheduling	DCH/DSCH (kbps)	HSDPA(kbps)
Macro-cell	Vehicular A	Fair throughput	675	930
Macro-cell	Vehicular A	Fair time	915	1520
Micro-cell	Pedestrian A	Fair throughput	990	3210
Micro-cell	Pedestrian A	Fair time	1260	5810

some value. The average user data rate coverage follows the I_{or}/I_{oc}-distribution of the cell and the amount of time dispersion. The user data rate coverage for a macro-cell scenario including significant AMC errors is illustrated in Figure 8.34. Compared to cell throughput capacity, the single-user data rate coverage is significantly lower since there is no gain of switching between users with favourable channel conditions. As there is no soft handover gain at the cell edge, the user data rates provided with HS-DSCH are much lower than for DSCH. With 100% cell area coverage, the average user data rate is limited to around 250 kbps supported. The maximum bit rate at the cell edge with Release 1999 is shown to be typically 384 kbps in Section 12.2.3. Hence, for guaranteed bit rates and 100% coverage, DCH/DSCH is still a competitive solution due to the soft handover gain discussed in Chapter 9.

Another issue of interest is the effective use of the HARQ technique. The experienced average BLER of the first transmissions is given in Figure 8.35 assuming (1) that the fair time scheduling is employed and (2) that each link is optimised towards the highest possible

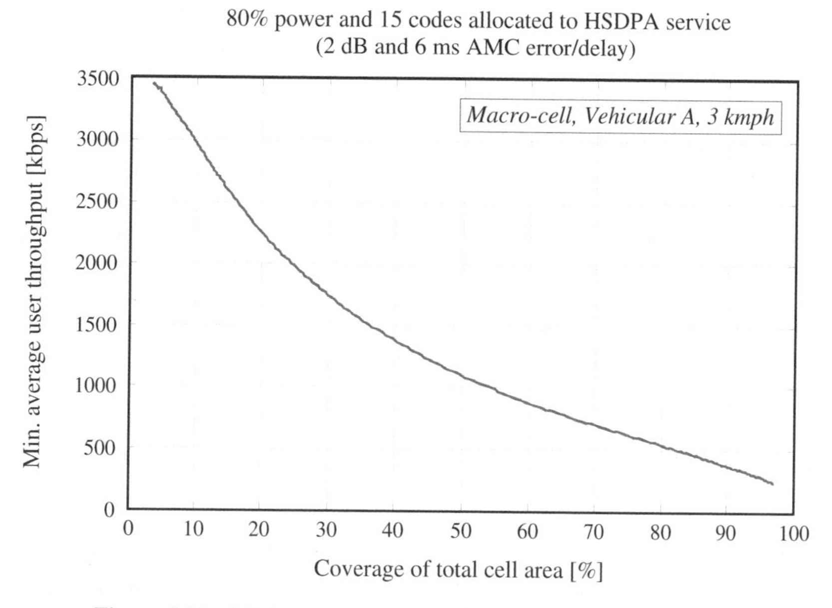

Figure 8.34 Minimum average user throughput vs. cell coverage.

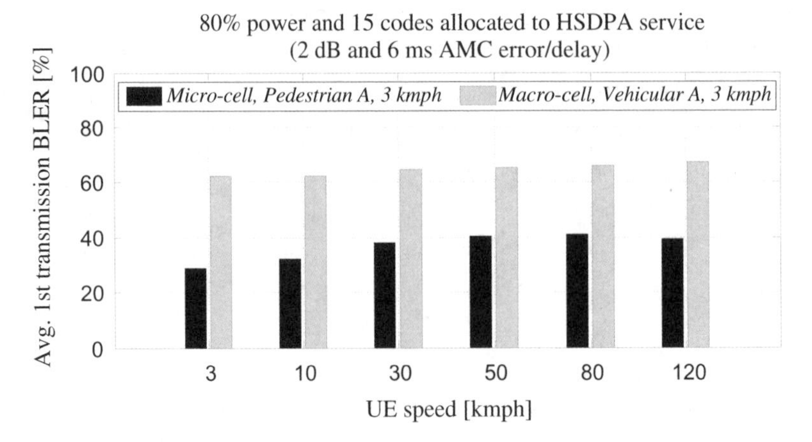

Figure 8.35 Average BLER of first transmission for macro-cell and micro-cell environments. Fair time packet scheduling is assumed.

instantaneous throughput. The average BLER of the first transmission is on the order of 30% for micro-cell and 60% for macro-cell scenarios, respectively. The experienced block error rate on average depends on whether the HSDPA users generally experience a quality in the lower or upper end of the dynamic range of the AMC. At the lower end, an extensive use of fast HARQ is employed which explains why the first transmission BLER is significantly higher for the macro-cell than for the micro-cell case. It should be noted, that the short-time BLER is hard to control since modulation and coding parameters are issued from imperfect E_s/N_o estimates. In practice, its value fluctuates between 0% and 100% because the E_s/N_o-to-BLER mapping curves have a very steep slope in the 0–1 transition region. Compared to DCH/DSCH packet services, the average BLER of the first transmission for HS-DSCH is increased from 10–30% to 30–60%. This is mainly due to the short re-transmission delay and the spectrally efficient re-transmission strategy employed in HSDPA. However, it should be noted that in terms of hardware utilisation, a very high re-transmission rate may be undesirable. Ultimately, the target BLER for the network will be set as a trade-off between the hardware utilisation and the spectral efficiency. Further, some highly delay sensitive services may require a lower target BLER. With the HSDPA concept, these delays and the target BLER are simply controlled by making the link adaptation more or less conservative.

8.7.3 Delay and QoS

While the DSCH packet scheduling is controlled in the RNC, the HSDPA packet scheduling is conducted directly in the Node B. When the scheduling is brought closer to the air interface, some limitations regarding Iub signalling delays are alleviated. A faster scheduling allows guaranteed bit rate services using packet scheduling without the need for a dedicated channel. This can lead to an improved utilisation of hardware and air interface resources, especially when the traffic is bursty. Finally, this approach enables a high spectral efficiency by extensive use of the HARQ technique and also an improved QoS control.

Table 8.12 Typical UMTS QoS classes with DSCH and with HSDPA

QoS class	DSCH	HSDPA
Conversational	No[a]	No[a]
Streaming	Yes/no[a]	Yes
Interactive	Yes	Yes
Background	Yes	Yes

[a]DCH is used.

Typical UMTS QoS classes that are transmitted with the Release 1999 DSCH and expected with the Release 5 HS-DSCH are shown in Table 8.12. The DSCH is used for non-real-time interactive and background services that do not require a guaranteed minimum bit rate. As discussed in Chapter 10, the streaming classes require a guaranteed minimum bit rate and the HS-DSCH is able to provide this service. The conversational services have a fairly constant bit rate and the HSDPA type scheduling does not provide considerable gains over the dedicated channel. Hence, it is assumed in Table 8.12 that the conversational services are carried on a dedicated channel while other services utilise the HS-DSCH. Due to low round trip delays and an accurate control of re-transmission delays over the Iub interface, the delay jitter due to re-transmission is reduced and HS-DSCH is well suited for the TCP protocol.

8.8 CONCLUSIONS

In this section the HSDPA concept was introduced and its performance was considered. The main aspects discussed can be summarised as follows:

- The HSDPA concept utilises a distributed architecture where the processing is closer to the air interface at Node B for low delay link adaptation.
- The HSDPA concept provides a 50% higher cell throughput than the Release 1999 DCH/DSCH in macro-cell and more than a 100% gain in micro-cell scenarios. For micro-cell, the HS-DSCH can support up to 5 Mbps per sector per carrier, i.e. 1 bit/Hz/cell.
- The HSDPA concept offers over 100% higher peak user bit rates than Release 1999. HS-DSCH bit rates are comparable to DSL modem bit rates. The mean user bit rates in large macro-cell environment can exceed 1 Mbps and in small micro-cells 5 Mbps.
- The HSDPA concept is able to support efficiency not only non-real-time UMTS QoS classes but also real-time UMTS QoS classes with guaranteed bit rates.

8.8.1 Multiple Receiver and Transmit Antenna Techniques

Using several transmitter antennas in the Node B and several receiver antennas in the terminal can increase the HSDPA bit rates. Such approaches are commonly denoted as

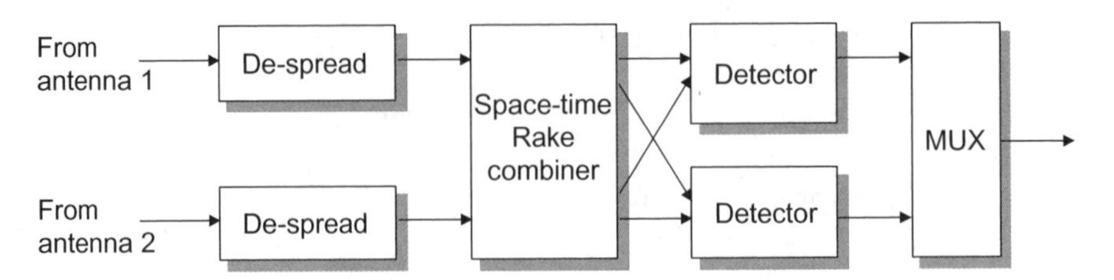

Figure 8.36 Example MIMO receiver.

Multiple Input Multiple Output (MIMO) techniques. Higher data rates can be achieved either (1) by an improved antenna transmit and receive diversity leading to better channel quality or (2) by re-using the spreading code on different antennas (higher throughput per code due to data layering). To distinguish between several sub streams sharing the same code, the terminal uses multiple antennas and spatial signal processing. An example of a MIMO receiver with two antennas is shown in Figure 8.36. The space–time rake combiner is the multiple antenna generalisation of the conventional rake combiner. As seen from the micro-cell results that were shown earlier, up to 20–30% of the users may have a channel quality which exceeds the requirements for 10 Mbps. For these users, MIMO schemes could potentially increase the user data rate further. However, inherent complexity and sensitivity issues must be considered in this context as well. As such, the MIMO technology will be studied further for future releases.

8.8.2 Fast Cell Selection

The concept of Fast Cell Selection (FCS) has been proposed to supplement hard handover and provide both a decreased interference and an improved system capacity. With the FCS technique, the terminal determines which cell is best for downlink service through radio propagation measurements, and makes a report to the network. Only one Node B at a time can be included in the active set. The selection of the most suitable cell may also be based on the available power and code resources for the cells in the active set. The FCS technique is conceptually similar to the Site Selection Diversity Transmission (SSDT) technique, which is already included in Release 1999, but applied on the downlink DCH. The FCS will be studied for further releases along with other potential enhancements.

REFERENCES

1. ITU-T H.323, Packet Based Multimedia Communications Systems, 1998.
2. 3GPP, Technical Report (TR) 25.858, High Speed Downlink Packet Access: Physical Layer Aspects, Release 5, V1.0.4, (2001–2002).
3. 3GPP TR 25.858 V1.0.4, High Speed Downlink Packet Access: Physical Layer Aspects, R5, 2001–2002.
4. 3GPP TR 25.855, V0.0.5, UTRA High Speed Downlink Packet Access.
5. 3GPP, Technical Specification Group, QoS Concept (3GPP TS 23.107, V5.5.0, 2002).
6. ITU-T H.324, Terminal For Low Bit-rate Multimedia Communication, 1998.
7. 3GPP, Mandatory Speech Codec Speech Processing Functions, AMR Speech Codec; General Description, 3G TS 26.071, 1999.

8. 3GPP, Technical Specification Group Services and System Aspects, Services and System Aspects, Location Services (LCS), Service description, Stage 1, 3G TS 22.071, 1999.
9. Handley, M., *et al.*, SIP: Session Initiation Protocol, RFC2543, IETF, 1999.
10. 3GPP, Technical Specification Group (TSG) RAN, Working Group 2 (WG2), Stage 2 Functional Specification of Location Services in URAN, 3G TR 25.923, 1999.
11. G. Feige, Cisco, 3GSM World Congress, Cannes, France, February 2002.
12. 3GPP, Technical Specification Group (TSG) SA, Transparent End-to-end Packet Switching Streaming Service (PSS) General Description, Release 5, V5.0.0, 2002–2003.
13. 3GPP, Architecture Principles for Release 2000, 3G TR 23.821, V1.0.1, 2000–2007.
14. 3GPPTS 23.207, End-to-End QoS Concept and Architecture, V5.8.0, 2003–2006.
15. 3GPP TS 23.002, Network Architecture.
16. RFC 2475, An Architecture for Differentiated Services (Diffserv).

9

The UTRA[1] Transmission System

9.1 UMTS SPECTRUM ALLOCATION

The UMTS frequency ranges are part of the worldwide spectrum allocation for 3rd or evolving 2nd generation systems. Figure 9.1 illustrates the representation of the spectrum from major regions (e.g. Europe, Japan, Korea and USA).

The distribution of the frequency bands from the allocated spectrum for the UTRA system is covered next. We present the ranges for the FDD and the TDD in parallel in order to unveil a complete view of the UMTS frequency assignment.

9.1.1 UTRA Frequency Bands

Table 9.1 summarises the frequency bands for the TDD and FDD modes, as well as the frequency distribution for the User Equipment (UE) and the Base Station (BS). Although, in some cases the frequency ranges may be the same for both UE and BS, these are noted separately for the completeness.

Additional spectrum allocations in ITU region 2 are FFS, and deployment of UMTS in existing and other frequency bands is not precluded. Furthermore, co-existence of TDD and FDD in the same bands (now under study) may be possible.

9.2 RADIO TRANSMISSION AND RECEPTION ASPECTS

After the allocation of the frequency ranges for the UTRA modes in the preceding section, in the following we present the transceiver parameters from the technical specifications [1–4]. These parameters will set the necessary background to consider equipment and network design, including traffic engineering issues.

9.2.1 Transmit to Receive (TX-RX) Frequency Separation

While the TDD mode does not need Frequency Separation (FS), the FDD mode does in both the EU and the BS (Table 9.2).

[1]The UMTS terrestrial radio access.

All IP in 3G CDMA Networks J. Castro
© 2004 John Wiley & Sons, Ltd ISBN: 0-470-85322-0

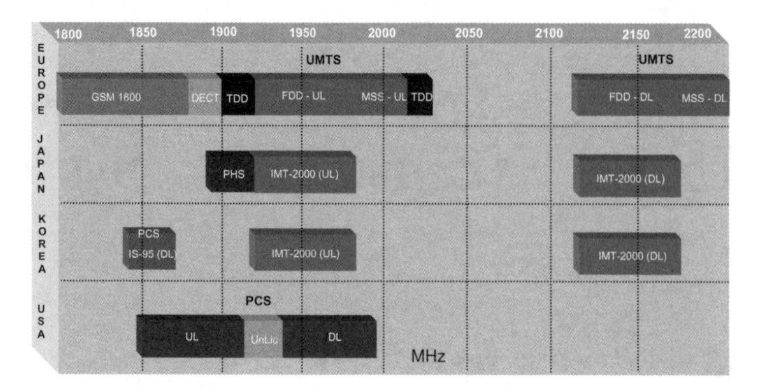

Figure 9.1 Spectrum allocation representation for 3G systems.

Table 9.1 UTRA frequency bands in the MS and Base Station (BS) side

	FDD (MHz)			TDD (MHz) up- and down-link	
Cases		UE	BS	UE	BS
Case (a)	Uplink (MS to BS)	1920–1980	1920–1980	1900–1920	1900–1920
	Downlink (BS to MS)	2110–2170	2110–2170	2010–2025	2010–2025
Region 2 (e.g. Europe)					
Case (b)	Uplink (MS to BS)	1850–1910	1850–1910	1850–1910	1850–1910
	Downlink (BS to MS)	1930–1990	1930–1990	1930–1990	1930–1990
Case (c)				1910–1930	1910–1930

Table 9.2 UTRA TX-RX frequency separation

	FDD	TDD
	User Equipment (UE) and Base Station (BS)	UE and BS
1.	Minimum value = 134.8 MHz Maximum value = 245.2 MHz All UE(s) shall support 190 MHz FS in case (a)[a]	No TX-RX frequency separation is required
2.	All UE(s) shall support 80 MHz FS in case (b)[a]	Each TDMA frame has 15 time-slots
3.	FDD can support both fixed and variable TX-RX FSs	Each time-slot can be allocated to either transit (TX) or receive (RX)
4.	Use of other TX-RX FSs in existing or other frequency bands shall not be precluded	

[a]When operating within spectrum allocations of cases (a) and (b) Table 9.1, respectively.

Table 9.3 UTRA channel configurations

	FDD (MHz)		TDD (MHz)
Channel	UE and BS		UE and BS
Spacing	5 MHz		5 MHz
Raster	200 kHz		200 kHz
Number	UL	$N_u = 5 \times (F_{uplink}\ \text{MHz})$ $0.0\ \text{MHz} \le F_{uplink} \le 3276.6\ \text{MHz}$	$N_t = 5 \times (F)$ $0.0\ \text{MHz} \le F \le 3276.6\ \text{MHz}$ (F is the carrier frequency in MHz)
	DL	$N_d = 5 \times (F_{downlink}\ \text{MHz})$ $0.0\ \text{MHz} \le F_{downlink} \le 3276.6\ \text{MHz}$	

Note: F, F_{uplink} and $F_{downlink}$ are the carrier uplink and downlink frequencies in MHz, respectively.

9.2.2 Channel Configuration

The channel spacing, raster and numbering arrangements aim to synchronise in both FDD and TDD modes as well as keep certain compatibility with GSM, in order to facilitate multi-mode system designs. This applies, e.g. to the raster distribution where 200 kHz corresponds to all (UE and BS in FDD and TDD modes). Table 9.3 summarises the specified channel configurations.

The nominal channel spacing (i.e. 5 MHz) can be adjusted to optimise performance depending on the deployment scenarios; and the channel raster (i.e. 200 kHz) implies the centre frequency which must be an integer multiple of 200 kHz.

In the case of the channel number, the carrier frequency is designated by the UTRA Absolute Radio Frequency Channel Number (UARFCN), Table 9.3 shows those defined in the IMT-2000 band.

9.3 TRANSMITTER CHARACTERISTICS

As in the UE or otherwise stated, we specify transmitter characteristics at the BS antenna connector (test port A) with a full complement of transceivers for the configuration in normal operating conditions. When using external apparatus (e.g. TX amplifiers, diplexers, filters or a combination of such devices), requirements apply at the far-end antenna connector (port B).

9.3.1 Maximum Output Power

9.3.1.1 User Equipment (UE)

At this time detailed transmitter characteristics of the antenna connectors in the UE are not available; thus, a reference UE with integral antenna and antenna gain of 0 dBi is assumed.

Table 9.4 UL reference measurement channel physical parameters (12.2 kbps)

FDD		TDD	
Parameter	Level	Parameter	Level
Information bit rate (kbps)	12.2	Information data rate	12.2 kbps
DPDCH (kbps)	60	RUs allocated	2 RU
DPCCH (kbps)	15	Mid-amble	512 chips
DPCCH/DPDCH (dB)	−6	Inter-leaving	20 ms
TFCI	On	Power control	2 bits/user
Repetition (%)	23	TFCI	16 bits/user
		Inband signalling DCCH	2 kbps
		Puncturing level at code rate 1/3: DCH/DCCH	5%/0%

For the definition of the parameters to follow we use the UL reference measurement channel (12.2 kbps) illustrated in Table 9.4, other references can be found in [1,2].

About four UE power classes have been defined (Table 9.5). The tolerance of the maximum output power is below the suggested level even when we would use multi-code transmission mode in the FDD and TDD modes.

Other cases applying to the TDD mode from [2] are:

- maximum output power refers to the measure of power while averaged over the useful part of transmit time-slots with maximum power control settings;
- in multi-code operation, the maximum output power decreases by the difference of the peak to average ratio between single- and multi-code transmission;
- UE using directive antennas for transmission, will have a class-dependent limit placed on the maximum Equivalent Isotropic Radiated Power (EIRP).

9.3.1.2 BS Output Power

In the TDD mode, *BS output power*, P_{out} represents the one carrier mean power delivered to a load with resistance equal to the nominal load impedance of the transmitter during one

Table 9.5 UE power classes

	FDD		TDD	
Power class	Maximum output power (dB m)	Tolerance (dB)	Maximum output power	(dB m)
Tolerance (dB)	1	+33	+1/−3	2
+27	+1/−3	+24	+1/−3	3
+24	+1/−3	+21	+2/−2	4
+21	±2			

slot. Likewise, BS *rated output power*, P_{RAT}, indicates the manufacturer declared mean power level per carrier over an active time-slot available at the antenna connector [4].

In FDD or TDD *BS maximum output power*, P_{max}, implies the mean power level per carrier measured at the antenna connector in specified reference conditions. In normal conditions, BS maximum output power remains within +2 dB and −2 dB of the manufacturer's rated output power. In extreme conditions, BS maximum output power remains within +2.5 dB and −2.5 dB of the manufacturer's rated output power.

9.3.2 Frequency Stability

Here frequency stability applies to both FDD and TDD modes. The required accuracy of the UE-modulated carrier frequency lies within ±0.1 ppm when compared to the carrier frequency received from the BS. The signals have apparent errors as a result of BS frequency error and Doppler shift; hence signals from the BS need averaging over sufficient time.

The BS-modulated carrier frequency is accurate to within ±0.05 ppm for RF frequency generation.

9.3.3 Output Power Dynamics

9.3.3.1 User Equipment (UE)

In the FDD as well as TDD we use power control to limit interference. The *Minimum Transmit Output Power* is better than −44 dB m measured with a Root-Raised Cosine (RRC) filter having a roll-off factor $\alpha = 0.22$ and a bandwidth equal to the chip rate.

9.3.3.1.1 Open-Loop Power Control

Open-loop power control enables the UE transmitter to set its output power to a specific value, where in normal conditions it has tolerance of ±9 and ±12 dB in extreme conditions. We defined it as the average power in a time-slot or ON power duration depending on the availability. The two options are measured with a filter having an RRC response with a roll-off factor $\alpha = 0.22$ and a bandwidth equal to the chip rate.

9.3.3.1.2 Uplink Inner-Loop Power Control

Through the uplink inner-loop power control, the UE transmitter adjusts its output power according to one or more TPC command steps received in the downlink. The UE transmitter will change the output power in step sizes of 1, 2 and 3 dB, depending on derived Δ_{TPC} or $\Delta_{RP\text{-}TPC}$ values in the slot immediately after the TPC_cmd. Tables 9.6 and 9.7 illustrate the transmitter power control range and average output power, respectively.

We define the inner-loop power as the relative power differences between averaged power of original (reference) time-slot and averaged power of the target time-slot without transient duration. The UE has minimum controlled output power with the power control set to its minimum value. This applies to both inner-loop and open-loop power control, where the minimum transmit power is better than −50 dB m [1]. These are measured with a filter that

Table 9.6 Transmitter power control range

TPC_cmd	1 dB step-size		2 dB step-size		3 dB step-size	
	Lower	Upper	Lower	Upper	Lower	Upper
+1	+0.5	+1.5	+1	+3	+1.5	+4.5
0	−0.5	+0.5	−0.5	+0.5	−0.5	+0.5
−1	−0.5	−1.5	−1	−3	−1.5	−4.5

has an RRC filter response with a roll-off factor $\alpha = 0.22$ and a bandwidth equal to the chip rate.

9.3.3.1.3 Uplink Power Control TDD

Through the uplink power control, the UE transmitter sets its output power taking into account the measured downlink path loss, values determined by higher layer signalling and filter response α. This power control has an *initial error accuracy* of less than ±9 dB under normal conditions and ±12 dB under extreme conditions.

From [2] we define the *power control differential accuracy* as the error in the UE transmitter power step, originating from a step in SIR_{TARGET} when the parameter $\alpha = 0$. The step in SIR_{TARGET} is rounded to the closest integer dB value. The error does not exceed the values illustrated in Table 9.8.

9.3.3.2 Base Station (BS)

In FDD, the transmitter uses a quality-based power control on both the uplink and the downlink to limit the interference level. In TDD, the transmitter uses a quality-based power control primarily to limit the interference level on the downlink.

Through *inner loop power control* in the downlink the FDD BS transmitter has the ability to adjust the transmitter output power of a code channel in accordance with the

Table 9.7 Transmitter average power control range

TPC_cmd	Transmitter power control range after 10 equal TPC_cmd groups				Transmitter power control range after 7 equal TPC_cmd groups	
	1 dB step-size		2 dB step-size		3 dB step-size	
	Lower	Upper	Lower	Upper	Lower	Upper
+1	+8	+12	+16	+24	+16	+26
0	−1	+1	−1	+1	−1	+1
−1	−8	−12	−16	−24	−16	−26
0,0,0,0,+1	+6	+14	N/A	N/A	N/A	N/A
0,0,0,0,−1	−6	−14	N/A	N/A	N/A	N/A

Table 9.8 Transmitter power step tolerance in normal conditions[a]

ΔSIR_{TARGET} (dB)	Transmitter power step tolerance (dB)
$\Delta SIR_{TARGET} \leq 1$	±0.5
$1 < \Delta SIR_{TARGET} \leq 2$	±1
$2 < \Delta SIR_{TARGET} \leq 3$	±1.5
$3 < \Delta SIR_{TARGET} \leq 10$	±2
$10 < \Delta SIR_{TARGET} \leq 20$	±4
$20 < \Delta SIR_{TARGET} \leq 30$	±6
$30 < \Delta SIR_{TARGET}$	±9[1]

[a]For extreme conditions the value is ±12 dB.

corresponding TPC symbols received in the uplink. In the TDD, inner-loop control is based on SIR measurements at the UE receiver and the corresponding TPC commands are generated by the UE, although the latter may or does also apply to the FDD.

9.3.3.2.1 Power Control Steps

The *power control step change* executes step-wise variation in the DL transmitter output power of a code channel in response to a corresponding power control command. The *aggregated output power change* represents the required total change in the DL transmitter output power of a code channel while reacting to multiple consecutive power control commands corresponding to that code channel. The BS transmitter will have the capability of setting the inner-loop output power with a step-size of 1 dB mandatory and 0.5 dB optional [3]. The power control step and the aggregated output power change due to inner-loop power control shall be within the range illustrated in Table 9.9.

In TDD, power control steps change the DL transmitter output power in response to a TPC message from the UE in steps of 1, 2 and 3 dB. The tolerance of the transmitter output power and the greatest average rate of change in mean power due to the power control step will remain within the range illustrated in Table 9.10.

Table 9.9 FDD transmitter power control steps and aggregated output power change range

Power control commands in the DownLink (DL)	Transmitter power control step range			
	1 dB step-size		0.5 dB step-size	
	Lower	Upper	Lower	Upper
Up (TPC command "1")	+0.5	+1.5	+0.25	+0.75
Down (TPC command "0")	−0.5	−1.5	−0.25	−0.75
	Transmitter aggregated output power change range after 10 consecutive equal commands (up or down)			
Up (TPC command "1")	+8	+12	+4	+6
Down (TPC command "0")	−8	−12	−4	−6

Table 9.10 TDD power control step-size tolerance

Step size (dB)	Tolerance (dB)	Range of average rate of change in mean power per 10 steps	
		Minimum (dB)	Maximum (dB)
1	±0.5	±8	±12
2	±0.75	±16	±24
3	±1	±24	±36

9.3.3.2.2 Power Control Dynamic Range and Primary CPICH–CCPCH Power

We refer to the difference between the maximum and the minimum transmit output power of a code channel for a specified reference condition as the *power control dynamic range*. This range in DL has a maximum power → BS maximum output power of −3 dB or greater, and minimum power → BS maximum output power of −28 dB or less.

By *total power dynamic range*, we mean the difference between the maximum and the minimum total transmit output power for a specified reference condition. In this case, the upper limit of the dynamic range is the BS maximum output power and the lower limit the lowest minimum power from the BS when no traffic channels are activated. The DL total power dynamic range is 18 dB or greater [3].

We call *Primary CPICH* power to the transmission power of the common pilot channel averaged over one frame and indicated in a BCH. This power is within ±2.1 dB of the value indicated by a signalling message [3].

In TDD, the power control dynamic range, i.e. the difference between the maximum and the minimum transmit output power for a specified reference condition has a DL minimum requirement of 30 dB. The minimum transmit power, i.e. the minimum controlled BS output power with the power control setting set to a minimum value, has DL maximum output power of −30 dB. The primary CCPCH power is averaged over the transmit time-slot and signalled over the BCH. The error between the BCH-broadcast value of the primary CCPCH power and the primary CCPCH power averaged over the time-slot does not exceed the values illustrated in Table 9.11. The error is a function of the total power averaged over the time-slot, P_{out}, and the manufacturer's rated output power, P_{RAT} [4].

Table 9.11 Errors between primary CCPCH power and the broadcast value (TDD)

Total power in slot (dB)	PCCPCH power tolerance (dB)
$P_{RAT} - 3 < P_{out} \leq P_{RAT} + 2$	±2.5
$P_{RAT} - 6 < P_{out} \leq P_{RAT} - 3$	±3.5
$P_{RAT} - 13 < P_{out} \leq P_{RAT} - 6$	±5

9.3.4 Out-of-Synchronisation Output Power Handling

The UE monitors the DPCCH quality to detect Line 1 (L1) signal loss. The thresholds Q_{out} and Q_{in} specify at which DPCCH quality levels the UE shall shut its power off and when it

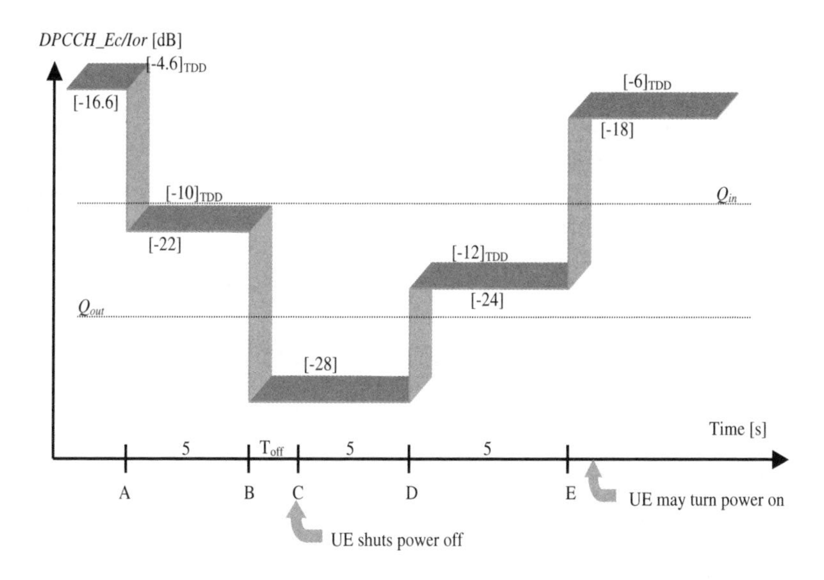

Figure 9.2 UE out-of-synch handling. Q_{out} and Q_{in} thresholds are for reference only [1].

may turn its transmitter on, respectively. The thresholds are not defined explicitly, but are defined by the conditions under which the UE shuts its transmitter off and turns it on.

Figure 9.2 illustrates the DPCH power level and the shutting 'off' and 'on', where the requirements for the UE from [1,2] are that:

- the UE shall not shut its transmitter off before point B;
- the UE shall shut its transmitter off before point C, which is $T_{\text{off}} = [200]$ ms after point B;
- the UE shall not turn its transmitter on between points C and E.

The UE may turn its transmitter on after point E.

9.3.5 Transmit ON/OFF Power

Transmit OFF power state occurs when the UE does not transmit, except during UL DTX mode (see Figure 9.3). We define this parameter as the maximum output transmit power within the channel bandwidth when the transmitter is OFF. The requirement for transmit OFF power shall be better than −56 dB m for FDD and −65 dB m for TDD, defined as an averaged power within at least one time-slot duration measured with an RRC filter response having a roll-off factor $\alpha = 0.22$ and a bandwidth equal to the chip rate.

The time-mask for transmit ON/OFF defines the UE ramping time allowed between transmit OFF power and transmit ON power. This scenario may include the RACH, CPCH or UL slotted mode. We define ON power as one of the following cases [1]:

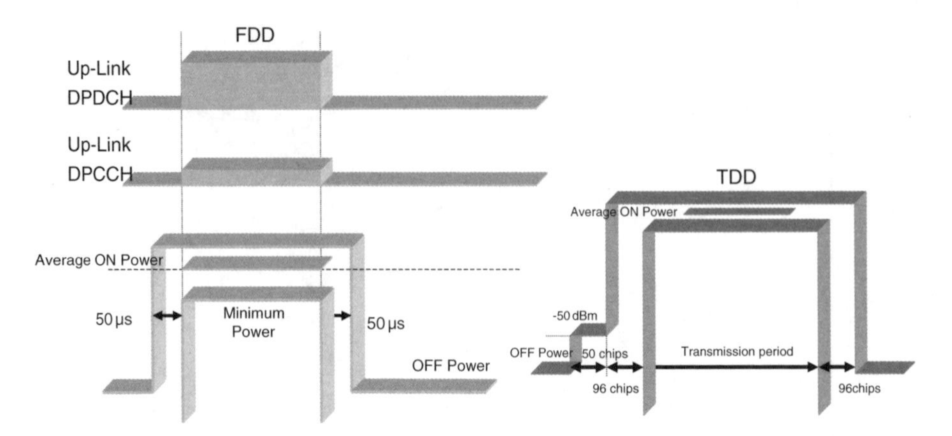

Figure 9.3 Transmit ON/OFF template.

- first preamble of RACH: open-loop accuracy;
- during preamble ramping of the RACH and compressed mode: accuracy depending on size of the power step;
- power step to maximum power: maximum power accuracy.

Specifications in [1] describes power control events in Transport Format Combination (TFC) and compressed modes.

9.3.5.1 BS Transmit OFF Power (TDD)

When the BS does not transmit, it remains in *transmit off power* state, which we defined as the maximum output transmit power within the channel bandwidth when the transmitter states OFF. Its required level shall be better than -79 dB m measured with an RRC filter response having a roll-off factor $\alpha = 0.22$ and a bandwidth equal to the chip rate.

The time-mask transmit ON/OFF defines the ramping time allowed for the BS between transmit OFF power and transmit ON power. The transmit power level versus time meets the mask illustrated in Figure 9.4.

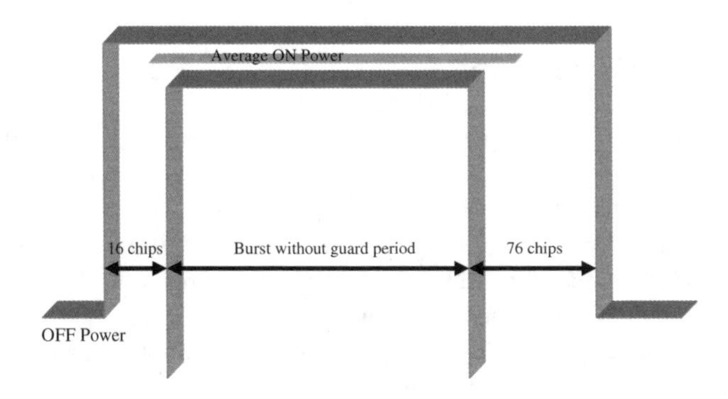

Figure 9.4 BS transmit ON/OFF template (TDD).

9.3.6 Output RF Spectrum Emissions

9.3.6.1 Occupied Bandwidth and Out of Band Emission

Occupied bandwidth implies a measure of the bandwidth containing 99% of the total integrated power of the transmitted spectrum, centred on the assigned channel frequency. In the TDD as well as FDD, the occupied channel bandwidth shall be less than 5 MHz based on a chip rate of 3.84 Mcps.

Out of band emissions are unwanted emissions immediately outside the nominal channel originating from the imperfect modulation process and non-linearity in the transmitter but excluding spurious emissions. A spectrum emission mask and adjacent channel leakage power ratio specify out of band emission limits.

9.3.6.2 Spectrum Emission Mask

The UE spectrum emission mask applies to frequencies that are between 2.5 and 12.5 MHz away from the UE carrier frequency centre. The out of channel emission is specified relative to the UE output power measured in a 3.84 MHz bandwidth. Table 9.12 illustrates UE power emission values, which shall not exceed specified levels.

Table 9.12 Spectrum emission mask requirement

Frequency off-set from carrier Δf (MHz)	Minimum requirement (dBc)	Measurement bandwidth (MHz)
2.5–3.5	$-35-15\ (\Delta f-2.5)$	30 kHz
3.5–7.5	$-35-1\ (\Delta f-3.5)$	1
7.5–8.5	$-39-10\ (\Delta f-7.5)$	1
8.5–12.5	−49	1

The first and last measurement position with a 30 kHz filter is 2.515 and 3.485 MHz.
The first and last measurement position with a 1 MHz filter is 4 and 12 MHz.
The lower limit shall be −50 dB m/3.84 MHz or whichever is higher.

The BS spectrum emission mask illustrated in Figure 9.5 and outlined in Table 9.13 may be mandatory in some regions and may not apply to others. Where it applies, BS transmitting on a single RF carrier and configured according to the manufacturer's specification shall meet specified requirements. The mask basically applies to the FDD and TDD.

For example, emissions for the appropriate BS maximum output power, in the frequency range from $\Delta f = 2.5$ MHz to f_offset_{max} from the carrier frequency, shall not exceed the maximum level specified in Table 9.13 [3–4], where:

- Δf = separation between the carrier frequency and the nominal −3 dB point of the measuring filter closest to the carrier frequency;

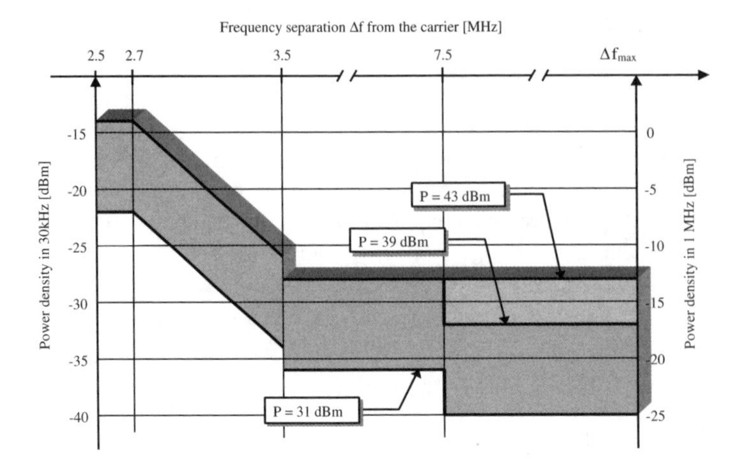

Figure 9.5 BS spectrum emission mask [3].

Table 9.13 BS spectrum emission mask values

Δf of measurement filter −3 dB point (MHz)	Δf of filter measurement at centre frequency (MHz)	Maximum level (dB m)	Measurement bandwidth
BS maximum output power $P \geq 43$ dB m			
$2.5 \leq \Delta f < 2.7$	$2.515 \leq \Delta f < 2.715$	-14	30 kHz
$2.7 \leq \Delta f < 3.5$	$2.715 \leq \Delta f < 3.515$	$-14-15\cdot(\Delta f - 2.715)$	30 kHz
	$3.515 \leq \Delta f < 4.0$	-26	30 kHz
$3.5 \leq \Delta f$	$4.0 \leq \Delta f < \Delta f_{max}$	-13	1 MHz
BS maximum output power $39 \leq P < 43$ dB m			
$2.5 \leq \Delta f < 2.7$	$2.515 \leq \Delta f < 2.715$	-14	30 kHz
$2.7 \leq \Delta f < 3.5$	$2.715 \leq \Delta f < 3.515$	$-14-15\cdot(\Delta f - 2.715)$	30 kHz
*	$3.515 \leq \Delta f < 4.0$	-26	30 kHz
$3.5 \leq \Delta f < 7.5$	$4.0 \leq \Delta f < 7.5$	-13	1 MHz
$7.5 \leq \Delta f$	$7.5 \leq \Delta f < \Delta f_{max}$	$P - 56$	1 MHz
BS maximum output power $31 \leq P < 39$ dB m			
$2.5 \leq \Delta f < 2.7$	$2.515 \leq \Delta f < 2.715$	$P - 53$	30 kHz
$2.7 \leq \Delta f < 3.5$	$2.715 \leq \Delta f < 3.515$	$P - 53 - 15\cdot(\Delta f - 2.715)$	30 kHz
*	$3.515 \leq \Delta f < 4.0$	-26	30 kHz
$3.5 \leq \Delta f < 7.5$	$4.0 \leq \Delta f < 7.5$	$P - 52$	1 MHz
$7.5 \leq \Delta f$	$7.5 \leq \Delta f < \Delta f_{max}$	$P - 56$	1 MHz
BS maximum output power $P < 31$ dB m			
$2.5 \leq \Delta f < 2.7$	$2.515 \leq \Delta f < 2.715$	-22	30 kHz
$2.7 \leq \Delta f < 3.5$	$2.715 \leq \Delta f < 3.515$	$-22 - 15\cdot(\Delta f - 2.715)$	30 kHz
*	$3.515 \leq \Delta f < 4.0$	-26	30 kHz
$3.5 \leq \Delta f < 7.5$	$4.0 \leq \Delta f < 7.5$	-21	1 MHz
$7.5 \leq \Delta f$	$7.5 \leq \Delta f < \Delta f_{max}$	-25	1 MHz

*This frequency range ensures that the range of values of Δf is continuous.

- F_offset = separation between the carrier frequency and the centre of the measuring filter
- f_offset_{max} = 12.5 MHz or is the offset to the UMTS Tx band edge, whichever is higher.

9.3.6.3 *Adjacent Channel Leakage Power Ratio (ACLR)*

The ratio of the transmitted power to the power measured in an adjacent channel corresponds to the Adjacent Channel Leakage Power Ratio (ACLR). Both the transmitted and the adjacent channel power measurements use an RRC filter response with roll-off factor $\alpha = 0.22$ and a bandwidth equal to the chip rate. If the adjacent channel power greater than −50 dB m then the ACLR shall be higher than the value specified in Table 9.14 [1].

Table 9.14 UE Adjacent Channel Leakage Power Ratio (ACLR)

Power class	Adjacent channel relative to UE channel (MHz)	ACLR limit (dB)
3	±5	33
3	±10	43
4	±5	33
4	±10	43

9.3.6.4 *Spurious Emissions*

Spurious emissions or unwanted transmitter effects result from harmonics emission, parasitic emission, Inter-Modulation (IMD) products and frequency conversion products, but not from band emissions. The frequency boundary and the detailed transitions of the limits between the requirement for out band emissions and spectrum emissions are based on ITU-R Recommendations SM.329. These requirements illustrated in Table 9.15 apply only to frequencies which are greater than 12.5 MHz away from the UE carrier frequency centre [1].

Table 9.15 General spurious emissions requirements

Frequency bandwidth	Resolution bandwidth (kHz)	Minimum requirement (dB m)
9 kHz $\leq f >$ 150 kHz	1	−36
150 kHz $\leq f >$ 30 MHz	10	−36
30 MHz $\leq f >$ 1000 MHz	100	−36
1 GHz $\leq f >$ 12.75 GHz	1 MHz	−30

Note: Measurements integer multiples of 200 kHz.

9.3.6.5 Transmit Modulation and Inter-Modulation

The transmit modulation pulse has an RRC shaping filter with roll-off factor $\alpha = 0.22$ in the frequency domain. The impulse response of the chip impulse filter $RC_0(t)$ is

$$RC_0(t) = \frac{\sin\left(\pi \frac{t}{T_C}(1-\alpha)\right) + 4\alpha \frac{t}{T_C} \cos\left(\pi \frac{t}{T_C}(1+\alpha)\right)}{\pi \frac{t}{T_C}\left(1 - \left(4\alpha \frac{t}{T_C}\right)^2\right)} \tag{9.1}$$

where the roll-off factor $\alpha = 0.22$ and the chip duration is $T = 1/\text{chip rate} \approx 0.26042^{\mu}$.

9.3.6.5.1 Vector Magnitude and Peak Code Domain Error

The Error Vector Magnitude (EVM) indicates a measure of the difference between the measured waveform and the theoretical modulated waveform (the error vector). A square root of the mean error vector power to the mean reference signal power ratio expressed in % defines the EVM. One time-slot corresponds to the measurement interval of one power control group. The EVM is less or equal to 17.5% for the UE output power parameter (≥ -20 dB m) operating at normal conditions in steps of 1 dB.

The *code domain error* results from projecting the *error vector power* onto the code domain at the maximum spreading factor. We define the *error vector* for each power code as the ratio to the mean power of the reference waveform expressed in dB, and the *peak code domain error* as the maximum value for the *code domain error*. The measurement interval is one power control group (time-slot). The requirement for the peak code domain error applies only to multi-code transmission, and it shall not exceed -15 dB at a spreading factor of 4 for the UE output power parameter having a value (≥ -20 dB m) and operating at normal conditions [1].

9.3.6.5.2 Inter-Modulation (IMD)

By transmit Inter-Modulation (IMD) performance we meant the measure of transmitter capability to inhibit signal generation in its non-linear elements in the presence of wanted signal and an interfering signal arriving to the transmitter via the antenna. For example, user equipment(s) transmitting in close vicinity of each other can produce IMD products, which can fall into the UE, or BS receive band as an unwanted interfering signal.

We define UE IMD attenuation as the output power ratio of wanted signal to the output power of IMD product when an interfering CW signal adds itself at a level below a wanted signal. Both the wanted signal power and the IM product power measurements use an RRC filter response with roll-off factor $\alpha = 0.22$ and a bandwidth equal to the chip rate. Table 9.16 illustrates IM requirement when transmitting with 5 MHz carrier spacing.

Table 9.16 Transmit Inter-Modulation (IM)

Interference signal frequency off-set (MHz)	5	10
Interference CW signal level (dB c)	−40	
IM product (dB c)	−31	−41

9.4 RECEIVER CHARACTERISTICS

We specify receiver characteristics at the UE antenna connector, and for UE(s) with an integral antenna only, we assume a reference antenna with a gain of 0 dBi. Receiver characteristics for UE(s) with multiple antennas/antenna connectors are FFS.

9.4.1 Diversity

We assume appropriate receiver structure using coherent reception in both channel impulse response estimation and code-tracking procedures. The UTRA/FDD includes three types of diversity:

1. *Time diversity* → channel coding and inter-leaving in both up- and down-link;
2. *Multi-path diversity* → rake receiver or other appropriate receiver structure with maximum combining;
3. *Antenna diversity* → occurs with maximum ratio combining in the BS and optionally in the MS.

9.4.2 Reference and Maximum Sensitivity Levels

Reference sensitivity implies the minimum receiver input power measured at the antenna port at which the Bit Error Ratio (BER) does not exceed a specific value, e.g. BER = 0.001, the DPCH_E_c has a level of −117 dB m/3.48 MHz and the $\hat{I}_{\text{or}}$ a level of −106.7 dB m/3.84 MHz.

For the maximum input level, also with BER not exceeding 0.001, $\hat{I}_{\text{or}} = -25$ dB m/3.84 MHz and DPCH_$E_c/\hat{I}_{\text{or}} = -19$ dB.

In the TDD mode reference sensitivity levels for ΣDPCH_$E_c/\hat{I}_{\text{or}}$ and $\hat{I}_{\text{or}}$ are 0 dB and −105 dB m/3.84 MHz, respectively, while the maximum sensitive level requirements are −7 dB and −25 dB m/3.84 MHz.

9.4.3 Adjacent Channel Selectivity (ACS)

Adjacent Channel Selectivity (ACS) refers to the measure of a receiver's ability to receive a WCDMA signal at its assigned channel frequency in the presence of an adjacent channel signal at a given frequency offset from the centre frequency of the assigned channel. We define the ACS as the ratio of receive filter attenuation on the assigned channel frequency to the receive filter attenuation on the adjacent channel(s) [1].

The ACS shall be better than 33 dB in Power Class 2(TDD), 3 and 4 for the test parameters specified in Table 9.17, where the BER shall not exceed 0.001.

The $(\Sigma(\text{DPCH\_}E_c/\hat{I}_{\text{or}})_{\text{TDD}}$ has 0 dB as test parameter for (ACS).

Table 9.17 Test parameters for Adjacent Channel Selectivity (ACS)

Parameter	Unit	Level
DPCH_E_c	dB m/3.84 MHz	−103
$\hat{I}_{or}$	dB m/3.84 MHz	−92.7
I_{oac}(modulated)	dB m/3.84 MHz	−52
F_{uw}(off-set)	MHz	±5

9.4.4 Blocking

The blocking characteristic indicates the measure of the receiver's ability to receive a wanted signal at its assigned channel frequency in the presence of an unwanted interference on frequencies other than those of the spurious response or the adjacent channels. The unwanted input signal shall not cause a degradation of the performance of the receiver beyond a specified limit, and the blocking performance shall apply at all frequencies except those at which a spurious response occur.

The BER shall not exceed 0.001 for the parameters specified in Tables 9.18 and 9.19. For Table 9.19 up to (24) exceptions are allowed for spurious response frequencies in each assigned frequency channel when measured using a 1 MHz step-size.

The TDD out of band blocking differs from the FDD because they do not have the same frequency range allocation (see Table 9.20).

Table 9.18 In-band blocking FDD and TDD

Parameter	Unit	Offset	Offset
Wanted signal$_{TDD}$	dB m/3.84 MHz	⟨RefSens⟩ + 3 dB	⟨RefSens⟩ + 3 dB
DPCH_E_c	dB m/3.84 MHz	−114	−114
$\hat{I}_{or}$	dB m/3.84 MHz	−103.7	−103.7
$I_{blocking}$(modulated) applies to FDD and TDD	dB m/3.84 MHz	−56	−44
F_{uw}(off-set) FDD and TDD	MHz	±10	±15

Table 9.19 Out of band blocking FDD

Parameter	Unit	Band 1	Band 2	Band 3
DPCH_E_c	dB m/3.84 MHz	−114	−114	−114
$\hat{I}_{or}$	dB m/3.84 MHz	−103.7	−103.7	−103.7
$I_{blocking}$(CW)	dB m	−44	−30	−15
F_{uw}	MHz	$2050 < f < 2095$ $2185 < f < 2230$	$2025 < f < 2050$ $2230 < f < 2255$	$1 < f < 2025$ $2255 < f < 12\,750$
F_{uw}	MHz	$1870 < f < 1915$ $2005 < f < 2050$	$1845 < f < 1870$ $2050 < f < 2075$	$1 < f < 1845$ $2075 < f < 12\,750$

Table 9.20 Out of band blocking TDD

Parameter	Unit	Band 1	Band 2	Band 3
Wanted signal	dB m/3.84 MHz	⟨RefSen⟩ + 3 dB	⟨RefSen⟩ + 3 dB	⟨RefSen⟩ + 3 dB
Unwanted signal level (CW)	dB m	−44	−30	−15
F_{uw}	MHz	$1840 < f < 1885$ $1935 < f < 1995$ $2040 < f < 2085$	$1815 < f < 1840$ $2085 < f < 2110$	$1 < f < 1815$ $2110 < f < 12\,750$
F_{uw}	MHz	$1790 < f < 1835$ $2005 < f < 2050$	$1765 < f < 1790$ $2050 < f < 2075$	$1 < f < 1765$ $2075 < f < 12\,750$
F_{uw}	MHz	$1850 < f < 1895$ $1945 < f < 1990$	$1825 < f < 1850$ $1990 < f < 2015$	$1 < f < 1825$ $2015 < f < 12\,750$

9.4.5 Spurious Response

Through the spurious response, a receiver has the ability to receive a desired signal on its assigned channel frequency, without exceeding a given degradation originating from an undesired CW interfering signal. The latter occurs at any other frequency at which the blocking limit is not met. Table 9.21 illustrates the spurious responses, where the BER does not exceed 0.001.

Table 9.21 Spurious response FDD and TDD

Parameter	Unit	Level
Wanted signal $_{TDD}$	dB m/3.84 MHz	⟨RefSens⟩ + 3 dB
DPCH_E_c $_{FDD}$	dB m/3.84 MHz	−114
$\hat{I}_{or\ (FDD)}$	dB m/3.84 MHz	−103.7
$I_{blocking}$ (CW) $_{(FDD\ and\ TDD)}$	dB m	−44
$F_{uw\ (FDD\ and\ TDD)}$	MHz	Spurious response frequencies

9.4.6 Inter-Modulation (IMD)

Inter-Modulation (IMD) response rejection enables the receiver to receive a wanted signal on its assigned channel frequency in the presence of two or more interfering[2] signals, which have a specific frequency relationship to the wanted signal. Table 9.22 illustrates the IMD characteristics, where BER does not exceed 0.001.

In the notation of tables, the TDD subscript implies that it applies to the TDD mode. If there is not a TDD subscript or an FDD subscript exist, it applies to the FDD mode.

[2]Two interfering RF signals of 3rd and higher order mixing can produce interfering signal in the desired channel band.

Table 9.22 Receive IMD characteristics FDD and TDD

Parameter	Unit	Level
DPCH_E_c	dB m/3.84 MHz	−114
$\hat{I}_{or}$	dB m/3.84 MHz	−103.7
$\hat{I}_{or\ (TDD)}$	dB m/3.84 MHz	⟨RefSens⟩ + 3 dB
(ΣDPCH_$E_c/\hat{I}_{or}$) $_{(TDD)}$	dB	0
I_{ouw1}(CW)	dB m	−46
I_{ouw2}(modulated)	dB m/3.84 MHz	−46
F_{uw1} (off-set)	MHz	10
F_{uw2} (off-set)	MHz	20

9.4.7 Spurious Emissions Power

We refer to the power of emissions generated or amplified in a receiver and appearing at the UE antenna connector as *spurious emissions power.* The spurious emission shall be (see [1] and Table 9.23):

- less than −60 dB m/3.84 MHz at the UE antenna connector, for frequencies within the UE receive band. In URA_PCH-, Cell_PCH- and IDLE-stage, the requirement applies also for the UE transmit band;
- less than −57 dB m/100 kHz at the UE antenna connector, for the frequency band from 9 kHz to 1 GHz;
- less than −47 dB m/100 kHz at the UE antenna connector, for the frequency band from 1 to 12.75 GHz.

Specifications in [1,2] describe the performance for the transmitter and receiver characteristics.

Table 9.23 TDD receiver spurious emission requirements [2]

Band	Maximum level (dB m)	Measurement bandwidth
9 kHz–1 GHz	−57	100 kHz
1–1.9 GHz, 1.92–2.01 GHz and 2.025–2.11 GHz	−47	1 MHz
1.9–1.92 GHz, 2.01–2.025 GHz and 2.11–2.170 GHz	−60	3.84 MHz
2.170–12.75 GHz	−47	1 MHz

The UE uses the last carrier frequency, except for frequencies between 12.5 MHz (below the first carrier frequency) and 12.5 MHz (above the last carrier frequency).

9.5 UTRA RF PERFORMANCE EXAMPLES

In the sequel we provide RF system scenarios based on the studies reported in [5]. Here we aim primarily to illustrate the principles outlined in the preceding sections in order to present practical applications of the recommended parameters. The examples may not strictly apply to actual designs; however, these could serve as reference for initial analysis.

9.5.1 Co-existence FDD/FDD: ACIR

Before we describe a methodology, we first define some of the essential terminology as in [5] for the context of the examples to follow:

Outage – in this context an outage occurs when, due to a limitation on the maximum TX power, the measured E_b/N_o of a connection is lower than the E_b/N_o target.
Satisfied User – a user is satisfied when the measured E_b/N_o of a connection at the end of a snapshot, is higher than a value equal to E_b/N_o target -0.5 dB.
ACIR – the Adjacent Channel Interference Power Ratio (ACIR) is defined as the ratio of the total power transmitted from a source (BS or UE) to the total interference power affecting a victim receiver, resulting from both transmitter and receiver imperfections.

9.5.1.1 Overview of Simulation Assumptions

Simulations use snapshots where we place subscribers randomly in a pre-defined deployment scenario; each snapshot simulates a power control loop until it reaches a target E_b/N_o; a simulation is made of several snapshots. We obtain the measured E_b/N_o by the measured C/I multiplied by the processing gain.

UEs do not reach the target E_b/N_o *at the end* of a PC loop in the outage state. We consider satisfied users those able to reach at least (E_b/N_o -0.5 dB) at the end of a Power Control (PC) loop. Statistical data related to outage (satisfied users) are collected at the end of each snapshot.

We model soft handover allowing a maximum of 2 BTS in the active set, where we set the window size of the candidate to 3 dB, and the cells in the active set are chosen randomly from the candidate set. We use selection combining in the uplink and maximum ratio combining in DL, and simulate uplink and downlink independently.

9.5.1.2 Simulated Scenarios

We have already outlined the background of the simulated scenarios in Chapter 2. Nonetheless, here we briefly describe them again to introduce the proper context of the different environments considered, e.g. macro-cellular and micro-cellular environments with their respective cases, i.e. macro-to-macro multi-operator case and macro-to-micro case.

9.5.1.3 Macro-to-Macro Multi-operator Case

In a *single operator* layout, we place BS on a hexagonal grid with distance of 1000 m; the cell radius is then equal to 577 m (see, e.g. Figure 9.6). We assume BSs with omni-directional antennas in the middle of the cell. In practice, we use either 3 or 6 sector antennas. We also assume 19 cells (or higher) for each operator in the macro-cellular environment. This number appears suitable when using the wrap around technique.

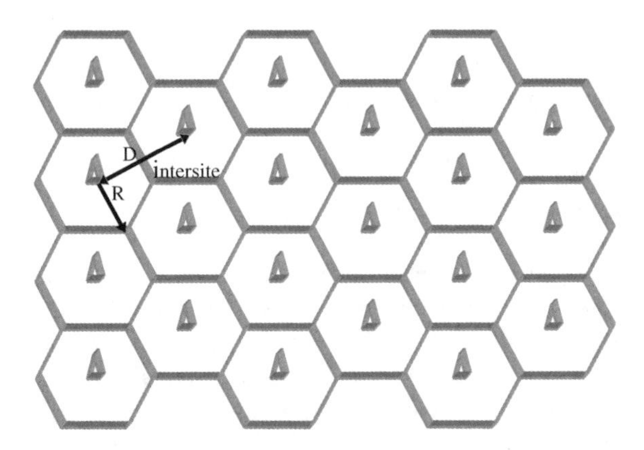

Figure 9.6 Macro-cellular deployment.

In the *multi-operator* case, we consider two shifting BSs shifting two operators, e.g. (worst case scenario) 577 m BS shift, and (intermediate case) 577/2 m BS shift. We do not consider the best case scenario (i.e. 0 m shifting = co-located sites).

9.5.1.4 Macro-to-Micro Multi-operator Case

9.5.1.4.1 Single Operator Layout, Micro-Cell Layer

For the micro-cell deployment in a Manhattan deployment scenario, we place the BSs so that they stand at street crossings in every second junction as illustrated in Figure 9.7 [6]. Although the model does not reflect efficient planning, it does provide sufficient amount of inter-cell interference generation with reasonably low number of micro-cell BSs. The parameters of the micro-cells are thus: block size = 75 m, road width = 15 m, inter-site distance between line of sight = 180 m, and the number of micro-cells in the micro-cellular scenario is 72.

9.5.1.4.2 Multi-operator Layout

In this micro-cell layout, we use the parameters proposed earlier (i.e. 72 BSs in every second street junction, block size 75 m, road width 15 m). We also apply a macro-cell radius of 577 m with a distance of 1000 between BSs.

Figure 9.8 illustrates the cellular layout to simulate Hierarchical Cell Structures (HCS). This layout allows large enough macro-cells and a low number of micro-cells so that computing simulation times remain reasonable. Furthermore, we select macro-cell BS positions to observe handovers and many other conditions (e.g. border conditions).

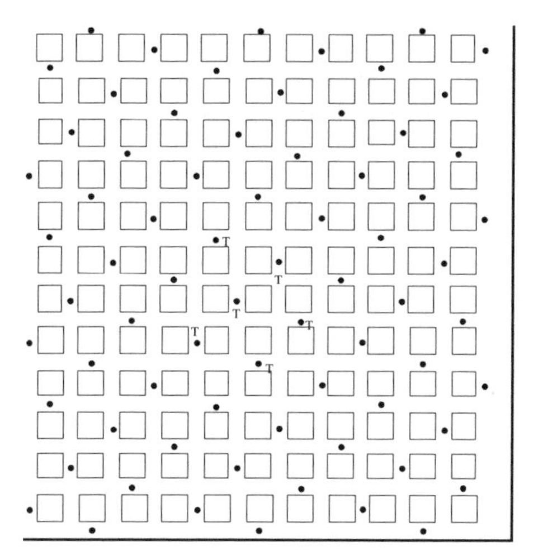

Figure 9.7 Micro-cell deployment.

When measuring interference at macro-cell BSs in UL, we measure same channel interference *only* from those users connected to the observed BS. Then we multiply the measured same channel interference by $1/F$, where F is the ratio of intra-cell interference to total interference, i.e.

$$F = I_{\text{intra}}(i)/(I_{\text{intra}}(i) + I_{\text{inter}}(i)) \tag{9.2}$$

F depends on the assumed propagation model; earlier studies suggest a typical value of around 0.6. However, an appropriate value for F can also be derived from specific macro-cell only simulations. We measure interference from micro-cells to macro-cells by using the wrap-around technique. We can then define the interference that a macro-cell BS receives as

$$I = \text{ACIR} \times I_{\text{micro}} + (1/F) \times I_{\text{macro}} \tag{9.3}$$

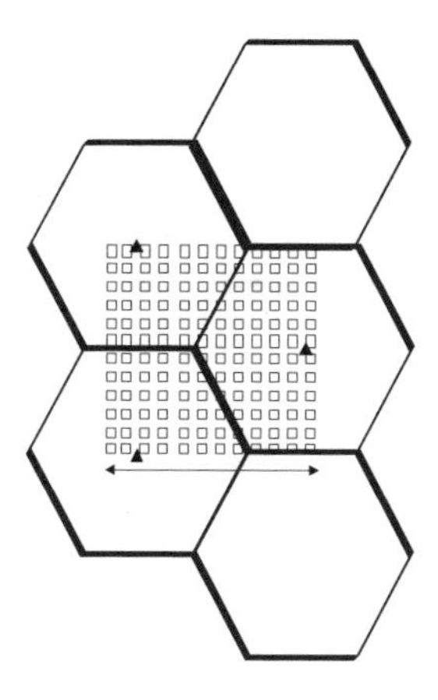

Figure 9.8 Macro-to-micro deployment.

where ACIR is the adjacent channel interference rejection ratio, and I_{macro} is same channel interference measured from users connected to the BS.

When we measure DL interference, same channel and adjacent channel interference gets measured from all BSs. To measure interference from micro-cells we use the wrap-around technique. When measuring interference at micro-cells in UL and DL, same channel and adjacent channel interference gets measured from all BS, and when measuring same channel interference we apply the wrap-around technique.

When measuring simulation results we consider all micro-cell users and those macro-cell users that are in the area covered by micro-cells. We also need to plot figures depicting the position of bad quality calls, in order to see how these are distributed in the network. In addition, noise rise should be measured at every BS and from that data, a probability density function should be generated [5].

9.5.1.5 Simulated Services

The following services were considered:

- Speech 8 kbps
- Data 144 kbps

Speech and data services were simulated in separate simulations, i.e. no traffic mix was simulated.

9.5.2 Description of the Propagation Models

Two propagation environments were considered in the ACIR analysis, i.e. macro-cellular and micro-cellular environments. For each environment, a propagation model was used to evaluate the propagation path loss due to the distance. As noted earlier, these propagation models are described in Chapter 2 and are also briefly presented in the forthcoming sections.

9.5.2.1 Received Signal

Before describing the propagation environments, a key parameter to be defined is the Minimum Coupling Loss (MCL), i.e. the minimum distance loss including antenna gain measured between antenna connectors. This represents the minimum loss in signal due to fact that the BSs are always placed much higher than the UE(s). The following values are assumed in our example for MCL:

- 70 dB for the macro-cellular environment
- 53 dB for the micro-cellular environment

With the above definition, the received power in DL or UL can be expressed for the macro-environment as:

$$\text{RX\_PWR} = \text{TX\_PWR} - \text{Max}(\text{pathloss\_macro} - \text{G\_Tx} - \text{G\_RX}, \text{MCL} \tag{9.4}$$

and for the micro-environment as

$$\text{RX\_PWR} = \text{TX\_PWR} - \text{Max(pathloss\_m1cro} - \text{G\_Tx} - \text{G\_RX, MCL} \qquad (9.5)$$

where

- RX_PWR: the received signal power
- TX_PWR: the transmitted signal power
- G_Tx: the Tx antenna gain
- G_RX: the Rx antenna gain

For the set of simulations in this section, we have assumed 11 dB antenna gain (including cable losses) in BS and 0 dB in UE [5]. In Chapter 7 we use other assumptions.

9.5.2.2 Macro-Cell Propagation Model

The macro-cell propagation model serves here to test scenarios in urban and sub-urban areas outside the high rise core where the buildings are of nearly uniform height as discussed in Chapter 2 and [7].

$$L = 40(1 - 4 \times 10 - 3\text{Dhb}) \ \log_{10}(\text{R}) - 18 \log_{10}(\text{Dhb}) + 21 \log_{10}(f) + 80 \text{ dB} \qquad (9.6)$$

where R is the BS–UE separation in kilometres, f is the carrier frequency of 2000 MHz and Dhb is the BS antenna height, in metres, measured from the average rooftop level.

When the BS antenna height gets fixed at 15 m above the average rooftop (i.e. Dhb = 15 m) and when considering a carrier frequency of 2000 MHz, the macro-cell propagation model formula becomes

$$L = 128.1 + 37.6 \log_{10}(R) \qquad (9.7)$$

Once we calculate L we add a log-normally distributed shadowing ($\log F$) with Standard Deviation (SD) of 10 dB to obtain the path loss as follows:

$$\text{Pathloss\_macro} = L + \log F \qquad (9.8)$$

To complete the definition of the path loss for our analysis we should note from [5] that:

- L shall in no circumstances be less than free-space loss, and this model applies only to the Non-Line-of-Sight (NLOS) case and describes worse case propagation for the examples in this section;
- the path-loss model is valid for a range of Dhb from 0 to 50 m;
- this model concerns designs mainly for distance from a few hundred metres to kilometres. Thus, it may not accurately apply to short distances.

9.5.2.3 *Micro-Cell Propagation Model*

We use the micro-cell propagation model (also covered in Chapter 2) for spectrum efficiency evaluations in urban environments modelled through a Manhattan-like structure. It allows us to appropriately evaluate the performance in micro-cell situations that will be typical, for example, in European like cities at the time of UMTS deployment.

In this case, this recursive model calculates the path-loss as a sum of LOS and NLOS segments. We find the shortest path along streets between the BS and the UE within the Manhattan environment. The path loss in dB is thus given by the well-known formula

$$L = 20 \log_{10} \frac{4\pi d_n}{\lambda} \tag{9.9}$$

where d_n is the 'illusory' distance, λ is the wavelength and n is the number of straight street segments between BS and UE (along the shortest path).

The illusory distance is the sum of these street segments and can be obtained by recursively using the expressions: $k_n = k_{n-1} + d_{n-1}c$ and $d_n = k_n s_{n-1} + d_{n-1}$, where c is a function of the angle of the street crossing. For a 90° street crossing, the value c should be set to 0.5. Further, s_{n-1} is the length in metres of the last segment. A segment is a straight path. The initial values are set according to: k_0 is set to 1 and d_0 is set to 0. The illusory distance is obtained as the final d_n when the last segment has been added.

The model is extended to cover the micro-cell dual-slope behaviour, by modifying the expression to

$$L = 20 \log_{10} \left(\frac{4\pi d_n}{\lambda} D \left(\sum_{j=1}^{n} s_{j-1} \right) \right)$$

where

$$D(x) = \begin{cases} x/x, & x > x_{\text{br}} \\ 1, & x \leq x_{\text{br}} \end{cases} \tag{9.10}$$

Before the break point x_{br} the slope is 2, after the break point it increases to 4. The break point x_{br} is set to 300 m. x is the distance from the transmitter to the receiver.

To take into account effects of propagation going above roof-tops, the path loss according to the shortest geographical distance must also be calculated. This is done by using the commonly known COST Walfish–Ikegami Model and with antennas below roof-tops

$$L = 24 + 45 \log(d + 20) \tag{9.11}$$

where d is the shortest physical geographical distance from the transmitter to the receiver (m).

The final path-loss corresponds to the minimum value between the path-loss value from the propagation through the streets and the path-loss based on the shortest geographical distance, plus the log-normally distributed shadowing ($\log F$) with SD of 10 dB.

$$\text{Pathloss\_micro} = \min\ (\text{Manhattan path loss, macro-path loss}) + \log F \tag{9.12}$$

The above path-loss model applies only to the micro-cell coverage with antenna located below roof-top. For the urban structure covered by the macro-cells, the path-loss defined in the preceding section *applies*.

9.5.3 The Simulation Process

Only one link gets considered in a single simulation, i.e. we simulate UL and DL independently. A simulation, aiming to cover most of the possible UEs of single placement in the network, consists of several simulation steps (snapshot).

9.5.3.1 Single Step (Snapshot) Simulation

In general, a simulation step (snapshot) constitutes placement of MSs, path-loss calculations, handover, power control and collection of statistics.

In particular:

- each simulation step begins with the uniform random distribution of UE(s) across the network;
- for each UE (e.g. in the case of macro-to-macro simulation), we randomly select an operator, so that the number of users per BS is the same for both operators;
- after the placement of UEs, we calculate the path-loss between each UE and BS, then add the log-normal fading and stored to a so-called gain matrix (G-matrix).

We keep constant the distance attenuation and log-normal fading during the execution of a snapshot

- using the G-matrix and based on the HO algorithm, we select the transmitting BSs for each UE;
- then a power control loop[3] stabilisation period gets started. During this stabilisation span, we execute power control as long as the used powers reach the level required to meet the expected quality;
- the acceptable number of power control commands in each power control loop, can exceed 150;
- we collect statistical data at the end of a power control loop. UE(s) with quality below the target remain in outage state; UE(s) with quality higher than the target –0.5 dB are in satisfied state.

[3]During the power control loop, the gain matrix (G-matrix) remains constant.

9.5.3.2 *Multiple Steps (Snapshots) Execution*

Multiple steps occur when a single step (snapshot) finishes, UE(s) are re-located to the system and the above processes get executed again. During a simulation, we execute as many simulation steps (snapshots) as required in order to achieve the ideal amount of local-mean-SIR values. The ideal number of snapshots for 8 kbps speech service amounts to 10 000 values or more. For data service, we require a higher number of snapshot, e.g. 10 times the value used for 8 kbps speech.

During one simulation step (i.e. snapshot), we obtain as many local-mean-SIR values as UE(s) in the simulation. The outputs from a simulation include SIR-distribution, outage probability, capacity figures, etc.

9.5.4 Modelling of Handover and Power Control

9.5.4.1 *Handover Modelling*

Here we model non-ideal soft handover, where an active set for the UE gets selected from a pool of candidate BSs for handover. The candidate set consists of BSs whose path-loss is within handover margin, i.e. BSs whose received pilot is stronger than the received pilot of the strongest BS subtracted by the handover margin. We select the active set of BS randomly from the candidate BSs, where a single UE may be connected to maximum of two BSs simultaneously. We assume 3 dB as a soft handover margin.

9.5.4.1.1 UpLink (UL) and DownLink (DL) Combining

In UL, selection combining among active BSs takes place to use the frame with the highest average SIR for statistics collecting purposes, while the other frames get discarded.

In DL, we model macro-diversity to sum together the signal received from active BSs. Thus, we realise maximal ratio combining by summing measured SIR value, i.e.

$$\mathrm{SIR} = \frac{C_1}{I_1 + N} + \frac{C_2}{I_2 + N} \tag{9.13}$$

9.5.4.2 *Power Control (PC) Modelling of UpLink (UL) Traffic Channels*

In these simulations, Power Control (PC) corresponds to the SIR-based fast inner-loop power control. Here we assume perfect PC, i.e. during the PC loop, each UE achieves perfectly the E_b/N_o target, assuming that the maximum TX power is not exceeded. Assuming perfect PC, we imply that PC error equals 0%, and PC delay equals 0 s.

As noted earlier, UE(s), which cannot achieve the target E_b/N_o at the end of a PC loop, are in outage. We base the initial TX power for the PC loop of the UL traffic channel on path-loss, thermal noise and 6 dB noise rise. However, the initial TX power should not affect the convergence process (PC loop) to the target E_b/N_o [5].

9.5.4.2.1 Simulation Parameters

UE Max TX power. The maximum UE TX power is 21 dB m (both for speech and data), and the UE power control range is 65 dB m; the minimum TX power is thus −44 dB m. For UL E_b/N_o target, based on [7] we assume

- macro-cellular environment: speech 6.1 dB, data 3.1 dB
- micro-cellular environment: speech 3.3 dB, data 2.4 dB

9.5.4.2.2 SIR Calculation in UL

We calculate the local mean SIR by dividing the received signal by the interference, and multiplying by the processing gain. Signals from the other users are summed together and seen as interference. SIR for our analysis is thus

$$\mathrm{SIR}_{\mathrm{UL}} = \frac{G_{\mathrm{P}} S}{(1-\beta) I_{\mathrm{own}} + I_{\mathrm{other}} + N_{\mathrm{o}}} \tag{9.14}$$

where S is the received signal, G_p is processing gain, I_{own} is interference generated by those users who are connected to the same BS as the observed use, I_{other} is interference from other cells, N_{o} is thermal noise and β is an interference reduction factor due to the use of, e.g. Multi-User Detection (MUD) in UL. However, MUD is NOT included in these simulations, therefore $\beta = 0$.

We calculate thermal noise for the 4.096 MHz band by assuming 5-dB system noise. Thermal noise power is thus equal to −103 dB m. In the multi-operator case, I_{other} also includes the interference coming from the adjacent operator, which is decreased by ACIR (dB).

9.5.4.3 *PC Modelling of Traffic Channels in DL*

As in the UL case, DL power control corresponds to the SIR-based fast inner-loop PC. Here too we assume perfect PC, i.e. during the PC loop, each DL traffic channel achieves perfectly the E_b/N_o target, assuming that the maximum TX power is not exceeded. Assuming perfect PC, we imply that PC error equals 0%, and PC delay equals 0 s. The UE(s) whose DL traffic channel is not able to achieve the E_b/N_o target at the end of a PC loop are considered also in outage as in the UL.

We choose randomly the initial TX power for the PC loop of the DL traffic channel in the TX power range; however, the initial TX power should not affect the convergence process (PC loop) to the target E_b/N_o.

9.5.4.3.1 Reference Simulation Parameters

Traffic channel TX power. We assume 25 dB m for DL traffic channel PC-range, and for the maximum power for each DL traffic channel (i.e. both speech and data) we assume 30 and 20 dB m in the macro-cellular and micro-cellular environment, respectively.

The DL E_b/N_o target following directives in [7] assumes 7.9 and 2.5 dB for speech and data, respectively, with DL TX or RX diversity[4] in the macro-cellular environment. In the micro-cellular environment, the DL E_b/N_o target assumes 6.1 dB and 1.9 dB for speech and data, respectively, with DL TX or RX diversity.

9.5.4.3.2 DL SIR Calculation

The DL Signal-to-Interference Ratio (SIR) can be expressed as

$$\mathrm{SIR}_{\mathrm{DL}} = \frac{G_{\mathrm{P}}S}{\alpha I_{\mathrm{own}} + I_{\mathrm{other}} + N_{\mathrm{o}}} \tag{9.15}$$

where S is the received signal, G_p is processing gain, I_{own} is the interference generated by those users linked to the same BS that the observed user (it includes also interference caused by perch channel and common channels), I_{other}[5] is interference from other cells, α is the orthogonality factor and N_o is thermal noise, which is calculated for the 4.096 MHz band by assuming 9 dB system noise figure. Thermal noise power is then equal to −99 dB m. Transmission powers for them are in total 30 and 20 dB m for macro-cells and micro-cells, respectively.

As mentioned in earlier chapters, the orthogonality factor takes into account the fact that the DL is not perfectly orthogonal due to multi-path propagation. An orthogonality factor of 0 implies perfectly orthogonal intra-cell users, while the value of 1 implies that the intra-cell interference has the same effect as inter-cell interference. Here the orthogonality factor α assumes 0.4 and 0.06 for macro-cells and micro-cells, respectively.

9.5.4.3.3 DL Maximum TX Power

For the maximum BS TX power, i.e. when the sum of all DL traffic channels in a cell exceeds the maximum BS TX power, here we assume 43 and 33 dB m in the macro-cell and micro-cell environments, respectively. Thus, during simulations if in the PC loop of each snapshot the overall TX power of each BS gets higher than the maximum power allowed, we record the event and validate it to guide a future DL approach. The scheme used to maintain the output level of the BS equal or below the maximum BS TX power, is similar to an analog mechanism to protect the power amplifier. At each iteration, the MSs request more or less power, depending on their C/I values. A given BS will be requested to transmit the common channels and the sum of the TCHs for all the MSs it is in communication with. If this total output power exceeds the maximum allowed for the PA, an attenuation gets applied in order to set the output power of the BS equal to its maximum level. As an RF variable attenuator would operate, this attenuation gets applied on the output signal with the exception of common channels, i.e. all the TCHs are reduced by this amount of attenuation. The power of the TCH for a given mobile will be [5]

$$\mathrm{TCH}(n+1) = \mathrm{TCH}(n) \pm \mathrm{Step} - \mathrm{RF\_Attenuation} \tag{9.16}$$

[4] 4.5 dB without diversity.

[5] In the multi-operator case, I_{other} also includes adjacent operator interference, which is decreased by ACIR (dB).

9.5.5 System Loading

9.5.5.1 Uplink

The steps for *single operator loading* in these simulation examples can be outlined from [5] as follows: we define the number of users in the UL of the single operator case as N_UL_single. The latter gets evaluated according to a 6 dB noise rise over the thermal noise in the UL (6 dB noise rise is equivalent to 75% of the pole capacity of a CDMA system).

We measure a simulation run with a pre-defined number of users at the end of the average noise rise (over the thermal noise). If lower than 6 dB, we increase the number of users until we reach the 6 dB noise. Thus, we define here the number of users corresponding to a 6 dB noise rise as N_UL_single.

9.5.5.1.1 Multi-operator Scenario with Macro-to-Macro Cellular Environment

We define the number of users in the UL of the multi-operator case as N_UL_multi. It gets evaluated, as in the single case, according to a 6 dB noise rise over the thermal noise in the UL. We run a simulation with a pre-defined number of users, and measure it at the end of the average noise rise (over the thermal noise). If lower than 6 dB, we increase the number of users until we reach the 6 dB noise rise. Thus, here we also define the number of users corresponding to a 6 dB noise rise as N_UL_multi.

Then, for a given value of ACIR, the obtained N_UL_multi gets compared to N_UL_single to evaluate the capacity loss due to the presence of a second operator [5].

9.5.5.1.2 Multi-operator Scenario with Macro-to-Micro Cellular Environment

In general, the noise rise does not change by the same amount for micro- and macro-cell layers if the number of users do change in the system. Thus, [5] proposes that loading in this case takes the following steps with two different numbers of input of users included in the simulator, i.e. N_users_UL_macro and N_users_UL_micro. Then the steps are:

1. selection of an ACIR value;
2. begin a simulation (made of several snapshots) with an arbitrary number of N_users_UL_micro and N_users_UL_macro;
3. system loading measurement;
4. run a 2nd simulation (made of several snapshots) by increasing the number of users (i.e. N_users_UL_macro or micro) in the cell layer having lower noise rise than the layer-specific threshold, and decreasing number of users ((i.e. N_users_UL_micro or macro) in the cell layer in which noise rise is higher than the layer-specific threshold, etc.;
5. redo phases 1 and 2 until noise rise is equal to the specific threshold for both layers;
6. when each layer reaches on average the noise rise threshold, the input values of N_UL_users_UL_macro and micro are taken as an output and compared to the values obtained in the single operator case for the ACIR value chosen at step 1.

We investigate two options (e.g. Option A and Option B) in relation to the noise rise threshold. In the first option, the noise rise threshold for the macro-layer is equal to 6 dB whilst the threshold for the micro-layer is set to 20 dB. The noise rise results from the combination of interference coming from the micro- and the macro-cell layers. Micro- and macro-cell layers interact, i.e. micro-cell interference affects the macro-cell layer and vice versa. In the second option, we set the noise rise threshold to 6 dB for both the macro- and the micro-layer, but the micro-cells are de-sensitised at 14 dB.

9.5.5.2 Downlink

9.5.5.2.1 Single-Operator Loading

As in the UL, the number of users in the DL for the single-operator case gets defined as N_DL_single. Then DL simulations occur in a way that a single-operator network gets loaded so that 95% of the users achieve an E_b/N_o of at least (target E_b/N_o −0.5 dB) (i.e. 95% of users are satisfied) and the supported number of users N_DL_single is then measured [5].

9.5.5.2.2 Multi-operator for Macro-to-Macro and Macro-to-Micro

In the *macro-to-macro* multi-operator case, the networks get loaded so that 95% of users are satisfied and the obtained number of users is defined as N_DL_multi. For a given value of ACIR, the measured N_DL_multi is obtained and compared to the N_DL_single obtained in the single-operator case. The multi-operator case (macro-to-micro) follows similar reasoning to that of the UL case.

9.5.5.3 Simulation Output

Finally, the expected outputs include: capacity figures (N_UL and N_DL), DL and UL capacity versus ACIR in the multi-operator case, as well as outage (non-satisfied users) distributions.

9.5.6 BTS Receiver Blocking and Simulation Assumptions

The simulations are static Monte Carlo simulations, using a methodology consistent with the ACIR approach described in the preceding sections. For our examples, we construct the simulations using two un-coordinated networks at different frequencies. The frequencies assume a separation by 10–15 MHz or more so that the BS receiver selectivity will not limit the simulation, and so that the UE spurious and noise performance will dominate over its adjacent channel performance. These are the factors that distinguish a blocking situation from an adjacent channel situation in which significant BS receiver degradation can be caused at very low levels due to the poor ACP from the UE [5].

During each trial of the simulations, we make uniform drops of the UE, adapt power levels and record data.[6] From these results, we plot CDF of the total signal appearing at the receivers' inputs to be covered in the results sections.

[6]A thousand such trials are made.

9.5.6.1 Simulation Scenario Assumptions for 1 and 5 km Cell Radius

The assumptions here are extracts from [5] to be consistent with the example results. The primary simulation assumptions for the 1 km radius are then:

1. Both networks are operated with the average number of users (50) that provide a 6 dB noise rise
2. The two networks have maximal geographic offset (a worst case condition)
3. Cell radius is 1 km
4. Maximum UE power is 21 dB m
5. UE spurious and noise in a 4.1 MHz bandwidth is 46 dB
6. BS selectivity is 100 dB (to remove its effect)
7. C/I requirement is −21 dB
8. BS antenna gain is 11 dB
9. UE antenna gain is 0 dB and
10. Minimum path loss is 70 dB excluding antenna gains.

The primary assumptions that are common to all simulations in the 5 km radius are:

1. The two networks have maximal geographic offset (a worst case condition)
2. Cell radius is 5 km
3. UE spurious and noise in a channel bandwidth is 46 dB
4. BS selectivity is 100 dB (to remove its effect)
5. BS antenna gain is 11 dB
6. UE antenna gain is 0 dB
7. Minimum path loss is 70 dB including antenna gains; in addition
8. For the speech simulations, maximum UE power is 21 dB m and the C/I requirement is −21 dB and
9. For the data simulations, maximum UE power is 33 dB m and the C/I requirement is −11.4 dB.

Note that this is different from the basic assumption in the ACIR section, since its data power level is 21 dB m, just like the speech level.

9.5.6.2 Simulation Parameters for 24 dB m Terminals

9.5.6.3 Uplink

The only difference with respect to the parameters listed in the previous sections are:

- 3.84 Mcps chip rate considered;

- 24 dB m Max TX power for the UE (results provided for 21 dB m terminals as well);
- 68 dB dynamic range for the power control;
- number of snapshots per simulation (3000).

Therefore, the considered parameters are shown in Table 9.24.

Table 9.24 Simulation parameters 24 dB m terminals

MCL	70 dB
BS antenna gain	11 dBi
MS antenna gain	0 dBi
Log normal shadowing	SD of 10 dB
No. of snapshots	3000
Handover threshold	3 dB
Noise figure of BS receiver	5 dB
Thermal noise (NF included)	−103.16 dB m@3.84 MHz
Max TX power of MS	21 dB m/24 dB m
Power control dynamic range	65 dB/68 dB
Cell radius	577 m (for both systems)
Inter-site distance	1000 m (for both systems)
BS off-set between two systems (x, y)	Intermediate: (0.25 km, 0.14425 km) → 0.289 km shift, Worst: (0.5 km, 0.2885 km) → 0.577 km shift
User bit rate	8 and 144 kbps
Activity	100%
Target E_b/I_o	6.1 dB (8 kbps), 3.1 dB?(144 kbps)
ACIR	25–40 dB

9.5.6.4 *Summary of Simulation Parameters*

For completeness in Table 9.25, we list the same assumptions simulation parameters as in [5] to be consistent with the example results in the forthcoming sections.

9.5.7 Example Results FDD/FDD

Here we illustrate example results primarily for the FDD to FDD mode. In [5] we can see additional cases. The goal of the preceding section and this section is simply to practically visualise some of the procedures covered in the first part of the chapter. Thus, this section aims to collect results on carrier spacing evaluations to illustrate deployment co-ordination, and multi-layer deployment considerations.

Table 9.25 Simulation parameters

Parameter	UL value	DL value
Simulation type	Snapshot	Snapshot
Propagation parameters		
MCL macro (including antenna gain)	70 dB	70 dB
MCL micro (including antenna gain)	53 dB	53 dB
Antenna gain (including losses)	11 dBi	0 dBi
	0 dBi	11 dBi
Log-normal fade margin	10 dB	10 dB
PC modelling		
No. of snapshots	>10 000 for speech >10 × no. of snapshots for speech for 144 kbps service	>10 000 for speech >10 × no. of snapshots for speech in the 144 kbps case >20 000 for data
No. of PC steps per snapshot	>150	>150
Step size PC	Perfect PC	Perfect PC
PC error	0%	0%
Margin with respect to target C/I	0 dB	0 dB
Initial TX power	Path loss and noise, 6 dB noise rise	Random initial
Outage condition	E_b/N_o target not reached due to lack of TX power	E_b/N_o target not reached due to lack of TX power
Satisfied user		Measured E_b/N_o higher than E_b/N_o target −0.5 dB
Handover modelling		
Handover threshold for candidate set	3 dB	
Active set	2	
Choice of cells in the active step	Random	
Combining	Selection	Maximum ratio combining
Noise parameters	Noise figure	5 dB
9 dB	Receiving bandwidth	4.096 MHz proposed
4.096 MHz proposed	Noise power	−103 dB m proposed
−99 dB m proposed	TX power	
	Maximum BTS power	
43 dB m macro		
33 dB m micro	Common channel power	
30 dB m macro		
20 dB m micro	Maximum TX power speech	21 dB m
30 dB m macro		
20 dB m micro	Maximum TX power data	21 dB m
30dB m macro		
20dB m micro	Power control range	65 dB
25 dB		
Handling of DL maximum TX power		
Admission control	Not included	Not included
User distribution		Random and uniform across the network

Table 9.25 (*Continued*)

Parameter	UL value	DL value
Interference reduction		
MUD	Off	N/A
Non-orthogonality factor macro-cell	N/A	0.4
Non-orthogonality micro cell	N/A	0.06
Common channel orthogonality		Orthogonal
Deployment scenario		
Macro-cell		Hexagonal with BTS in the middle of the cell
Micro-cell		Manhattan (from 30.03)
BTS type		Omni-directional
Cell radius macro		577 macro
Inter-site single operator		1000 macro
Cell radius micro		Block size = 75 m, road 15 m
Inter-site single micro		Inter-site between LoS = 180 m
Inter-site shifting macro		577 and 577/2 m
No. of macro-cells		>19 with wrap-around technique)
Inter-site shifting macro–micro		See scenario
Number of cells per each operator		See scenario
Wrap-around technique		Should be used
Simulated services		
Bit-rate speech	8 kbps	8 kbps
Activity factor speech	100%	100%
Multi-path environment macro	Vehicular macro	Vehicular macro
E_b/N_o target	6.1 dB	7.9 dB
Multi-path environment macro	Out-door micro	Out-door micro
E_b/N_o target	3.3 dB	6.1 dB
Data rate	144 kbps	144 kbps
Activity factor speech	100%	100%
Multi-path environment macro	Vehicular macro	Vehicular macro
E_b/N_o target	3.1 dB	2.5 dB with DL TX or RX diversity, 4.5 dB no diversity
Multi-path environment macro	Out-door micro	Out-door micro
E_b/N_o target	2.4 dB	1.9 dB with DL TX or RX

9.5.7.1 ACIR for 21 dB m Terminals

Figure 9.9 illustrates the UL speech ACIR for intermediate and worst case scenarios for the Macro-to-Macro cellular environment, while Figure 9.10 shows the DL case.

The examples include UL and DL 8 kbps speech service with the following characteristics:

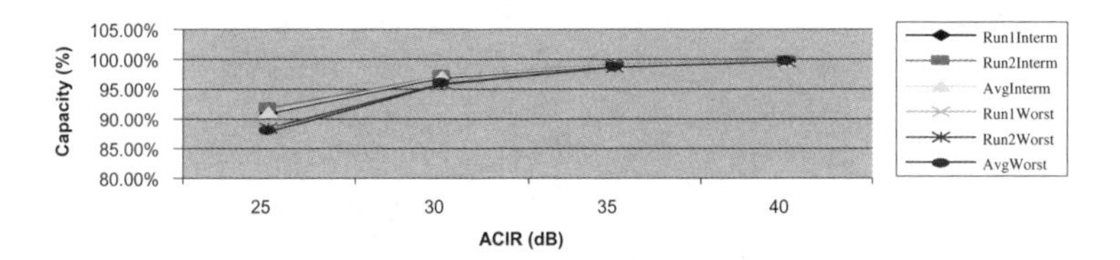

Figure 9.9 ACIR intermediate and worst case for macro-to-macro UL speech 8 kbps.

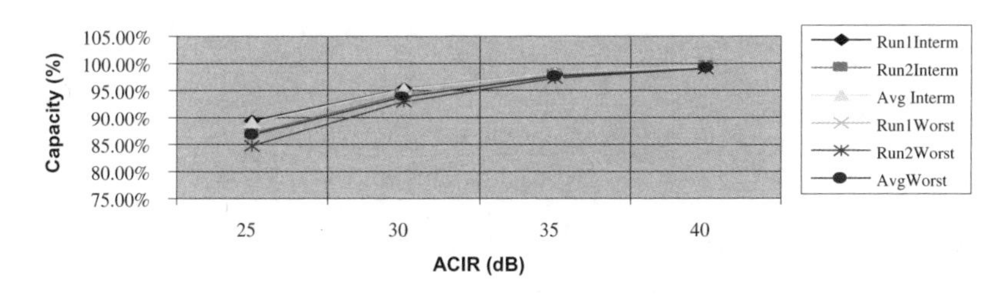

Figure 9.10 ACIR intermediate and worst cases for macro-to-macro DL speech 8 kbps.

- intermediate case scenario where the 2nd system is located at a half-cell radius shift;
- worst case scenario where the second system BSs are located at the cell border of the first system;
- average results for intermediate and worst case.

We can see clearly in Figure 9.9 that as the ACIR increases the capacity also increases. The impact applies to the intermediate as well as to the worst case situations. Likewise for the DL case, as the ACIR value increase the capacity keeps high. The worst case run still has impact at 32.5 dB but remains at minimum at 35 dB.

9.5.7.2 ACIR for 24 dB m Terminals

Other simulation results also following [5] include outputs for UL ACIR with 24 dB m terminals, for both speech (8 kbps) and data (144 kbps). We compare the results with those obtained with 21 dB m terminals. Figure 9.11 illustrates the UL ACIR in 24 dB m terminals.

When comparing the results illustrated in Figures 9.11 and 9.12, we can see that with lower ACIR values in speech, the capacity degrades much more than with the same ACIR values in data.

9.5.8 BTS Receiver Blocking

9.5.8.1 Simulation Results for 1 km Cell Radius

Figure 9.13 shows the overall Cumulative Distribution Function (CDF) of the input signals to the receivers using 21 dB m terminals. Based on the preceding simulation assumptions

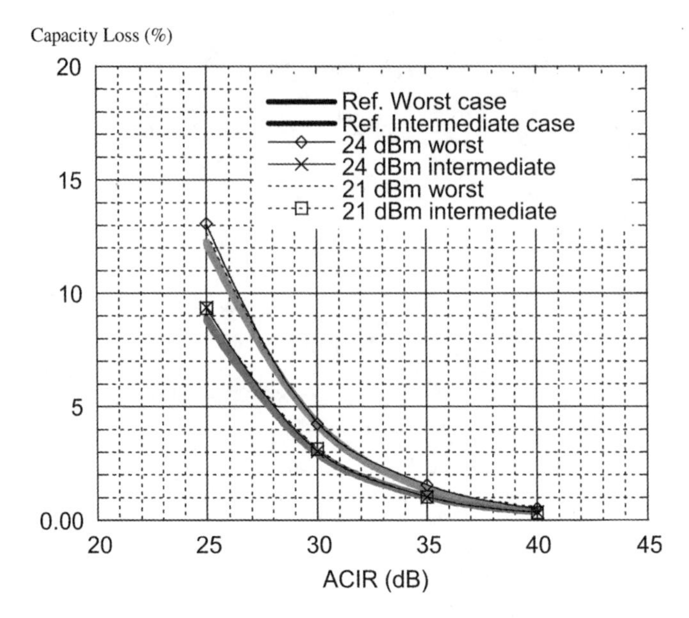

Figure 9.11 UL ACIR speech 24 dB: macro-to-macro.

and parameters, we can perceive that the largest signal appears at −54 dB m amplitude while occurring in less than 0.01% of the cases. Although simulations have not been done for a higher power terminal, according to [5] it is reasonable to assume approximated scaling of the power levels by 12 dB (i.e. from 21 to 33 dB m). Thus, it is proposed that

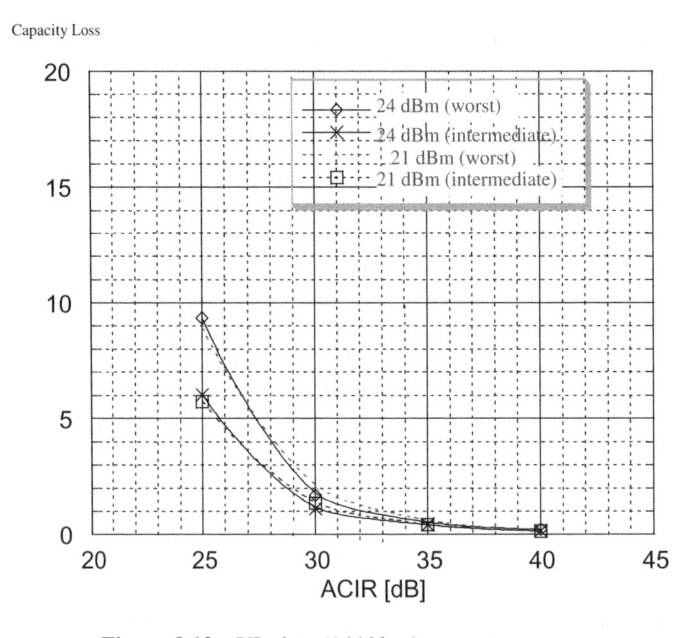

Figure 9.12 UL data (144 kbps) macro-to-macro.

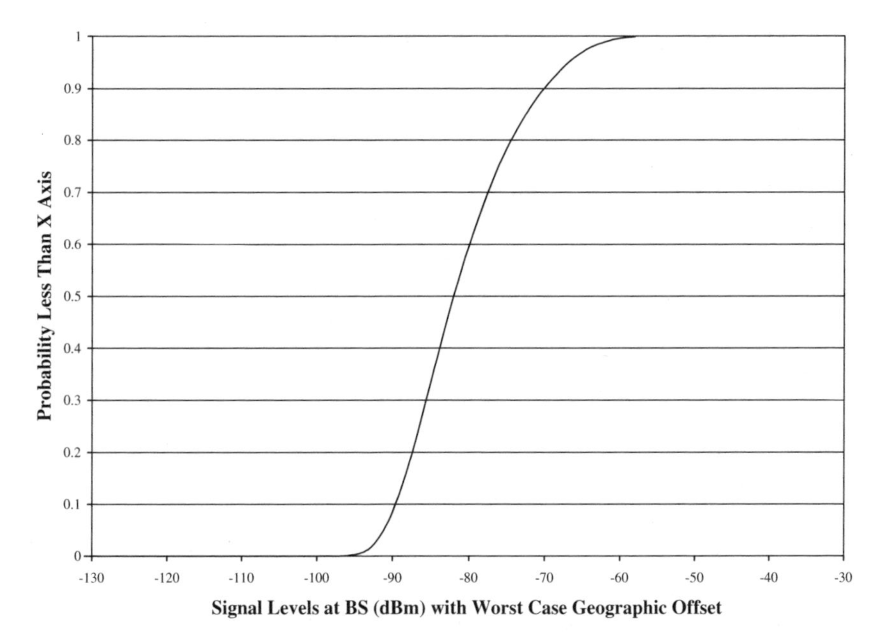

Figure 9.13 BS signals levels [5].

$-54 + 12 = -42$ dB m should be considered a reasonable (if not slightly pessimistic) maximum value for the largest WCDMA blocking signals.

9.5.8.2 Simulation Results for 5 km Cell Radius

Figure 9.14 illustrates the overall CDF of the input signals to the receivers using speech only. Discontinuity occurs, e.g. at -49 dB m input level because in large cells there are a few

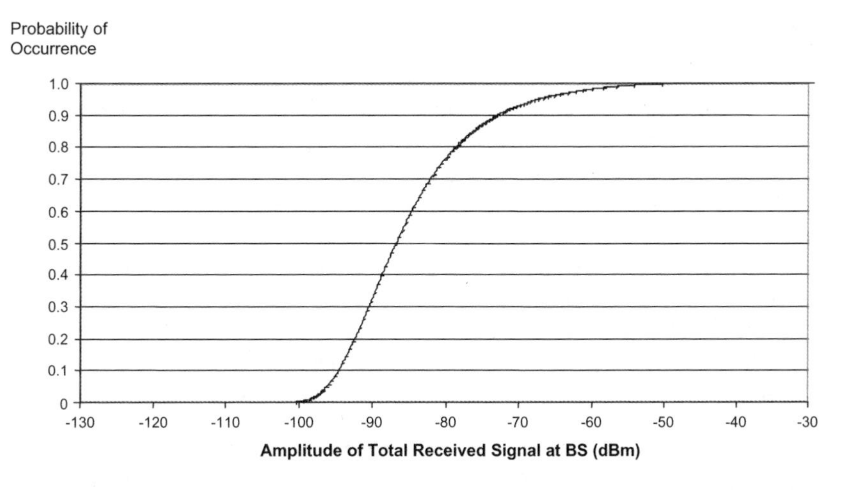

Figure 9.14 CDF of total signal for speech only system with 5 km cells and worst case geographic offset (after [5]).

occurrences of users operating at their maximum transmitted power level of 21 dB m while remaining close enough to another network's cell to produce a minimum coupling loss condition.

Therefore, for this large cell, the received signal power level corresponding to 99.99% of the occurrences is very close to the level dictated by MCL and is about −49 dB m (= 21 dB m − 70 dB) [5]. The preceding event may show the same phenomenon with mixed speech and data systems, i.e. it would produce approximately the same result if the maximum power level for a data terminal were also 21 dB m.

Figure 9.15 illustrates the CDF of the input signals to the receivers in mixed speech and data systems. This indicates that 99.99% of occurrences of the input signals to the receivers are about −40 dB m or less. Because of the large cell, the MCL dictates the absolute maximum signal; and it is only a few dB higher (i.e. 33 dB m − 70 dB = −37 dB m). Discussion in [5] indicates that it may be desirable to allow more than the 3 dB degradation in sensitivity to that which is typically used in the measurement of a blocking specs. This can be justified because:

- the interfering UEs spurious and noise are going to dominate the noise in the victim cell in a real system and
- the measurement equipment is approaching the limit of its capability in the performance of this test.

The first reason seems evident from observing that the interfering UEs noise two channels from its assigned frequency is probably typically in the range of −90 dB m (=−40 dB m − 50 dB), which is much larger than the typical noise floor of the receiver at −103 dB m. The 2nd reason appears evident from observing that the typical noise floor of the most high quality signal generators equals 65–70 dB c with a WCDMA signal. The latter results in test equipment generated noise of −105 to −110 dB m, which can produce a significant error in the blocking measurement.

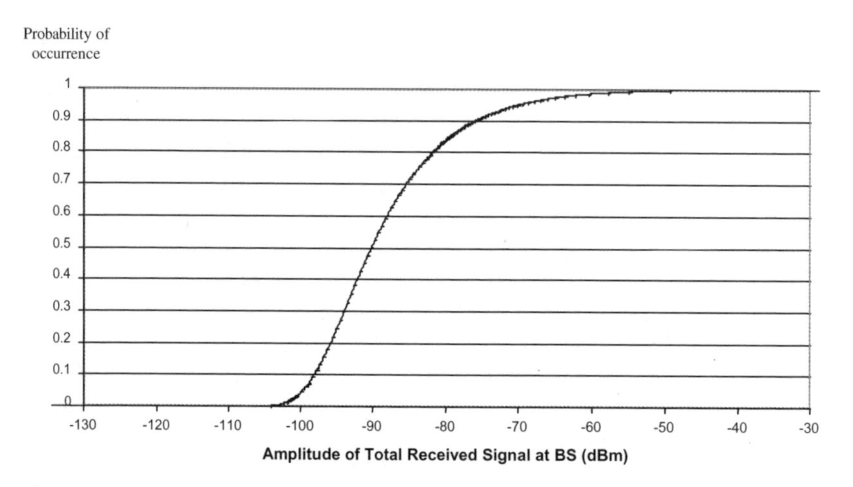

Figure 9.15 CDF of total signal for mixed speech and data system with 5 km cells and worst case geographic offset (after [5]).

In view of these concerns, it is probably reasonable to allow more than a 3 dB increase in the specified sensitivity level under the blocking conditions. In conclusion, it seems reasonable to assume that the in-band blocking specification for UTRA should be −40 dB m (considering that 33 dB m terminals will exist), and the interfering (blocking) test signal should be an HPSK carrier. A 6 dB degradation in sensitivity under the blocking condition should be allowed [5].

9.5.9 Transmit Inter-Modulation (IMD) for the UE

User equipment(s) transmitting in close vicinity of each other can produce Inter-Modulation (IMD) products, which can fall into the UE, or BS receive band as an unwanted interfering signal. The transmit IMD performance indicates the ability of a transmitter to inhibit the generation of signals in its non-linear elements caused by presence of the wanted signal and an interfering signal reaching the transmitter via the antenna.

We define the UE IMD attenuation by the ratio of the output power of the wanted signal to the output power of the IMD product when an interfering CW signal gets added at a level below the wanted signal. We measure, both the wanted signal power and the IMD product power with a filter that has a Root-Raised Cosine (RRC) filter response with roll-off factor of 0.22 and a bandwidth equal to the chip rate. Such test procedure is identical to the ALCR requirement with the exception of the interfering signal. Thus, when performing the aforementioned test, we cannot separate the ACLR impact due to the wanted signal, which would fall into the 1st and 2nd adjacent channel from the IMD product as a result of the interfering signal. Consequently, the IMD cannot be specified to be of the same value as the ALCR. It has to be of a lower value (e.g. 2 dB) to account for the worst case ALCR contribution [5].

9.6 CONCLUSIONS

In this chapter we outlined the UTRA transmission system characteristics and also provided reference examples quantifying some of the recommended parameters or threshold values. The examples use the principles described in Chapter 2 through simulation techniques carried out during the specifications. Thus, these serve primarily as illustrations to visualise some of the impacts while designing actual UMTS networks.

REFERENCES

1. 3GPP, Technical Specification Group (TSG) RUN WG4, UE Radio Transmission and Reception (FDD), 3G TS 25.101, V3.1.0, 1999–2012.
2. 3GPP, Technical Specification Group (TSG) RUN WG4, UTRA (UE) TDD; Radio Transmission and Reception, 3G TS 25.102, V3.1.0, 1999–2012.
3. 3GPP, Technical Specification Group (TSG) RUN WG4, UTRA (BS) FDD; Radio Transmission and Reception, 3G TS 25.104, V3.1.0, 1999–2012.
4. 3GPP, Technical Specification Group (TSG) RUN WG4, UTRA (BS) TDD; Radio Transmission and Reception, 3G TS 25.105, V3.1.0, 1999–2012.

5. 3GPP, Technical Specification Group (TSG) Radio Access Networks; RF System Scenarios, Release 1999, 3G TS 25.942, V2.1.3, 2000–2003.
6. Pizarrosa, M. and Jimenez, J. (eds.), Common Basis for Evaluation of ATDMA and CODIT System Concepts, MPLA/TDE/SIG5/DS/P/001/b1, September 1995.
7. Universal Mobile Telecommunications System (UMTS); Selection Procedures for the Choice of Radio Transmission Technologies of the UMTS, TR 101 112 V3.1.0, 1997–2011, UMTS30.03 V3.1.0.

10

3G Services Enablers

10.1 INTRODUCTION

Just as mentioned in Chapter 2, UMTS is all about services; despite the fact that complex technology may be behind it, the ultimate objective lies on enhancing the wireless user experience through rapid service response and higher choice of applications.

As illustrated in Figure 10.1, increasing channel bandwidth affords faster transmission rates and higher capacity. For both 2G and 3G there exist bearer independent services, but there are or will be also 3G specific services, which will exploit broadband wireless transmission [1–9].

Therefore, we cannot underestimate the potential of specific 3G services when deploying 'service enabler platforms' and specifying terminals. Thus, 3G services, whether they are enabled by independent or specific, must intrinsically maximise broadband bandwidth availability.

Current trends clearly indicate the adoption of higher combination of media, i.e. text, audio, video in non-real-time and real-time settings or environments. Figure 10.2 [10], e.g. illustrates the progress in terms of putting together the different media enablers and the impact on traffic growth as the applications and services get more sophisticated. It does also illustrate the uptake of kilo Bytes (kB) per user as the combined media applications become prevalent among users. The latter without taking into account the progress in build-in cameras in newer handsets, i.e. more mega-pixel, will yet increase the kB uptake as a result of more perfected pictures, and thereby also augment the network traffic.

In conclusion, to better enable wireless services we need enablers that efficiently combine media and networks that rapidly transmit information to enhance the user experience and meet growing multimedia communication expectations.

Likewise, network infrastructure needs to maximise response time to match broadband demands on multimedia services. Figure 10.3 illustrates Nokia's indicative and generic benchmarking examples, where detailed network or application delays are not explicitly described. Nevertheless, it does provide a comparative view of what an infrastructure can do today. For example, without a doubt UMTS-WCDMA stands as the fastest global wireless system delivering as much as nine times faster than a current GPRS, where EDGE does a barely three times more.

All IP in 3G CDMA Networks J. Castro
 ISBN: 0-470-85322-0

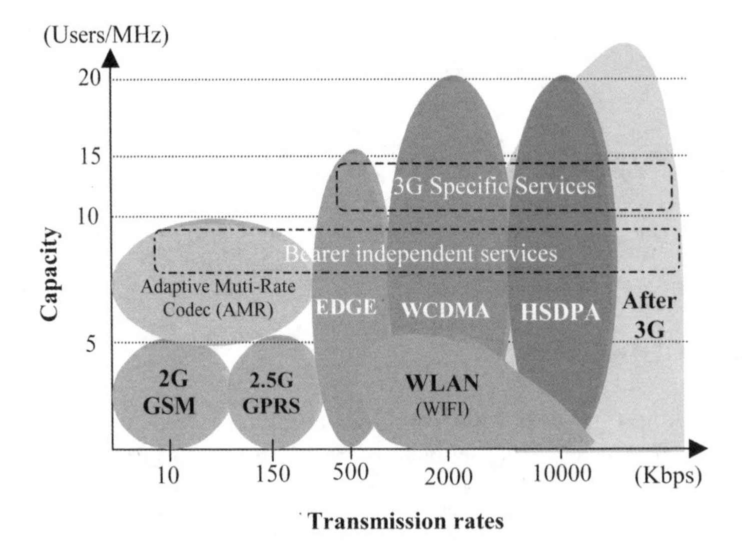

Figure 10.1 Transmission rate and capacity evolution.

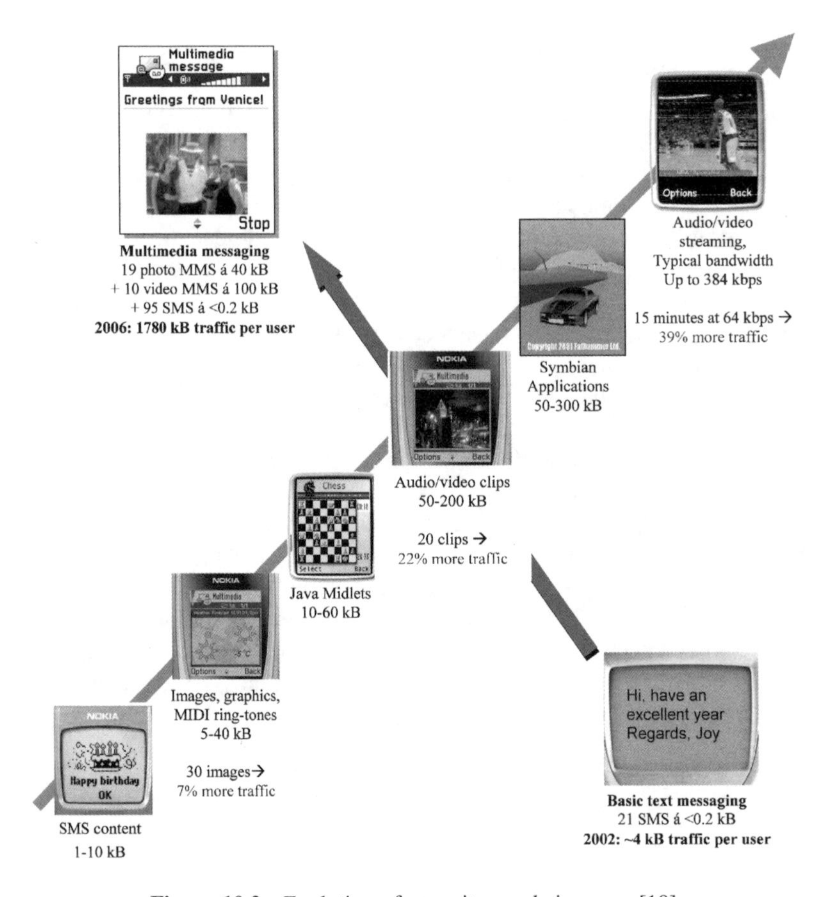

Figure 10.2 Evolution of capacity needs increase [10].

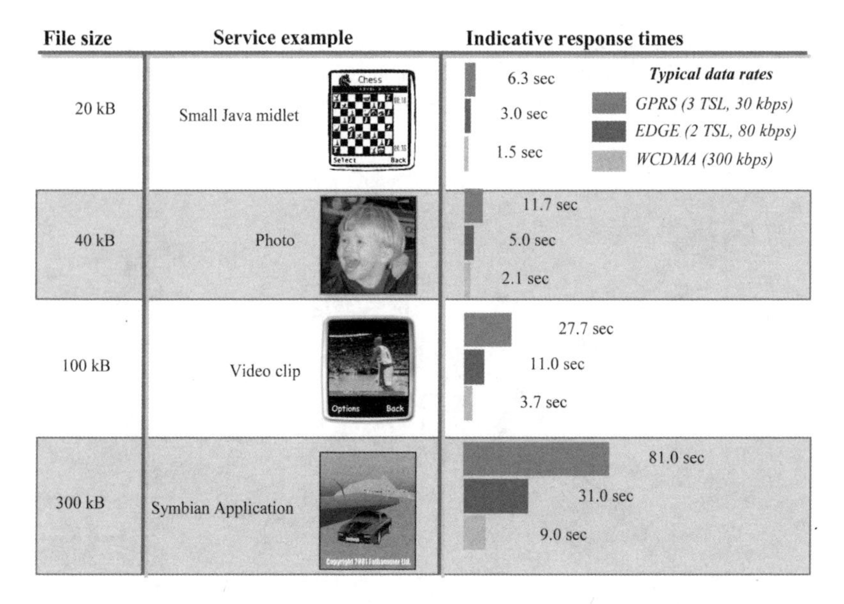

Figure 10.3 Fastest response time with UMTS-WCDMA [10].

From the above, it is evident that multimedia applications, or better yet in the context here, 3G services will benefit the most from the best of combination of media enablers and fastest infrastructure.[1]

With this introduction, in the following we will highlight the application enabler platforms and devices, which will facilitate the 3G-service enabling and wireless user experience enhancement.

10.2 CONTENT DISTRIBUTION PLATFORMS

If we equate the above in general terms, we may say that multimedia applications are primarily about generating content and content distribution, though broadband wireless is about 3G services.

Therefore, optimising content distribution whether it is in real- or non-real-time stands as the greatest challenge to maintain response time and thereby maximise the user experience.

10.2.1 The Streaming Solution

10.2.1.1 Streaming Solution Aspects and Usage Scenarios

In applications, *streaming* refers to the ability to play synchronised media streams like audio and video streams in a continuous manner as these streams originate from a server through a network. These applications can be demand (e.g. music, news, etc.) and live (radio, television, etc.)

[1]WLANs here are only complementary solutions to global coverage potential of UMTS.

In the context UMTS, Packet-switched Streaming Service (PSS) fills the gap between 3G MMS (e.g. downloading and conversational) services and streaming over fixed-IP networks highly successful in many application today. Figure 10.4 illustrates a generic overview of the 3GPP solution [11].

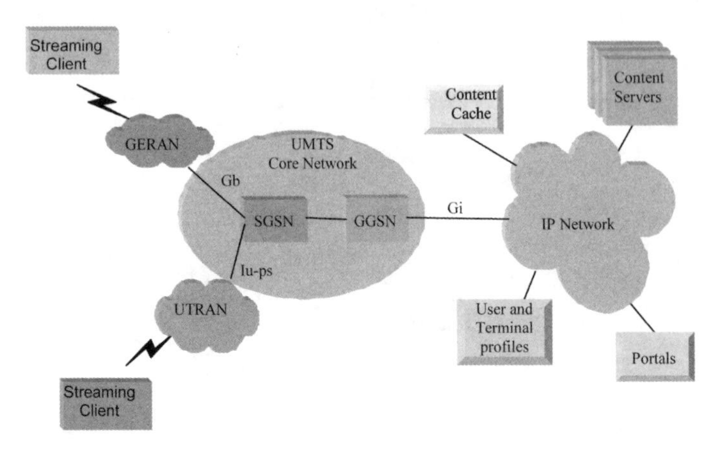

Figure 10.4 Building blocks of the 3G packet-switched streaming service [11].

Figure 10.4 shows the key service specific entities involved in a 3G PSS. A PSS requires at least content or streaming server and a streaming client, where the first resides behind the Gi interface. Other elements like portals, profile servers, caching servers and proxies also located behind the Gi interface might be involved to provide additional services or to improve the overall service quality. Portals are also servers, which allow access to streamed media content, e.g. they might offer content browse and search facilities (Web/WAP page) to reach network-content storing servers. User and device profile servers store user preferences and device capabilities used to control presentation of streamed media content to a mobile user.

10.2.1.2 A Logical Architecture

As described above, PSS technical specifications from 3GPP [11] provide the framework for commercial streaming solutions today.

Figure 10.5 illustrates a logical architecture showing the main streaming building blocks. In this architecture the trigger for the streaming service starts from the user side starting with the handset and flows all the way through the GGSN to the streaming server or the streaming proxy. In the process authentication verification occurs through the profiling event, where MSISDN information gets compiled for content authorisation. The streaming administration centre co-ordinates all the events and manages the flow to reply the user with the right content. The flow has also access check through the Fire Walls (FW) to provide end-to-end service security.

The 3G RAN offers the necessary radio-access bearers to meet the QoS requests from the core side. The service network's session cache would allow communication to the Streaming

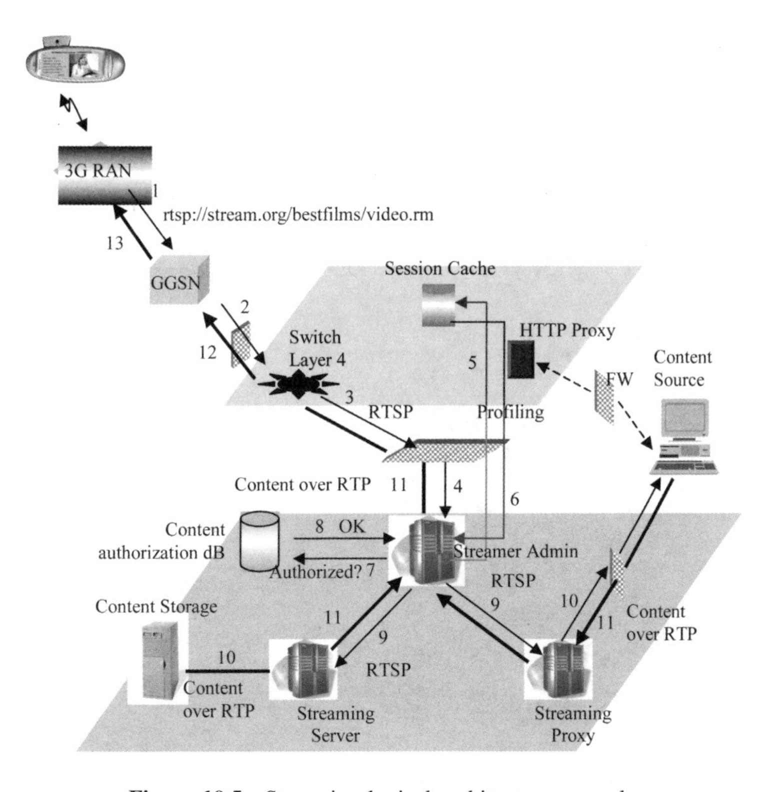

Figure 10.5 Streaming logical architecture example.

Centre (SC), assuming that the client's RTSP[2] requests arrives to the SC with the same IP address as the one stored in the session cache. The level 4 (L4) switch redirects RTSP requests transparently to the SC and the FireWalls update the streaming (RTSP/RTP/RTCP) requests and API (HTTP) requests.

Figure 10.6 illustrates a basic streaming flow, where the QoS event appears as part of the call set-up. In this example [11], the mobile user gets a Universal Resource Identifier (URI) for a desired content, which complies with his/her terminal. This URI, which specifies a streaming server and the address of the content on that server, may originate from a WWW browser, a WAP browser or be entered manually. A PSS application that establishes a streaming session does need to understand a Session Description Protocol (SDP) file.

Sessions containing only non-streamable content, e.g. SMIL file, still images or text to form a time-synchronised presentation, do not require use of a SDP file for session establishment. Instead, they use the HTTP protocol for receiving the presentation files. On the other hand, PSS SMIL sessions can also include URIs to streamable content, requiring parsing a SDP file and/or RTSP signalling [11].

We can obtain the SDP file a link inside the HTML page that the user downloads, via an embed tag directly by typing it as a URI, or through RTSP signalling via the DESCRIBE method. In case of streaming delivery option by MMS service (Figure 10.7), we get the SDP file via the MMS user agent that receives a modified MMS message from the MMS relay or

[2]Real time service protocol.

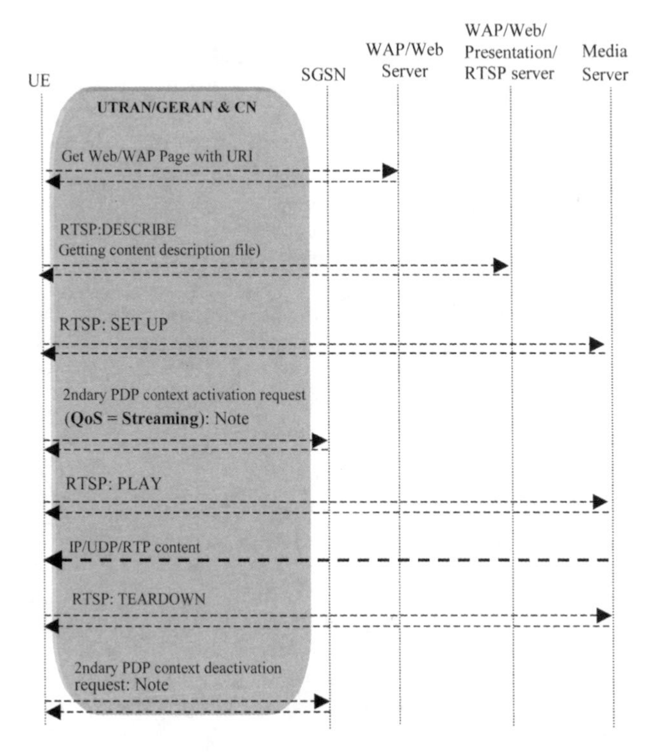

Figure 10.6 Basic streaming session example, after [12].

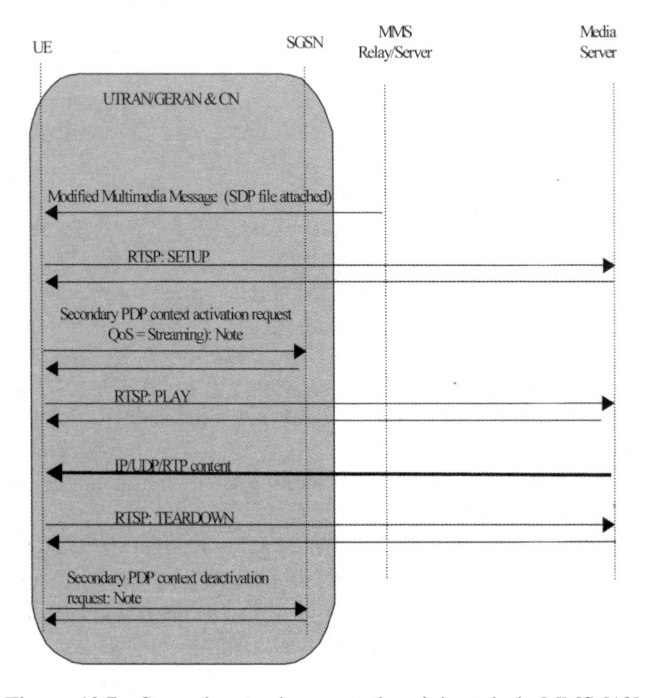

Figure 10.7 Streaming session example originated via MMS [12].

server. The SDP file contains the description of the session (session name, author, etc.), media type to be presented and media bit rate [12].

A session starts through a process in which the browser or the mobile user invokes a streaming client to set up the session against the server. The UE will have an active PDP context in accordance with [13] or other type of radio bearer that enables IP packet transmission before the session signalling establishment. Once the client gets the necessary information about the content, it initiates bearer provisioning with the appropriate streaming media QoS.

In Figure 10.7 a streaming service set-up starts with an RTSP SETUP message for each media stream chosen by the client. This returns the UDP and/or TCP port etc. for usage by the respective media stream. The client sends a RTSP PLAY message to the server, which starts to send one or more streams over the IP network [11].

10.2.1.3 Functional Building Blocks

Figure 10.8 and Table 10.1 illustrate the main functional elements for streaming solution and describe the main actions each unit may execute, since they are non-exhaustive.

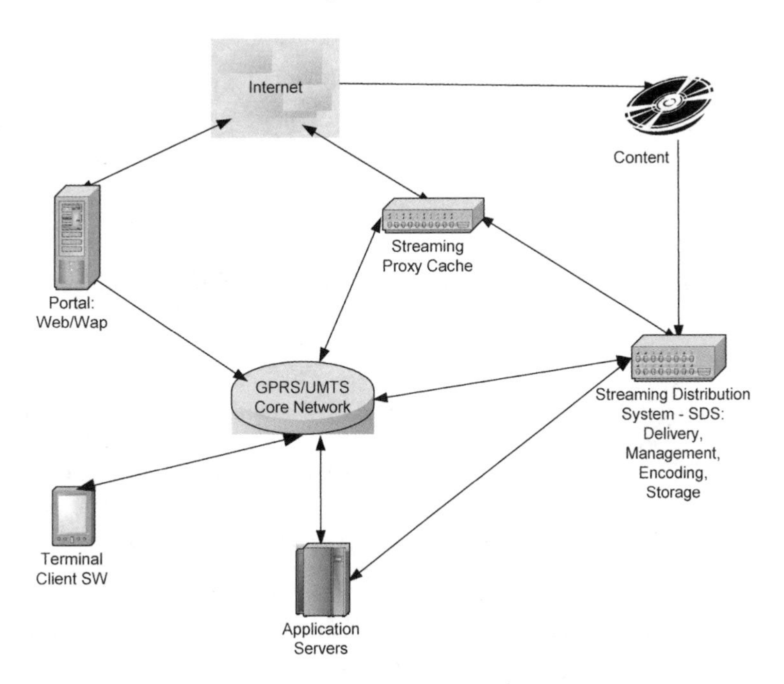

Figure 10.8 Streaming functional elements.

10.2.1.3.1 The Streaming Distribution System—SDS

As illustrated in Table 10.1, the SDS incorporates four main functions, i.e. delivery, management, encoding and storage.

A *Delivery* function with open standard interfaces and widespread technologies will at least:

Table 10.1 High level description of streaming functional elements

Elements	Main Feature
Streaming Distribution System—SDS	
Delivery	e.g. MMS, Download and Streaming + Dynamic Rate Control,
Management	Load Balancing, Logging, O&M, Session Control and Transport
Encoding	Dynamic encoding/Transcoding and Real-time encoding
Storage	Content indexing and Monitoring/Scheduling, Metadata management
Streaming Proxy Cache	Content Caching and Internet Content Proxying
Application Servers	Interfaces: Billing, Java apps, exchange, Logging, O&M, Security, Authentication and Accounting
Portals (Web/Wap)	Browsing, Logging, Security, Authentication and Personalisation
Terminal Client Software	Content Playback: Java MIDP, Mobile streaming player

1. enable real-time content download to a handset;
2. embody a set of conveyance mechanisms to meet different user profiles, while allowing personalised content discovery options and perceiving the user's terminal or device;
3. support Java technology as one of the major enablers for downloading and interactive content interface;
4. map to different charging options to comply with business requirements and service characteristic.

The *Delivery* constituents will incorporate one or more streaming media server(s) delivering at least 50 data types, including MPEG-4, audio, video and MP3, and operate either in 2.5G or 3G environments, but optimised for the latter.

Management in streaming implies semantic administration of content, i.e. automatic configuration of workflows by processing content based on its metadata features. The latter based on World Wide Web consortium (W3C) standards and operating on diverse content types, which enable reception, aggregation and distribution of content objects consistently in forward compatible manned.

The *Encoding* function permits creation of digital (both live and on-demand), media content condition for delivery through mobile devices[3] optimised for specific user profiles automatically configured, i.e. according to target codec, audio/video transmission bit rate and video characteristics, e.g. video frame size and video frame rate.

Through a set of databases the *Storage* function meets the management and content distribution demands.

[3]In 2.5G/3G wireless environments.

10.2.1.3.2 The Streaming Proxy Cache—SPC
The SPC manages media critical tasks in 2.5G/3G environments enabling thus content pull from the Core Network (CN) application servers or simply the Internet through third parties and retransmit with the lowest latency based on user's demands.

10.2.1.3.3 Application Servers—APS
The APS incorporates an integrated set of software enablers to provide an inclusive streaming media solution, which comprises, e.g.:

1. a robust operating system, i.e. Unix or Linux carrier class clusters;
2. a web server, e.g. Apache, which establishes the web-type internal interaction;
3. SunOne or BEA WebLogic Server supporting Java messaging services, XML (extensible Markup Language) as well as J2EE standards;
4. Oracle 9i enabling database management assistance;
5. file system/volume/backup suites affording data storage, file interaction and data integrity management; typical suites, e.g. Veritas.

10.2.1.4 Streaming Content Staging

The sequence or content streaming stages include at least re-sourcing, customisation, encoding, administration and dissemination, where the order simply implies a logical delivery structure. Ideally, the content may be as illustrated in Figure 10.9.

Figure 10.9 illustrates a content staging flow, which can adapt to all types of sources depending on user requests, i.e. internal or external to an operator's service offering.

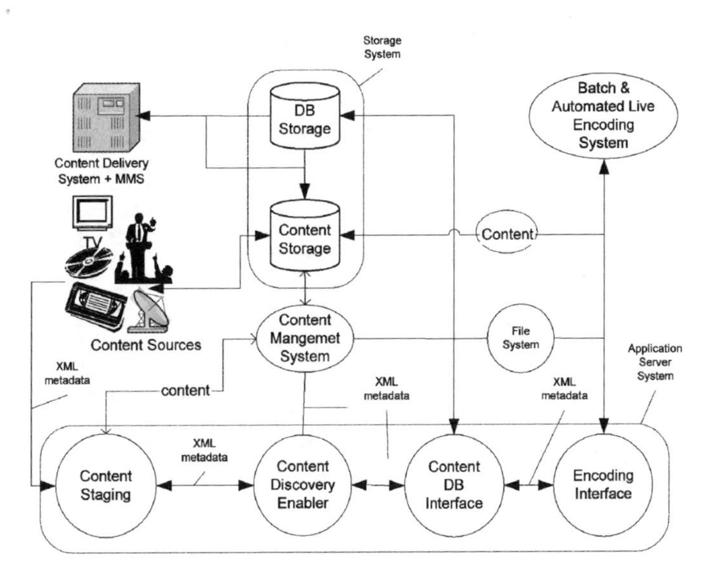

Figure 10.9 Streaming content sourcing solution.

Sourcing

Requests can be on demand from stored sources or on real-time from live events. In the first case, the process may be as follows:

- A content discovery application looks XML control files and new content in an input box. If available, it passes it to the Web Server (WS) using the open APIs.
- Then the WS registers the content and stores its metadata,[4] which if encoded, the file mover transfers it directly to the content management system. Otherwise, non-encoded content passes through the encoding system.

Real time content process involves a live scheduler and an automated encoder, where:

- an XML control will contain the metadata defining the live event coming into the scheduler and interact through the necessary APIs with the live encoder and manages the encoder and source;
- the automated live encoder inquiries the live scheduler through APIs and fetches tasks at high speed and encodes as scheduled in the XML control file by setting the encoding properties, i.e. formats, QoS, capture device, transmission rate, segment level, etc. Table 10.2 illustrates typical formats.

Table 10.2 File formats and standard encoding features

File Format Examples	Standard Features
MPEG-1	IETF RFC 2429 to support RTP Payloads format for H.263
MPEG-2	3GPP file format support
MPEG-4	IETF RFC 1889 for RTP/AVP over UDP
Apple QuickTime	ITU-T H.263 Profile 0 Level 10 Video codec
Real Networks (Media, Audio, Video)	3GPP TS 26.071 AMR Narrowband audio codec
WAV	IETF RFC 3267 support for RTP payload format for AMR
Microsoft Windows Media	IETF RFC 2326 support for RTSP over TCP
H.263, AMR-NB, AAC	MPEG-4 Visual Simple Profile Level 0 codec
AVI	MPEG-4 AAC Low Complexity Audio codec
MOV	IETF RFC 3016—RTP payload in MPEG-4 visual and audio codecs
MP3	IETF RFC 2327 support for SDP for RTSP

[4]*Metadata*—descriptive information providing content properties.

Encoding features and standards
Table 10.2 illustrates typical coding standards and features, which would logically be supported by streaming solutions.

Authoring

- It refers to the customisation of content for flexible and adaptive delivery through multiple service scenarios.
- Authoring can take place through tools, which allows modification, alignment, branding and new presentation formats of content.

Content Management and DRM
The context here implies end-to-end administration of content flows throughout the HW and SW environments, which will depend on the service specifications and operational functionalities. Since the latter will vary among all streaming service providers due to differentiation and content distribution targets, we will not detail them here. However, we will cover briefly Digital Rights Management (DRM), because it will influence all solutions regardless the implementation model.

DRM
The most representative specs for DRM are those from OMA, which are meant to supervise dedicated delivery of mobile content types coming from multifarious channels, e.g. WAP and MMS downloading media ringing all the way from ring tones, MIDI, screensavers, GIF, JPEG, Java content, etc. including video clips. OMA-DRM v1.0 it enables transactional services with the following three features: Forward-lock, Combined delivery and separate delivery.

1. *Forward-lock* enables delivery of subscription-based services such as news, imaging, sports, info, etc; i.e. media without media modification.
2. *Combined delivery* provides regulations for media delivery interaction, e.g. previews, device recognition, delivery rights, etc.
3. *Separate delivery* enables media discrimination based on content rights through different channels. These functions use content encryption into DRM Content Format (DCF) applying symmetric encryption, where
   - usage rights contain the Content Encryption Key (CEK) available for the DRM User Agent (UA) in the device to decrypt the content;
   - the device may forward the media (DCF) object but not the rights of usage.

Content Dissemination
Has everything to do with distributing streaming content to all subscriber segments on demand from stored sources or in real-time from live events. Here again the characteristics of the solution will depend on the approach of delivery and strategy to reach the user. It will also depend on the number of building blocks involved, i.e. on the content sources and channels enabling the streaming chain.

Figure 10.10 illustrates a typical content distribution flow with the following logical steps.

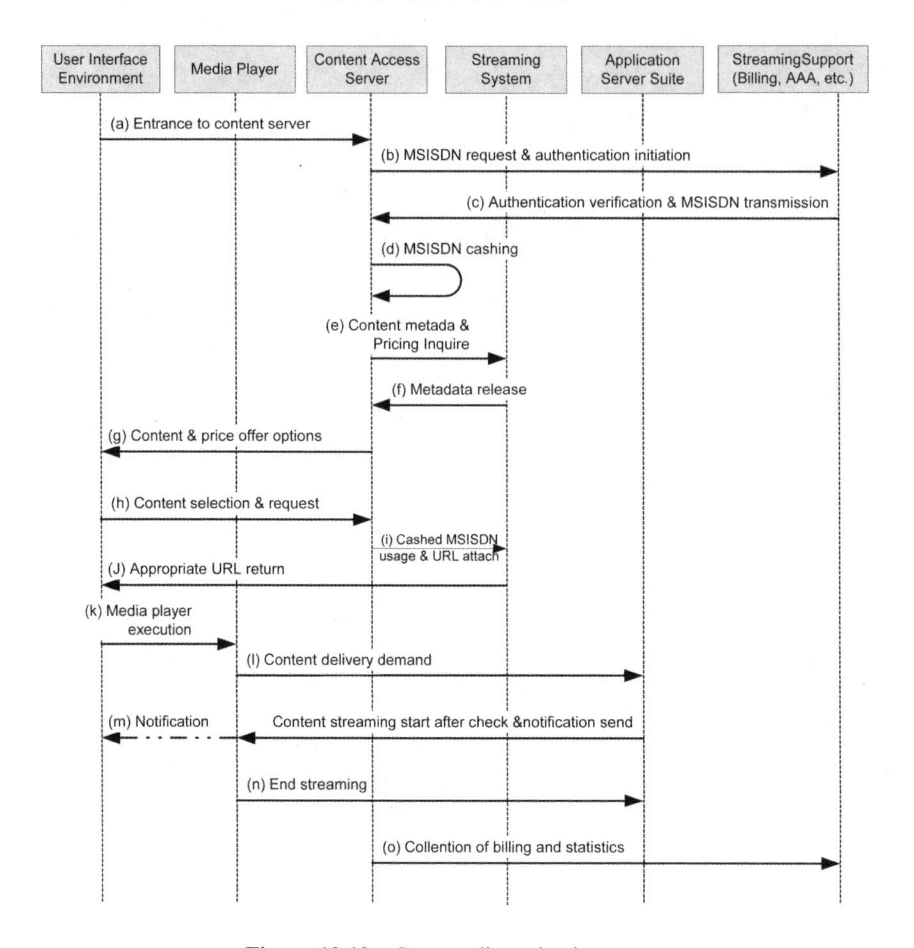

Figure 10.10 Content dissemination steps.

The distribution flow begins with access to the content portal and authentication process through the terminal browser. Then the encrypted MSISDN value returns and cashed to speed up forthcoming steps and minimise response delays. Subsequently, the portal retrieves the metadata info, which provides the content characteristics, including pricing and the selection, and content delivery requests begins. After selection verification, streaming goes on to the end and billing/statistic recorded.

The streaming distribution process will vary from provider to provider; but regardless of the approach, the objectives are to minimise the number of steps to maximise the user experience through a fastest response. Thus, all internal maintenance steps should be transparent to the user and with the lowest latency.

10.2.1.5 Other Streaming Supporting Features

Supporting features include security, performance and charging, which incorporates 'Event Based Billing'.

These features will depend on the deployment architecture and will be cover in the forthcoming sections along the downloading system.

10.2.2 The Content Downloading Solution

10.2.2.1 Function Synthesis

Today there are many downloading implementation solutions and the architecture depends on the number of components integrated or the types of elements deployed. In any case, most solutions will have at least the following key building blocks:

- content provisioning interface
- content type manager
- content delivery manager
- content user selection interface

The delivery and storage sequence may very well align with that of the streaming solution presented earlier. However, for completeness here we will illustrate the downloading principles by illustrating a practical commercial implementation developed by Sun Microsystems, i.e. a J2EE application running on top of SUN ONE Application Server [14].

Figure 10.11 illustrates the main building blocks mentioned above, where:

1. the whole downloading sequence begins with content providers and developers submitting *content* to a *delivery server* through a *developer portal*;

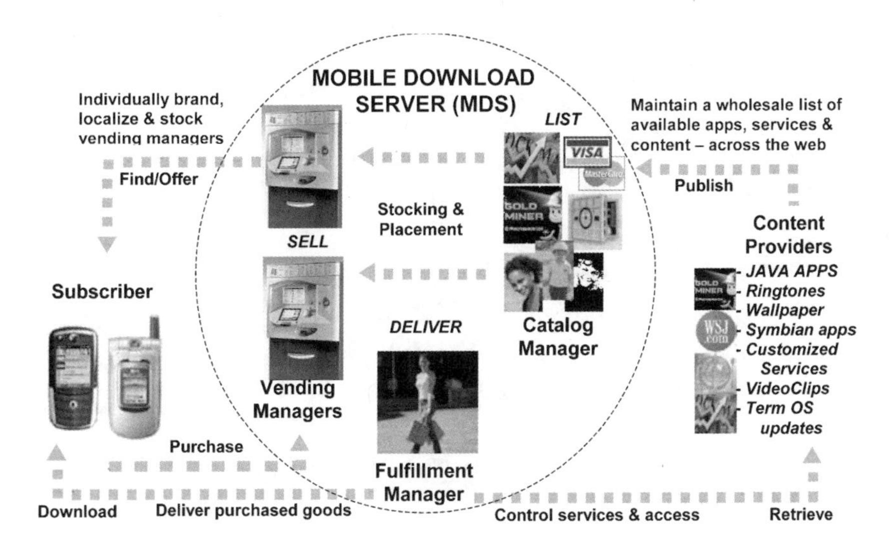

Figure 10.11 Architecture overview of a delivery system, after Sun [14].

2. a centralised or distributed *catalogue manager* gets the accepted[5] content before it is channelled to a subscriber;
3. from *catalogue manager* authorised content gets stored into the *vending managers*, which can be published to one or more *vending managers*;
4. customising *vending managers* enables flexible and dedicated content delivery addressing all types of segments and demands;
5. the *fulfilment manager* delivers content according to selection characteristics while taking care of DRM and billing functions.

10.2.2.2 Integration Interfaces

The integration of a Mobile Download Server (MDS) to a cellular network infrastructure may take place through common interfaces as illustrated in Figure 10.12. The actual configuration will vary from network to network and from service approach to delivery approach. Current and forthcoming network elements will play an active role in content downloading. As already presented in the streaming section, all depends on what is already in the wireless environment, or fixed network for that. In this particular case, we illustrate

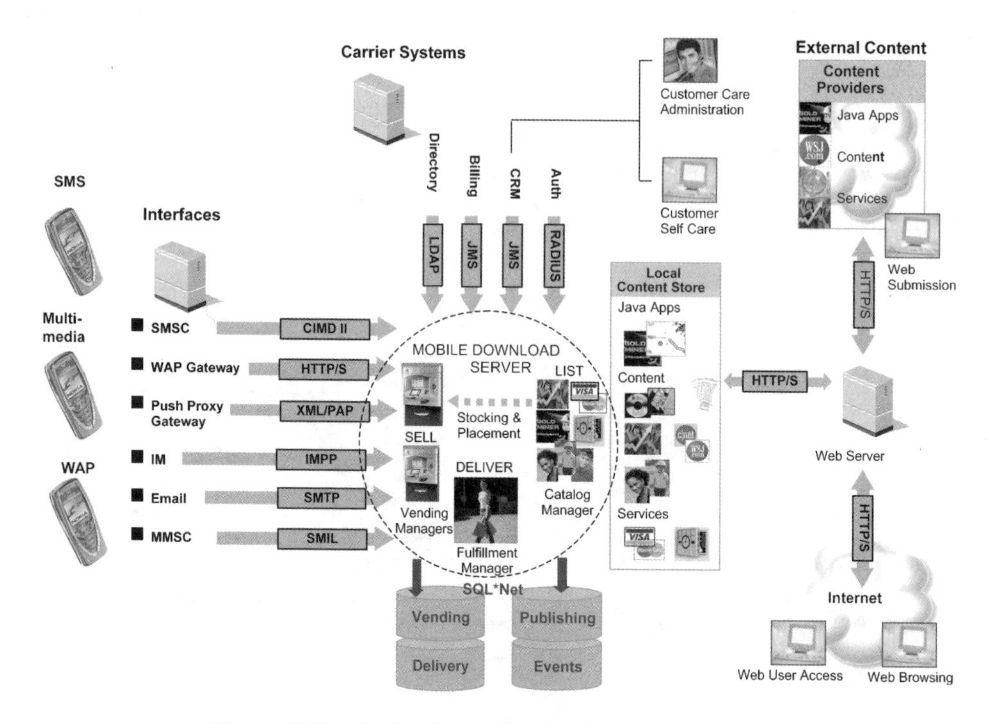

Figure 10.12 Typical integration interfaces, after Sun [14].

[5]Acceptance implies complete review and discrimination of the content and its compliance for service.

integration with a working product simply to present a realistic multi-purpose platform, which will be supporting 2.5G and 3G infrastructures concurrently. What is then the difference between the last two? In 3G, availability of wider bandwidth will be non-negligible to the user through faster content transfer or faster execution of applications and thereby overall rapid response.

Thus, when it comes to integration in practice, we may see the same elements for 2.5G and 3G at least at the beginning of the latter. In the following, we illustrate typical integration interfaces [14] with the following features:

1. *Authentication*—elucidation of conventional HTTP information coming from the WAP/WEB gateway for subscriber recognition/verification through an MSISDN or a unique subscriber identifier.
2. *Subscriber profile*—providing a plug-in interface to integrate with an existing subscriber database hosted by the service provider, e.g. LDAP integration.
3. *Billing APIs*
   a. *Postpaid*—allowing MDS asynchronous (e.g. to JMS clients) interaction with billing solutions for reporting, charging and notifying billable download events to external systems, including flexible configuration of CDR record formatting.
   b. *Pre-pay*—real-time synchronous credit authorisation and funds *clearance* for already paid accounts.
4. *Web portal functions*—facilitating sale, DRM, customer-care and reporting interfaces for direct or remote system access.
5. *Push interfaces*—content independent customised or by default, e.g. WAP and SMS push messaging implementations containing content object embedded data.
6. *Pull interfaces*—low level (PULL-subscriber initiated) downloading APIs for both java and non-java content delivery onto a device.
7. *Content delivery flow*—suites automatically validating content deposits while assuring content quality (e.g. virus checking, data compression, signatures, etc.).
   a. *Content adjusters*—facilitate content (e.g. new MIMEs) submission, management and download without code modification for a target device.
   b. Support of new content types and OTA delivery formats while inter-operating with third-party content deliver adapters in standard web servers.
8. *Statistic event collection*—interfaces allowing concurrent system and subscriber activity monitoring through multiple JMS event channels.
   a. Enables dynamic status report gathering on system performance and content delivery processing for the user's download-profile tracking.
   b. Facilitates system maintenance by supervising system failure logs and alarm reporting, including customer support monitoring and assistance.
9. Parallel or multi-delivery vending, catalogue and fulfilment managers to support subscriber segmentation or distributed provisioning scenarios.

a. Enabling highly customisable environments for standalone or hosted operation to different consumer sectors and/or enterprise services.

b. Large MDS server configuration options for multiple service providers, ASPs or wholesale content aggregators.

10. *Third Party Software (SW) Optimisation*—enabling faster downloads through structured internal/external SW interaction and eliminations of superfluous steps and efficient de-compilation, e.g. JODE.[6]

10.2.2.3 A Deployment Example

Figure 10.13 illustrates a very logical simplified view of practical downloading solution deployment. Notice that the sophistication will depend on the level of elements involved. Here we only point out that in practice most of the building blocks for application platforms

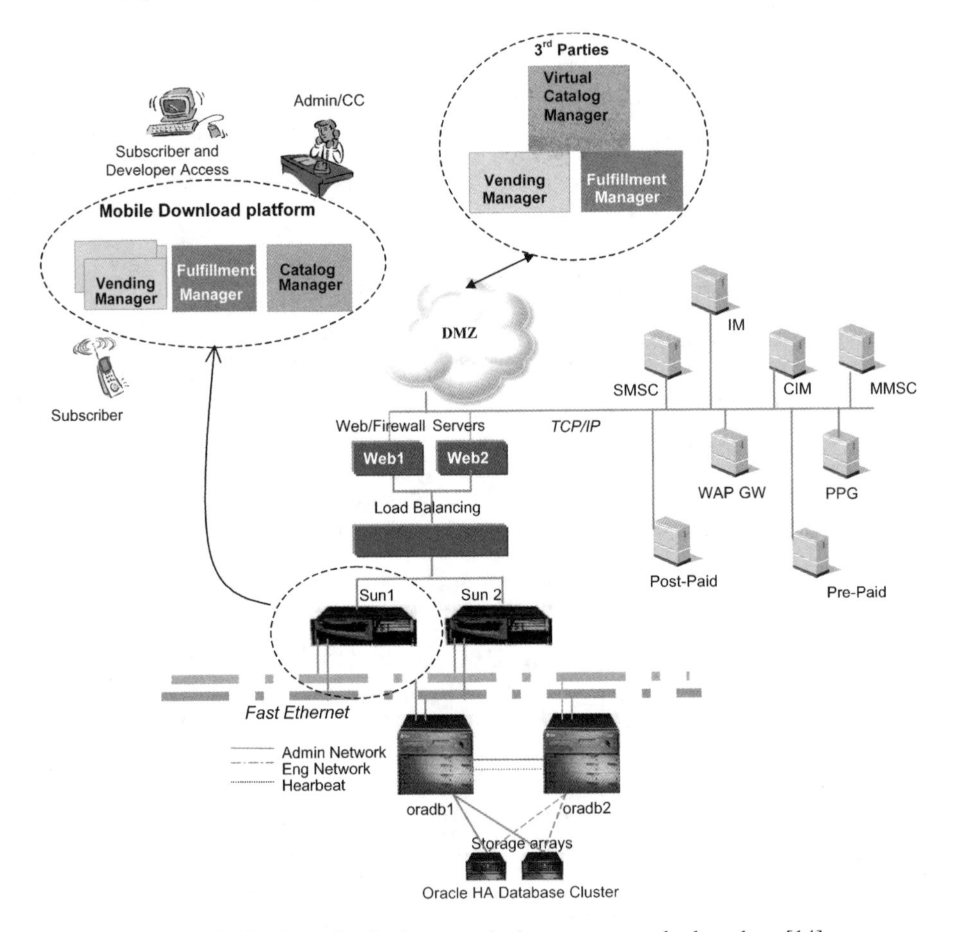

Figure 10.13 Downloading system deployment example, based on [14].

[6] Java optimise and decompile environment.

such as downloading already exist in 2.5G mobile networks like, which 3G will exploit directly and enhance response times and performance capabilities by providing broadband wireless bearers and better QoS features.

For example, as illustrated in Figure 10.12 integration with Value Added Service (VAS) elements become a natural step, including the billing, where legacy solutions still separate in post- and pre-paid.

In addition, databases, storage units and Web servers are also an integral part of mobile Internet solutions. Likewise, load-balancing techniques are part of the standard procedures on web environments. Thus, enabling content download boils downs to the downloading server and optimised IP based integration of building blocks to ensure minimise dead milliseconds while responding to usage.

Nonetheless, there remain critical efficient features such as 'virtual content list managers' and 'multiple content distribution managers', which adapt automatically to external/internal inputs and demands to serve a variety of subscriber segments and content deposit/distribution scenarios such as third party developers and virtual service providers. Not to loose the response gains due to broadband bearers in UMTS, it is imperative that such features be fully optimised and transparent to all concerned, i.e. users, providers and developers. In conclusion, end-to-end optimisation of application platforms or content distribution enablers providing rich multimedia content in 3G, will be absolutely essential for seamless data services adoption.

10.3 FACILITATING REAL TIME IP SERVICES

On the one hand, today's predominant mobile voice services still depend primarily on conventional circuit-switched based calls, where radio resources are reserved for the duration of the call.

On the other hand, with the advent of IP-Multimedia Subsystem (IMS) and ALL IP based Radio-Access Networks (RAN), real-time services like voice will/use new enablers. For example, Figure 10.14 illustrates how projected packet traffic will begin to grow rapidly starting 2005.

In fact today, IP base traffic volume has already surpassed that of circuit switched in fixed networks and this trend will follow in the wireless cellular environment as person-to-person communications becomes widespread.

10.3.1 Enabling Push-to-Talk over Cellular-PoC

PoC stands as new IP based real-time voice service, which will start exploiting the packet-switched environment. 2.5G networks facilitate efficacious use of radio and system resources. For example, GPRS with 1/2 duplex VoIP technology and Session Initiation Protocol (SIP) enable an effective deployment of call processing and application servers for PoC. It brings a direct one-to-one and one-to-many voice communication service in wireless networks, thus making two-way radio service accessible through mobile handsets. As special voice offering and not a substitute for legacy cellular voice services, PoC will become a natural IMS application.

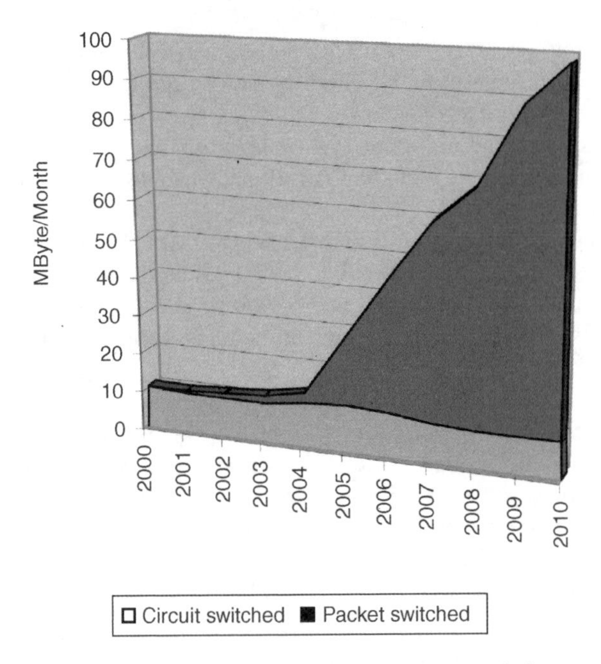

Figure 10.14 Packet-switched growth extrapolation.

10.3.1.1 PoC Generic Architecture Overview

Earlier PoCs known as Push-to-Talk (PTT) are available with proprietary architectures even though they stand as open and capable to integrate to current GSM/GPRS networks. An evolution of these solutions is the OMA base architecture illustrated in Figure 10.15, where industry defined functionalities aim to provide optimal end-to-end solution independent of IMS but fully inter-operating with it.

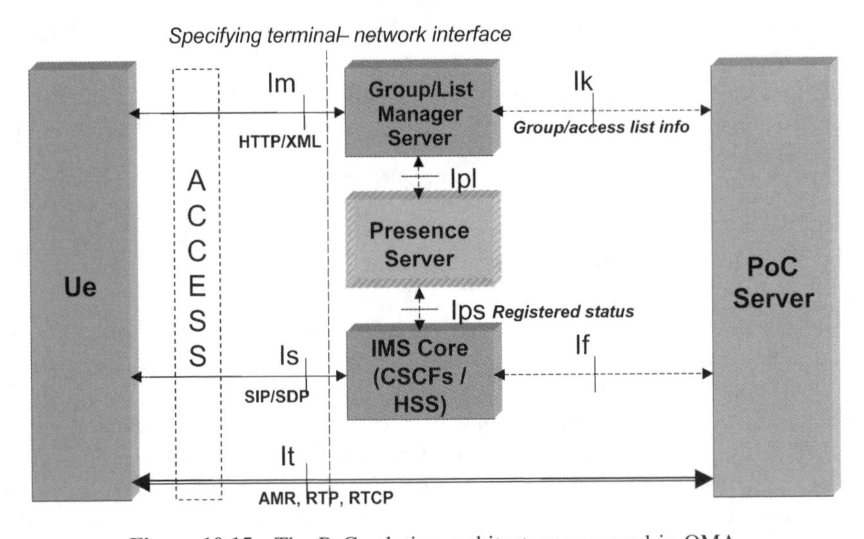

Figure 10.15 The PoC solution architecture proposed in OMA.

IMS network elements carry out functions such as SIP routing, authentication, QoS control, real-time charging and overall application-server interface management. The PoC servers embody specific push-to-talk functions and necessary interaction data, and generate dedicated CDRs to correlate them through the IMS charging functions. IMS thus provides a standardised approach of IP connectivity between the terminals a sole registering point and network authentication for PoC as it would for other application platforms.

In Figure 10.16 a typical *PoC processor* will perform internal system management, control and user planes functions, e.g. speech packet administration, i.e. multiplication, filtering, forwarding, speech quality optimisation, single and group session control, as well as login access. The *Register* function comprises provisioning, API interfaces to users, O&M, PoC processor, network monitoring, etc.

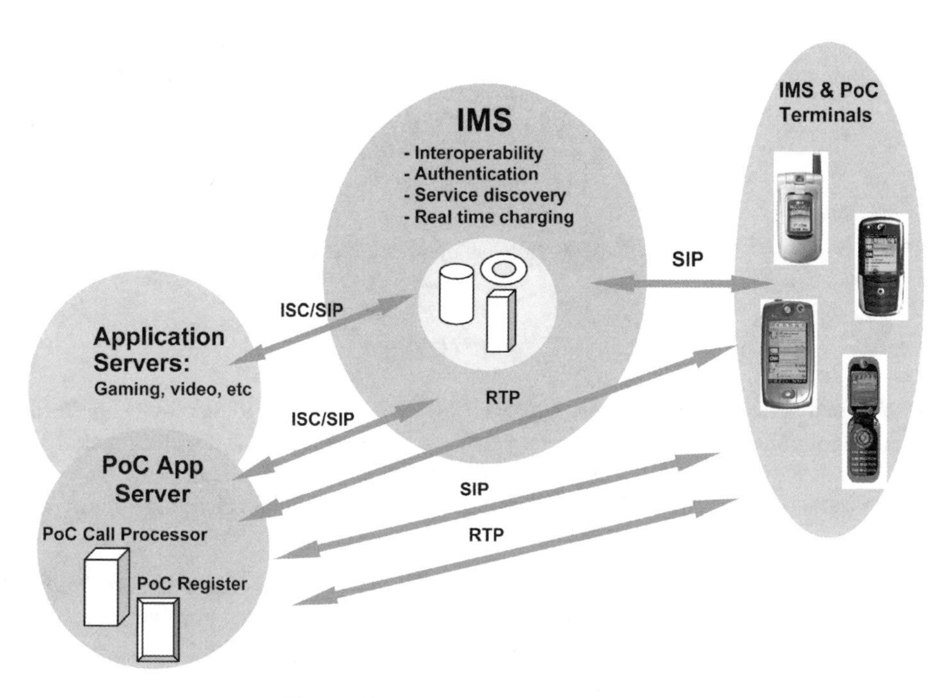

Figure 10.16 IMS based PoC solution.

Practically, a PoC solution would be core network through the PS domain with the following logical planes: provisioning, management, control and user planes. Where as illustrated by Figure 10.16, in reference solution we would include the provisioning plane in the *PoC register* and remaining planes in the *PoC call processor*.

In Figure 10.17 the user plane treats dynamic time critical operations, e.g. voice packets, the provisioning plane carries static data, the management plane comprises static user and group data, and finally the control plane deals administrates the SIP signalling of the terminal the PoC solution.

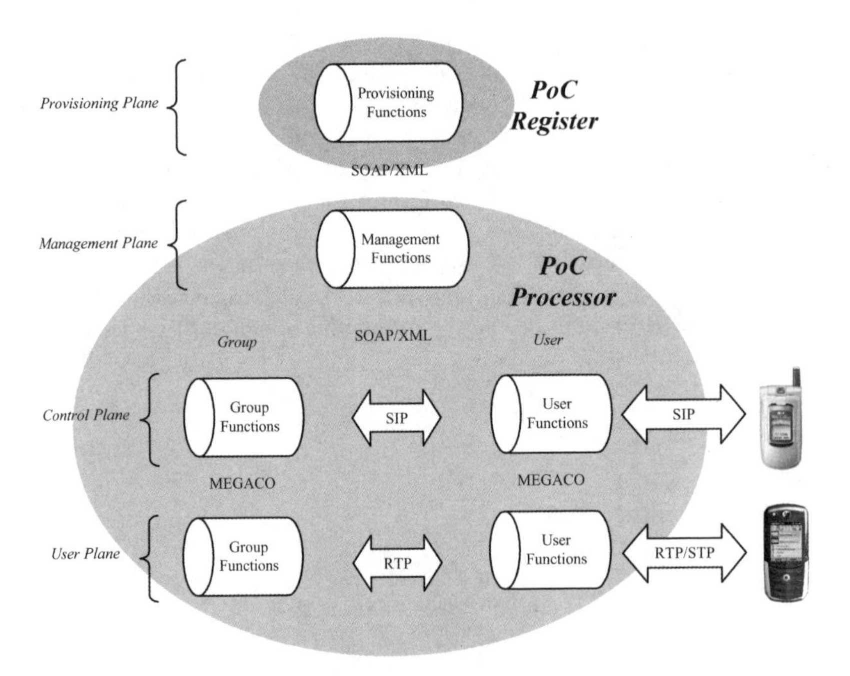

Figure 10.17 PoC architecture protocol stack example.

10.3.1.2 PoC Charging and Provisioning Options

While commercial implementation will vary from deployment to deployment, intrinsic charging functions within IMS solutions will have most likely the following functions:

- CDR gathering for charging correlation with existing PS domain charging solutions, e.g. charging gateway.
- 3GPP specified *diameter online* charging interface, which is transferred as *diameter Attribute Value Pair* (AVP) and *diameter accounting* requests. For real-time or hot billing, a *online diameter interface* would enable account query, credit control for event and session transactions and price indication.
- Billing field to accommodate types of billing, e.g. post-paid, pre-paid and hot billing to discriminate user profiles, and IMS charging ID (for IMS subscribers).

Provisioning would take place based on user profile configuration through the HLR/HSS for existing users and HSS for new IMS only users.

10.4 ENABLING LOCATION COMMUNICATION SERVICES (LCS)

Here we assume that User Equipment (UE) positioning is a network provided feature, which enables the provision of *service provider* or *third party* location applications. With UMTS, location applications take a new dimension because they can exploit broadband bearers capable to combine with reach multimedia characteristics for higher user experience. Thus, LCS will play a key role in the implementation of innovative 3G services.

In LCS, by applying radio signals we determine the subscriber's UE (geographic) location, which may be requested by and reported to a client (application) associated with the UE, or by a client within or attached to the core network. The location information may also be applied internally, e.g. for location assisted handover or to support other features such as home location billing [15]. Position information gets reported in standard, i.e. geographical co-ordinates, together with the time-of-day and the estimated errors (uncertainty) of the location of the UE according to specification TS 23.032 [16]. The position information accuracy will depend on the method used, i.e. the UE position and activity within the coverage area. Key UTRAN system design options, e.g. cell size, adaptive antenna technique, path loss estimation, timing accuracy and Node B surveys will allow a network operator to deploy the most appropriate and cost effective UE positioning method. Key assumptions for 2.5G and 3G networks before coming into the solution architecture include [15] the following:

1. Positioning methods are access network specific.
2. Commercial location services apply only to UEs with a valid SIM or USIM.
3. LCS availability in the access network is optional as is in the MSC and SGSN. However, it is applicable to any target UE whether the latter supports LCS or not.
4. LCS applies to both circuit-switched and packet-switched services, which may also be used for internal system operations to improve system performance.
5. It is a logical expectation from users/operators and a compelling necessity to support LCS signalling between separate access networks via the core network.

10.4.1 UTRAN UE Positioning Architecture

Figure 10.18 illustrates the general arrangement of the UE positioning feature in UTRAN. The UTRAN UE Positioning entities interact through the messaging and signalling capabilities of the UTRAN interfaces (Iub, Iur, Iupc). The communications can be summarised as follows [17]:

- The SRNC receives authenticated requests for UE positioning information from the CN across the Iu interface.
- The RNCs manage the UTRAN resources (including Node Bs, LMUs and the SAS), the UE and calculation functions, to estimate the position of the UE and return the result to the CN.
- The SRNC may also make use of the UE positioning function for internal purpose, e.g. position based handover.

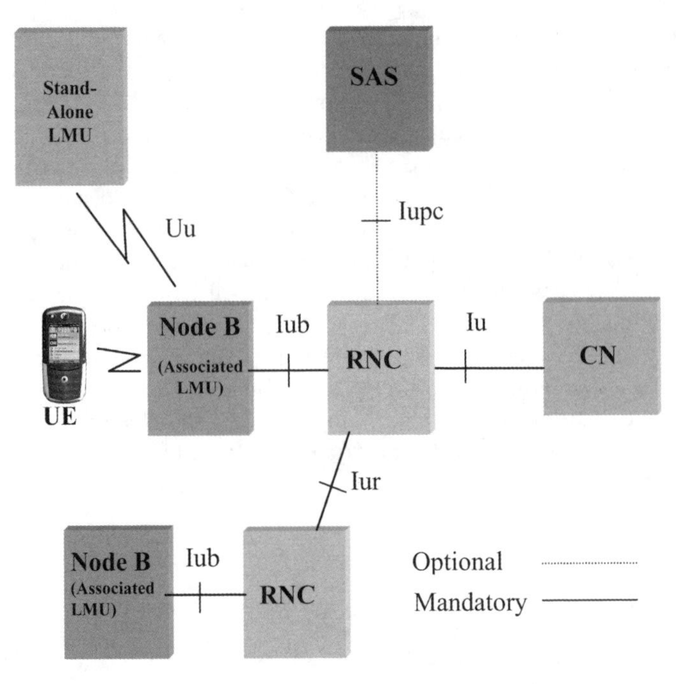

Figure 10.18 The UE positioning in UTRAN, after [17].

10.4.2 LCS Categories and UE Positioning Methods

The four LCS categories include [15] the following:

a. *The commercial LCS (or value-added services)*—associated with an application that provides a value-added service to the subscriber of the service.
   - For example, a directory of restaurants in the local area of the UE, incorporating directions for reaching them from the current UE location.

b. *The internal LCS*—developed to make use of the UE location information for access network internal operations.
   - For example, location assisted handover and traffic and coverage measurement, as well as O&M support related tasks, supplementary services, IN related services, etc.

c. *The emergency LCS*—part of a service provided to assist subscribers who place emergency calls. For example, mandatory in some jurisdictions like US.

d. *The lawful intercept LCS*—used for location information to support various legally required or sanctioned services.

The UTRAN may use one or more positioning methods to determine the position of an UE, which may involve two main steps, i.e. *signal measurements* and *position estimate computation* based on the measurements. The UE, the Node B or a Location Measurement

Unit (LMU) can make the signal measurements, which are basic UTRA radio transmissions or radio navigation signals. UTRAN's standard positioning methods comprise:

1. *Cell ID* based method.
2. *Observed Time Difference of Arrival* (OTDOA) method.
3. Network-assisted GPS methods.

10.4.3 The Cell ID based Positioning Method

In the cell ID based (i.e. cell coverage) method, the UE's position gets estimated with the knowledge of its serving Node B, which may be obtained by paging, locating area update cell update, UTRAN Registration Area (URA) update or Routing Area (RA) update.

This positioning information can be indicated as the *cell identity of the used cell, the service area identity* or as the *geographical co-ordinates* of a position related to the serving cell. The position information will consist of a QoS estimate (e.g. achieved accuracy) and, if available, the positioning method(s) used to obtain the position estimate. In a geographical co-ordinates method the UE estimated position is a fixed geophysical position within the serving cell (e.g. centre position of the serving Node B). The latter can also be obtained by combining information from the signal RTT in FDD [18] or Rx timing deviation measurement and knowledge of the UE timing advance, in TDD [19].

10.4.3.1 Determining the Cell ID

In the Cell ID method, the Serving Radio Network Controller (SRNC) identifies the cell providing coverage for the target UE following procedures illustrated in Figure 10.19. Here when the SRNC receives the LCS request from the Core Network (CN), it checks the state of

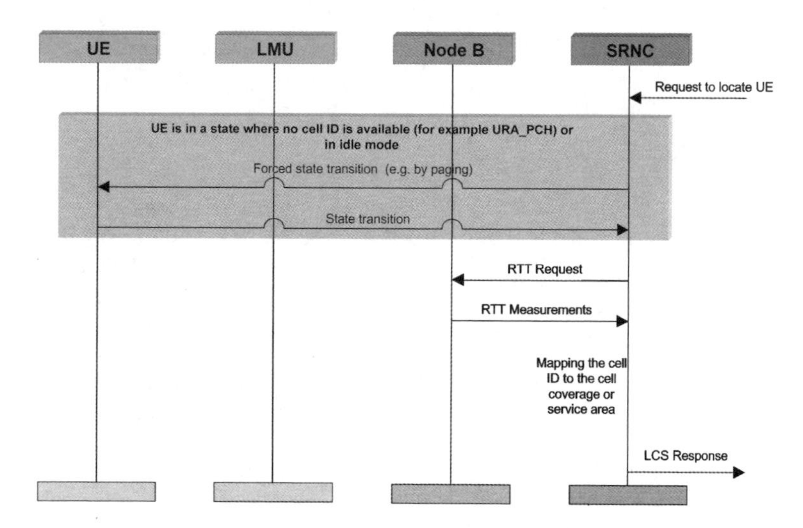

Figure 10.19 The cell ID determination process [17].

the target UE. If the latter is in a state where the cell ID is available, then the target cell ID is chosen as the basis for the UE positioning. In the absence of the cell ID, the UE is paged, so that SRNC can establish the cell with which the target UE is associated. To increase the accuracy of the LCS response, the SRNC may also request RTT (FDD only) measurements from the Node B or LMU associated with the cell ID. The SRNC may also map the cell ID to a corresponding *Service Area Identifier* (SAI) to match the service coverage information available in the CN (see more details in). The cell ID based method determines the position of the UE regardless of the UE RRC mode (i.e. connected or idle) [17].

10.4.4 The Observed Time Difference of Arrival (OTDOA) Method

OTDOA has as primary measurements the 'SFN[7]-SFN observed time difference' perceived at the UE [20,21], which together with other information concerning the surveyed geographic position of the transmitters and the Relative Time Difference (RTD) of the actual downlink transmission signals may be used to calculate an estimate of the position of the UE. As illustrated in Figure 10.20, each OTDOA measurement for a pair of downlink transmissions describes a line of constant a hyperbola difference along which the UE may be located. We determine the UE's position by the intersection of these lines for at least two pairs of Node Bs, where the position accuracy estimates made with this technique depends on:

1. the precision of the timing measurements;
2. the relative position of the Node Bs involved;
3. it is also subject to the effects of multi-path radio propagation.

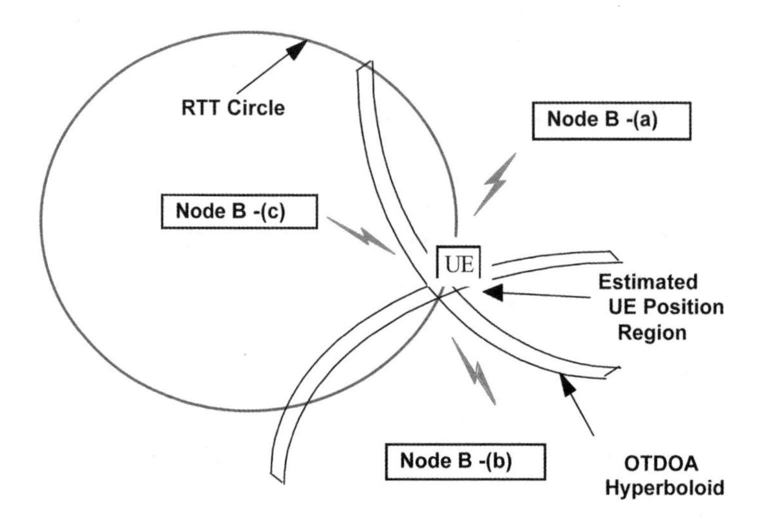

Figure 10.20 The OTDOA positioning method [17].

[7]System frame number.

The primary OTDOA measurements go to the SRNC through signalling over the Uu, Iub (and Iur) interfaces between the UE and the SRNC. In networks with a Stand-Alone SMLC (SAS), the SRNC may forward the measurement results over the Iupc interface to the SAS, where they are used in conjunction with the known positions of the transmitter sites and the RTD of the transmissions to estimate the UE's position [17].

10.4.5 Network Assisted GPS Positioning Method

As illustrated in Figure 10.21, when GPS is designed to inter-work with the UTRAN, the network assists the UE GPS receiver to improve its performance as follows [17]:

1. It decreases the UE GPS start-up and acquisition times; the search window can be limited and the measurements speed up significantly.
2. It increases the UE GPS sensitivity because we obtain positioning assistance messages via UTRAN so that the UE GPS can operate also in low Signal-to-Noise Ratio (SNR) situations when it is unable to de-modulate UE GPS signals.
3. It reduces UE power consumption due to rapid start-up times since the GPS can be in idle mode when not active.

This positioning method relays on signalling between UE GPS receivers and a continuously operating GPS reference[8] receiver network, which has clear line of sight to satellites in a GPS constellation as it does the assisted UEs.

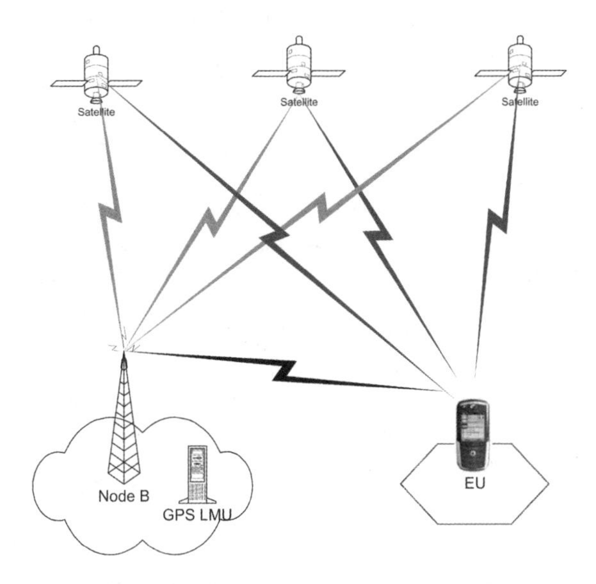

Figure 10.21 Network assisted GPS positioning.

[8]GPS reference receivers may be connected to the UTRAN to enable derivation of UE assistance signals.

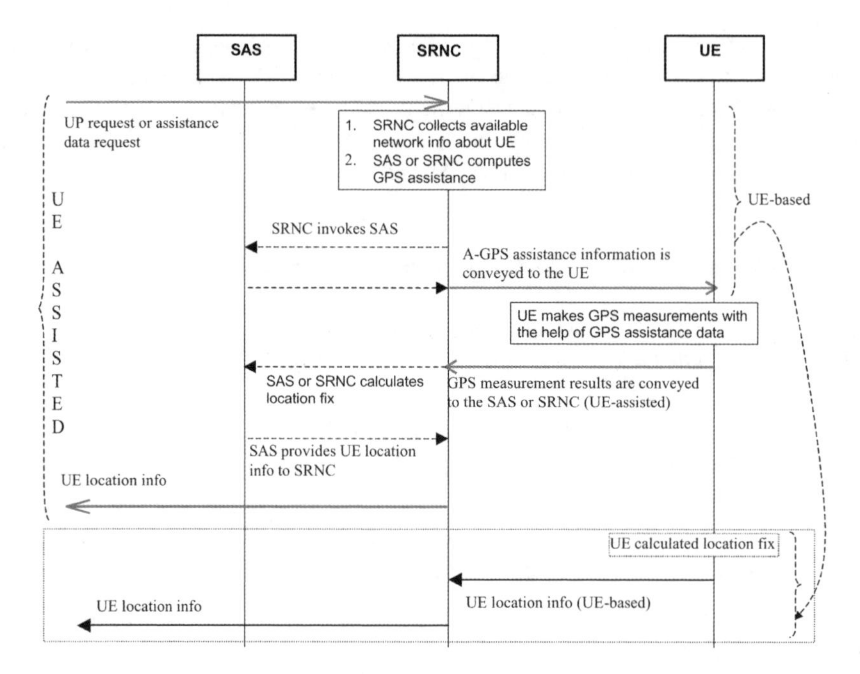

Figure 10.22 Network assisted GPS positioning: UE-assisted and UE-based, after [17].

Figure 10.22 illustrates end-to-end call flows for UE-assisted and UE-based positioning established from [17]. The SAS is an optional network element and the call segments do not apply in a network where the UE positioning resides within the SRNC.

These two types of network-assisted GPS methods differ according to where the actual position calculation is carried out. Computation of the position fix can either be performed in UTRAN (i.e. SRNC or SAS) for UE-assisted or in the UE for UE-based.

The *UE-based* method maintains a full GPS receiver functionality in the UE, and the position calculation is carried out by the UE, thus allowing stand-alone position fixes.

While in the *UE-assisted* method, the UE employs a reduced complexity GPS receiver functionality, which carries out the pseudo-range (code phase) measurements. These are signalled, using higher layer signalling, to the specific network element that estimates the position of the UE and carries out the remaining GPS operations. Here, accurately timed code phase signalling [19,20] is required on the downlink. Thus, when we perform downlink GPS in the UE, we should also signal differential corrections to it. On the other hand, we can apply these corrections to the final result in the network to improve the position accuracy without extra signalling to the UE [17].

10.4.6 The Location Communications Services Architecture

Figure 10.23 illustrates the general location service building blocks for GSM and UMTS, which shows the relation of LCS clients and servers in the core network with the GERAN

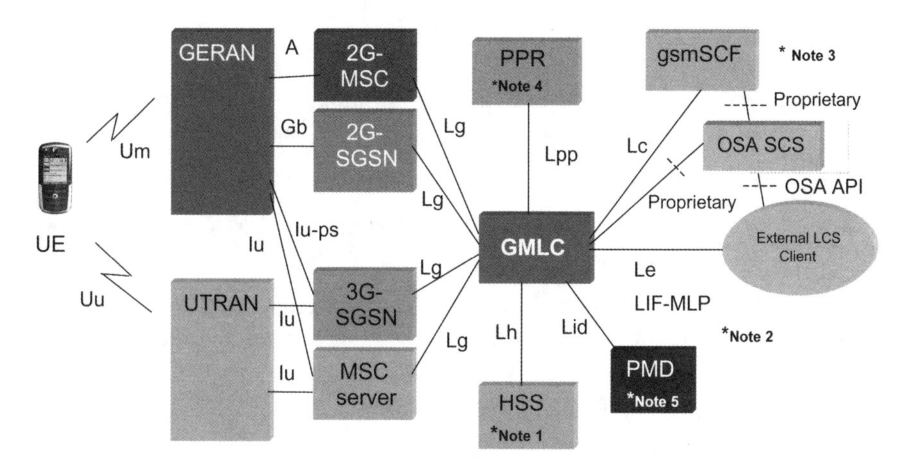

Figure 10.23 The Location Communications Architecture functional building blocks [17].

*Note 1: HSS includes both 2G-HLR/3G-HLR functionality. LCS appears on all overall architecture in [21].

*Note 2: LIF-MLP may be used on the Le interface.

*Note 3: As one alternative the LCS client may get location information directly from GMLC, which may contain OSA Mobility SCS with support for the OSA user location interfaces [22–26].

*Note 4: The PPR functionality may be integrated in GMLC.

*Note 5: The PMD functionality may be integrated in GMLC or PPR.

and UTRAN access networks. Here the LCS entities within the access network communicate with the Core Network (CN) across the A, Gb and Iu interfaces. Interaction between the access network and LCS entities take place through the messaging and signalling capabilities of the access network [15]. The key interactions can be summarised as follows:

- One or more LCS clients may request the location information of UE, which may be associated with the GSM/UMTS networks or the access networks operated as part of a UE application or accessed by the UE through its access to an application (e.g. through the Internet).
- Clients make their requests to a LCS server, where there could be more than one LCS server.
- The client must be authenticated and network resources co-ordinated including the UE and the calculation functions, to estimate the location of the UE and result returned to the client.
- Information from other systems (e.g. other access networks) can be used.
- The LCS server may provide 'estimates accuracy' and 'timing measurement' as part of the location information returned to the client.

10.4.6.1 Functional Description of LCS Operations

The operation starts with a LCS client requesting location information for a UE from the LCS server, the LCS server passes the request to the LCS functional entities in the core network, which do the following [15]:

1. Verify that the LCS client is authorised to request the location of the UE or subscriber and that the UE supports the LCS.
2. Establish if the LCS is allowed to locate the UE or subscriber, for service or operational reasons and which network element in the access network should receive the location request.
3. Request the access network (through the A, Gb or Iu interface) to provide location information for an identified UE, with the corresponding QoS.
4. Receive information about the location of the UE from the access network and forward it to the client.
5. Send appropriate accounting information to an accounting function.

The access network LCS functional entities determine the position of the target UE according to methods described in the preceding section.

10.4.6.2 LCS Function Allocation to Network Elements

From direct extracts [15], Table 10.3 summarises location services functional groups and functional blocks, and Figure 10.24 illustrates the generic configuration for the LCS and its functional blocks to network elements. As described in the preceding section, different positioning methods, i.e. network-based, mobile-based, mobile-assisted and network-assisted positioning methods can apply, where both the network and the mobiles can measure the timing of signals and compute the mobile's location estimate and get configured accordingly. For example, when applying a network-based positioning, the entities involved in measuring the mobile's signal and location-estimate calculation are allocated in the network elements of the access stratum. And when we apply a mobile-based or network-assisted method the entities reside in the UE [15].

We implement the LCS through an additional node in the network, i.e. the Mobile Location Centre (MLC), which will incorporate a number of new interfaces, and influence the generic LCS architecture variants.

The allocation of functional entities in the reference configuration of LCS, described in Table 10.4 and Figure 10.24, assumed that the CS and PS have either their own independent mobility management or use the joint mobility management through the optional Gs interface. On the other hand, the LCS may benefit also from the Iur interface between RNCs, when uplink radio information and measurement results are collected.

The functional model illustrated in Figure 10.24 includes functional entities for both CS and PS related LCS. In addition, as well as all the entities needed for different positioning methods, i.e. network based, mobile based, mobile assisted and network assisted positioning, exploiting either uplink or downlink measurements. It is clear that even if the UE may use,

Table 10.3 Summary of functional groups and functional blocks for location services [15]

Functional Group	Functional component	Full name of Functional Block	Abbreviation
Location Client	Location Client Component	(External) Location Client Function	LCF
		Internal Location Client Function	LCF-internal
LCS Server in PLMN	Client Handling component	Location Client Control Function	LCCF
		Location Client Authorization Function	LCAF
		Location Client Co-ordinate Transformation Function	LCCTF
		Location Client Zone Transformation Function	LCZTF
	System Handling component	Location System Control Function	LSCF
		Location System Billing Function	LSBF
		Location System Operations Function	LSOF
		Location System Co-ordinate Transformation Function	LSCTF
	Subsystem Handling Component	Location Subscriber Authorization Function	LSAF
		Location Subscriber Privacy function	LSPF
	Positioning Component	Positioning Radio Control Function	PRCF
		Positioning Calculation Function	PCF
		Positioning Signal Measurement Function	PSMF
		Positioning Radio Resource Management	PRRM

e.g. the GPS positioning mechanism, it will still demand, e.g. auxiliary measurements from the serving network [15,17].

10.4.6.3 Functional Description of Related LCS Network Elements

Access network depends on the positioning method applied as described in the preceding section.

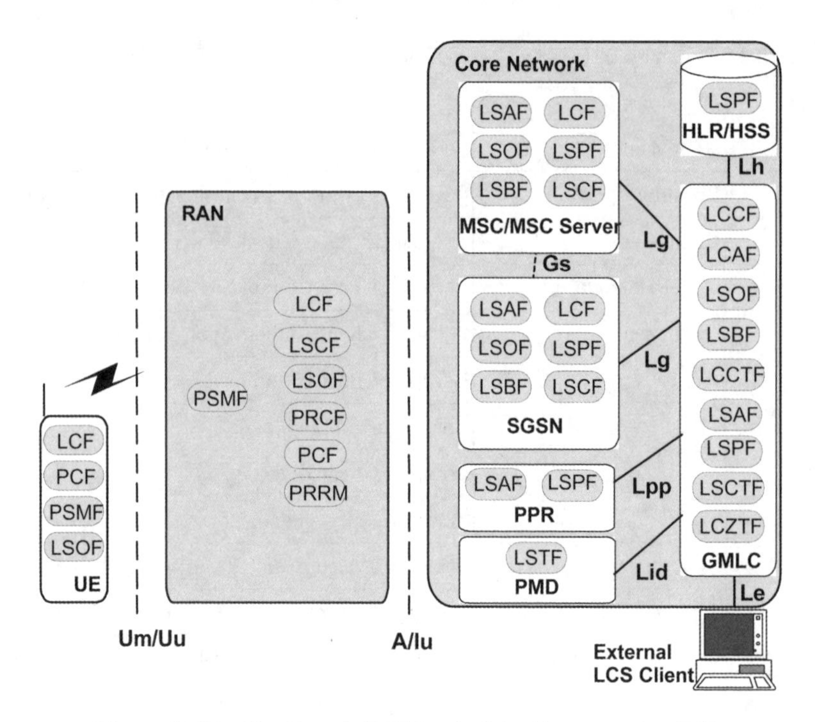

Figure 10.24 The Generic LCS Logical Architecture, after [15].

Internal LCS applications represent entities internal to UMTS, which make use of location information to improve or enhance network operation.

Internal LCS clients, which can be identified an internal ID, include the broadcasting-location client, HPLMN-O&M client, VPLMN-O&M client, anonymous-location recording client bearer service supporting client and UE supplementary service client.

External LCS applications represent entities (e.g. commercial or emergency services), which make use of location information for operations external to the mobile network.

External LCS client, which can also be identified by an external ID, includes the LCS applications interface to the LCS entities through their Location Client Functions (LCF).

Requestors may trigger location requests from external LCS clients, which should be appropriately authenticated.

LCS client may indicate the type of the Requestor [27] identity in the LCS service request as illustrated in Table 10.5.

10.4.6.3.1 Gateway Mobile Location Centre—GMLC

The GMLC contains functionality required to support the LCS, where there may be more than one GMLC in one PLMN.

A GMLC *Le* reference point is the first node an external LCS client accesses in a PLMN. The GMLC may request routing information from the HLR or HSS through the *Lh* interface, which after performing registration authorisation, it sends positioning requests to either VMSC, SGSN or MSC server and receives final location estimates from the corresponding

Table 10.4 LCS functional entities allocation to network elements [15]

	UE	RAN	GMLC	SGSN	MSC/MSC Server	HLR/HSS	PPR	PMD	Client
Location client functions									
LCF	X			X	X				X
LCF Int		X							
Client handling functions									
LCCTF			X						
LCCF			X						
LCAF			X						
LCZTF			X						
System handling functions									
LSCF		X		X	X				
LSBF			X	X	X				
LSOF	X	X	X	X	X				
LSCTF			X						
Subscriber handling functions									
LSAF			X	X	X		X		
LSPF			X	X	X	X	X		
LSTF								X	
Positioning functions									
PRCF		X							
PCF	X	X							
PSMF	X	X							
PRRM		X							
	UE	RAN	GMLC	SGSN	MSC/MSC Server	HLR/HSS	PPR	PMD	Client

Table 10.5 Type of requestor identity

1. Logical name	4. URL [29]
2. MSISDN [28]	5. SIP URL [30]
3. E-mail address [29]	6. IMS public identity [31]

entity through the *Lg* interface. Authorisation information for location service requests and location data exchange between GMLCs, located in the same or different PLMNs, occurs through the *Lr* interface. The target UE's privacy profile settings gets always verified in the UE's home PLMN prior to remit a location estimate. We use the *Lr* interface in order to allow location request from a GMLC outside the HPLMN while having privacy recognition take place in the HPLMN. Now we denote other functions as follows:

- '*Requesting GMLC*'—the one which receives solicitation from a LCS client.
- '*Visited GMLC*'—the one which is associated with the serving node of the target mobile.
- '*Home GMLC*'—the one residing in the target mobile's home PLMN, which is responsible for the control of privacy checking of the target mobile.
- '*Requesting GMLC*'—the visited GMLC, and either one or both of which can be the *Home GMLC* at the same time.

10.4.6.3.2 LCS Support in the UE

The UE, which utilises either of the positioning methods described earlier, interacts with the measurement co-ordination functions to transmit the needed signals for uplink based LCS measurements and to make measurements of downlink signals. It may also contain LCS applications, or access a LCS application through communication with a network accessed through the UE or an UE application, and may include the needed measurement and calculation functions to determine the UE's location with or without assistance of the GSM/UMTS LCS entities [15].

While in GSM the positioning methods supported by the UE are signalled by the UE to the core network and radio-access network using Classmark3 in CS mode [32], in UMTS the UE capability to support different positioning methods is only communicated within UTRAN [17]. More specifically, the UE may also contain an independent location function [e.g. an Assisted Global Satellite Positioning Service (A-GPS)] and thereby be able to report its location, independent of the RAN transmissions, or use RAN broadcast information.

10.4.6.3.3 The MSC/VLR and MSC Server

The *MSC/VLR* contains UE subscription authorisation, managing call-related and non-call-related positioning requests functions of LCS. It is accessible to the GMLC through the *Lg* interface. Some of the MSC's key functions for the LCS include:

- charging and billing
- LCS co-ordination
- location request
- authorisation
- operation of LCS services

When connected to SGSN through the Gs interface, it checks whether the UE is GPRS attached to decide whether to page the UE on the A/Iu or Gs interface [15].

The MSC/VLR may also inform HLR/HSS about the UE's LCS capabilities while including the IP address of the V-GMLC associated with the MSC/VLR in the MAP UPDATE LOCATION message, during registration and inter-MSC update location procedures [15].

The *MSC server* handles the same functionalities as the MSC/VLR including charging and billing, LCS co-ordination, location request, authorisation and operation of the LCS services. However, it communicates wit the GMLC through the *Lg* interface.

10.4.6.3.4 SGSN

The SGSN incorporates functions for UE subscription authorisation and LCS managing positioning requests in the PS domain. It communicates with the GMLC through the *Lg* interface and has the same functions as the MSC.

Like an MSC/MSC server, a SGSN may inform a HLR/HSS on the UE's LCS GPRS capabilities and may include the VE-GMLC IP address associated with the SGSN in the 'MAP UPDATE GPRS LOCATION' message, during Attach and Inter SGSN routing area update procedures. Finally, the SGSN forwards the CS paging request received from the Gs interface to the BSS/RNC [15].

10.4.6.3.5 HLR and HSS

The HLR and HSS contain LCS subscription data and routing information and they are accessible from the GMLC via the *Lh* interface, which for a roaming UE, they may be in a different PLMN [15].

10.4.6.3.6 gsmSCF and Privacy Profile Register—PPR

The *Lc* interface supports CAMEL access to LCS applicable in CAMEL Ph3. The procedures and signalling associated with it are defined in [33] and [34], respectively.

Privacy check may be done in the Privacy Profile Register (PPR). The HLR or HSS contains the address to the PPR, which is accessible from the H-GMLC via the Lpp interface. PPR may be a stand-alone network entity or the PPR functionality may be integrated in H-GMLC [15].

10.4.6.3.7 Pseudonym Mediation Device—PMD

The PMD functionality maps or decrypts the pseudonym into the corresponding verinym (i.e. IMSI or MSISDN). This functionality may be a stand-alone network entity or it may be integrated in the PPR, GMLC or other network entity, where it is not part of GMLC, it may be accessed using the *Lid* interface.

10.4.6.3.8 Target UE Verinyms and Pseudonyms

It is possible to address and indicate the target UE using MSISD, or in certain cases to address the target UE using IP address when a static or dynamic IP address (IPv4 or IPv6) has been allocated for the UE. However, in the PS domain mobile terminated location request procedures (as well as in the CS domain), the target UE is identified using either MSISDN or IMSI, because target UE IP-addressing is only possible when there is an active PDP context established between the target UE and the external LCS client. When using an established PDP context, the LCS client can request the target UE, as identified with the IP address it currently uses, to initiate a mobile originated location request.

10.4.6.3.9 Target UE Pseudonyms
To comply with national regulations requiring anonymity of the target mobile user in some countries, it is possible to address and indicate the target UE using a pseudonym, which may be the IMSI or MSISDN of the target UE encrypted, e.g. using the public key of the home operator. In addition, the network element that issued the pseudonym, i.e. the PMD address, gets attached to the pseudonym, if required, or it can be deduced from the pseudonym; likewise for the H-GMLC address [15].

10.4.6.3.10 Non-dialable Callback Numbers
In case of a SIM-less emergency call, we may use a non-dialable callback number to identify the target UE with a format according to national or regional regulations. The non-dialable callback number in North America, e.g. complies with the J-STD-036 recommendations, i.e. the digits 911 + the last seven digits of IMEI expressed in decimal numbers [35].

10.5 VHE/OSA

The Virtual Home Environment (VHE) as a key component in 3G mobile systems enables end users to bring with them their personal service environment whilst roaming between networks independent of terminals used [36]. Furthermore, in the VHE context, a Personal Service Environment (PSE) empowers users to manage and interact with their communication services, i.e. dynamically set preferences in their list of subscribed services depending on their device capabilities. In addition, PSE also allows users to manage multiple subscriptions, e.g. business and private, multiple terminal types and location preferences based on User profiles [37].

OSA makes possible to implement forthcoming applications exploiting network functionalities and Service Capability Features (SCFs) while enhancing present services. It aims to provide standardised, extendible and scalable interface that allows inclusion of new functionality in the network with a minimum impact on the applications.

The standardised OSA API (using, e.g. CORBA, SOAP based architecture capability servers operating as gateways between the network entities and the applications) is independent of vendor specific solutions, of programming languages, operating systems and other enablers used in the service capabilities. In addition, it is independent of the location within the home environment where service capabilities are implemented, and independent of supported service capabilities in the network [36].

The OSA API, found on lower layers using mainstream information technology and protocols (i.e. CORBA/IIOP, SOAP/XML and other XML based protocols), and lower layer protocols (e.g. TCP, IP, etc.) provides security mechanisms (e.g. TLS, IP sec, etc.) to encrypt data.

In new environments, e.g. IMS, while supporting existing CAMEL based services, allows the development of new services independent of network domain through the home CSCF supporting a CAP interface. As a result, service providers have the option to register their subscribers to the v-CSCF in networks that also support CAMEL.[9] On the other hand, when the visited network does not support CAMEL or the capabilities to support a given service, it will be possible to provide call control through the h-CSCF. This implies that either the

[9]As in GSM today.

user is registered with the h-CSCF directly, or if the user is registered in the v-CSCF, then the v-CSCF will be capable of forwarding the call set-up request to the h-CSCF. The latter will also apply to service providers or operators that choose to have all of their subscribers (including roamers) supported by the h-CSCF [36].

10.5.1 The OSA Main Characteristics

Applications: Corresponds to all applications platforms, e.g. VPN, video conferencing and location-based services, which could be implemented in one or more servers.

Framework: to exploit service capabilities in the network, e.g. authentication and discovery, which can be accessed by methods defined in the OSA interfaces.

Service capability servers: i.e. abstractions from underlying network functionality, e.g. call control and user location.

The OSA service capability features get specified in terms of a number of interfaces and their methods divided into two groups, i.e. *framework interfaces* and *network interfaces.*

The OSA interfaces provides the Service Capability Servers (SCS) as functional entities distributed across one or more physical nodes, e.g. user location interfaces and call control interfaces implemented on a single physical entity or distributed across different physical entities (Figure 10.25). For more details in OSA characteristics see [36].

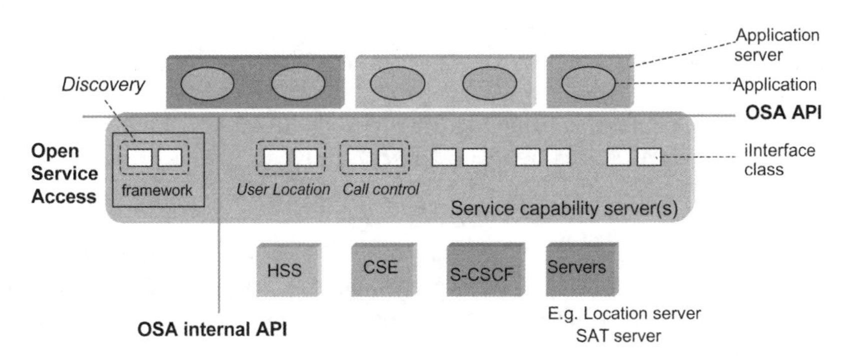

Figure 10.25 Main OSA characteristics, after [36].

10.5.2 Implementation Options of the VHE/OSA

Figure 10.26 illustrates the main components to realize the Virtual Home Environment (VHE). Notice that the applications servers need a set of enablers and management.

The two key options to follow include the following:

- Personalisation—Subscribers can customise their own service set and its corresponding interaction.

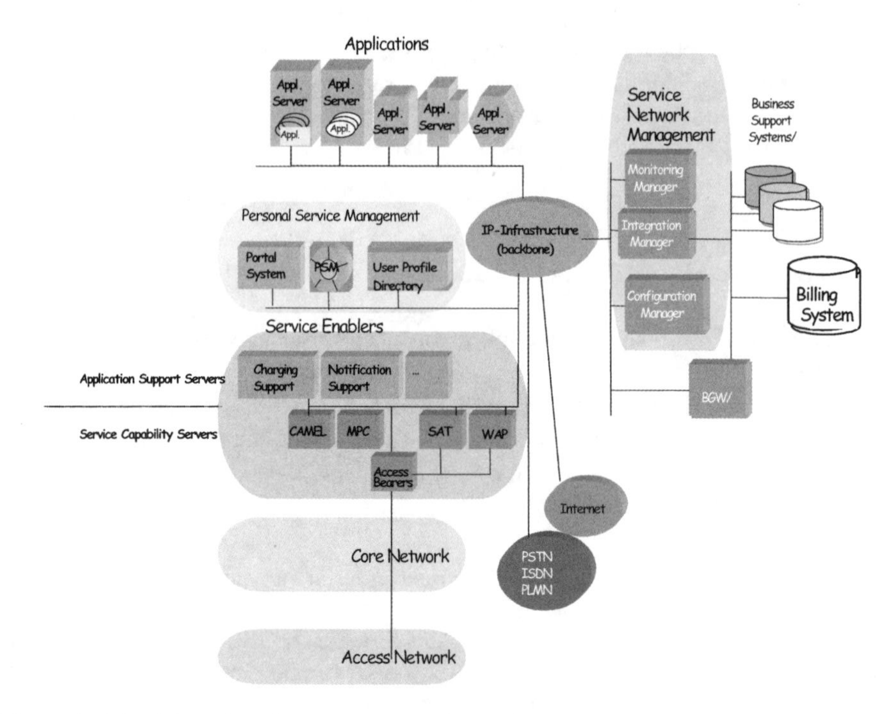

Figure 10.26 IP based service network overview and its relation to core and access networks.

- Transparency—Subscribers can see seamlessly the same service features across network boundaries and between terminals regardless of their location.

To further illustrate the implementation, we can divide the IP-infrastructure service network into functional areas as illustrated in Figure 10.27. Clearly, we have the service enablers as shown in Figure 10.26, then personal service management service network management and applications.

The service platform consists of service network access products, IP infrastructure servers, service enablers consisting of service capability servers, application support servers and other gateways. It also includes service network management, which in turn consists of the personal service management, application servers and applications and hardware/service platforms for the servers.

Thus, a launch service network will incorporate the VHE concept as a scenario where one can consistently access telecom and datacom services, through personalised user interface, from any network, on any terminal, anywhere.

10.5.3 The SIM Application Toolkit (SAT)

The SAT browser on the (U) SIM card and the SIM Application Toolkit Service Capability Server (SAT SCS) allow subscribers to access standard Web applications on the Internet, exploiting the widespread Web technology and the inherent security of smart cards, and opening up a whole new range of applications, such as wireless electronic commerce.

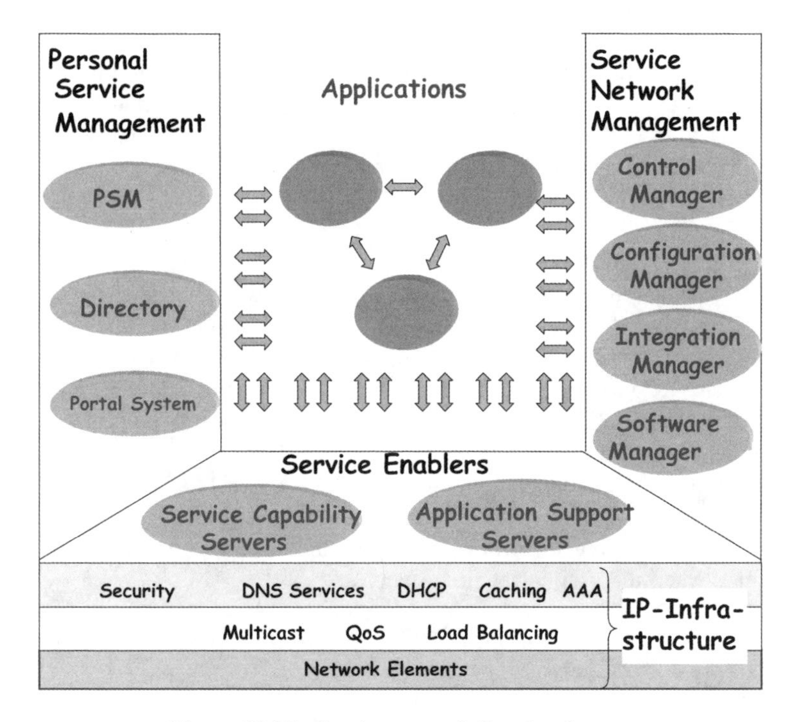

Figure 10.27 Service network functional areas.

A SAT Service Capability Server (SAT-SCS) solution affords service providers or network operators, as well as content providers to supply advanced services using standard tools and either HTML or the Wireless Markup Language (WML), where the Wireless Application Protocol (WAP) defines the latter. Thus, 3G terminals complying with the SIM application toolkit can co-exist with WAP terminals accessing the same services. The SAT SCS does not depend on the SIM card supplier.

Coupling the SIM's inherent potential to solve security issues and Over-The-Air (OTA) provisioning for device management, SAT enables an Internet interface for an application provider to create services based on SMS, OTA SIM File management and SIM application toolkit. On the wireless network side, the SAT SCS uses the SIM toolkit messaging,[10] i.e. for secure transport of the web pages. On the Internet side the SAT SCS connects to one or more applications through HTTP. On the terminal side, the SIM card includes a browser that receives web pages from the applications and converts them into SIM application toolkit commands, thus interacting with the user.

10.5.3.1 SAT Service Capability Server

Further realisation of the SAT will include, e.g. a Service Capability Server (SCS), which we can logically divide into two parts, i.e. the request and the push modules. The SCS receives web pages from the application, converts them into byte code and sends them to the browser

[10]GSM 03.48.

using GSM 03.48 for the transport. The difference lies in how the web page is fetched from the application. The request module waits for requests from the browser for a certain web page, fetches the page and returns it to the browser. The push module waits for the application to send a web page to a certain browser, thus the user or the browser does not have to take the initiative but it rather comes from the application. A web page could either be written in HTML or WML.

10.5.3.1.1 Browser and Menu

The browser resides on the SIM card of the terminal. It runs byte code strings and converts them into SAT commands. The byte code could be fetched from two places, i.e. they can reside on the SIM card or they could be sent from the SAT SCS. When the browser receives byte code it starts to interpret and convert it into SAT commands. The byte code was originally written in HTML/WML and thus the interpretation can be seen as conversion from HTML/WML to SAT commands.

A key part of SIM application toolkit constitutes the menu inserted into the standard menu structure of the terminal. The menu becomes the starting point for the user to access the web through the terminal. Each menu item belonging to the browser points to a byte code string that is executed when the user selects that menu item.

10.6 CONCLUSIONS

The set of service components for UMTS will continue to evolve and their numbers increase continuously. Therefore, the solutions outlined in this chapter are only representations of key platforms, which will/or already facilitate the delivery of 3G services.

The massive investment in license fees for UMTS spectrum[11] in the past, has without a doubt delayed UMTS service providers from practical dedication to the creation of services in closer collaboration with content providers, developers and manufactures. However, today the momentum is back to engage in the implementation of innovative applications and services to take full advantage of the potential of UMTS and its forthcoming technology.

As a result, exploiting streaming, downloading and location based service applications enabling platforms incorporated or independent of OSA solutions is already very high in the agenda for 3G services creation. In fact, we are already coming into a new phase of combining as much as possible these platforms to maximise efficiency and operational costs. For example, nothing should prevent combining streaming and downloading enablers, which in practice are still separate within service environments. After all, both are in reality content distribution channels, where the first operates on real-time live streaming and the second off-line on demand streaming.

On the service provider side, it does not matter within what segment subscribers are, at the end, with the penetration of mobile services, free-Internet and the choice[12] of service provider, users will only care about quality, price, choice and value. Only multimedia applications offering through efficient enabler platforms can sustain these demands profitably.

[11]UK and German operators invested more than 5 billion per UMTS license.

[12]The widespread of 3G service providers will create higher competition in terms of service options.

On the other hand, great challenges also remain for manufactures on the terminal side, to produce rapidly intelligent multi-functional terminals with efficient power consumption. For example, terminals supporting IMS should be rapidly available to developers align applications exploiting the aforementioned platforms in this chapter.

In conclusion, there exist expanding possibilities for service innovation and technology utilisation in applications development to keep up or motivate multimedia services penetration. Thus, broadband wireless pipe along with high performing application enablers and platforms should be the focus for future mobile services development. To focus on only one or the other will not fulfil user expectations or revenue oriented service providers.

REFERENCES

1. Drafting Team. Vision and Road Map for UMTS Evolution, TSGS#8(00)0337, 2000.
2. 3GPP, Study on Release 2000 SERVICES and Capabilities (3G TS 22.976), V2.0.0, 2000–2006.
3. TS 22.003. 3rd Generation Partnership Project, Technical Specification Group Services and System Aspects, Circuit Tele-services Supported by a Public Land Mobile Network (PLMN).
4. TS 22.002. 3rd Generation Partnership Project, Technical Specification Group Services and System Aspects, Circuit Bearer Services (BS) Supported by a Public Land Mobile Network (PLMN).
5. TS 22.004. 3rd Generation Partnership Project, Technical Specification Group Services and System Aspects, General on Supplementary Services.
6. TS 22.060. General Packet Radio Service (GPRS) Stage 1.
7. 3GPP, Architecture Principles for Release 2000 (3G TR 23.821), V1.0.1, 2000–2007).
8. 3GPP, Combined GSM and Mobile IP Mobility Handling in UMTS IP CN 3G TR 23.923 version 3.0.0, 2000–2005.
9. IETF RFC 2002 (1996): IP Mobility Support, C. Perkins.
10. Internet draft, Johson and Perkins, Mobility Support in IPv6, October 1999. http://www.ietf.org/internet-drafts/draft-ietf-mobileip-ipv6-09.txt
11. Source Nokia, 3GSMA Cannes, Feb 2003.
12. 3GPP, Technical Specification Group (TSG) SA, Transparent end-to-end Packet Switching Streaming Service (PSS) General Description, Release 5, v5.0.0, 2002–2003.
13. ITU-T H.323, Packet Based Multimedia Communications Systems, 1998.
14. Sun™ Content Delivery Server 3.6, Monetize, Manage, and Securely Deliver Content October 2003, Sun Microsystems.
15. 3GPP, Functional stage 2 description of Location Services (LCS), TS 23.271 V6.6.0, 2003–2012.
16. 3GPP TS 23.032: Universal Geographical Area Description (GAD).
17. 3GPP, Stage 2 functional specification of UE positioning in UTRAN, 3G TS 25.305, V6.0.0, 2003–2012.
18. 3GPP TS 25.214: Physical layer procedures (FDD).
19. 3GPP TS 25.215: Physical layer—Measurements (FDD).
20. 3GPP TS 25.225: Physical layer—Measurements (TDD).
21. 3G TS 23.002: Network architecture.
22. 3G TS 23.127: Virtual Home Environment /Open Service Access.
23. 3G TS 29.198-1: Open Service Access (OSA); Application Programming Interface (API); Part 1; Overview.
24. 3G TS 29.198-2: Open Service Access (OSA); Application Programming Interface (API); Part 2; Common Data.
25. 3G TS 29.198-3: Open Service Access (OSA); Application Programming Interface (API); Part 3; Framework.

26. 3G TS 29.198-6: Open Service Access (OSA); Application Programming Interface (API); Part 6: Mobility.
27. LIF TS 101: Mobile Location Protocol Specification (Location Interoperability Forum) [Available at http://www.openmobilealliance.org/tech/LIF/]
28. 3G TS 23.003: Numbering, addressing and identification.
29. RFC 2396: Uniform Resource Identifiers.
30. RFC 3261: SIP: Session Initiation Protocol.
31. 3G TS 23.228: IP multimedia subsystem (IMS).
32. 3G TS 24.008: Mobile Radio Interface—Layer 3 MM/CC Specification.
33. 3G TS 23.078: Customised Applications for Mobile network Enhanced Logic (CAMEL)—stage 2.
34. 3G TS 29.002: Mobile Application Part (MAP) Specification.
35. ANSI J-STD-036A: Enhanced Wireless 9-1-1 Phase 2.
36. Virtual Home Environment/Open Service Access, 3GPP TS 23.127 V5.2.0, (2002–2006).
37. 3G TS 22.121: Service Aspects; The Virtual Home Environment.

11

Resource and Network Management

11.1 INTRODUCTION

Operating a 3G network involves managing resources and Network Elements (NE). This chapter covers these two aspects to complete the deployment issues started in Chapter 7. Resources here refer primarily to the radio resources and NE refers to the 3G building blocks, i.e. elements in the CS, PS and radio-access networks.

11.2 RADIO RESOURCE MANAGEMENT AND SIGNALLING

Power control constitutes one of the major tasks of Radio Resource Management (RRM). Other tasks such as admission control, load control and packet scheduling also correspond to RRM; however, we will not emphasise them in this section. Power control aims to minimise interference levels in order to maintain an expected transmission quality in the air-interface. The UTRA FDD mode depends on soft blocking to efficiently manage multi-rate services. This takes place according to appropriate RRM algorithms covered in Chapter 4.

11.2.1 Managing Power

Power control becomes more critical in the FDD than in the TDD mode. Thus, this section concentrates primarily on managing power in WCDMA. The impacts on handover are also presented.

In WCDMA all users share the same RF band separated by spreading codes. As a result, each user appears as a random noise to other users. Non-controlled individual power can therefore interfere un-necessarily with those sharing the same frequency band. To illustrate the need for power control Figure 11.1 shows two MSs in the UL. MS_1 gets closer to the BS than MS_1, now if there was no power control, both MSs would transmit at their fixed power P_T. But since MS_1 is closer, it would have higher power than that of MS_2 if we assume that the distance of the latter is three times greater than that of MS_1. Thus, if the required SNR (S/N_{required}) is 1/3, then $S/N_1 = 3$ and $S/N_2 = 1$. Thus, MS_2 will suffer the classical near-far effect and may not satisfy the quality of service required in the link. Furthermore, any third MS coming into the cell will not get the required *S/N* either, and may even cause

All IP in 3G CDMA Networks J. Castro
© 2004 John Wiley & Sons, Ltd ISBN: 0-470-85322-0

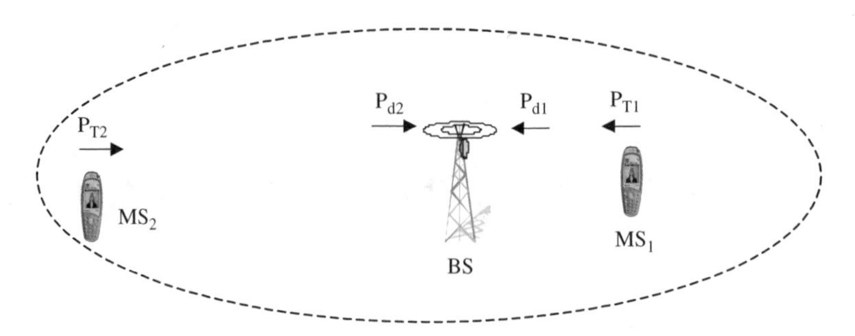

Figure 11.1 Power control to prevent near-far effect.

MS_2 to drop its *S/N* even lower. Power control will thus aim to overcome near-far effects and thereby increase capacity with acceptable link quality.

11.2.1.1 Fast Power Control (FPC)

The FDD mode uses fast power control with 1.5 kHz frequency (i.e. 1500 times/s) in both UL and DL. It operates at a faster rate than any path loss change. The FPC uses the closed-loop option as noted in Chapter 4. We see higher gains of FPC in low mobile speeds than for high mobile speeds, and in received powers than in transmitted powers. At speeds above 50 km/h, e.g. FPC does not contribute much due to the higher multi-path gains. We can find more information about fast power control in [1].

Other gains of FPC depend on diversity, e.g. multi-path diversity, receive, transmit antenna diversity and macro-diversity. Less diversity implies more variations in the transmitted power. Thus, we get smaller *power rise*[1] in the presence of more multi-path diversity.

In DL macro-cell coverage with WCDMA, *power rise* gets critical because it directly intervenes in the required transmission power, which determines the transmitted interference. Hence, to maximise the DL capacity, we should select the quantity of diversity, such that it minimises the transmission power required by a link, since the received power level does not affect the capacity in the DL.

In the UL, the level of transmission power from the different MSs does have direct impact on the interference to the adjacent cells, and the received power determines the level of interference to other users in the same cell. Diversity in this case does not have much impact, which means that UL capacity of a cell would be maximised by minimising the required received powers, and the amount of diversity would not affect the UL capacity.

When MSs move at high velocities, the FPC does not follow fast fading; we would require higher received power level to obtain the expected quality. Thus, in this scenario diversity does help to maintain the received power level constant, thereby allowing a lower average received power level to provide the required quality of service.

[1]If we define *power rise* as the relative average transmission power in a fading channel compared to the non-fading, while the received power level is the same both in fading and in non-fading channels with ideal power control.

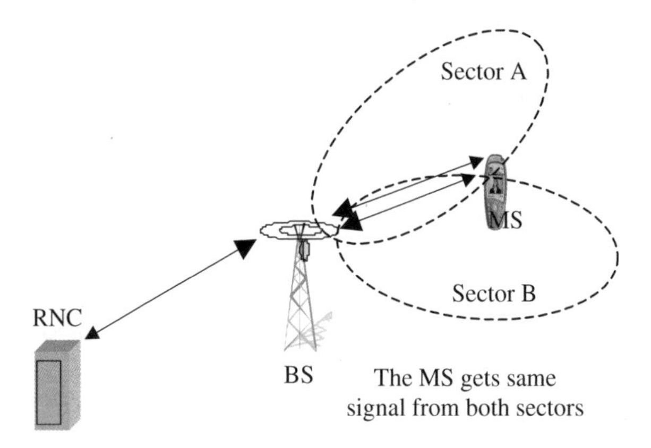

Figure 11.2 Softer handover event.

11.2.1.2 Power Control in Handover (HO)

Before we discuss power control in HO, we briefly review the HO types. The two types of HO in our FDD mode include softer and soft HO.

11.2.1.2.1 Softer Handover

As illustrated in Figure 11.2, softer HO occurs when an MS passes through the overlapping coverage of two adjacent sectors of a BS. Communications between the BS and MS take place concurrently through two channels (i.e. one to each sector or cell). The concurrent links use two separate DL codes, so the signals are perceived by the RAKE receiver and processed as in multi-path reception, but with the RAKE fingers generating the corresponding code for each sector.

A similar process occurs in the UL, each BS sector receives the MS code, which gets routed to the same RAKE receiver for maximal ratio combining. In softer HO we have only one power-control loop active per connection.

Softer HO events do not exceed 16% of established links, and in the UL we do not use additional resources except for the extra RAKE fingers. Neither does the BS need to provide additional DL transmission power to complete the softer HO process.

11.2.1.2.2 Soft Handover

In soft handover, an MS passes through the overlapping cell coverage area of two sectors, which correspond to different BSs, e.g. BS-a and BS-b as illustrated in Figure 11.3. Communications between the MS and BS occurs concurrently through two different channels, i.e. one from each BS. The MS receives both signals by maximal ratio combining RAKE processing.

While in the DL *softer* and *soft* HO behave basically in the same way[2] and the MS does not see any difference between them, in the UL *soft* HO behaves differently. For example, the MS receives the code channel from both BSs. This information then gets routed to the RNC for macro-diversity combining thereby to obtain the same frame reliability indicator

[2]Thus, soft and softer HO can also take place in combination with each other.

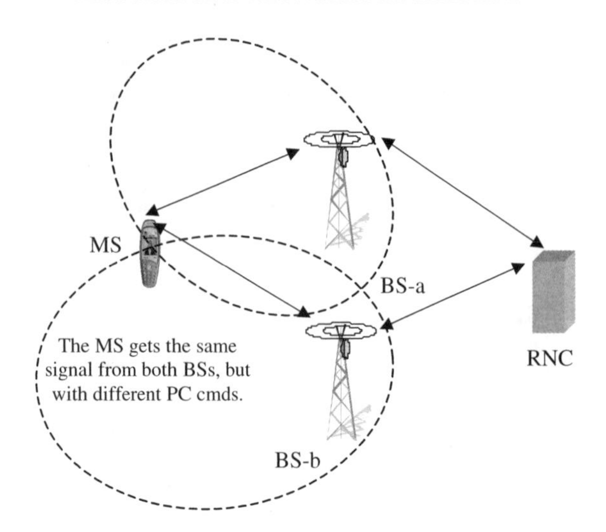

Figure 11.3 Soft handover event.

provided for outer loop PC, i.e. to select the best frame after each inter-leaving period within 10–80 ms.

In general, *soft* HO will not exceed 40% of the links. However, it will not go below 20% either. Thus, we cannot neglect soft HO overhead when dimensioning. For example, we must allocate extra transmission power in the BS, extra BS RAKE receiver channels, extra RAKE fingers in the MS and extra transmission links between the BSs and the RNCs.

An appropriate provision and/or an efficient FPC management in WCDMA will maintain most of its total capacity[3] during HO. In FPC we need to deal effectively with the BS power drifting and the accurate detection of UL power control commands from the MS.

Inaccurate reception of power-control commands in the BS due to propagation impacts, such as delay or shadowing, will trigger undesired power events from the BSs, e.g. increasing power when expecting power decrease. This power drifting will degrade *soft* HO. On the other hand, the RNC can control such drifting by limiting the power-control dynamics or by obtaining DL reference transmission power levels from the BSs. Then send this reference value for the DL transmission powers to the BSs.

In the UL all BSs send independent power-control commands to the MS to control its transmission power. The MS can then decrease its power if one BS demands so, and apply maximal ratio combining to the data bits in soft HO since the same data is sent from all soft HO BSs.

11.2.1.3 Outer-Loop Power Control

We use outer-loop power control to keep the quality of the FPC communication at the required level. An excessive high FPC quality will waste capacity. Outer-loop power control

[3]Otherwise, up to 40% of the total capacity can decrease.

applies to both UL and DL, since FPC also applies to both.[4] While FPC has a frequency of 1.5 kHz, the outer loop power control has a frequency range of 10–100 Hz.

11.2.1.4 Conclusions

In the preceding sections, we have highlighted power control and handover aspects primarily to indicate their importance when planning for capacity and coverage. Other sources such as [2–6,7,8] cover more in depth power-control issues.

Other related areas of radio resources for the FDD mode, e.g. admission control, are found in [9–12]. Sources that apply to the resource management of the TDD mode are found in [13–16].

11.3 NETWORK MANAGEMENT

11.3.1 Introduction

Forthcoming 3G systems such as UMTS will serve as multi-technology platforms[5] for new and innovative services. These services will appear within a highly competitive market demanding uniqueness at the best price. To meet the demands, it will be imperative to maintain efficient operational costs through an appropriate NE management system. We will obtain the ideal NMS only through the right combination of NE element control techniques. On the other hand, because of the widespread 2G networks evolving into 3G, managing UMTS NE will not be the only challenge. We also need integrated 2G/3G systems.

11.3.2 Network Management Characteristics

Considering the items in the preceding section, an NMS will have at least the following characteristics:

- Capabilities to integrate and manage 2G NE besides 3G building blocks.
- Support advanced functions and techniques to cope with the multi-farious UMTS technology and maintain diverse service functionality as well as quality of service provision.
- Have an inherent easy-to-use man–machine interface to minimise personnel training requirements.
- Support a multiple set of protocols and open interfaces to interact with multi-vendor equipment.

[4]In IS-95 outer-loop power control applies only to the UL because there is no fast power control in DL.
[5]For example, IP, ATM; WCDMA, etc.

In the context of GSM as a 2G system, a basic set of capabilities will include network management applications in combination with technology specific features to appropriately deploy and operate all components of a complex GSM/GPRS/UMTS network.

11.3.3 A Generic Functional View of a 3G NMS System

Figure 11.4 illustrates a reference architecture of an integrated NMS system capable of managing a combined 2G/3G network. A layered approach allows us to address the complex hybrid system to monitor, i.e. GSM and UMTS NEs and performance.

At the network management level the essential functions would include the following:

- *Fault control* – control and monitor the function and performance of allocated network resources.
- *Ticketing and reporting* – trouble reporting and service assignment to the operations team.
- *Set-up and configuration* – assist in complex system parameter configuration.
- *Resource management* – data and inventory tracking to provide visibility of available physical resources in the network.

At the sub-network management layer, the integrated architecture will aim to gather different sub-domains into one domain. This blending of different control technologies will provide a unified management process. The result will afford a consolidated view of alarm surveillance, performance and configuration access to all related nodes of the integrated domain. The sub-domains include but are not limited to the following:

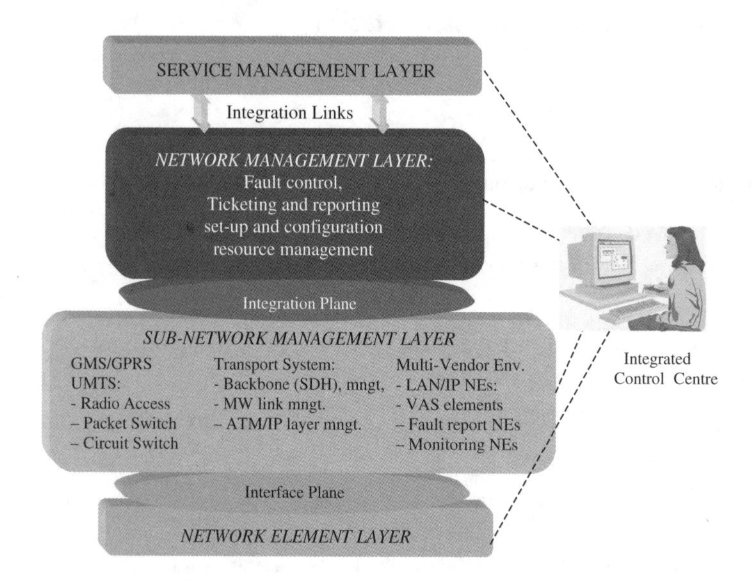

Figure 11.4 A layered NMS architecture reference.

- *GMS/GPRS and UMTS sub-domains* – incorporating radio access, packet and circuit-switching network elements.
- *The transport system* – has to do primarily with the core transport network incorporating, e.g. a SDH backbone, a set of micro-wave links and overlay ATM/IP network running on the SDH ring.
- *The multi-vendor environment set-up* – stands to support NE from different vendors, which will continue as part of a common element to 2G/3G or evolve through upgrade from 2G to 3G. The set-up may incorporate LAN or IP, VAS and fault report or monitoring NEs.

11.3.4 Main 3G Network Elements for Management

In the following, we describe the components of the network element layer illustrated in Figure 11.4. We start by outlining the elements corresponding to the radio-access network. However, because our interest lies primarily with the 3G elements, we describe mainly the elements corresponding to the UTRAN.

11.3.4.1 The UTRAN Building Blocks

The main components of UTRAN (illustrated in Figure 11.5), which would be managed by the integrated management system proposed in the preceding section, include:

- 3G Base Stations (BS, in 3GPP called Node B);
- site solution products, e.g. antennas and power systems;

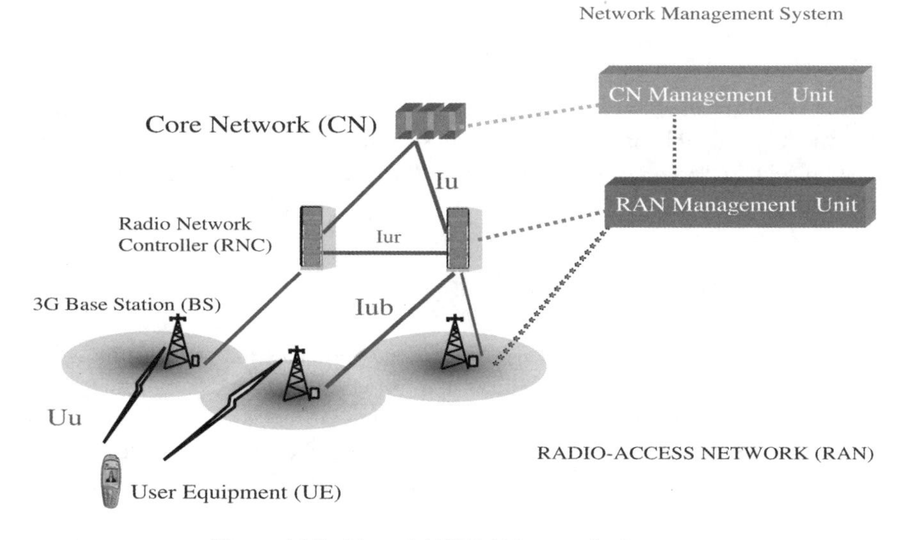

Figure 11.5 Essential UTRAN network elements.

- Radio Network Controllers (RNC);
- UTRAN functions (software for RNC and BS);
- radio-access network management.

Briefly reviewing from Chapter 3, the RNC takes care of the radio-access bearers for user data, the radio network and mobility. The 3G BS provides the radio resources. The main interfaces are Iu interface between RNC and CN and Uu between User Equipment (UE) and Node B or 3G BS. Within UTRAN, the RNCs communicate with each other over Iur and with 3G-BSs over Iub. The key functions to manage are thus as follows:

- *The Radio-Access Bearer* (RAB) functionality provides the CN with a set of services between the core network and the UE. It offers RABs appropriate for voice, CS data and PS data, including required information processing and signalling. It also supports multiple RAB connections to one UE, e.g. both voice and packet-switched services concurrently to one MS.
- *Link control functions*, i.e. paging, signalling channel management, RAB services and allocation and control of radio and other RAB resources.
- *Mobility functions* include handover, cell re-selection, macro-diversity combining and location update management.
- *Capacity management functions,* i.e. control the trade-off[6] between capacity, quality and coverage. The essential tasks are:

  capacity control handling allocation of the radio resources, which depends upon resource information from involved cells and neighbouring cells,

  admission control managing access of new users into the network based; it depends on network load status, subscriber priorities and resource availability,

  congestion control reducing load in high load situations, e.g. by queuing or delaying packet or best effort traffic,

  quality control based on power-control features.
- *Transmission and interface control* will aim to manage the logical interfaces, Iu, Iur and Iub, which can flexibly be mixed on the physical transport. For example, we can use the same links for access to the CN to carry Iur, or concentration of traffic to several 3G BSs on one physical link.

11.3.4.2 The Core Network (CN) Building Blocks

The management of the CN components in this example takes into account the horizontal integration of functional elements. As illustrated in Figure 11.1, the architecture has a total separation of the payload transport and traffic control into the user plane and the control plane, respectively. Here the media gateways constitute the centre components in the first plane, and switching servers (e.g. MSC; SGSN servers) in the database platforms (e.g. HLR)

[6]A precise management and control of the trade is critical to for the FDD mode or WCDMA.

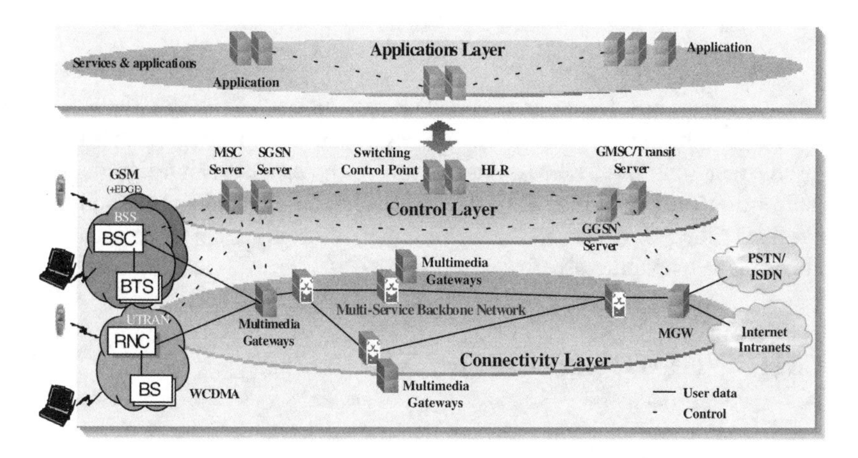

Figure 11.6 3G-CN elements for integrated management.

in the second plane. In the user plane, we aim to manage the traffic flow; and in the control plane, we will make sure that the traffic intensity does not overwhelm system boundaries.

11.3.4.2.1 Media Gateway Nodes (MGW)

The MGW nodes as constituents of the user plane handle CS and PS information and connect to the fixed network for CS traffic (ISUP) and PS traffic (internet/corporate LANs, etc.), and to the RAN through the RNC. Various traffic control nodes connecting through H.248 links (Chapter 6) manage the MGW.

11.3.4.2.2 Traffic Control Servers

The traffic control servers include CS and PS servers.

The MSC Server Nodes The MSC server controls the CS traffic in the MGW, including traffic transported on an IP/ATM backbone. The NMS will need thus to capture MSC server functions such as typical MSC functions, GMSC, VLR and signalling functions.

Packet Traffic Control Nodes The two PS servers include a Serving GPRS Support Node (SGSN Server) and a Gateway GPRS Supporting Node (GGSN). These server nodes maintain and update contexts for all attached users of packet data services. In the case of the SGSN server, the contexts focus primarily on macro-mobility, while in the GGSN the contexts deal with the type of network connections.

The Subscriber Database (HLR)

The HLR serves as a common platform for CS traffic servers (i.e. MSC servers) and the PS traffic servers (SGSN servers and GGSN nodes). It stores subscriber data downloaded to the nodes, from a domain where a subscriber presently roams.

11.3.4.3 Conclusions

In the preceding sections, we have outlined mainly the types of 3G functions that an integrated NMS will have to capture. Thus, we assume that a new 3G NMS will incorporate the typical 2G functions from GSM systems, for example, and seamlessly integrate them into its control mechanism. Many of the 3G logical functions will have the same operation principle as that of the 2G. However, the separation of the control and users planes will bring a new dimension to managing a network.

11.4 UMTS NETWORK OPTIMISATION

Network optimisation will depend on the operating environment, the loads for which we design the network and the appropriated allocation of resources.

The operating environment cannot neglect interference from adjacent networks, assuming the internal network interference is under control. Thus, in the following before we address or review capacity or load enhancing options, and efficient ways to allocate resources, we deal briefly with multi-operator interference issues.

To maximise the performance of the FDD (i.e. WCDMA) system, we need a minimum spectrum mask for a transmitter and highest selectivity for a receiver in the MS and BS, in order to minimise adjacent channel interference. In this context, we define the Adjacent Channel Interference Power Ratio (ACIR) as the ratio of the transmission power to the power measured after a receiver filter in the adjacent channel(s). Here we measure both the transmitted and the received power with a root-raised cosine filter response with roll-off 0.22 and a bandwidth equal to the chip rate as described in Chapter 4. ACIR occurs due to imperfect receiver filtering and a non-ideal transmitter. In the UL we get ACIR from the non-linearities of the MS power amplifier, where inter-modulation originates adjacent to channel leakage power. In the DL, the receiver selectivity of the FDD terminal will have great impact on ACIR. Technical specifications in [17] recommend for both UP/DL are 33 dB for adjacent carriers with 5 MHz separation, and 43 dB for the second adjacent carrier with 10 MHz separation.

11.4.1 ACIR Impacts in a Multi-operator Environment

Non-co-located BSs of two different operators can originate near-far effects; in particular, when an MS closer to another operator's BS stays far from its own BS. Despite the usage of different carriers, total interfering signal suppression will not be possible. Thus, the BS receiving the interference cannot control the output power of the interfering MS because it belongs to another operator.

As a result, there exists a need for Adjacent[7] Channel Protection (ACP), which is the ratio of the transmitted power and the power measured after a receiver filter in the adjacent channel. The ACP results from the combination of out-of-band emission and receiver selectivity, where these two quantities need balance to prevent over-specification.

[7]Adjacent channel may refer to the channel closest to the assigned channel and the second adjacent channel.

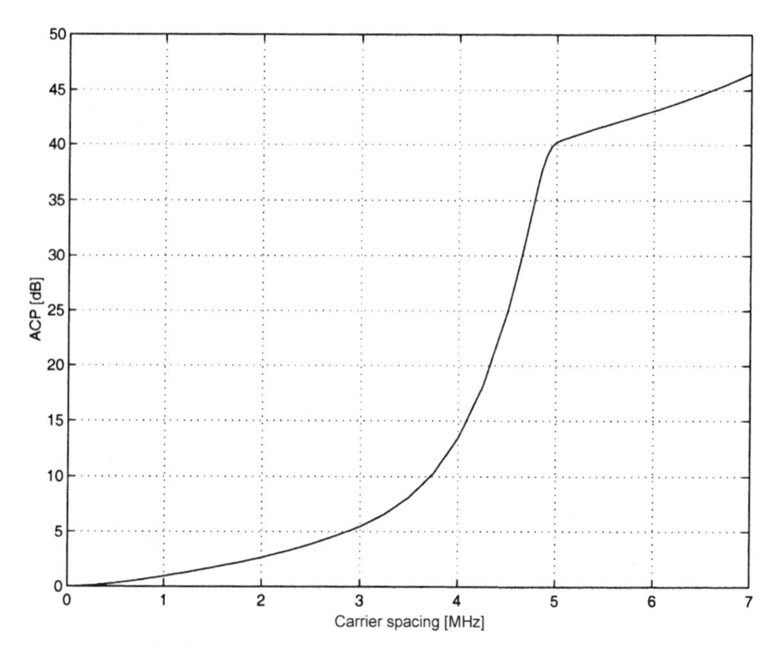

Figure 11.7 ACP as a function of carrier spacing [18].

In [18] measurements have been made on a commercial PA, and a model has been derived. Integrating the power spectrum over a receiver filter gives the ACP. For different offsets of the filter, the ACP as a function of carrier spacing was obtained. Assuming a receiver filter matched to the transmitter pulse, Figure 11.7 illustrates the ACP, where the curve gets steep just below 5 MHz, and where it becomes flatter thereafter. This implies that the Adjacent Channel Interference (ACI) increases when the carrier spacing falls below 5 MHz and that the ACI marginally decreases for carrier spacing larger than 5 MHz.

It also implies that capacity loss depends on ACP, cell radius and relative location of the BSs. Thus, co-location of BSs will ease the near-far problem, and that location of a base station on another operator's cell border gives the worst case. Consequently, site sharing with other operators in WCDMA will be imperative. In conclusion, about 35–40 dB ACP gives a worst case capacity loss of 5–10% for the worst case located base stations.

In UTRA, e.g. FDD, the nominal carrier spacing of 5.0 MHz can get adjusted in steps of 200 kHz according to the needs of the adjacent channel interference. Figure 11.8 illustrates this possibility.

Therefore, the process of optimising the network to minimise ACIR may involve maximum site sharing between operators, a dynamic adjustment of the carrier spacings within a network as well as inter-networks, de-sensitisation of the BS receiver, inter-frequency handovers when excess ACIR occurs and ideal antenna location or co-location.

11.4.2 Enhancing and Managing Capacity

Efficient network dimensioning and optimised BS deployment will complement a very good admission control system to manage and maximise capacity successfully.

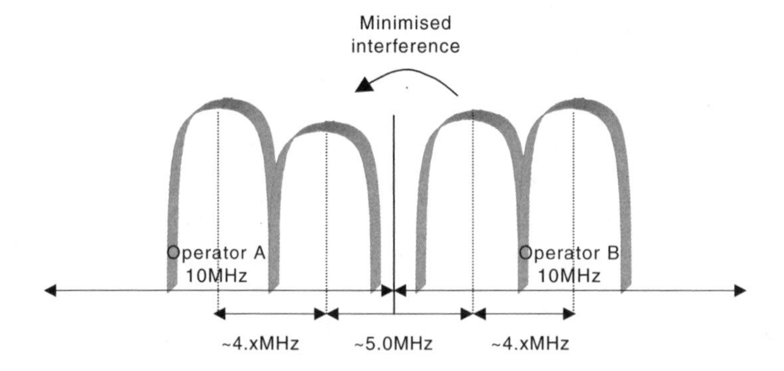

Figure 11.8 Carrier spacing example between operators.

Admission control has to do much with the RNC to maintain the requested quality of service of the radio links. It monitors the load (i.e. downlink power and UL interference level) on the cell carrier. RNC functions desiring to allocate radio resources on given cell carrier request to the admission control for the target cell resources. The functions triggering the admission control tasks may include:

- signalling connection set-up (allocating for radio resources administration);
- soft/softer HO;
- inter-system HO (e.g. GSM to UMTS or vice versa);
- radio link co-ordination (i.e. setting up radio resources);
- channel type switching/selection (e.g. common to dedicated channel);
- paging;
- code allocation/re-allocation.

The admission control also prioritises depending on the type of calls, e.g. emergency calls will precede a normal (voice or data) call. System control tasks such as handovers will also have priority over normal data calls. The admission control will thus discriminate between administrative tasks such as soft/softer HO, service calls and/or emergency calls.

Traffic levels representing the load of service request to a network will start the operation of the admission control for management of congestion control and link quality management. Thus, in the following we briefly discuss the factors affecting load control.

11.4.2.1 Load Control Analysis

We measure traffic intensity generating a certain load in Erlangs. We can define the latter as the average number of simultaneous calls in a given period. For example, we can measure a series of calls lasting x within 1 h in Erlangs, i.e.

$$\frac{\sum \text{calls} \times x\,\text{s}}{3600\,\text{s}} = \text{Erlangs} \tag{11.1}$$

If what we can measure of the generated traffic in a network is the carried load in Erlangs, then the offered load can be expressed as

$$\text{offered\_load} = \frac{\text{carried\_load}}{(1 - \text{blocking\_rate})} \tag{11.2}$$

where the blocking rate refers to the measured blocking probability measure in a given BS. The blocking probability, or grade of service, can be quantified through the Erlang-B formula, i.e.

$$P_{\text{blocking}} \frac{\rho^C / C!}{\sum_{i=0}^{C} \rho^i / i!} \tag{11.3}$$

where the C is the number of channels and ρ is the offered load. Based on the typical assumptions [19] for the offered traffic we can formulate the offered load as

$$\rho = \frac{\lambda}{\mu} \tag{11.4}$$

where λ is the Poisson arrival rate of λ calls/s, and $1/\mu$ is the exponential call service time of $(1/\mu)$ s/call. Notice that the logic applies mainly to CS calls; for data calls we need to use other distributions described in Chapter 2. Applying the blocking principle to the WCDMA we can now define soft blocking and hard blocking.

Soft blocking assuming UL limitation can be defined from the following assumptions: perfect power control, each subscriber requiring the same E/N_o, constant number of users in the cell. However, none of the three assumptions will hold in practice specially with mixed services. Nevertheless, we use them here. Then, we can say that soft blocking occurs when the total interference level exceeds the background noise level by a given quantity. Then from the logic for capacity started in Chapter 7 we define the total interference as

$$I_{\text{total}} = ME_bR(1 + \eta) + N \tag{11.5}$$

where M is the number of users in the same cell, E_b equals the energy per bit of the signal, R is the baseband data rate, N is the thermal noise power and η is the loading factor defined in Chapter 7 [20]; i.e. the interference ratio brought in by the MSs served by other cells to interference generated by MSs served by the home cell. Then, if we get soft blocking when I_{total} exceeds the background noise level by a given quantity $1/x$, we can prevent soft blocking if

$$I_{\text{total}} \geq ME_bR(1 + \eta) + N \tag{11.6}$$

If we define $x = N/I_{\text{total}}$, we can define the soft capacity as a function of the maximum allowable interference level for our simple model, which would not necessary apply in practice, but can serve to understand the concept of load control

$$M \leq \left(\frac{W/R}{E_b/I_o}\right)\left(\frac{1 - x}{1 + \eta}\right) \tag{11.7}$$

To conclude the load control, we next define hard blocking. Here we assume that the actual ratio of I_{total} to background noise is negligible, i.e. there exists very low probability for soft blocking.

To define hard blocking we first define the actual load and the conditional load. The first one refers to the real load passing through a network in action, while the second one refers to the actual load plus the administrative traffic such as soft handovers. Then we introduce a traffic factor as

$$\text{traffic\_factor} = \frac{\text{actual\_load}}{\text{conditional\_load}} \tag{11.8}$$

Then, the actual load is

$$\text{actual\_load} = \text{traffic\_factor} \times \text{conditional\_load} \tag{11.9}$$

Thus, we have a simple way to quickly determine the number of voice channels, e.g. we can use the actual load in conjunction with Erlang's C formula for the blocking rate. This formula implies that a subscriber who is blocked will re-originate calls until he/she goes through. This blocking probability is equivalent to the probability that the call has been delayed, and it can be expressed as

$$P_{\text{delayed}} = \frac{\rho^C/C!}{\frac{\rho^C}{C!} + \left(1 - \frac{\rho}{C}\right)\sum_{i=0}^{C-1} \rho^i/i!} \tag{11.10}$$

11.4.2.2 Conclusions

Enhancing and managing capacity has a lot to do with the condition of the projected traffic and the state of the network. The admission control and traffic congestion control in UTRA has more than voice traffic to deal with. Thus, the preceding formulation needs to be extrapolated to the mixed traffic before it can be applied directly. The traffic factors need to take into account background traffic.

REFERENCES

1. Sipilä, K., Laiho-Steffens, J., Wacker, A. and Jäsberg, M. Modelling The Impact Of The Fast Power Control On The WCDMA Uplink, Proc. VTC'99, 1999, pp. 1266–1270.
2. Viterbi, A.J., *et al.*, Soft handoff extends CDMA cell coverage and increases reverse link capacity, *IEEE J. Selected Areas Commun.*, **12**(8), 1994.
3. Viterbi, A.J., CDMA *Principles of Spread Spectrum Communications*, New York: Addison-Wesley, 1995.
4. Viterbi, A.J., Viterbi, A.M. and Zehavi, E., Performance of power-controlled wideband terrestrial digital communications, *IEEE Trans. Commun.*, **41**(4), 1993.
5. Viterbi, A.J. and Padovani, R., Implications of Mobile Cellular CDMA, *IEEE Commun. Mag.*, 1992.

6. Salonaho, O. and Laakso, J., Flexible Power Allocation for Physical Control Channel in Wideband CDMA, Proc. VTC'99, Houston, TX, 1999, pp. 1455–1458.
7. Huang, C.Y. and Yates, R.D., Call Admission in Power Controlled CDMA Systems, VTC'96, 1996.
8. Zander, J., Performance of optimal power control in cellular radio systems, IEEE Trans Vehic. Tech., **1**, 1992.
9. Knutsson, J., Butovitsch, T., Persson, M. and Yates, R., Evaluation of Admission Control Algorithms for CDMA System in a Manhattan Environment, Proc. 2nd CDMA Int. Conf., CIC '97, Seoul, South Korea, 1997, pp. 414–418.
10. Knutsson, J., Butovitsch, P., Persson, M. and Yates, R., Downlink Admission Control Strategies for CDMA Systems in a Manhattan Environment, Proc. VTC'98, Ottawa, 1998, pp. 1453–1457.
11. Liu, Z. and Zarki, M., SIR based call admission control for DS-CDMA cellular system, *IEEE J. Selected Areas Commun.*, **12**, 638–644, 1994.
12. Holma, H. and Laakso, J., Uplink Admission Control and Soft Capacity with MUD in CDMA, Proc. VTC'99, Amsterdam, 1999, pp. 431–435.
13. Berg, M., Petterson, S. and Zander, J., A Rado Resource Management Concept for 'Bunched' Personal Communication System, Proc. Multiaccess, Mobility and Teletraffic for Personal Communications Workshop, MMT'97, Melbourne, 1997.
14. Zander, J., Radio Resource Management – An overview, Proc. IEEE Vehicular Technology Conf., VTC'96, Atlanta, GA, 1996, pp. 661–665.
15. Chen, L., *et al.*, A Dynamic Channel Assignment Algorithm for Asymmetric Traffic in Voice/Data Integrated TDMA/TDD Mobile Radio, Int. Conf. on Information, Communications and Signal Processing ICICS'97, 1997, pp. 215–219.
16. Mihilescu, C., Lagrange, X. and Godlewski, Ph., Dynamic Resource Allocation in Locally Centralized Cellular Systems, Proc. IEEE Vehicular Technology Conf., Ottawa, 1998, pp. 1695–1700.
17. 3GPP Technical Specification 25.101, UE Radio Transmission and Reception (FDD).
18. Ericsson, Tdoc SMG2 UMTS-L1 101/98, UTRA FDD Guard Band Analysis, 1998.
19. Hess, G.C., *Land-Mobile Radio System Engineering*, Norwood, MA: Artech House, 1993.
20. Yang, S.C., *CDMA RF System Engineering*, Norwood, MA: Artech House, 1998.

12
Complementary UMTS Technologies

12.1 WLANs AND 3G NETWORKS

In the following, we will present Wireless Local Area Networks (WLAN) in the context of the evolution of GSM/GPRS systems towards UMTS, i.e. 2G & 2.5G evolving into universal 3G networks [1].

12.1.1 Background

Today despite controversial positions in the Telecom industry, WLANs stands as good complementary technology to 2.5G and 3G data solutions. Thus, premature speculations that WLANs will dilute the 3G business models will not have a place here.

Compared to 3G, in WLAN we still need to address many factors before it gets up to the level of 3G. These include:

- consolidated business models that meets all user segments needs;
- integrated (communication & applications) roaming options for both national and international environments;
- coverage as large as that of 3G;
- voice over IP optimised for mobile systems;
- seamless inter-system handover;
- wide area coverage with WLAN supporting handhelds;
- large profit margins to subsidise WLAN business as we subsidise terminals in cellular mobile markets.

Today WLANs do offer faster online data access than mobile networks would. For example, in an airport lounge it may be economic and allow faster Laptop or Pocket PC

All IP in 3G CDMA Networks J. Castro
 ISBN: 0-470-85322-0

connectivity than through a classical mobile data network. However, as soon as the user is on the move, WLAN advantages disappear.

Some may say in the future there will be large WLAN coverage both from public and private sectors. If so then other problems will arise, e.g. if you have a private WLAN network that others nearby can access - how can you ensure that they are not downloading illegal content? Or that your ISP charges you, e.g. 0.07 ∈ pr. Mb downloaded while other people tap in to your WLAN access?

Thus, WLANs can fill gaps for now, but in the longer term, their two basic current advantages, i.e. 'speed and price' will not be sufficient to ensure security and privacy. Besides, progress on download channel bandwidths for 3G will reach WLAN performance levels or bypass it (e.g. HSDPA), especially, if many users feed from the same WLAN hub. Of course, WLAN will also evolve, but in any case today there are few real indications that in the future WLAN will endanger a well planned and implemented 3G business models.

As a result, WLAN can only be a complementary technology to UMTS or 3G and not replace it. Furthermore, if anyone will or are contributing to the expansion and adding value to WLANs, are mobile operators who are also exploiting the cellular network infrastructure arrangements.

This section will present WLAN as a complement to cellular networks and emphasise the need of seamless inter-operability between the two systems.

12.1.2 Why WLANs Now?

Today there are already more than 30 million mobile laptop users with GSM devices, and will grow to about 130 millions including PDA users by 2006[1] capable to access Wireless LAN hot spots. Highly mobile business professionals and non-negligible segment of PDA users already welcome efficient, high-speed wireless broadband access for their laptops and PDAs.

In the mobile environment, subscribers, regardless of their type, business or average consumer, demand higher data speeds for fast connections to their entire office or favourite application/information server sites through broadband Internet links in hot spots (e.g. airports, hotels and convention centres).

The rapid adoption of ADSL in the consumer market has already changed the perception of remote connection (the faster the better). Hence, we can no longer change user expectations when it comes down to wireless Internet access. Today 128 and 256 kbps Internet access habits are getting common everywhere in the western world. Current trends indicate that all users will or expect similar speeds in the wireless environment. As a result, the idea that broadband wireless will mainly appeal the business segment is an obsolete view altogether.

In addition, mobile operators are watching nervously, how small Wireless ISPs are entering into the corporate market, providing fast Wireless Internet services with the potential to diminish their market potential. Thus, mobile operators do not only need to protect their market share but also offer more competitive solutions to differentiate their corporate offering from the competition.

[1]Estimate according to BWCS 2001.

To meet the needs of mobile subscribers for broadband data connectivity as well as stand the challenge, cellular network service providers need Operator Wireless LAN (OWLAN) solutions. These solutions need to maximise reuse of their current and future data infrastructures, e.g. GPRS [2] and UMTS. Popular solutions providing such offering are based on 11 Mbps IEEE 802.11 wireless LAN access technology.

12.2 WLAN DEPLOYMENT SCENARIOS

Mobile Operator (MO)-WLANs as a wireless extension of Ethernet networks, cover a radius of about 50 m indoor at transmission rates of 11- 0.x Mbps, compared to GSM cells (where x = 9 to 0), which cover a radius of about 10–35 km. All current solutions aim to maximise usage of GPRS (later UMTS) infrastructures applying IEEE 802.11b WLAN standard known as WIFI operating at 2.4 MHz ISM band.

Figure 12.1 illustrates an overview of the MO-WLAN example. Notice the additional nodes, which to a good extent represent the solution of most popular vendors (e.g. Nokia).

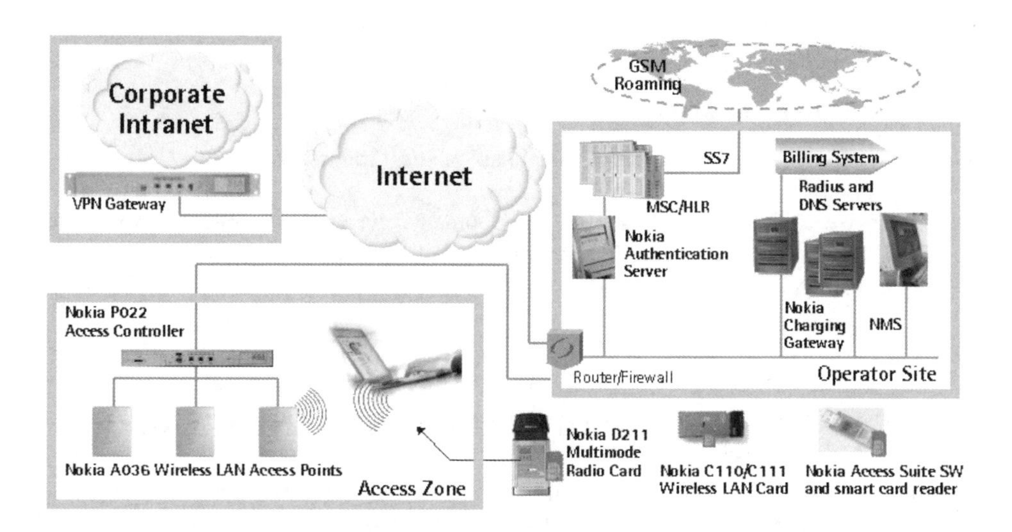

Figure 12.1 Mobile Operator (MO) Wireless LAN architecture example (after Nokia).

Thus, deploying MO-WLAN boils down to adding a few elements and reusing existing infrastructure. The reusable building blocks are:

1. GPRS backbone including routing and switching elements;
2. charging gateway or mediation device;
3. core application elements (e.g. SMSC, MMSC, downloading, streaming servers, etc.);
4. user profile and authentication, HLR and AuC, Radius server (AAA) solutions.

The new elements, as illustrated in Figure 12.1, include mainly:

- *the Access Points* (AP)→interaction ports to the WLAN user devices;
- *the Access Controller*→manages the APs;
- WLAN devices with classical authentication systems (e.g. PC cards, SIM, smart card reader, etc.).

12.2.1 WLAN Service Environment

The service environment will vary according to the place where business travellers or subscribers could take advantage of mobile broadband services for their laptops, PDAs, etc. A non-exhaustive list may include:

- hotels
- airport lounges
- business parks, various campus areas
- corporate buildings and meeting rooms
- residential private networks
- moving passenger vehicles, e.g. Cruise boats, planes, trains, etc.

Here since we are dealing with mobile environments, we will illustrate WLANs within trains. Therefore, in the following we shall describe the prospective deployment of APs in passenger in wagons.

12.3 TRAIN WLANs AND OTHER DEPLOYMENT SCENARIOS

Offering WLAN Access Points (APs) within passenger trains would simply be an expansion of '*access zones*' introduced in Figure 12.1 and illustrated within Figure 12.2 with case 5.

The Access Controller (AC) can be physically at the 'access zone', or centrally at the Mobile Operator core site handling multiple access zones. For our new 'train-access zone' represented with case 5, the centralised architecture mode would apply as it would also for case 6. These 'access zones' can be connected to the Access Controller both over a switched network and a routed network as shown for zones 5 and 6 in Figure 12.2.

The centralised Access Controller (AC) solution would be cost-effective, for small access zones like trains (i.e. 1–2 wireless LAN access points per train wagon).

Today, the integration between the WLAN infrastructure and the train Access Points (APs) would be done through a GPRS modem and an inter-working unit, which boils down to SW and a mini-portable using a GPRS PC card available in the market now. This, until the new generation of APs and ACs do directly support GPRS integration; projected dates for integrated GPRS solutions have started already in 1H03.

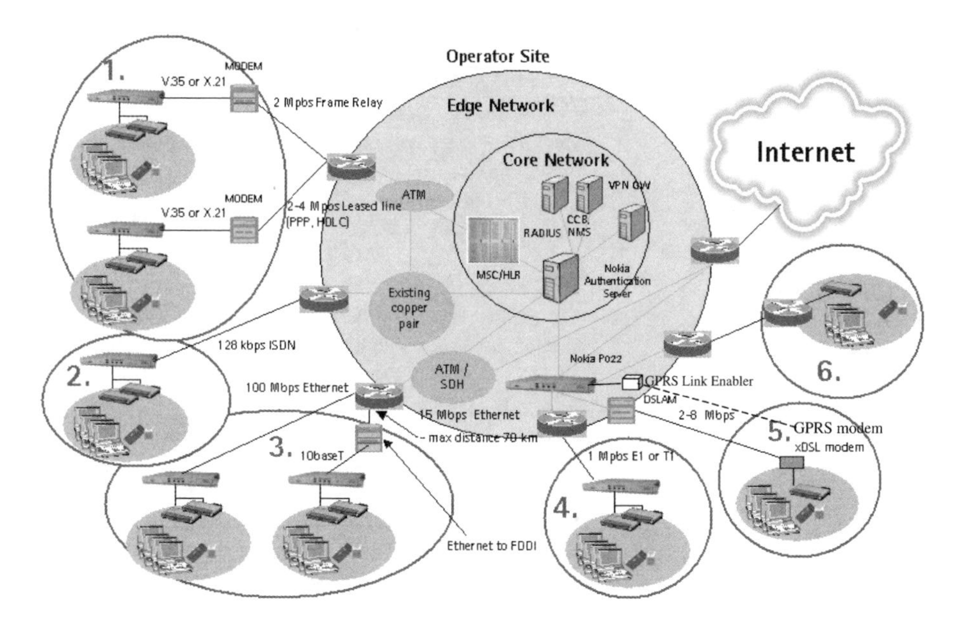

Figure 12.2 Access zone examples connecting the operators site (based on Nokia's access controller).

The Inter-working unit, for the initial phase, would be a 'Portable Office' product from 'red-M', which has the essential characteristics for the GPRS link, besides, e.g. ISDN port, 10/100 Ethernet Internet port, 10/100 Ethernet Office LAN port.

Table 12.1 illustrates the key parameters of the Indoor-Train WLAN link through GPRS. This inter-working unit becomes the essential bridge to integrate the moving WLAN, i.e. the train-access zone[2] with the supporting WLAN building blocks and a mobile network infrastructure.

Logically, we would use similar gateway to interconnect a dedicated UMTS packet domain environment.

12.3.1 Propagation Aspects and Capacity

The train wagons stand as indoor environments hosting the solution where we would also have to minimise degradation from propagation effects and consider the following:

- Appropriate train floor plans to maximise transmission by strategic position of antennas.
- Consider RF systems already in use. Most interference will occur primarily from elements in the vicinity, e.g. MW ovens, existing WLANs BSs. We assume the latter will not be the case.

[2]For the record, there also exists GPRS supporting modems, which work with satellite links to integrate WLANs in passenger planes and cruise boats, if so desired.

Table 12.1 Key characteristics of inter-working unit for GPRS-WLAN link (portable office red-M)

ISDN Internet port	The BRI Euro ISDN (2B) port provides an Internet connection, which can be always on or dial-up. It allows data rates of up to 128 kbps by using both of the ISDN B channels and the built-in MLPPP software. The port uses an RJ45 connector with a status LED.
10/100 Ethernet Internet port	This allows the Portable Office to be connected to a WAN modem (e.g. a DSL modem), or to a corporate network. The port uses an RJ45 connector with a status LED.
10/100 Ethernet office LAN port	This provides an interface to a standard 10/100 Mbps Ethernet Local Area Network. The port uses an RJ45 connector with a status LED.

GPRS uplink	The GPRS uplink provides a wireless alternative to ISDN or Ethernet-based connectivity from the Access Server. Using the GPRS network, download speeds of up 53.6 kbps and upload speeds of 26.8 kbps are possible.
	Other GPRS specifications are
	• GPRS multi-slot class 10
	• GPRS coding schemes CS-1,2,3,4
	• GPRS maximum data rates for CS-2:
	  ○ 53.6 kbps downlink, 26.8 kbps uplink.
	• Transparent and non-transparent GSM circuit-switched data up to 14.4 kbps
	• Standard radio frequency bands for EGSM 900, DCS 1800 and PCS 1900
	• RF output power:
	  ○ EGSM 900-2 W max. (Power Class 4)
	  ○ DCS 1800 and PCS 1900-1 W max. (Power Class 1)
	• TCP/IP header compression
	• Air interface data encryption
	• standard GSM error correction algorithms

User storage	8.8 GB is available on the Portable Office hard disk for user data.
Auto switching 110/240V, 50/60 Hz power supply	The maximum power consumption is 120 W. There is also an LED power indicator.
Physical Height 334 mm, Width 245 mm, Depth 145 mm, Weight 3.9 kg	The Portable Office can be placed on a desk or mounted on a wall. If it is mounted on a wall the base can be removed to reduce the dimensions to a height of 312 mm, width of 245 mm and depth of 125 mm.

- Interfering metal fire-breaks or metal walls or floor structures with heavy metal components may be an issue when trying to minimise the number of APs. Thick concrete drastically degrading and reducing transmission rates elements may not be present here.
- Empty passages or doorways, which cause RF propagation effects and thereby reduce transmission efficiency, may also be minimum in trains.
- Overcrowded spectrum—which also degrade transmission rates, as a result of the success of 1st generation WLANs operating in the 2.4 MHz spectrum—does not appear to be a problem at the moment.

As illustrated in Figure 12.3, the position of the user with regard to the antenna will determine the information rate he/she gets. However, degradation due to interference, e.g. from the aforementioned constraints will affect the user individually and will not bring the system down.

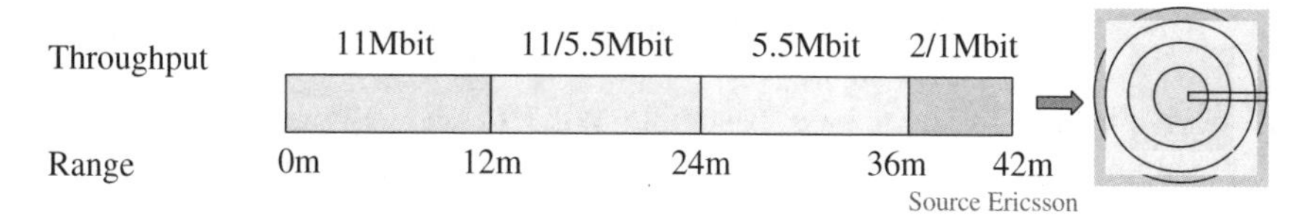

Figure 12.3 Transmission degradation based on distance.

The location of the APs will thus have a great impact on the distribution of bandwidth (11 Mbps) among the users within the train. We can see this transition of rates in Figure 12.4, which also affects the number of users sharing the pipe provided by the AP. On the other hand, the distance would not be a problem if we put one AP per wagon. E.g. an Inter-city wagon in Switzerland is about 25 m and a regional train wagon about 19 m. However, recent WLAN technology is allowing larger coverage, e.g. 50–100 m or more; in this case optimisation will be imperative.

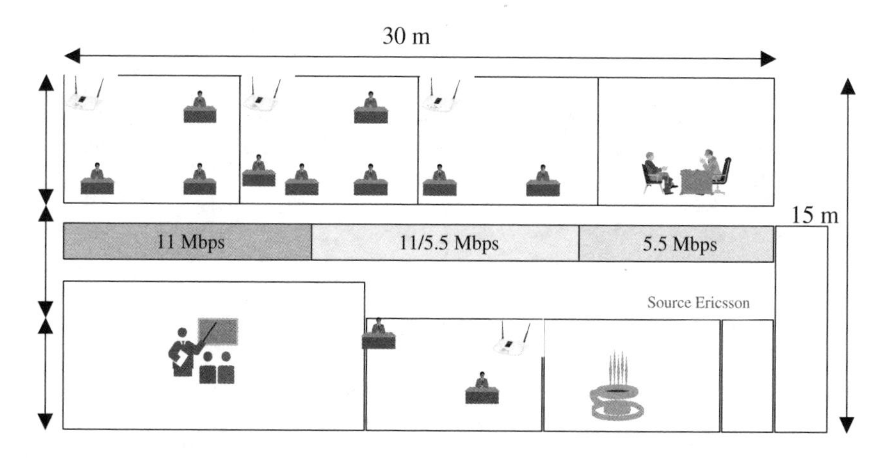

Figure 12.4 Mobile-office like environment scenario.

Figure 12.4 gives also a window to the type of Quality of Service (QoS)we would want to offer based on the number of APs and users sharing the available bandwidth. Here it should be noted that not all users will need to access simultaneously. Thus, in general about 15 users per AP may be possible; however, it will depend on the type of services offered.

MO-WLAN access sites can thus be classified[3] as

Small	containing up to 3 APs
Medium	covering from 4–10 APs
Large	containing 10–50 APs

The APs can be daisy-chained towards one Access Controller in the train if necessary for better performance.

12.3.2 Bluetooth-WLAN Combined Deployment Options

Recent developments for combined Bluetooth-WLAN Access Points are on their way, and commercial products have been available since 1H03. This hybrid APs would expand access availability to a larger set of terminals. As new multimedia terminals, e.g. Nokia, Ericsson (7650, P800 and others, and without a doubt UMTS terminals), become popular, capabilities to access the Bluetooth-WLAN combined network will be higher.

The infrastructure to support the combined solution will be the same as for WLAN alone, except that the range coverage for the Bluetooth will remain shorter, but still up to 20 m range. This means that coverage optimisation would have to be done in a little different manner. Support links to the infrastructure would take place in similar way to that of Train-WLANs

The same inter-working administration work will apply to both. Thus, with the combined solution we would meet the points mentioned above, i.e.:

- Internal costs would not go any higher due to development or additional infrastructure.
- External costs would not increase much either, because we would use similar infrastructure support.
- The combined solution would not deviate from our initial strategy of Personal Area Networks through WLANs, now underway within CSS or engineering, or plans within OCH.

[3]For example, Ericsson.

12.4 DESCRIPTION OF MO-WLAN BUILDING BLOCKS

Other detailed solution descriptions would cover more in depth any dedicated environment WLAN (e.g. that for trains), here we want to illustrate other building blocks MO-WLAN. This includes the: transmission system, authentication, security and charging/billing.

12.4.1 Maximising Transport Network Usage

The Control Servers, attached to the *GPRS/UMTS primary or core sites*, can be reached by connecting to a GPRS/UMTS backbone through the Core Transport system based on fibre optic or point-to-point (e.g. MW Mini Link E) as illustrated in Figure 12.5.

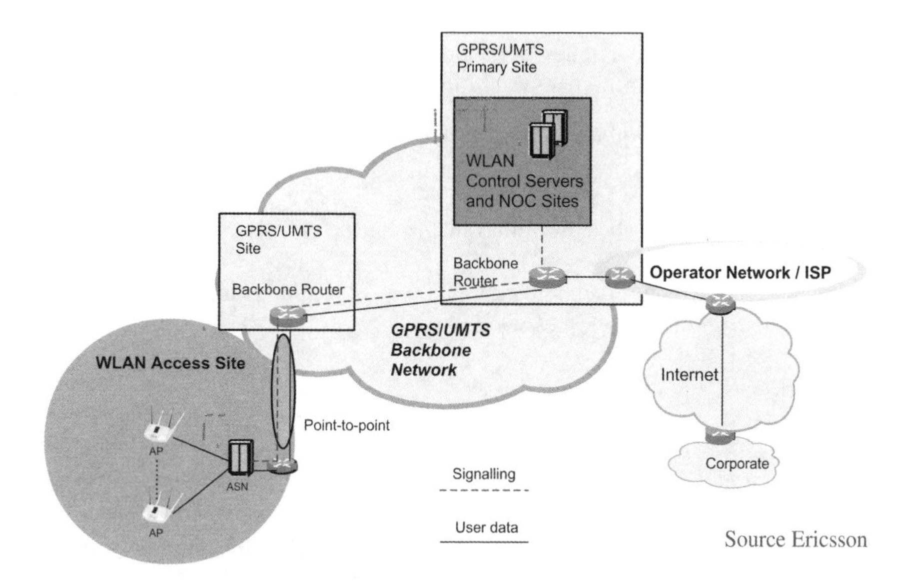

Figure 12.5 Transport over GPRS/UMTS backbone network.

An advanced R99 GPRS/UMTS core solution and GSM/UMTS radio will afford the use of some types of QoS to maximise transmission rate to given MSISDN or modem running with a SIM/USIM [3]. This would allow dedicated links to the IP infrastructure from the moving vehicles, and thereby assure the maximum bandwidth possible for APs connecting the Access controllers in the supporting network. Then,

1. the MO-WLANs infrastructure at a mobile network site would connect through an L2 connection to the nearest GPRS/UMTS backbone network site or transport access and from there, reach the primary or core site where its servers are;
2. the data traffic would use the existing access to the Internet without much impact on the existing backbone capacity;
3. the access process would benefit from the Security set-up for the GPRS/UMTS network.

As reported earlier, the multiple use of the cellular transport network would motivate the building of a flexible and intelligent transport network, which can discriminate circuit and packet-switched traffic.

This means that we can go, e.g. from two simple POPs in the MSC sites to up to as many POPs is required by equipping the necessary BSC/RNC sites or some BS/Node B sites with intelligent switch/routers capable to route traffic generated in the IP world.

12.5 SECURITY AND AUTHENTICATION

As in any of the 'access zones', security threats in the 'Train-access zone' will also be critical. Thus, we would have to apply at least the same type of security schemes as in the classical zones. Some of the commonly known threats are:

- *Masquerading* → attacker pretends to be an authorised user to gain access to the system or the attacker pretends to be the system to which end users try to gain access,
- *Identity or Data Interception* → subscriber ID or user data gets captured for later use,
- *Info Manipulation* → data content is read, replaced, inserted or deleted in the system,
- *Service Denial* → network becomes unavailable for the authorised users (info distorted).

The minimum security requirements to confidently deploy the 'Train-access zone' services and for end users using these services would be:

- the network must authenticate the user;
- the end user must authenticate the network;
- a man-in-the-middle attack shall not be possible.

By deploying an access zone system that implements the requirements above, we can be confident that at least the Masquerading and Identity Interception threats are not possible.

To increase the security to also cope with the Manipulation of data threat, we and/or end user can use data encryption, i.e. not only that data is payload but also authentication, authorisation and accounting data.

The Wireless LAN standard IEEE 802.11i specifies major enhancements both to authentication and radio interface encryption. The IEEE 802.11i standard addresses all the concerns expressed by the Wireless LAN user community.

12.5.1 SIM Authentication

Today the authentication based on EAP/SIM that can be applied to the 'Train-access zone' is even more secure than standard GSM authentication/authorisation technology.

The IETF standard EAP/SIM, specifies enhanced GSM authentication by using multiple GSM triplets per authentication, i.e. stronger challenges, responses and session keys [4]. In addition, the terminal also authenticates the network.

- Challenges are accompanied with a message authentication code.
- The terminal's nonce (random number) is included in session key derivation and network authentication.
- Attacker cannot build a rogue network based on the authentication data.
- It is not vulnerable to 'man-in-the-middle' attacks.

12.5.2 Forced VPN and End-to-End Security

An encrypted VPN tunnel is the most secure technology available to protect data over any network. Hence, many corporations already provide their employees with VPN software to create an encrypted end-to-end tunnel between the user's terminal and the corporate network. To ensure security and service trust in the 'Train-access zone', we could therefore enforce VPN traffic as follows:

- Corporate users shall use the VPN provided by their corporation. Only encrypted user traffic is processed at the mobile network (no compromise on the end user's data security).
- A mobile network would provide VPN to non-corporate users, e.g. by basing VPN authentication on the cellular identity (SIM/USIM).
- If desired, the corporate users may also use the mobile network VPN, and user traffic can be further VPN tunnelled to the user's company.

12.6 ENHANCED DATA RATE FOR GSM EVOLUTION (EDGE)

Enhanced Data Rate for Global Evolution (EDGE) enhances GSM networks' data capabilities towards Third Generation (3G) services. EDGE raises the Air interface data throughput threefold over today's GSM and increases both Circuit-Switched (CS) and Packet-Switched (PS) services. US mobile networks use EDGE as their evolution to 3G-radio technology and Europeans as a complement to WCDMA-based UMTS 3G networks.

EDGE applies to 900-, 1800- and 1900-MHz GSM frequencies and has become interesting to service providers to rapidly and cost-effectively increase capacity and boost transmission rates. It provides significantly higher data rates on the current 200 kHz GSM carrier, i.e. ECSD rates up to 43.2 kbps/timeslot and EGPRS rates up to 59.2 kbps/timeslot, where the data throughput per carrier increases to even up to 473 kbps.[4] ECSD may enable to support a 64 kbps real time service with a low Bit Error Ratio (BER) by allocating two time slots of 32 kbps each. The EDGE modulation adapts to radio conditions, thereby offering the highest data rates in good propagation environments, while ensuring wider area coverage at lower data speeds per timeslot.

[4] In practice, transmission rates may stay around 150 kbps while supporting bandwidth intensive services.

Table 12.2 Comparison between EDGE and GSM modulation parameters

	EDGE	GSM
Modulation	8-PSK	GMSK
Symbol rate	270.8 kbps	270.8 kbps
Number of bits/symbol	3 bits/symbol	1 bit/symbol
Payload/burst	342 bits	114 bits
Gross rate/time slot	68.4 kbps	22.8 kbps

12.6.1 EDGE Modulation and Coding

The circumvention to triple the data rates as illustrated in Table 12.2 is the introduction of 8-PSK (Phase Shift Keying), a linear higher order modulation in addition to the existing GMSK (Gaussian Minimum Shift Keying). An 8-PSK signal carries 3 bits per modulated symbol over the radio, while a GMSK signal carries only 1 bit per symbol. EDGE keeps the GSM carrier symbol rate (270.833 ksps) for 8-PSK, and the burst length remains identical to the current GMSK, i.e. the same 200 kHz carrier spacing (Figure 12.6).

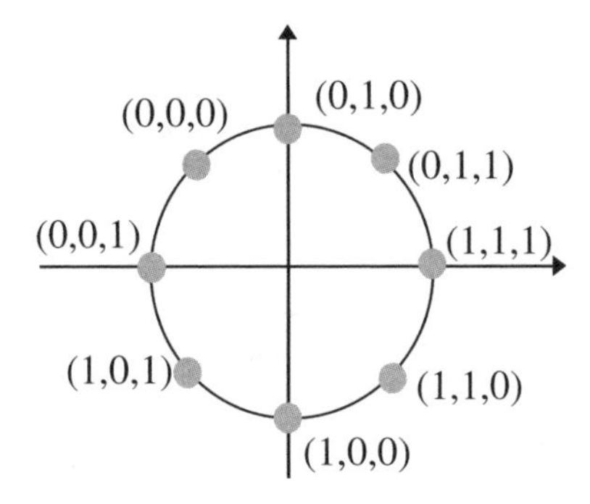

Figure 12.6 8-PSK modulation in EDGE.

Thus, subscribers can send more bits per radio time slot in the same airtime without a need of new frequency band or license to obtain higher data rate services.

12.6.2 Enhanced Circuit-Switched Data (ECSD)

ECSD does not necessarily substitute HSCSD (≈64 kbps) when it comes to transmission rates, but enables achieving these rates with less time slots (e.g. 64 kbps transparent[5] with 2×32 kbps instead 4 TS of 14.4 kbps providing 57.6 kbps with HSCSD), which are sufficient to offer transparent and non-transparent CS services as illustrated in Table 12.3.

[5]Available throughput is constant and the transmission delay is fixed.

Table 12.3 ECSD transmission data rates

Service	Code rate	Modulation	Gross rate (kbps)	User rate (kbps)	Radio interface rate (kbps)
E-TCH/28.8 NT/T	0.42	8-PSK	69.2	28.8	29.0
E-TCH/32 T[a]	0.46	"	"	32.0	32.0
E-TCH/43.2 NT	0.63	"	"	43.2	43.5

[a]Available only with 2 radio slots yielding to 64 kbps for maximum service rate in multi-slot.

The 28.8 kbps data rate will be available for *transparent* and *non-transparent*[6] services for both single and multi-slot configurations. The 32.0 kbps service will be available only in multi-slot configuration (two-slot) and will be used for offering 64 kbps transparent service, whilst 43.2 kbps will be available for non-transparent services only.

ECSD exploits the HSCSD transmission and signalling architecture, where the same Transcoding and Rate Adaptation Unit (TRAU) frame formats and 16 kbps sub-channels apply. For example, we get 28.8 kbps services from one radio interface timeslot based on 2×16 kbps sub-channels and 14.4 kbps TRAU frames.

12.6.3 Enhanced General Packet Radio Service (EGPRS)

EGPRS builds on GPRS, i.e. the GSM packet-switched data domain enabling Internet/ Intranet type services. It uses several Modulation and Coding Scheme (MCS) varying from 8.8 kbps up to 59.2 kbps in the radio interface, as illustrated in Table 12.4.

Table 12.4 Single slot EGPRS peak data rates

MCS	Modulation	Code rate	Family	User rate (kbps)
1	GMSK	0.53	C	8.8
2	GMSK	0.66	B	11.2
3	GMSK	0.80	A	14.8
4	GMSK	1	C	17.6
5	8-PSK	0.37	B	22.4
6	8-PSK	0.49	A	29.6
7	8-PSK	0.76	B	44.8
8	8-PSK	0.92	A	54.4
9	8-PSK	1	A	59.2

[6]Available throughput varies with the quality of transmission → higher error probability = lower throughput.

In Table 12.4, GMSK modulation applies to wide area coverage while 8-PSK suits better higher rates. The different MCSs enable re-segmentation of the data block for link adaptation, where we choose the protection that best fits the channel condition to offer maximum throughput, since higher protection implies lower throughput.

12.6.4 EDGE Network Pre-requisites

Increasing transmission rates through the EDGE air interface puts new demands in the existing GSM network, e.g. EDGE capable transceivers including transmission requirements in the transport and core networks. Higher transmission capacity implies also more capacity in the A-bis interface (the link between the BTS and BSC) as illustrated in Figure 12.7.

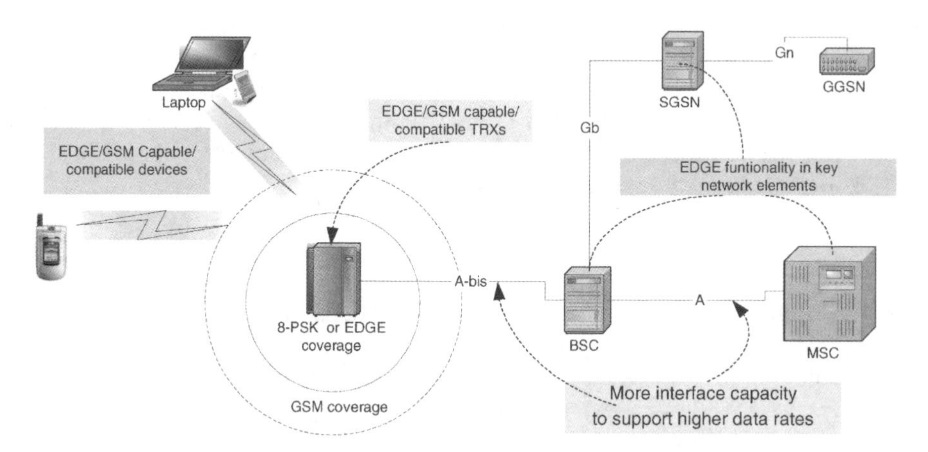

Figure 12.7 GSM/GPRS/EDGE network minimum requirements.

12.6.4.1 ECSD Network Link Control Characteristics

The *Link Adaptation* (LA) mechanisms work by adjusting to the protection of the info transmitted depending on the link attribute perceived from forthcoming channel quality measurement.

In EDGE Link Adaptation (LA) features enable switching between 8-PSK and GMSK modulated channels, which are not mandatory features in GSM networks, to provide the ideal service continuation over large coverage areas with high data rate *transparent* mode calls. For example, a two slot 28.8 kbps transparent connection in unstable GMSK channels will go near the centre to only one slot 8-PSK channel.

On the other hand, for *non-transparent* services, link adaptation is not imperative because Backward Error Control methods and variable bit rate mitigate instability.

12.6.4.2 EGPRS Link Quality Control Characteristics

The link adaptation schemes in EGPRS vary according to the propagation channels, e.g. urban, rural area, hilly terrain, etc., where a finite (limited) set of info protection may be

Table 12.5 EGPRS protection options

Modulation/scheme	Data rate (kbps)	Protection
8-PSK/MCS-9	59.2	Low
8-PSK/MCS-8	54.4	↓
8-PSK/MCS-7	44.8	
8-PSK/MCS-6	29.6 (27.2)	
8-PSK/MCS-5	22.4	
GSMK/MCS-4	17.6	
GSMK/MCS-3	14.8 (13.6)	
GSMK/MCS-2	11.2	
GSMK/MCS-1	8.8	High

defined as 20 ms blocks, e.g. EGPRS uses nine protection schemes, i.e. MCS-1[7] to MCS-9 (high protection/low bit rate to no protection/high bit rate Table 12.5), where we encode info to preserve channel degradation and modulated it before over the air interface transmission.

All MCSs in EDGE may bi-directionally switch from one data block to another between them as in GPRS, but in the latter occurs only in one direction; thus, link adaptation does not take place.

12.6.5 EDGE Network Deployment Aspects

Edge supports ECSD data rates through the High-Speed Circuit-Switched Data (HSCSD) infrastructure elements and standard A/A-bis interfaces. We choose rates as multiples of each other for link adaptation between different modes of modulation and number of time slots. Likewise, we introduced EGPRS through the GPRS infrastructure without major modifications.

8-PSK modulation, which does not have a constant envelope, puts requirements on the linearity of the power amplifier, in particularly for high-output power equipment. It has an Average Power Decrease (APD) of about 2–3 dB.

12.6.5.1 Coverage Options

Higher bit rates over standard GSM/GPRS decreases system's resistance to time dispersion for users moving and vehicle speeds. Thus, EDGE will have lower coverage span than GSM and presumably used mainly in quasi-stationary environments. However, if speeds and time dispersion exceed EDGE thresholds operating in 8-PSK, users will be switched to GMSK modulation mode through the link adaptation feature and get lower transmission rates but still having coverage.

[7]GPRS coding schemes CS1 to CS4 are different to EGPRS GMSK MCSs.

12.6.5.2 Capacity Options

EDGE cells enable simultaneously users with different transmission rates. So EGPRS subscriber closer to the base station will get high bit rates seamlessly, while those at cell border will pick up GPRS type rates according to the link adaptation and switch to GMSK modulation from 8-PSK as indicated in the preceding section. Now, high usage of traffic channels at the border of an EDGE cell may increase the probability of interference, decreasing thereby the user bit rates and increasing C/I. Edge mitigates the latter effects by using lower amount of time slots, reducing thereby the overall C/I.

While future services such as Voice over IP (VoI) demand real-time packet traffic channels, current GPRS type serve primarily non-real-time services, where 'best effort' is acceptable and discontinuous connection or transmission brake does not cause drop calls. Thus, the enhancement of GPRS, i.e. EGPRS exploits also different amounts of radio resources allocated to users by enabling dynamic multi-slot connections without changes to the frequency plans. Figure 12.8 estimates how the preceding requirements imply that average user data rates in *capacity limited* environments would reach 40 kbps per time slot, and in good C/I the rates would exceed 40 kbps. However, in *coverage-limited* cells average rates are lower due to lower C/I.

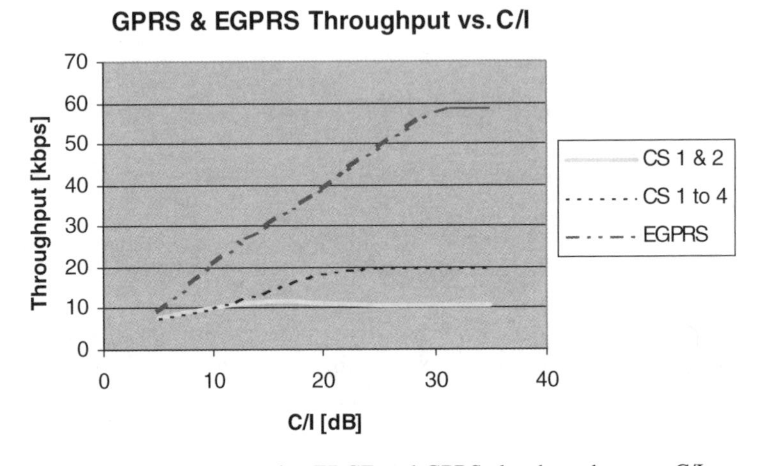

Figure 12.8 Comparative EDGE and GPRS slot throughput vs. C/I.

Likewise, in ECSD design margins need to be taken into account to operate in capacity or coverage environments. For example, to warrant constant throughput we add a fading margin the link budget and increase the frequency re-use factor (15–18) to obtain sufficient network C/I to operate with 8-PSK modulation in EDGE.

12.6.5.3 Frequency Planning and BCCH Utilisation

In GSM, if we apply frequency hopping we can tighten the average frequency re-use factor of 9 and 12, which EDGE also supports quite well through its link adaptation feature.

Thus, EDGE can apply random frequency plan and benefit from high C/I closer to the cell centre.

Finally, since EDGE transmit power can often be limited at the edges, 8-PSK applies only on the traffic channels. Thus, the BCCH carrier, utilised for cell selection transmitting with constant power, use GMSK modulation on the signalling channels.

12.7 ALL IP RADIO-ACCESS NETWORK CHARACTERISTICS

As already indicated in preceding chapters, for example 10th, current mobile network architectures designed to support predominantly circuit-switched traffic still rely on conventional Radio-Access Network (RAN) structures. But the rapid growth of IP traffic, i.e. IP-based services as illustrated in Figure 12.9, demand already more flexible RAN today to address rapidly access enhancements such as High Speed Downlink Packet Access (HSDPA) and High Speed Uplink Packet Access (HSUPA), and will grow higher yet in the remaining part of this decade.

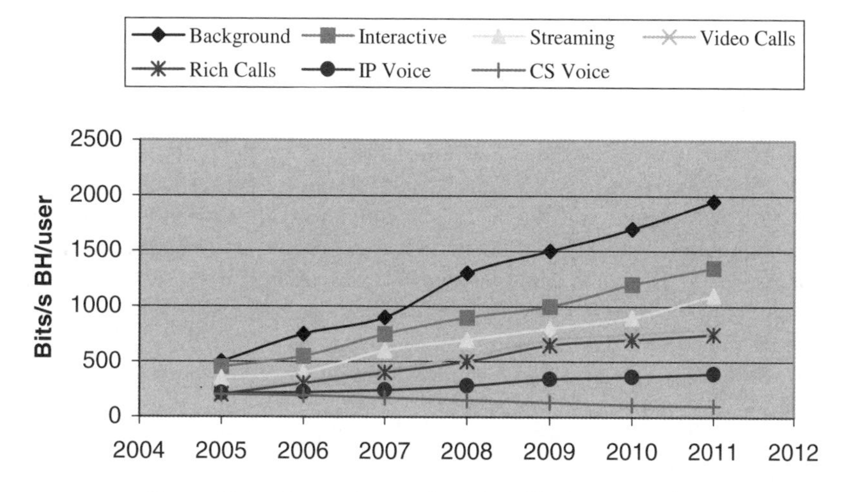

Figure 12.9 Extrapolation of multi-traffic data projection.

On the other hand, the new variety of person-2-person (p2p) IP-based services requires different QoS levels, which must be provided by the RAN, it is difficult to predict when compared to voice dominated traffic. The QoS requirements in the radio access are non-negligible and it can only meet demands if it is optimised to carry IP traffic. The latter can be achieved better through an efficient end-2-end (e2e) IP network architecture, where the radio too is optimised to handle packet traffic without adaptation.

12.7.1 IP-Based Radio Architecture

Logically, an IP-based radio will work simultaneously with IPv4/IPv6 and operate at least with the following features [5]:

1. distributed control tasks by shifting functions from the RNC to the Node B;
2. intelligently splitting the control and user planes to dynamically allocate capacity based on service demands;
3. maximise synergies between transport infrastructure elements, e.g. gateways, routers, etc.;
4. multi-radio[8] access capabilities from core and service networks with appropriate QoS features through an optimised air-interface and packet scheduler.

12.7.1.1 Spread Control Functions

An optimised packet access approach would not depend on hierarchical tasks, which traditionally take place, e.g. on the RNC of conventional 3G mobile network, because these would reside on the Node Bs as discussed briefly in Chapter 1, where the multi-node to multi-node communications aim for an efficient shared utilisation of resources through the whole network.

12.7.1.2 Split User and Control Planes

This split is already part of current solutions defined by the IETF and applied to the core segment by 3GPP tech specs, e.g. R5; thus its application to a packet optimised radio access will be a customised expansion and a normal evolution [6–11]. Independent high-speed data forwarding and routing control functionality will then become also a natural feature as it has been in packet-based core networks. As a result, we can utilise finest routers in the user plane and optimum SW functions in the control plane and assure all QoS levels to meet all types of service requirements.

When we take into account the co-existing diversity of 2G access technologies such as GSM/EDGE, UMTS and its complements like WLAN, Bluetooth Gateways a single packet based access network can support all demands and facilitate operations based on multi-mode Base Station Transceivers.

12.7.1.3 Optimum Quality of Service Performance

Non-packet access optimised 3G air interfaces (e.g. A, Iu) work in real time regardless of the type of service. On the other hand, optimised solutions would address service requirements on case by case basis following QoS demands. Logically, packet optimisation functions, such as packet scheduling, would reside on the Node B elements instead the RNC as indicated in the preceding sections. This distributed function will in addition minimise latency times for services due to efficient routing responses and will apply by default to the evolution of radio features, e.g. HSDPA and HSUPA.

[8] For example, WCDMA, GSM, etc.

12.7.1.4 All IP Radio-Access Building Blocks

The packet optimised Node Bs would not only hold optimised packet scheduling functions but also WCDMA features like macro-diversity combining and soft handover to afford dynamic mobility functions, which would use the transport network effectively. Other elements, multi-radio and core gateways, routers and universal OSS/O&M solutions to manage even the IT or IP backbone and transport-elements, will also naturally evolve to support the new Node B functions.

12.7.1.5 Maximising Transport Resources

Physical transport solutions, i.e. fibres, micro-wave, etc., will continue to support transport protocols, like ATM, IP, etc., and co-exist with the evolved 'All IP' functions in the access segment. Then where is the gain of all IP?

The added value rests on optimising the usage of the transport means by identifying the proportions utilised in the access segment. Conventional configurations in mobile networks indicate that about 85% of the transport resources go to the access segment, e.g. the integration of the Node Bs. Thus, an access element like a Node B, which contains intelligent mobility functions, will have lower traffic latency response to services sensitive to packet transmission jitter, will exploit efficient separation of traffic classes, will use single layer and direct transport layout thereby minimising packet overhead and maximising location for macro diversity combining functions.

12.7.1.6 Effective End-to-End Quality of Service Execution

Practical packet optimisation in an 'All IP' access segment will most likely make use of DiffServ (DS) [12] as the QoS control mechanism because of its flexibility and scalability. Thus, UMTS traffic classes would be mapped into DS codepoints at the Node B or radio gateway elements. Then each packet marked with DS codepoints would be handled in association with the configuration parameters of the radio-access element, i.e. the Node Bs, radio GW or the RNC itself following the transport plan and the standard interfaces.

12.8 WiMAX – ENHANCING BROADBAND HOTSPOTS

The *Worldwide Inter-operability Micro-wave Access (WiMAX)* industry Forum is promoting the 802.16x wireless metropolitan area network (WMan) standard as the next broadband transport bearer for hotspots now getting filled primarily by WLANs -WiFi. The 802.16 specs use OFDM modulation from Wi-Land, Canada.

Although there are claims that WiMAX will extend the potential of Wi-Fi as far as 30 miles, it may not stand the ubiquitousness and intelligence of UMTS. *Thus, in the full mobile environment it may remain primarily as complementary access technology despite all the industry hype.*

Nokia and Intel have been driving and dominating its standardisation; however, recently the first has left the consortium. Thus, we are yet to see how fast will evolve at the commercial level. Nevertheless, as a technology, which has created great interest we will next summarise its key features and some deployment options.

12.8.1 Brief WiMAX Technical Overview

As articulated by the WiMAX Forum [13], 'the IEEE 802.16 Air Interface Standard is truly a state-of-the-art specification for *fixed* broadband wireless access systems employing a Point-to-Multipoint (PMP) architecture'. The mobile version is still in the beginning process. To incorporate mobility functions may not be as trivial as would appear; henceforth, non-negligible time would be expected before the completion of stable mobile standards excluding implementation for commercial use.

The original specs were developed to meet a large scope of Broadband Wireless Access (BWA) options operating between 10 and 66 GHz, but to exploit specific markets an amendment is almost under conclusion to operate between 2 and 11 GHz, for which IEEE does not necessarily provide conformance statements and testing performance. So here the WiMAX technical Working Groups (WGs) are striving to make it inter-operable between suppliers competing in the same market. The WGs set system profiles and produce Protocol Implementation Conformance Statements (PICS) proforma, Test Suite Structure and Test Purposes specifications and Abstract Test Suite specifications according to the ISO/IEC 9464 [13,14].

The standard covers both the Medium Access Control (MAC) and the physical (PHY) layers. For the latter, at lower end (2–11 GHz) frequencies, a line of sight is not required. Instead they utilise *adaptive* modulation and forward error correction to further increase the typical capacity of 802.16 systems. While the MAC accommodates either Time Division Duplex (TDD) or Frequency Division Duplex (FDD) deployments, enabling for both full and half-duplex terminals in the FDD case, the OFDM-based physical layer is designed for TDD.

As indicated above, the MAC aimed explicitly for PMP wireless access, endeavours by design to seamlessly carry any higher layer or transport protocol, e.g. ATM, Ethernet or Internet Protocol (IP), and any future protocol for that. This MAC conceived for very high bit rates (i.e. duplex up to 268 Mbps), stands for truly broadband physical layer, while delivering ATM compatible Quality of Service (QoS) to ATM as well as non-ATM (MPLS, VoIP, etc.) service. Which on the other hand, this may very well be a drawback first before is a plus, i.e. the IEEE 802.16 Air Interface Specification may just be very well too large specification.

12.8.2 Consolidating Inter-operability

The IEEE 802.16 was designed to cover the *fixed* broadband wireless access needs for large variety of scenarios, thus allowing for different physical layers options and frequency bands to accommodate country-by-country frequency use restrictions. For example, there are features that allow one to build an IP centric system or an ATM centric system depending upon the needs of customers. Such flexibility had made total inter-operability a daunting task, which now the IEEE 802.16 WG started to address by using the 'System Profiles' in the IEEE 802.16 specification. These system profiles allow inclusion of features as mandatory or optional for the different MAC or PHY scenarios that are most likely to arise in the deployment of real systems. Thus, vendors would focus on a variety of market demands while ensuring inter-operability but without implementing absolutely every possible feature [13].

12.8.3 WiMAX Application Options

Initial solutions appeared focusing mainly on broadband last mile in unwired areas, and on backhaul for 'Hotspots', although the latter stands amid all the doubts about the sustainability of its boom. Therefore, WiMAX will be an alternative mainly to DSL and cable, not to cellular.

For some mobile service providers WiMAX may become a risk first, before it brings some profits. As a result, a careful complementary usage is must. By operating in a mixture of spectrum, e.g. 2.5 GHz/3.5 GHz licensed and 5.8 GHz unlicensed bands, it opens to wide range of opportunities. Rural areas or developing regions may benefit faster from licensed solutions than large city conglomerates.

A pragmatic UMTS operator may thus control unlicensed solutions to strengthen its 3G offering and maximise its service synergies before independent wild attempts of Wi-Fi providers pretend to offer what an advanced UMTS network can deliver.

12.9 CONCLUSIONS

Without a doubt UMTS remains as the mainstream and future proof solution for mobile networks. It exploits the GSM footprint to the uttermost and benefits to the maximum of the deployment synergies. With its evolution building blocks such as HSDPA (+ forthcoming HSUPA) in the access side, and IMS in the core side, the economy of scales will be non-negligible to maintain and promote growth in the very near future, i.e. in next 2 years.

EDGE may appear as quick win in the immediate future, yet it may not sustain the same evolution path foreseen by UMTS. Besides, the terminal penetration may not warrant the same boon and speeds of GSM/GPRS volumes. In the mean time UMTS will continue to progress and expand thus reducing the gaps even for the terminals.

As a result, UMTS co-existing with the complementary technologies, such as EDGE, WLANs and even WiMAX (in the future), may be the ideal model to establish integrated mobile solutions without neglecting the prospect fixed/mobile convergence through wireless. In addition, served by "All IP" features UMTS and its complement stands out as the solution to foster, master it and evolve it appropriately in the cellular environment in years to come.

Finally, it is hoped that this chapter, even though it did not cover in depth some of the outlined UMTS-complementary technologies, has still provided a window to the opportunities and potential of combined mobile broadband wireless solutions.

REFERENCES

1. Drafting Team, Vision and Road Map for UMTS Evolution, TSGS#8(00)0337, 2000.
2. TS 22.060, General Packet Radio Service (GPRS) Stage 1.

3. 3GPP, Combined GSM and Mobile IP Mobility Handling in UMTS IP CN 3G TR 23.923, Version 3.0.0, 2000–05.
4. IETF RFC 2002 (1996): IP Mobility Support, C. Perkins.
5. Internet draft, Johson and Perkins, Mobility Support in IPv6, October 1999. http://www.ietf.org/internet-drafts/draft-ietf-mobileip-ipv6-09.txt
6. 3GPP, Study on Release 2000 SERVICES and Capabilities (3G TS 22.976), V2.0.0, 2000–06.
7. TS 22.003, 3rd Generation Partnership Project, Technical Specification Group Services and System Aspects, Circuit Teleservices Supported by a Public Land Mobile Network (PLMN).
8. TS 22.002, 3rd Generation Partnership Project, Technical Specification Group Services and System Aspects, Circuit Bearer Services (BS) Supported by a Public Land Mobile Network (PLMN).
9. TS 22.004, 3rd Generation Partnership Project, Technical Specification Group Services and System Aspects, General on Supplementary Services.
10. 3GPP, Architecture Principles for Release 2000 (3G TR 23.821), V1.0.1, 2000–07.
11. 3GPPTS 45.005, Radio Transmission and Reception, 3GPP Technical Specification Group, GERAN.
12. IETF RFC 2474, The differentiated Services Architecture.
13. WiMAX Forum, http://www.wimaxforum.org/tech
14. G. Antonello, Conformance & WiMAX Certification, Jan. 21st, 2004. Source WiMAX Forum.

APPENDIX C: ALLOCATION OF SECONDARY SYNCHRONIZATION CODES (SSCs)

Table 4.48 Allocation of SSCs for secondary SCH [4]

Scrambling Code Group	Slot Number														
	#0	#1	#2	#3	#4	#5	#6	#7	#8	#9	#10	#11	#12	#13	#14
Group 0	1	1	2	8	9	10	15	8	10	16	2	7	15	7	16
Group 1	1	1	5	16	7	3	14	16	3	10	5	12	14	12	10
Group 2	1	2	1	15	5	5	12	16	6	11	2	16	11	15	12
Group 3	1	2	3	1	8	6	5	2	5	8	4	4	6	3	7
Group 4	1	2	16	6	6	11	15	5	12	1	15	12	16	11	2
Group 5	1	3	4	7	4	1	5	5	3	6	2	8	7	6	8
Group 6	1	4	11	3	4	10	9	2	11	2	10	12	12	9	3
Group 7	1	5	6	6	14	9	10	2	13	9	2	5	14	1	13
Group 8	1	6	10	10	4	11	7	13	16	11	13	6	4	1	16
Group 9	1	6	13	2	14	2	6	5	5	13	10	9	1	14	10
Group 10	1	7	8	5	7	2	4	3	8	3	2	6	6	4	5
Group 11	1	7	10	9	16	7	9	15	1	8	16	8	15	2	2
Group 12	1	8	12	9	9	4	13	16	5	1	13	5	12	4	8
Group 13	1	8	14	10	14	1	15	15	8	5	11	4	10	5	4
Group 14	1	9	2	15	15	16	10	7	8	1	10	8	2	16	9
Group 15	1	9	15	6	16	2	13	14	10	11	7	4	5	12	3
Group 16	1	10	9	11	15	7	6	4	16	5	2	12	13	3	14
Group 17	1	11	14	4	13	2	9	10	12	16	8	5	3	15	6
Group 18	1	12	12	13	14	7	2	8	14	2	1	13	11	8	11
Group 19	1	12	15	5	4	14	3	16	7	8	6	2	10	11	13
Group 20	1	15	4	3	7	6	10	13	12	5	14	16	8	2	11
Group 21	1	16	3	12	11	9	13	5	8	2	14	7	4	10	15
Group 22	2	2	5	10	16	11	3	10	11	8	5	13	3	13	8
Group 23	2	2	12	3	15	5	8	3	5	14	12	9	8	9	14
Group 24	2	3	6	16	12	16	3	13	13	6	7	9	2	12	7
Group 25	2	3	8	2	9	15	14	3	14	9	5	5	15	8	12
Group 26	2	4	7	9	5	4	9	11	2	14	5	14	11	16	16
Group 27	2	4	13	12	12	7	15	10	5	2	15	5	13	7	4
Group 28	2	5	9	9	3	12	8	14	15	12	14	5	3	2	15
Group 29	2	5	11	7	2	11	9	4	16	7	16	9	14	14	4

Table 4.48 (*Continued*)

Scrambling Code Group	Slot Number														
	#0	#1	#2	#3	#4	#5	#6	#7	#8	#9	#10	#11	#12	#13	#14
Group 30	2	6	2	13	3	3	12	9	7	16	6	9	16	13	12
Group 31	2	6	9	7	7	16	13	3	12	2	13	12	9	16	6
Group 32	2	7	12	15	2	12	4	10	13	15	13	4	5	5	10
Group 33	2	7	14	16	5	9	2	9	16	11	11	5	7	4	14
Group 34	2	8	5	12	5	2	14	14	8	15	3	9	12	15	9
Group 35	2	9	13	4	2	13	8	11	6	4	6	8	15	15	11
Group 36	2	10	3	2	13	16	8	10	8	13	11	11	16	3	5
Group 37	2	11	15	3	11	6	14	10	15	10	6	7	7	14	3
Group 38	2	16	4	5	16	14	7	11	4	11	14	9	9	7	5
Group 39	3	3	4	6	11	12	13	6	12	14	4	5	13	5	14
Group 40	3	3	6	5	16	9	15	5	9	10	6	4	15	4	10
Group 41	3	4	5	14	4	6	12	13	5	13	6	11	11	12	14
Group 42	3	4	9	16	10	4	16	15	3	5	10	5	15	6	6
Group 43	3	4	16	10	5	10	4	9	9	16	15	6	3	5	15
Group 44	3	5	12	11	14	5	11	13	3	6	14	6	13	4	4
Group 45	3	6	4	10	6	5	9	15	4	15	5	16	16	9	10
Group 46	3	7	8	8	16	11	12	4	15	11	4	7	16	3	15
Group 47	3	7	16	11	4	15	3	15	11	12	12	4	7	8	16
Group 48	3	8	7	15	4	8	15	12	3	16	4	16	12	11	11
Group 49	3	8	15	4	16	4	8	7	7	15	12	11	3	16	12
Group 50	3	10	10	15	16	5	4	6	16	4	3	15	9	6	9
Group 51	3	13	11	5	4	12	4	11	6	6	5	3	14	13	12
Group 52	3	14	7	9	14	10	13	8	7	8	10	4	4	13	9
Group 53	5	5	8	14	16	13	6	14	13	7	8	15	6	15	7
Group 54	5	6	11	7	10	8	5	8	7	12	12	10	6	9	11
Group 55	5	6	13	8	13	5	7	7	6	16	14	15	8	16	15
Group 56	5	7	9	10	7	11	6	12	9	12	11	8	8	6	10
Group 57	5	9	6	8	10	9	8	12	5	11	10	11	12	7	7
Group 58	5	10	10	12	8	11	9	7	8	9	5	12	6	7	6
Group 59	5	10	12	6	5	12	8	9	7	6	7	8	11	11	9
Group 60	5	13	15	15	14	8	6	7	16	8	7	13	14	5	16
Group 61	9	10	13	10	11	15	15	9	16	12	14	13	16	14	11
Group 62	9	11	12	15	12	9	13	13	11	14	10	16	15	14	16
Group 63	9	12	10	15	13	14	9	14	15	11	11	13	12	16	10

Index

All IP in 3G CDMA Networks J. Castro
 ISBN: 0-470-85322-0